HOUSEHOLD CHEMICALS AND EMERGENCY FIRST AID

HOUSEHOLD CHEMICALS AND EMERGENCY FIRST AID

CHARLES R. FODEN
JACK L. WEDDELL

CRC Press
Taylor & Francis Group
Boca Raton London New York

CRC Press is an imprint of the
Taylor & Francis Group, an **informa** business

CRC Press
Taylor & Francis Group
6000 Broken Sound Parkway NW, Suite 300
Boca Raton, FL 33487-2742

© 1993 by Charles R. Foden and Jack L. Weddell
CRC Press is an imprint of Taylor & Francis Group, an Informa business

First issued in paperback 2019

No claim to original U.S. Government works

ISBN 13: 978-0-367-45015-1 (pbk)
ISBN 13: 978-0-87371-901-8 (hbk)

This book contains information obtained from authentic and highly regarded sources. Reasonable efforts have been made to publish reliable data and information, but the author and publisher cannot assume responsibility for the validity of all materials or the consequences of their use. The authors and publishers have attempted to trace the copyright holders of all material reproduced in this publication and apologize to copyright holders if permission to publish in this form has not been obtained. If any copyright material has not been acknowledged please write and let us know so we may rectify in any future reprint.

Visit the Taylor & Francis Web site at
http://www.taylorandfrancis.com

and the CRC Press Web site at
http://www.crcpress.com

Library of Congress Card Number 92-26703

Library of Congress Cataloging-in-Publication Data

Foden, Charles R.
  Household chemicals and emergency first aid / Charles R. Foden and Jack L. Weddell.
      p. cm.
    ISBN 0-87371-901-8 (alk. paper)
    1. First aid in illness and injury—Handbooks, manuals, etc.
    2. Household products—Toxicology—Handbooks, manuals, etc.
    3. Household products—safety measures—Handbooks, manuals, etc.
    I. Weddell, Jack.   II. Title.
    [DNLM: 1. Emergencies—handbook. 2. First Aid—handbook.
    3. Household products—poisoning—handbook.   WA 39 F653h]
  RC86.8.F64 1992
  615.9'08—dc2(92-26703

TO MY WIFE BETTY

for
the happiest sixteen
years of marriage

PREFACE

This book was written for use by fire fighters, police, Environmental Protection Agencies, environmental health, emergency medical services and other responders to emergency incidents involving hazardous materials. It will assist the Incident Commander in making the initial evaluation about the possible potential of the hazard and in initiating actions to protect the property and lives of not only any victims involved, but those of the emergency personnel handling the incident.

The intent of this information is to provide guidance, primarily during the initial phase following an incident. To obtain additional expert assistance for the most effective methods to handle a long term incident, it is important that a call be made as soon as possible to:

U.S. Coast Guard National Response Center
1-800-424-8802

CHEMTREC TOLL FREE 1-800-424-9300

The Association of American Railroads
Bureau of Explosives 1-202-639-2222

U.S. Department of Transportation
Hot Line 1-202-426-2075

The National Agricultural Association
Class "B" Poison Pesticide Chemicals
1-513-961-4300
or
To the manufacturer of the material

Numerous schemes for assessing the hazard of materials have been developed by various organizations. Nevertheless, none of these completely fulfill the Fire Service and other agency requirements. We have compiled as much available information as possible to assist the first arriving officer at an incident in taking the initial action to reduce the loss of life and property.

The intent of this book is not to serve as a substitute for your knowledge or judgment. This distinction is important since the recommendations given are those most likely to apply in the majority of cases. We cannot anticipate every possible case.

The following material was assembled from authoritative sources. The author makes no warranty. The information is correct to the best of our knowledge.

HOUSEHOLD PRODUCTS

"Household product" is a very common term. It sounds so innocent and commonplace. However, in most cases throughout the United States, what can be found under the kitchen sink or in the broom closet is a chemical accident just waiting to happen.

Household products or consumer products range from band-aids to laundry detergent, dish washing detergents and floor stripping compounds. Additions to this list are house and garden insect sprays, herbicides, swimming pool chlorine and acids.

Mixing household cleaners of various types is a common practice among housewives. They are usually unaware of the dangers of breathing the fumes and vapors that are released by some of these common products.

A typical example of this is cleaning the stains in the toilet bowl. Common household bleach is a very good stain remover and so is Vanish toilet bowl cleaner. However, <u>DO</u> <u>NOT</u> <u>MIX</u> these two common products. The gas that is generated can be very hazardous to your health.

Ammonia is a very good product for walls, but, you do not want to mix it with chlorine bleach, wall or tile cleaners.

The gas, when inhaled into the lungs, causes severe irritation with coughing and labored breathing, the lungs also can fill up with fluid. If death does not occur, you also can turn blue, due to the lack of oxygen in your blood.

This mistake puts unsuspecting women in the hospital every year. Many have died from this practice.

Household products are safe to use if you read the label and <u>do</u> <u>not</u> <u>mix</u> <u>them!</u>

Child-proof containers are not child proof for all children. If there is a way to open or otherwise operate the container, some children will find it. So you must store household products out of the reach of children. You should lock them up so your child cannot get into them.

Most homes that have swimming pools in their back yard have muriatic acid, otherwise known as, swimming pool acid. There is also swimming pool chlorine (in granular form) that may be stored in the same area.

Swimming pool acid is very strong and will cause serious burns within a few minutes. The chlorine granules are normally 72 percent pure. They react quickly when spilled on the ground, forming chlorine gas. This may create a chlorine gas cloud.

Chlorine, if mixed with a light oil or a flammable liquid, will start a fire.

I know everyone likes to BBQ during the summer. I also know that it feels like it takes a long time for some charcoal to start. If you are using an electric-style charcoal starter, never use charcoal starter fluid to assist the electric starter. Also do not use any flammable liquid with this type of starter.

There are so many accidents just waiting to happen. I cannot cover them all here, so, the motto should be <u>"BE PREPARED."</u> In an emergency have the very important telephone numbers readily available next to the phone. Include in that list, the phone number for the nearest poison control center nearest to your home. You will find the one nearest to you listed in the front of this book with the toll-free emergency telephone number and the address. By all means, call the Poison Control Center if you have a medical emergency involving household products.

INTERNATIONAL HAZARD CLASS DESCRIPTION

CLASS 1 EXPLOSIVES
    Division 1.1 Explosives with a mass explosion hazard
    Division 1.2 Explosives with a projection hazard
    Division 1.3 Explosives with predominantly a fire hazard
    Division 1.4 Explosives with no significant blast hazard
    Division 1.5 Very insensitive explosives
    Division 1.6 Extremely insensitive explosive articles

CLASS 2 GASSES
    Division 2.1 Flammable gases
    Division 2.2 Nonflammable gases
    Division 2.3 Poison gases
    Division 2.4 Corrosive gases (Canadian)

CLASS 3 FLAMMABLE LIQUIDS
    Division 3.1 Flashpoint below - 18°C (0°F)
    Division 3.2 Flashpoint-18°C and above but less than 23°C (73°F)
    Division 3.3 Flashpoint of 23°C and up to 61°C (141°F)

CLASS 4 FLAMMABLE SOLIDS; SPONTANEOUSLY
             COMBUSTIBLE MATERIALS; AND MATERIALS
             THAT ARE DANGEROUS WHEN WET
    Division 4.1 Flammable solids
    Division 4.2 Spontaneously combustible materials
    Division 4.3 Materials that are dangerous when wet

CLASS 5 OXIDIZERS AND ORGANIC PEROXIDES
    Class 5.1 Oxidizers
    Class 5.2 Organic peroxides

CLASS 6 POISONOUS AND ETIOLOGIC MATERIALS
    Division 6.1 Poisonous Materials
    Division 6.2 Etiologic (infectious) materials

CLASS 7 RADIOACTIVE MATERIALS   CLASS 8 CORROSIVES   CLASS 9 MISCELLANEOUS HAZARDOUS MATERIALS

Explanation of (NAS) hazard ratings

The hazard classification system described within covers four main classes of hazards: fire, health, water pollution, and reactivity. Health, water pollution, and reactivity are subdivided further into sub-classes. A numerical rating indicates the relative degree of potential hazard. General guidelines are used to describe the five (5) levels of severity for each. Remember that these ratings relate to hazard situations that may arise from an accident involving these materials. The National Fire Protection Association hazard classification is also indicated when known.

Incidents involving more than one hazardous material at a time requires that the ON-SCENE/INCIDENT COMMANDER obtain expert advise as soon as the scope of the incident can be determined. The materials involved in an accident may, by themselves, be non hazardous. Nevertheless, any combination of different materials or the involvement of a single material in a fire may produce serious health, fire, or explosion hazards.

EXPLANATION OF THE NATIONAL ACADEMY
OF SCIENCES (NAS) HAZARD RATINGS

Health

Rating	Fire	Vapor Irritant	Poisons
0	No hazard	No effect	No effect
1	Flash point (CC) above 140°F	Slight effect	Slight toxicity
2	Flash point (CC) 100° to 140°F	Moderate irritation; temporary effect	Intermediate toxicity
3	Flash point (CC) below 100°F; boiling point above 100°F	Irritaton; cannot be tolerated	Moderate toxicity
4	Flash point (CC) below 100°F boiling point below 100°F	Severe effect; may do permanent injury	Severe toxicity

Water Pollution

Rating	Human Toxicity	Aquatic Toxicity
0	Nontoxic LD50 15 g/kg	Acute threshold limits above 10,000 ppm
1	Practically nontoxic; LD50 5 to 15 g/kg	Threshold limits 1,000 to 10,000 ppm
2	Slightly toxic; LD50 0.5 to 5 g/kg	Threshold limits 100 to 1,000 ppm
3	Moderately toxic; LD50 50 to 500 mg/kg	Threshold limits 1 to 100 ppm
4	Toxic; LD50 50 mg/kg	Threshold limits below 1 ppm

LD50 = Lethal Dosage 50% of a specified population

Reactivity

Rating	Other Chemicals	Water	Self-reaction
0	Inactive; may be attacked by materials rated 4	No reaction	No reaction
1	React only with materials rated 4	Mild reaction: unlikely to be hazardous	Mild self reaction under some conditions
2	React with materials rated 3 or 4	Moderate reaction	Will undergo self reaction if contaminated; do not require stabilizer
3	React with each other and with materials rated 2 or 4	More reactions; may be hazardous	Vigorous self-reaction; required stabilizer
4	React with each other and with materials rated 0-3	Vigorous reaction; likely to be hazardous	Self-oxidizing chemical; capable of explosion or detonation

Physical and Chemical Properties

Appearance - - APPEAR:	millimeters - - - mm	Specific Gravity - - - SG
At - - - - - - - @	Millimeters of Mercury - - mmHg	Temperatures
Chemical Abstract System - - CAS	Not Available - - - - - - NA	25F degrees in Fahrenheit
		25C degrees in Celsius
Department of Transportation - DOT	Not Established- - - - - - - Not Est.	Threshold Limit Values - TLV(TWA)
Flash Point - - - - - - - - - FP	Not Flammable- - - - - - - Not Flam	Number - - - - - - - NUMBER
Upper Explosive Limit - UEL	Ignition Temperature - - - Igni Temp	Physical State - - - PS
Vapor Density - - - - - VD	Lower Explosive Limit - - LEL	Pounds Per Square Inch Absolute-psia
Vapor Pressure - - - - - VP	Milligrams Per Cubic Meter - mg/m3	Pounds Per Square Inch Gage - psig
Reportable Quantity- - - RQ		

CONTAINMENT

Containment methods or techniques will include any actions taken that will keep the spilled chemical from moving to an area that will cause additional problems. These techniques may appear simple and easy, however, you must keep in mind that you are dealing with very dangerous substances. Simple containment does not eliminate the basic hazard nor does it bring the incident to a successful conclusion.

The simplest containment method is to construct a dike around the spill area. This can be constructed of soil, sand, or other inert materials. Soil may be the choice since it is readily available at the scene. Other inert materials may include absorbents that are on the market. If the spill is large, there may not be enough of the absorbent material readily available to do the job. You might want to direct the run off to a holding pit or pond in a safe location with a series of ditches or trenches. You must be aware that this method of containment will allow the spilled chemical to be absorbed into the soil and this may complicate the removal of all contaminated soil (including the diking material) under any circumstances.

Containment of dangerous fumes generated by a spill may be affected by a fog stream. This technique will remove most vapors and fumes from the air. However, whenever water is to be used, remember that some chemicals react violently when they come in contact with it. This water also must be contained or otherwise prevented from entering a sewer, stream or other waterway.

If the chemicals spill into a waterway, containment is much more difficult. Total damming of the stream or river is necessary to prevent contamination downstream. However, since this is rarely possible, all users of the water downstream must be notified of the contamination so they may be able to divert the flow. The city, county, or state public health service must be notified.

ABSORPTION

The spilled liquid chemicals may be absorbed into inert materials such as soil, sand, clay, powdered cement, fly ash, or commercial absorbents, i.e. 911, Cat litter, Safe Step or Dries It. If any of the liquid can be removed safely, this will reduce the amount of absorbent required. Remember, absorption, like containment, is not the final step in handling the incident. The hazardous matter is still present. All you have done is made it immobile by getting it to adhere to the surface of another material. Like the contaminated soil and/or diking material in the containment procedure, the absorbent that holds the chemical must now be removed.

DILUTION

A third method of handling a liquid spill of these chemicals is to dilute them with water (remember what happens when water comes in contact with some chemicals). This means that you may need a very large containment area or pit to hold the diluted chemical. Never allow the dilute solution to enter the sewers or waterways, or the contamination that you are trying so hard to prevent will spread.

It could be necessary to use water in a volume of 1,000 times the volume (or more) of the spilled material to get the chemical to a concentration where it no longer poses a hazard. Then a decision must be made concerning what to do with the diluted solution. This decision should be made by the environmental representative, who will surely have responded to the incident.

NEUTRALIZATION

Neutralization is a chemical reaction that will change the hazardous chemical to a usually non-hazardous material and water. For acids, there are several neutralization agents that may be used, but they all have drawbacks:

Sodium hydroxide (caustic soda) or potassium hydroxide (caustic potash), both in a concentrated solution in water, will be the most effective, but they are expensive, and are very hazardous materials themselves.

Calcium hydroxide (slaked lime) dissolved in water is not as dangerous to use as the sodium and potassium hydroxides, but it is expensive and not as effective.

All three hydroxides may be used in their dry, solid form, but they are very hazardous in this state and will not mix as rapidly as the solutions will.

Sodium Bicarbonate (baking soda), sodium carbonate (soda ash), and calcium carbonate (crushed limestone) are usually recommended because they are relatively inexpensive (calcium carbonate is the least expensive and the least effective, but works adequately) and they will do the job.

With the addition of these materials, there may be bubbling and gassing. This is the generation of carbon dioxide, a product of the neutralization process. In any event, a small sample of the spilled acid should be obtained in a small container, and the chosen neutralizer should be added slowly. This simple experiment will show what will happen when large amounts are added to the spill, and will eliminate any surprises. The use of litmus paper also will indicate the relative amount to be added to the spill. There are other neutralizing materials on the market, too numerous to list here.

Again, the ranking environmental representative should decide when the spilled acid has been rendered safe.

VAPOR CONTROL

AFFF has proven very effective for controlling vapors emitted from spilled hydrocarbons. It is also very good for hydrocarbon fires, including unleaded gasoline and gasohol. It may be effective on some polar solvents. AFFF has a shorter drop-out rate, but forms an aqueous film on the surface of the liquid.

AFFF has an unlimited shelf life and will work in sub-surface injection. Freezing temperatures do not affect the use of AFFF. Follow the manufacturer's recommendations for application devices. A fog nozzle gives greater reach and provides a protective curtain while advancing for a fast attack.

The use of the fog nozzle reduces the amount of suds and the thickness of the applied blanket, however this does not reduce the effectiveness of the aqueous film that is formed. An air aspirating nozzle loses the reach and protective curtain, but provides a long-lasting aerated foam blanket.

High expansion foam is good on hydrocarbon fires. It has also been used on insecticides and pesticides where restricted run-off is desired due to low water content. High expansion foam has been used on liquid chlorine and liquid natural gas (LNG). Upon contact with LNG, it freezes to form an insulating blanket, reducing vapors. High expansion foam also can be used on fuming acids to control vapor emissions. High expansion foam has little or no effect if the blanket is moved by the winds or thermal updrafts, or if you do not have enough to do the job.

EMERGENCY FIRST AID TREATMENT OF POISONING

Many household products in everyday use, such as medicines, certain cosmetics, cleaning agents, plant and insect sprays, bleaches, etc., when accidentally or intentionally misused, cause illness or death.

Furthermore, household appliances such as natural gas stoves, ovens, heating units, portable kerosene heaters, and the family automobile if not checked regularly, account for additional illness, and very often, death from carbon monoxide poisoning.

A general knowledge of the potentially harmful effects of these various items and an understanding of how to combat their toxic effects is important. The emergency care rendered within the first moments after exposure to a poison may mean the difference between life and death.

Because of the widespread incidences of poisonings of all kinds and the need for exceedingly prompt and effective action in even the most unusual cases, a nationwide network of Poison Control Centers (see P.C.C. listing) has been set up in or near all major population areas.

The centers are specially equipped to handle all types of poison cases. They maintain a detailed file of the ingredients of all new and potentially dangerous products, regardless of their nature. With such facilities and special knowledge, these centers offer the best chance of survival for any poisoning victim. If poisoning is suspected, call a center for assistance or get the victim to a center as quickly as possible. If there are no poison control centers nearby, take the victim to the nearest hospital emergency room, as soon as possible, after rendering first-aid assistance.

Check the Poison Control Center listing for the center nearest you and enter it in your phone book under emergency numbers.

PREVENTION OF POISONING

The most important factors in the prevention of accidental poisoning are reasonable care and common sense in using and storing potentially poisonous substances. Equipment may give off poisonous fumes during normal or defective operation so proper care and maintenance is important. This can be accomplished by following a few simple rules:

1. Do not keep medicines, cosmetics, cleaning fluids, rat poisons, or other insecticides where they can be reached by children.

2. Do not keep any medical or other type of chemical preparations that are not clearly labeled.

3. Do not store drugs or household chemicals in food containers.

4. Be certain that the purpose and dosage of a medicine is clearly stated on the label and that you understand them before you or anyone else takes any.

5. <u>Do</u> <u>not</u> take medicine in the dark. That means, never take a medicine without reading the label.

6. Never keep prescription drugs after they have served the purpose of the person for whom they were prescribed.

7. Do not exceed the recommended dosage of aspirin or any other medication containing salicylates for a very sick child with a high fever and dehydration exact under medical supervision.

8. <u>Do</u> <u>not</u> keep known poisons in a medicine cabinet, the kitchen, or any other location where food is stored or where they may inadvertently become accessible to a child.

9. <u>Do</u> <u>not</u> <u>use</u> any type of cleaning fluid near an open flame (e.g. gas water heater, gas stove or oven, or a lighted cigarette). Do not use where ventilation is not adequate because it allows vapors to collect.

10. Thoroughly wash the hands and other exposed body parts with plenty of soap and water after using rat poison, plant or insect spray.

11. <u>Do</u> <u>not</u> <u>stay</u> in a closed room in which an insect spray has just been used. Do not allow food to remain exposed in such a room.

12. <u>Do</u> <u>not</u> <u>operate</u> a gasoline motor in unventilated spaces such as a closed garage.

13. <u>Do</u> <u>not</u> <u>use</u> a gas or kerosene stove in unventilated rooms.

14. <u>Do</u> <u>not</u> <u>use</u> charcoal for cooking or heating in an enclosed area.

15. Have any appliance using natural, propane, butane gas, or kerosene inspected regularly. This will ensure there are no leaks and that it is working properly and is venting correctly.

Most poisonous substances are properly labeled with clear directions for their use. <u>Read</u> <u>the</u> <u>directions</u> on the container and follow them exactly.

HOW POISONS ENTER THE SYSTEM

By Mouth: The most common route by which a poison finds its way into the body is through the mouth. This is particularly true of young children, since almost everything they touch goes directly into their mouth. Although adults should know better, they often do not since by design or accident, they swallow corrosives such as acids, alkalies, or too many sleeping pills. Various disinfectants that contain carbolic acid and many other known or potential poisons sometimes are misused.

By Inhalation: Very often poisonous vapors or fumes from insecticides, cleaning fluids, leaky gas appliances, and even the family automobile, are inhaled.

By Absorption: Many substances can be absorbed directly through the skin in large enough quantities to be poisonous. This includes some insecticides, plant sprays, and cleaning fluids. These are only a few, but are the most common. These substances produce general toxic effects or severe local irritation that will increase the amount and rapidity of absorption.

By Injection: Poisons can be inadvertently injected into the system. You may never be shot with a poisoned arrow, but bees, wasps, and scorpions do sting. Tarantulas and black widow spiders bite, as do rattlesnakes, copperheads, and moccasins. Saltwater catfish, stingrays, and Portuguese man-of-war have various ways of getting their particular "brand" of poison into the unwary.

HOW POISONS ACT

Once in the system, poisons act in various ways. The acids and alkalies produce actual burning and corrosion of the tissue they meet. They can burn a hole in a person's stomach.

Sleeping pills and alcohol act as strong depressants of the central nervous system. Insecticides, like chlordane, or a drug, like strychnine, produce extreme stimulation of the central nervous system. Other compounds, such as cyanide and carbon monoxide, produce an asphyxial type of death. They combine with the oxygen-carrying system of the blood and thus prevent oxygen from being carried to the tissues.

Muscarine, the toxic substance in certain poisonous mushrooms, acts on a special part of the nervous system. It depresses breathing and heart action and increases the worm-like action of the intestinal tract.

If effective first-aid is rendered before medical aid becomes available, the doctor can then make use of drugs that specifically counteract the effects of a particular poison and save the life of the victim. It is the emergency aid to the victim that may determine the success of the medical treatment.

GENERAL PROCEDURES FOR TREATING POISONING VICTIMS

Often the specific poison taken by a person cannot be determined immediately. Under these circumstances, no time should be wasted in attempting to find a specific antidote. The poison should be neutralized as quickly as possible if the poison is a corrosive. This is usually indicated by the stains and burns about the victim's mouth and lips.

If the poison is determined to be a noncorrosive one (no stains or burns about the victim's mouth and lips), give the victim an emetic, such as two or three glasses baking soda solution (one tablespoon of baking soda to eight ounces of water), or syrup of ipecac (one half (1/2) ounce for a child over one (1) year of age), plus at least a cup (8 fluid ounces) of water to make the child vomit. This procedure serves to dilute the poison and helps the body to get rid of it. If no vomiting occurs after twenty (20) minutes, this dose may be repeated one (1) time only.

The best early treatment for any poison taken by mouth is evacuation of the stomach contents. The most common exception to this rule is the ingestion of a caustic poison or a petroleum distillate such as kerosene.

It is uncommon for the first dose of ipecac to fail. Ipecac is so effective as an emetic, both in children and adults, that many authorities feel that keeping it for emergency use in all homes is a necessity.

The victim should be taken to a hospital as soon as feasible after the stomach has been evacuated. Do not wait for vomiting to occur and transport the person to a medical facility immediately.

Be sure the stomach contents are saved for chemical analysis so the nature of the poison may be determined. This step may not only later prove to have been of lifesaving importance, but it may well have strong legal significance. If the package of the suspected poison can be found, any remaining contents should be sent to the hospital with the victim.

TYPES OF POISONS

Corrosive Acids and Alkalies

When the exact nature of the poison is known, specific steps are indicated. One of the frequently encountered groups of poisons likely to be swallowed comprises the corrosive acids and alkalies. Clues to these are the stains and burns about the mouth and lips, or the presence of a characteristic odor. The common ones and their common uses are listed below.

Corrosive Acids	Corrosive Alkalies
Sulfuric Acid (Car batteries)	Potash (lye; drain cleaners)
Nitric Acid (industrial strength cleaners)	Caustic Soda (soap making)
Oxalic Acid (cleaning solutions)	Lime (building trades)
Carbolic Acid or Crude Cresol (used in disinfectants)	Ammonia (household and industrial uses)

Do not cause vomiting if any of the products listed above has been swallowed. Do not use a stomach tube because of the danger of causing a perforation of the stomach or esophagus where the chemical may have caused severe tissue damage.

If a corrosive acid has been swallowed (EXCEPT CARBOLIC ACID or any other cresol disinfectant) give the victim lime water, chalk and water, small amounts of diluted milk of magnesia, or 1 tablespoon of baking soda in a glass of water. The objective is not to induce vomiting but simply to neutralize the acid. Try to keep the victim from straining if vomiting does occur.

Give large amounts of lemon juice, orange juice or equal parts of vinegar and water if a corrosive alkali has been swallowed.

After the first neutralization has been effected for either an acid or alkali, give the victim milk for the formation of a protective coating on the tissues.

The pain in these cases is usually severe and the victim may go into shock. Medical aid should be obtained quickly since some form of sedation will be required with medical supportive measures.

CARBOLIC ACID (PHENOL) POISONING

Usually the clue to carbolic acid poisoning is the very strong characteristic odor on the breath and from the vomited matter. When carbolic acid is taken orally, it produces extensive local burning and corrosion, whitening of the skin and severe burning pain in the mouth, throat and stomach. Phenol is absorbed from the gastrointestinal tract, skin, and lungs, and this causes a severe general reaction, the symptoms of which are slightly delayed. In general, there is a depression of the central nervous system and marked irritation of the kidneys. Headache may be severe and there is muscular weakness, dark bluish coloration of the face, rapid pulse, delirium or coma, and occasionally convulsions. The breathing rate is irregular, and death occurs as the result of respiratory failure.

The principal object of treatment is to remove the phenol from the stomach as quickly as possible before much absorption has occurred. The victim should be taken to a hospital. There, a stomach tube can be passed and the stomach washed out with a substance that will slow the absorption of the phenol as well as dissolve and get rid of it.

IRRITANT SUBSTANCES

There is a large and varied group of compounds that have in common the characteristics of causing irritation of the stomach when swallowed. They produce nausea, vomiting, and severe abdominal pain. Many of these irritants are found in combination with other chemicals as active ingredients in commonly used items in most households. This group includes the following:

Arsenic (rat poisons; weed killers.)	Copper (plant sprays; rat poisons)
Iodine (antiseptic)	Mercury (fireworks; plant sprays; and germicides)
Phosphorous (fireworks; matches; and rat poisons)	Silver Nitrate (inks; cleaning solutions)
Zinc (weed killers; soldering paste and metallurgy)	

Signs and Symptoms: The victim complains of a metallic taste in the mouth, the lips and tongue may appear white and shriveled. In the most serious cases shock is present. The victim should be transported to the hospital immediately.

POISONING BY SUBSTANCES COMMONLY FOUND ABOUT THE HOME

A wide variety of miscellaneous substances are included in this category. If one was to take the substances that might logically be found around the bathroom or in a medicine cabinet, for instance, the following possibilities should be considered:

Atropine and Belladonna: These drugs are commonly used in many prescriptions, and if taken in an overdose (particularly by children), produce marked symptoms.

Signs and Symptoms: Normally, the skin becomes hot, dry, and flushed. The mouth and throat are so dry that swallowing or talking is difficult and the pupils are widely dilated. The victim may be very restless and difficult to control. They also may have hallucinations and the pulse will be weak and rapid. The victim will usually complain of double vision.

<u>Treatment</u>

1. Induce vomiting and give strong tea if the drug has been taken by mouth.

2. Control fever by ice packs, place the victim in a tub with very cold water containing cracked ice, if possible. When the body temperature has been lowered to 100 degrees F., remove the victim, wrap in wet sheets and transport to medical facility.

3. If respiration is inadequate or failing, artificial respiration is indicated.

Medical aid should be obained promptly. There are specific antidotes that a doctor can give by injection.

Nitroglycerine: Nitroglycerine tablets are commonly used for the control of spasmodic suffocative attacks. An overdose may cause most disturbing symptoms.

Please note, however, that mild cases of nitroglycerine poisoning, resulting from absorption through the skin, are seen among workers who handle dynamite. Some persons are much more susceptible than others.

Signs and Symptoms: Symptoms consist of a full, throbbing headache, dizziness, tenseness of the head and neck muscles, dilated pupils, irregular pulse, pain in the chest, and weakness. In severe cases, there will be nausea, vomiting, unconsciousness, and convulsions.

Treatment

Where a sizable overdose has been taken by mouth, medical aid should be obtained quickly. In the meantime, empty and thoroughly clean the victim's stomach with a dose of ipecac or other emetic. This should be followed by repeated glasses of salt water. These victims easily go into shock so they should be kept warm, lying down and given stimulants to help raise the blood pressure. The latter is markedly lowered by the poison.

Salicylates: Salicylate poisoning is most commonly acquired in one of three ways: (1) by the swallowing of pleasantly flavored asprin tablets, usually by a child who believes it to be candy, (2) by swallowing oil of wintergreen rubbing compound (which smells like peppermint) and (3) by overdose of either asprin or sodium salicylate, given for rheumatic fever or for some other illness.

In recent years, the incidence of asprin intoxication has reached alarming proportions. This is attributed to the fact that the drug is so commonly used and because of its availability in pleasantly flavored forms for children. The highest mortality rate occurs in children between one (1) to four (4) years old.

Signs and Symptoms: The onset of symptoms from a single large overdose (such as a child eating a bottle of tablets) may not occur for 12 to 24 hours. It may take from 1 to 4 days for symptoms of poisoning to appear if extreme dosages are taken at regular intervals over a long period of time.

The first and most dependable sign of poisoning is rapid, deep breathing. This sign is so common that if there is an unexplained increase in rate and depth of breathing, the possibility of asprin poisoning should be borne in mind. Hyperventilation can eventually lead to dangerous or even fatal changes in the body chemistry.

Bleeding from the lining of the stomach is the result of corrosive action of the salicylate. This is what causes the coffee ground appearance in the vomitus that sometimes occurs in salicylate poisoning.

If salicylate poisoning occurs, the child will be irritable, restless, and in severe cases may go into convulsions. The child may be in a semiconscious state or a complete stupor. Older children and adults may complain of ringing in the ears or they may be delirious. There may be either a pallor or dusky blue color of the face and there often is a high fever and some evidence of acute abdominal pain.

Treatment

If aspirin poisoning does occur:

1. Take the child to a hospital as quickly as possible. To combat the successive stages of the poisoning, there are many highly technical procedures that must be followed.

2. However, if the child is found shortly after swallowing a single large amount of the salicylate (such as asprin or oil of wintergreen) the stomach should be emptied immediately. You can do this by giving the child a dose of ipecac that you can buy at the local pharmacy. At the hospital the child's stomach may further be cleansed by gastric lavage, using magnesium oxide as a neutralizing agent.

If, however, an increased rate of breathing is noted, absorption will already have occurred and stomach washing will be of little value. Therefore, treat the child for shock, and transport without delay to the hospital.

IRON: Iron in various forms is widely used as a tonic and for the treatment of anemia. It also is given during pregnancy. Several cases of poisoning in children who have had access to capsules of medicinal iron preparations have been reported. It is well to be aware of the danger and to be able to recognize the symptoms, since only about half the cases of those ill enough to attract attention recover. As little as one (1) gram (15 grain) capsule is dangerous to a child, and two capsules have caused death. As dispensed, such capsules are easily available to children. All iron-containing preparations should be kept from any possible access by children.

Signs and Symptoms: The child becomes pale, restless, and nauseated, and there may be vomiting and very severe bloody diarrhea. Drowsiness gradually merges into coma, the pulse becomes weak and rapid, and the skin bluish. The child begins to go into a state of profound shock. Curiously, some victims appear to improve rapidly for 12 to 24 hours. Then they die suddenly, with little warning, 24 to 48 hours after the iron has been taken.

Treatment: The child should be taken to a hospital as quickly as possible, as other technical supportive measures will be required. In addition, a new antidote, desferioxamine, has become available to most centers and apparently is lifesaving in many instances.

Boric Acid: Boric acid is often used in powdered form for antiseptic solutions for washing out the eyes and for wet dressings. The virtues of boric acid have been very overrated. It is not a good antiseptic, and it is a dangerous poison under certain conditions of misuse. It has no place in the household.

Pure boric acid is poisonous when taken accidentally by mouth or when applied to extensive raw surfaces of the body.

Signs and Symptoms: The symptoms of boric acid poisoning are a bright red appearance of the skin. A generalized rash, with shedding of the skin, rapid breathing and suppression of urine. This condition is difficult to diagnose with certainty and may be confused with illnesses having similar symptoms.

<u>Treatment:</u> Get the victim to a hospital as quickly as possible.

Camphor: Camphor is not seen in the family medicine chest as much as it used to be, but it is still an occasional cause of poisoning in children, particularly if camphorated oil is given by mistake for castor oil. Camphorated oil has little demonstrable use and should be banned from the family medicine cabinet. Tincture of camphor is sometimes used for "fever blisters."

Signs and Symptoms: The general symptoms are a burning sensation in the throat and stomach, excessive thirst, nausea and vomiting. Also included are blurred vision, dizziness, headache and convulsions. After a few hours the victim shows all the symptoms of being drunk, and breath or urine usually smells like camphor.

<u>Treatment:</u> Get the victim to a hospital as quickly as possible.

Strychnine: Strychnine is often found around the home in some form of rat poison, causing the death of many children each year.

Signs and Symptoms: Usually the symptoms come on rapidly after swallowing the poison though they may be delayed for more than an hour. They are unmistakable. There is apprehension, shuddering, and then, suddenly, a generalized spasm which grips the entire body. This spasm causes the back to arch so the victim is supported only by the back of the head and the heels of the feet. During the spasm the victim cannot breathe and will turn blue in the face. The pulse is rapid, weak and thready and the face assumes a horrible grin. Once witnessed you will not forget it. Muscular contractions are extremely painful. When the spasm passes the victim relaxes, exhausted, remaining so until the next spasm. Spasms may be brought on by a sudden noise, touch, or a strong light. Victim must be handled gently and moved slowly.

<u>Treatment:</u> Emergency aid consists of the emptying of the stomach, followed by large amounts of very strong tea. The tannic acid in the tea combines with and neutralizes the strychnine. The mixture is then removed by vomiting or lavage and the procedure repeated as necessary.

This is an acute medical emergency. Anesthesia may be required to control the convulsions. Medications must be given to carry the victim through and to support him until the effects of the drug wear off. This will be a matter of several hours in those cases that survive.

Chloral Hydrate: Chloral hydrate was widely used as a sleeping powder and still is. Under proper conditions, it produces normal sleep; combined with alcohol, it makes a "Mickey Finn." This rapidly produces unconsciousness, followed, upon awakening, by a severe headache and nausea.

Signs and Symptoms: When taken by itself in an overdose, chloral hydrate renders a person suddenly helpless and, for that reason, has been used as "knockout drops" for vicious purposes. The symptoms come on suddenly, with burning and dryness of the throat, collapse, shallow breathing, weak pulse, and coma. The victim will seem to be in shock.

<u>Treatment:</u> Emergency aid consists of emptying the stomach, rendering artificial respiration when required, and administering strong stimulants. Medical aid should be summoned quickly, and the victim taken to a hospital. Such a case may be serious enough that survival remains in doubt for days.

Lead Poisoning: The condition now ranks as one of the most common causes of childhood mortality as the result of poisoning. Children 1 to 4 years of age living in older type houses, where lead-containing paints were widely used, are most frequently the victims.

Signs and Symptoms: The earliest signs of trouble in children are vomiting, constipation and a peculiar ashy pallor, coupled with loss of appetite and irritability. The effect of lead on the central nervous system may produce very marked irritability if a child has taken in enough quantities of lead over a long period of time. In advanced cases, convulsions and loss of consciousness are the symptoms. These symptoms can accompany other types of disease. Lacking some other obvious explanation they are sufficient to raise suspicion in the mind of the physician of the possibility of lead poisoning.

POISONS FOUND ESPECIALLY IN THE GARAGE AND STORAGE AREAS

Carbon Monoxide: This gas is given off in large quantities by automobile exhausts. It may prove quickly fatal in unventilated spaces, such as garages or automobiles with closed windows. This is true even if the vehicle is moving.

Signs and Symptoms: It is of the utmost importance that every person familiarize themselves with the signs and symptoms of carbon monoxide poisoning. This is true especially to save their own life or the life of another. They are headache, yawning, dizziness, faintness, ringing in the ears, nausea, pounding of the heart, bright cherry-red color of the skin, lethargy, stupor, and coma.

<u>Treatment:</u> To treat carbon monoxide or any other gas poisoning:

1. Move the victim to fresh air. If breathing is labored, give the victim oxygen. If breathing has stopped, give artificial respiration.

2. Summon medical aid as quickly as possible.

Preferably the victim should be taken to a hospital by a well equipped rescue squad/ambulance.

Kerosene and Other Petroleum Distillates: Kerosene poisoning is common in certain rural and city area where kerosene is used for heating and lighting purposes and as a solvent for insecticides. Because "pop" bottles and coffee cans containing kerosene are carelessly left about, young children can easily get hold of it and drink large quantities.

Signs and Symptoms: The child may choke, cough, gasp, strangle, or vomit immediately upon drinking the kerosene. Often the child is found lying depressed or stuporous beside the container from which they drank. In about half of the victims there may be abdominal pain, and most will show some degree of fever.

<u>Treatment:</u> Take the child to the nearest hospital as quickly as possible, where the proper treatment will be undertaken. If the child does vomit spontaneously, an effort should be made to prevent the sucking of the vomited matter into the lungs. This is done by keeping the head lower than the rest of the body.

METHYL ALCOHOL (METHANOL or WOOD ALCOHOL)

Methyl alcohol is used around the home for cleaning windows, alcohol rubs, burning in spirit lamps, spot removing etc. In industry, it is used as a solvent.

Methyl alcohol is an extremely poisonous substance; the ingestion of a fraction of an ounce has been known to cause death.

Signs and Symptoms: The most characteristic symptoms of wood alcohol poisoning are some that may be due to simultaneous drinking of ethyl alcohol. These are intoxication, extreme thirst, visual disturbances, very severe abdominal pain, and occasionally convulsions.

<u>Treatment:</u> Get the victim to the nearest hospital as soon as possible.

POISONOUS BITES AND STINGS

The insects that are the highest source of danger in this regard are the honey bee, bumblebee, wasp, black hornet, and the yellow jacket. The venom of these stinging insects causes more deaths in the United States each year than are caused by rattlesnakes bites.

Signs and Symptoms: In non-sensitive individuals, a bee sting produces nothing more than a painful swelling and redness, aching and itching. Also, if several stings are received at one time, enough venom can be absorbed to make the victim quite ill. This can cause a case of severe hives to develop or a generalized swelling of all the tissue. If the victim is allergic to bee stings, they may become very ill and may require emergency treatment to save their life.

Unlike the honey bee, the wasp, black hornet, and yellow jacket retain their stingers and can sting repeatedly. This makes them capable of inflicting multiple injections of venom into the same victim. After stinging its victim, the honey bee disembowels itself, leaving behind the stinger and the venom sac. The muscles surrounding the venom sac continue to force venom into the sting site and can do so for 15 to 20 minutes.

Treatment: Unless the victim is in danger because of a basic reaction, remove the stinger immediately by scraping it out gently with a sharp object, such as a knife blade, the side of tweezers, or a finger nail. Never grasp the sac or stinger with finger nails or forceps since this simply forces the remaining venom into the skin. Treat by using an ice pack to reduce the swelling.

If the victim is extremely sensitive to bee venom because of previous stings, anaphylactic shock usually develops quickly and can be easily recognized when the victim experiences severe difficulty with breathing, becomes restless and may develop discolored areas; also develops livid blueness of the skin, begins to cough, complains of headache, and may become unconscious. These are warning signs of extreme sensitivity, and the victim must be taken to the hospital with all possible speed.

Black Widow Spiders: Bites by these spiders are normally serious, but are not, as many people suppose, fatal.

Signs and Symptoms: One of the symptoms caused by a black widow spider bite is severe abdominal pain due to muscle spasms (without nausea or vomiting). Other symptoms are a partial state of collapse, dilated pupils, generalized swelling of the face and extremities, and at times, convulsions.

Treatment: Place a constriction band above the bite site (tight enough to restrict blood flow in the veins but not the arteries). Also, place an ice pack on the bite site. Transport the victim to the nearest medical facility.

Brown Recluse Spider: The brown recluse spider is at least as poisonous as the black widow spider. Although many spiders are poisonous, the bites of these two spiders are to be feared the most.

Signs and Symptoms: There may be a mild stinging sensation at the time of bite or it may go unnoticed. In a few hours, the site may start to itch, become inflamed, be quite tender, and a blister will appear in its center. In about 12 hours a high fever develops, chills, grippe-like sensations, and often nausea and vomiting. In about two days, the area around the bite site turns white and the local pain will be so severe narcotics may not be effective.

Treatment: Transport the victim to the nearest hospital for proper treatment consisting of strong analgesics and sedative drugs.

Tarantulas: Tarantula bites are normally similar in their effects to those of the black widow spider. The highest danger arises when young children are bitten.

Treatment: The treatment can only consist of measures to relieve whatever symptoms appear, as there is no specific antivenom available.

Scorpions: Scorpion stings, although feared, are usually not very serious, although the stings of breeds in certain parts of the United States and other parts of the world are considered more dangerous.

Signs and Symptoms: There is a burning sensation at the site of the sting, followed by pain that spreads to the entire limb. General symptoms appear within about an hour with a headache, dizziness, nausea, vomiting, increased salivation, shock, and coma.

Treatment: Get the victim to a hospital as quickly as possible.

Snake Bite: There are four major types of poisonous snakes in the United States. These are Rattlesnakes (there are many varieties), Water moccasins (cottonmouths), Copperheads, and Coral Snakes.

Urban expansion into the areas where you find these snakes has led to a large increase in snake bites. The increase in outdoor recreation has also been a factor in the increase of reported cases.

Signs and Symptoms: There will be immediate severe pain accompanied by swelling and dark-purplish discoloration of the skin following the bite from one of the first three snakes mentioned. The puncture marks of the fangs are easily found. The general symptoms from these bites are growing weakness, shortness of breath, the increasing feeling of weariness leading to unconsciousness, dimness of vision, rapid pulse, nausea and vomiting.

The symptoms of coral snake bite are similar, with the addition of severe drowsiness. The victim will sink into unconsciousness as if they were asleep.

<u>Treatment:</u> 1. Immobilize the extremity and put a constricting band just above the bite, tight enough to stop the flow of blood in the veins. <u>Do</u> <u>not</u> <u>tighten</u> <u>the</u> <u>band</u> <u>enough</u> <u>to</u> <u>stop</u> <u>the</u> <u>arterial</u> <u>blood</u> <u>supply.</u> <u>Remember</u> <u>this</u> <u>is</u> <u>not</u> <u>a</u> <u>tourniquet.</u> If you have a snake bite kit, use the rubber suction cups to pull out some of the poison. If ice is available, place it over the bite. Transport the victim to a hospital.

POISON CONTROL CENTERS LISTED BY STATE
==

ALABAMA	ARKANSAS

Birmingham
The Children's Hospital of
Alabama Regional Poison
Control Center
1600 Seventh Ave. South
zip: 35233-1711
Phone: (800) 292-6678 (Alabama only)
       (205) 933-4050
             939-9201
             939-9202

Tuscaloosa
Alabama Poison Control System Inc.
809 University Ave. S.
zip: 35401
Phone: (800) 462-0800 (Alabama only)
       (205) 345-0600

ALASKA

Anchorage
Anchorage Poison Center
Providence Hospital
3200 Providence Dr.
P.O. Box 196604
zip: 99519-6604
Phone: (800) 478-3193 (Alaska only)
       (907) 261-3193

ARIZONA

Phoenix
Samaritan Regional Poison Center
Good Samaritan Medical Center
1130 E. McDowell Rd. Ste A-5
zip: 85006
Phone: (602) 253-3334

Tuscon
Arizona Poison and Drug Info. Center
University of Arizona Health Service
Center
1501 N. Campbell Ave. Rm 3204-K
zip: 85724
Phone: (800) 362-0101 (Arizona only)
       (602) 626-6016

Little Rock
Arkansas Poison and Drug Info. Center
University of Arkansas for Medical
Sciences College of Pharmacy
Slot 522 (internal mailing)
4301 W. Markham Street
zip: 72205
Phone: (501) 661-6161
       (800) 482-8948 (MD's & hospitals)
                      (Arkansas only)
       (501) 666-5532 (MD's & hospitals)

==
CALIFORNIA
==

Fresno
Fresno Regional Poison Control Center
Fresno Community Hospital & Med Center
Fresno and R Streets P.O. Box 1232
zip: 93715
Phone: (800) 346-5922 (Fresno,Kern,Kings
       Madera,Mariposa,Merced,Tulare only)
       (209) 445-1222

Los Angeles
Los Angeles County Medical Association
Regional Poison Control Center
1925 Wilshire Blvd.
zip: 90057
Phone: (800) 777-6476
       825-2722 (MD's & hospitals)
                (California only)
       (213) 484-5151
       664-2121 (MD's & hospital)

Orange
UC Irvine Regional Poison Center
University of Calif. Irvine Med. Center
101 The City Drive Rte 78
zip: 92668
Phone: (800) 544-4404 (Inyo,Mono,Orange
       San Bernardino and Riverside Cnty only)

       (714) 634-5988
       634-6665 (haz. mat hotline)

Sacramento
UC Davis Medical Center
Regional Poison Control Center
2315 Stockton Blvd. Rm-1511
zip: 95817
Phone: (800) 342-9293 (No. Calif. only)
       (916) 734-3692

San Diego
San Diego Regional Poison Center
University of California
San Diego Medical Center
225 Dickinson St. H-925
zip: 92103-1990
Phone: (800) 876-4766 (Imperial
       & San Diego Cnty only)
       (619) 543-6000

San Francisco
San Francisco Bay Area Regional
Poison Control Center
San Francisco General Hospital
1001 Potrero Ave. Rm 1 E 86
zip: 94110
Phone: (800) 523-2222 (415 & 707
                       area codes only)
       (415) 476-6600

San Jose
Santa Clara Valley Med Center
Regional Poison Center
751 So. Bascom Ave.
zip: 95128
Phone: (800) 662-9886 (Monterey,
       San Benito,San Luis Obispo,Santa
       Cruz,Santa Clara cnty only)

       (408) 299-5112

===================================|===================================|===================================
COLORADO | FLORIDA | HAWAII
===================================|===================================|===================================

Denver | Jacksonville | Honolulu
Rocky Mtn Poison and Drug Center | St. Vincent's Medical Center | Hawaii Poison Control Center
645 Bannock Street | 1800 Barrs St./ P.O. Box 2982 | Kapiolani Medical Center for Women
zip: 80204-4507 | zip: 32203 | and Children
Phone: (800) 332-3073 (Colorado) | Phone: (904) 387-7500 | 1319 Punahou Street
       (800) 525-5042 (Montana) | 387-7499 (TTY) | zip: 96826
       (800) 442-2702 (Wyoming) | | Phone: (800) 362-3585 (Outer Islands
       (303) 629-1123 | Tallahassee | of Hawaii only)
                                   | Tallahassee Memorial Regional Med Ctr. | (800) 362 3586
Interstate | 1300 Miccosukee Road | (808) 941-4411
The Poison Control Center | zip: 32308 |===================================
Omaha, Neb. | Phone: (904) 681-5411 | IDAHO
Phone: (800) 955-9119 | |===================================
===================================| Tampa |
CONNECTICUT | Florida Poison Information Center | Boise
===================================| Tampa General Hospital | Idaho Statewide Emergency Poison
                                   | Davis Island | Communication Center
Farmington | P.O. Box 1289 | St. Alphonsus Regional Medical Center
Connecticut Poison Control Center | zip: 33601 | 1055 North Curtis Road
University of Connecticut | Phone: (800) 282-3171 (Florida only) | zip: 83706
Health Center |===================================| Phone: (800) 632-8000 (Idaho only)
zip: 06030 | GEORGIA | (208) 378-2707
Phone:(800) 343-2722 (Connecticut) |===================================|===================================
       (203) 679-3473 (Admin) | Atlanta |
       (203) 679-4346 (TDD) | Georgia Regional Poison Control Center |===================================
===================================| Grady Memorial Hospital | ILLINOIS
DELAWARE | 80 Butler Street S.E. |===================================
===================================| Box 26066 |
                                   | zip: 30335-3801 | Chicago
Wilmington | Phone: (800) 282-5846 (Georgia only) | Chicago & Northeastern Illinois
Poison Information Center | (404) 589-4400 | Regional Poison Control Center
Medical Center of Delaware | 525-3323 (TTY) | Rush-Presbyterian-St Luke's Medical Ctr
Wilmington Hospital | Macon | 1753 West Congress Parkway
501 W. 14 St. | Regional Poison Control Center | zip: 60612
zip: 19899 | Medical Center of Central Georgia | Phone: (800) 942-5969 (North East
Phone: (302) 655-3389 | 777 Hemlock Street | Illinois only)
===================================| zip: 31208 | (312) 942-5969
DISTRICT of COLUMBIA | Phone: (912) 744-1427 |
===================================| (912) 744-1146 | Springfield
                                   | 744-1000 | Central & Southern Illinois Regional
Washington | Savannah | Poison Control Resource Center
National Capital Poison Center | Savannah Regional Poison Control Ctr | St. John's Hospital
Georgetown University Hospital | Department of Emergency Medicine | 800 East Carpenter Street
3800 Reservoir Road N.W. | Memorial Medical Center | zip: 62769
zip: 20007 | 4700 Waters Avenue | Phone: (800) 252-2022 (Illinois only)
Phone: (202) 625-3333 | zip: 31403 | (217) 753-3330
       784-4660 (TTY) | Phone: (912) 355-5228 (So East Georgia |
                                   | and So. West South Carolina Only) | Interstate Centers
                                   | | Cardinal Glennon Children's Hospital
                                   | Interstate Centers | Regional Poison Center
                                   | Chattanooga Poison Control Center | St. Louis, MO
                                   | (For Children Only) | Phone: (800) 366-8888 (Western
                                   | Chattanooga, TN | Illinois only)
                                   | Phone: (615) 778-6100 (Northern |
                                   | Georgia only) |

```
=================================| =================================| ==================================
INDIANA                          | KANSAS  Cont'd                    | MASSACHUSETTS
=================================| =================================| ==================================
                                 |                                   |
Indianapolis                     | Wichita                           | Boston
Indiana Poison Center            | HCA Wesley Medical Center         | Massachusetts Poison Control System
Methodist Hospital of Indiana    | 550 North Hillsdale Avenue        | 300 Longwood Avenue
1701 North Senate Blvd.          | zip: 67214                        | zip: 02115
zip: 46206                       | Phone: (316) 688-2277             | Phone: (800) 682-9211 (Massachusetts
Phone: (800) 382-9097 (Indiana   |                                   |                            only)
                        only)    | Interstate Centers                |        (617) 232-2120
       (317) 929-2323            | Cardinal Glennon Children's Hospital| =================================
             929-2336 (TTY)      | Regional Poison Center            | MICHIGAN
                                 | St. Louis,  MO                    | =================================
Interstate Centers               | Phone: (800) 366-8888 (Topeka only)|
Kentucky Regional Poison Center  | =================================|
of Kosair Children's Center      | KENTUCKY                          | Detroit
Louisville, KY                   | =================================| Poison Control Center
Phone: (502) 589-8222 (Southern  |                                   | Children's Hospital of Michigan
                Indiaria only)   | Louisville                        | 3901 Beaunien Blvd.
=================================| Kentucky Regional Poison Center of| zip: 48201
IOWA                             | Kosair Children's Hospital        | Phone: (800) 462-6642 (Michigan only)
=================================| P.O. Box 35070                    |        (313) 745-5711
                                 | zip: 40232-5070                   |
Des Moines                       | Phone: (800) 722-5725 (Kentucky only)| Grand Rapids
Variety Club Poison and Drug     |        (502) 589-8222 (Metropolitan| Blodgett Regional Poison Center
Information Center               |        Louisville and So.Indiana only)| Blodgett Memorial Medical Center
Iowa Methodist Medical Center    |                                   | 1840 Wealthy Street S.E.
1200 Pleasant Street             | Interstate Centers                | zip: 49506
zip: 50309                       | Knoxville Poison Control Center   | Phone: (800) 632-2727 (Michigan only)
Phone: (800) 362-2327 (Iowa only)| Knoxville,  TN                    |        356-3232 (TTY)
       (515) 283-6254            | Phone: (615) 544-9400 (Southern   |        (616) 744-7854
                                 |                Kentucky only)     |
Iowa City                        | =================================| Kalamazoo
Poison Control Center            | MAINE                             | Bronson Poison Center
University of Iowa Hospitals and | =================================| Bronson Methodist Hospital
Clinics                          |                                   | 252 East Lovell Street
zip: 52242                       | Portland                          | zip: 49007
Phone: (800) 272-6477 (Iowa only)| Maine Poison Control Center at Maine| Phone: (800) 442-4112 (Michigan only)
       (319) 356-2922            | Medical Center                    |        (616) 341-6409
                                 | 22 Bramhall Street                | =================================
Interstate Centers               | zip: 04102                        | MINNESOTA
McKennan Hospital Poison Center  | Phone: (800) 442-6305 (Maine only)| =================================
Sioux Falls, SD                  |        (207) 871-2381 (ER)        |
Phone: (800) 843-0505            | =================================| Minneapolis
=================================| MARYLAND                          | Hennepin Regional Poison Center
KANSAS                           | =================================| Hennepin County Medical Center
=================================|                                   | 701 Park Avenue
                                 | Baltimore                         | zip: 55415
Kansas City                      | Maryland Poison Center            | Phone: (612) 347-3141
Mid-America Poison Control Center| University of Maryland School of  |        337-7474 (TTY)
University of Kansas Med Center  | Pharmacy                          |
3900 Rainbow  Rm-B-400           | 20 North Pine Street              | St. Paul
zip: 66103                       | zip: 21201                        | Minnesota Regional Poison Center
Phone:(800) 332-6633 (Kansas only)| Phone: (800) 492-2414 (Maryland only)| St. Paul-Ramsey Medical Center
       (913) 588-6633            |        (410) 528-7701             | 640 Jackson Street
                                 |                                   | zip: 55101
                                 |                                   | Phone: (800) 222-1222 (Minnesota only)
                                 |                                   |        (612) 221-2113
```

```
===============================  =================================  ===============================
MINNESOTA Cont'd               | MONTANA                          | NEW MEXICO Cont'd
===============================  =================================  ===============================
                               |                                  |
Interstate Centers             | Interstate Centers               | Albuquerque
McKennan Hospital Poison Center| Rocky Mountain Poison Center     | New Mexico Poison and Drug Info Ctr
Sioux Falls, SD                | Denver, CO                       | University of New Mexico
Phone: (800) 843-0505          | Phone: (800) 525-5042            | zip: 87131
                               | =================================| Phone: (800) 432-6866 (New Mexico
No. Dakota Poison Information Ctr| NEBRASKA                        |                         only)
Fargo, ND                      | =================================|
Phone: (701) 234-5575          |                                  |    (505) 843-2551
   Northwestern Minnesota only | Omaha                            | ===============================
        (Call Collect)         | The Poison Control Center        | NEW YORK
                               | Children's Memorial Hospital     | ===============================
St. Luke's Midland Regional    | 8301 Dodge Street                |
Medical Center                 | zip: 68114                       | Buffalo
Poison Control Center          | Phone: (800) 955-9119 (Nebraska only)| Western New York, Regional Poison
Aberdeen, SD                   |        (402) 390-5555            | Control Center at Children's
Phone: (800) 592-1889          |                                  | Hospital of Buffalo
===============================| Interstate Centers               | 219 Bryant Street
MISSISSIPPI                    | McKennan Hospital Poison Center  | zip: 14222
===============================| Sioux Falls, SD                  | Phone: (800) 888-7655 (Western New
                               | Phone: (800) 843-0505            |                     York only)
Jackson                        | =================================|        (716) 878-7654
Regional Poison Control Center | NEW HAMPSHIRE                     |             878-7655
University Medical Center      | =================================|
2500 North State Street        |                                  | East Meadow
zip: 39216                     | Hanover                          | Long Island Regional Poison
Phone: (601) 354-7660          | New Hampshire Poison Information Ctr| Control Center
===============================| Dartmouth/Hitchcock Medical Center| Nassau County Medical Center
MISSOURI                       | 2 Maynard Street                 | 2201 Hempstead Tpk.
===============================| zip: 03756                       | zip: 11554
                               | Phone: (800) 562-8236 (New Hampshire| Phone: (516) 542-2323
Kansas City                    |                         only)    |        542-2324
Children's Mercy Hospital      |    (603) 646-5000                |        542-2325
2401 Gillham Road              | =================================| (911) (TTY)
zip: 64108-9898                | NEW JERSEY                       |
Phone: (816) 234-3000          | =================================| New York City
                               |                                  | New York City Poison Control Center
St. Louis                      | Newark                           | 455 First Avenue  Rm 123
Cardinal Glennon Children's    | New Jersey Poison Information and| zip: 10016
Hospital Regional Poison Center| Education System                 | Phone: (800) 545-1796 (New York City
1465 So. Grand Blvd.           | Newark Beth Israel Medical Center|                          only)
zip: 63104                     | 201 Lyons Avenue                 |    (212) 340-4494
Phone: (800) 366-8888 (Missouri| zip: 07112                       |         764-7667
    Western Illinois & Topeka, KS)| Phone:(800) 962-1253 (New Jersey only)|
                               |    (201) 923-0764               | Nyack
   (800) 392-9111 (Missouri only)|      926-8008 (TTY)           | Hudson Valley Regional Poison Center
   (314) 772-5200              | =================================| Nyack Hospital
        577-5336 (TTY)         | NEW MEXICO                       | North Midland Avenue
                               | =================================| zip: 10960
                               |                                  | Phone: (800) 336-6997 (from 518 and
                               | Interstate Centers               |             914 area codes only)
                               | El Paso Poison Control Center    |    (914) 353-1000
                               | El Paso, TX                      |
                               | Phone: (915) 533-1244 (Southern New|
                               |                     Mexico only) |
                               |                                  |
```

NEW YORK Cont'd	NORTH CAROLINA Cont'd	OHIO

NEW YORK Cont'd

Rochester
Life Line/ Finger Lakes Regional
Poison Control Center
University of Rochester Med. Ctr.
Box 777
zip: 14642
Phone: (800) 333-0542 (Ontario &
Wayne Counties only)
(716) 275-5151
275-2700 (TTY)

Syracuse
Central New York Poison
Control Center
University Hospital of Syracuse
750 East Adams Street
zip: 13210
Phone: (800) 252-5655 (outside of
Onondaga County)
(315) 476-4766

Interstate Centers
Northwest Regional Poison Center
Erie, PA
Phone: (800) 822-3232 (South-
Western New York only)

Vermont Poison Center
Burlington, VT
Phone: (802) 658-3456

NORTH CAROLINA

Asheville
Western N.C. Poison Control Ctr.
Memorial Mission Hospital
509 Biltmore Avenue
zip: 28801
Phone: (800) 542-4225 (North
Carolina only)
(704) 255-4490

Charlotte
Mercy Hospital Poison Control Ctr
2001 Vail Avenue
zip: 28207
Phone: (704) 379-5827

NORTH CAROLINA Cont'd

Durham
Duke University Regional Poison
Control Center
Duke University Medical Center
P.O. Box 3007
zip: 27710
Phone: (800) 672-1697 (No. Carolina
only)
(919) 684-8111

Greensboro
Moses H. Cone Memorial Hospital
Triad Poison Center
1200 North Elm Street
zip: 27401-1020
Phone: (800) 722-2222 (Alamance,
Forsyth, Guilford, Rockingham,
and Randolph Counties only)
(919) 379-4105

Hickory
Catawba Memorial Hospital
Poison Control Center
810 Fairgrove Church Road S.E.
zip: 28602
Phone: (704) 322-6649

Interstate Centers
Virginia Poison Center
Virginia Commonwealth University
Box 522, MCV Station
Richmond, VA 23298
Phone: (800) 552-6337 (804 Area code
only)

NORTH DAKOTA

Fargo
North Dakota Poison Information Ctr.
St. Luke's Hospitals
720 Fourth Street North
zip: 58122
Phone: (800) 732-3300 (North Dakota
only)
(701) 234-5575 (Local, North-
Western Minnesota only Call Collect)

Interstate Centers
St. Luke's Midland Regional Med. Ctr.
Poison Control Center
Aberdeen, SD
Phone: (800) 592-1889

OHIO

Akron
Akron Regional Poison Control Center
Children's Hospital Medical Center
of Akron.
281 Locust Street
zip: 44308
Phone: (800) 362-9922 (Ohio only)
(216) 379-8562
379-8446 (TTY)

Cincinnati
Regional Poison Control System and
Drug & Poison Information Center
University of Cincinnati
College of Medicine
231 Bethesda Avenue M.L. 144
zip: 45267-0144
Phone: (800) 872-5111 (Ohio only)
(513) 588-5111

Cleveland
Greater Cleveland Poison Control Ctr
2101 Adelbert Road
zip: 44106
Phone: (216) 231-4455

Columbus
Central Ohio Poison Center
Children's Hospital
700 Children's Drive
zip: 43205
Phone: (800) 682-7625 (Ohio only)
(614) 228-1323
228-2272 (TTY)

Dayton
Western Ohio Poison and Drug
Information Center
Children's Medical Center
1 Children's Plaza
zip: 45404-1815
Phone: (800) 762-0727 (Ohio only)
(513) 222-2227

Lorain
Poison Control Center of Lorain Cnty
Lorain Community Hospital
3700 Kolbe Road
zip: 44053
Phone: (800) 821-8972 (Ohio only)
(216) 282-2220

```
====================================   ====================================   ====================================
OHIO  Cont'd                          PENNSYLVANIA                          PENNSYLVANIA  Cont'd
====================================   ====================================   ====================================
```

OHIO Cont'd	PENNSYLVANIA	PENNSYLVANIA Cont'd
Toledo Poison Information Center of Northwestern Ohio Medical College of Ohio 3000 Arlington Avenue zip: 43614 Phone: (419) 381-3897	Allentown Lehigh Valley Poison Center Allentown Hospital 17th & Chew Streets zip: 18102 Phone: (215) 433-2311	Pittsburgh Pittsburgh Poison Center Children's Hospital of Pittsburgh 1 Children's Place 3705 Fifth Avenue at DeSoto Street zip: 15213 Phone: (412) 681-6669
Youngstown Mahoning Valley Poison Center St. Elizabeth Hospital Med. Ctr. 1044 Belmont Avenue zip: 44501 Phone: (800) 426-2348 (Ashtabula Columbiana, Mahoning, Trumbull Counties only) Pennsylvania: Mercer, Lawrence Counties only (216) 746-2222 746-5510 (TTY)	Altoona Keystone Region Poison Center Mercy Hospital 2500 Seventh Avenue zip: 16603 Phone: (814) 946-3711 Danville Susquehanna Poison Center Geisinger Medical Center North Academy Avenue zip: 17821 Phone: (800) 352-7001 (Pennsylvania only) (717) 275-6119	Interstate Centers Mahoning Valley Poison Center Youngstown, OH Phone: (800) 426-2348 (Lawrence and Mercer Counties only) ==================================== RHODE ISLAND ==================================== Providence Rhode Island Poison Center Rhode Island Hospital 593 Eddy Street zip: 02903 Phone: (401) 277-5727
Interstate Centers Northwestern Regional Poison Ctr. Erie, PA Phone: (800) 822-3232 (North- Western Ohio only) ==================================== OKLAHOMA ====================================	Erie Northwest Regional Poison Center Saint Vincent Health Center 232 West 25th Street zip: 16544 Phone: (800) 822-3232 (Northwestern Pennsylvania, Northeastern Ohio, and Southwestern New York only) (814) 452-3232	==================================== SOUTH CAROLINA ==================================== Columbia Palmetto Poison Center University of South Carolina College of Pharmacy zip: 29208 Phone: (800) 922-1117 (South Carolina only) (803) 765-7359
Oklahoma City Oklahoma Poison Control Center Children's Hospital of Oklahoma 940 N.E. 13th Street zip: 73104 Phone: (800) 522-4611 (Oklahoma only) (405) 271-5454 ==================================== OREGON ====================================	Hershey Capital Area Poison Center University Hospital Milton S. Hershey Medical Center University Drive zip: 17033 Phone: (800) 521-6110 (717) 531-6110 531-6039	Interstate Centers Savannah Regional Poison Control Center Savannah, GA Phone: (912) 355-5228 (Southwestern South Carolina only) ==================================== SOUTH DAKOTA ====================================
Portland Oregon Poison Center Oregon Health Sciences University 3181 S.W. Sam Jackson Park Road zip: 97201 Phone:(800) 452-7165 (Oregon only) (503) 494-8968	Philadelphia Delaware Valley Regional Poison Control Center 1 Children's Ctr. 34th & Civic Center Blvd. zip: 19104 Phone: (215) 386-2100	Aberdeen St. Luke,s Midland Regional Medical Center Poison Control Center 305 South State Street zip: 57401 Phone: (800) 592-1889 (South Dakota Minnesota, North Dakota, Wyoming) (605) 622-5678

SOUTH DAKOTA Cont'd	TEXAS Cont'd	VIRGINIA Cont'd

Sioux Falls
McKennan Hospital Poison Center
800 East 21st Street
P.O. Box 5045
zip: 57117-5045
Phone: (800) 952-0123 (South
Dakota only)
(800) 843-0505 (Iowa,
Minnesota, Nebraska only)
(605) 336-3894

TENNESSEE

Knoxville
Knoxville Poison Control Center
University of Tennessee Memorial
Research Center and Hospital
1924 Akoa Hwy.
zip: 37920
Phone: (615) 544-9400 (Eastern
Tennessee, Southern Kentucky only)

Memphis
Southern Poison Center, Inc.
848 Adams Avenue
zip: 38103
Phone: (901) 528-6048

Nashville
Middle Tennessee Regional Poison
Clinical Toxicology Center
1161 21st Avenue
501 Oxford House
zip: 37232
Phone: (800) 288-9999 (Mid-
Tennessee only)
(615) 322-6435
(Nashville only)

TEXAS

Dallas
North Texas Poison Center
Parkland Hospital
5201 Harry Hines Blvd.
P.O. Box 35926
zip: 75235
Phone: (800) 441-0040 (Texas only)
(214) 590-5000

El Paso
El Paso Poison Control Center
R.E. Thomason General Hospital
4815 Alameda Avenue
zip: 79905
Phone: (915) 533-1244 (Southwestern
Texas, Southern New Mexico only)

Galveston
Texas State Poison Center
University of Texas Medical Branch
Eighth & Mechanic Streets
zip: 77550-2780
Phone: (800) 392-8548 (MD's and
ambulance personnel only; Texas only)
(409) 765-1420
(713) 654-1701 (Houston only)
(512) 478-4490 (Austin only)

UTAH

Salt Lake City
Intermountain Regional Poison
Control Center
50 North Medical Drive Bldg 528
zip: 84132
Phone: (800) 456-7707 (Utah only)
(801) 581-2151

VERMONT

Burlington
Vermont Poison Center
Medical Center Hospital of Vermont
Colchester Avenue
zip: 05401
Phone: (802) 658-3456 (Vermont &
bordering parts of New York only)
(802) 656-2721 (Education
Program)

VIRGINIA

Charlottesville
Blue Ridge Poison Center
University of Virginia Health
Sciences Center
Blue Ridge Hospital
Box 67
zip: 22901
Phone: (800) 451-1428 (Virginia only)
(804) 924-5543

Richmond
Virginia Poison Center
Virginia Commonwealth University
Box 522, MCV Station
Richmond, VA 23298
Phone: (800) 522-6337 (804 area code
only)
(804) 786-9123

WASHINGTON

Seattle
Seattle Poison Center
Children's Hospital and Medical Ctr
4800 Sand Point Way N.E.
P.O. Box C-5371
zip: 98105-0371
Phone: (800) 732-6985 (Washington
only)
(206) 526-2121
(206) 526-2223 (TTY)

Spokane
Spokane Poison Center
St. Luke's Hospital
S 711 Cowley
zip: 99202
Phone: (800) 572-5842 (Northern Idaho,
Western Montana, Northwestern Oregon)
(509) 747-1077

Tacoma
Mary Bridge Poison Center
Mary Bridge Children's Hospital
317 South K Street
P.O. Box 5299
zip: 98405-0987
Phone: (800) 542-6319 (Washington
only)
(206) 594-1414

Yakima
Central Washington Poison Center
Yakima Valley Memorial Hospital
2811 Tieton Drive
zip: 98902
Phone: (800) 572-9176 (Washington
only)
(509) 248-4400

```
============================== | ============================== | ==============================
  WEST VIRGINIA               |                               |
============================== | ============================== | ==============================
                              |                               |
  Charleston                  |                               |
  West Virginia Poison Center |                               |
  West Virginia University    |                               |
  Health Science Center       |                               |
  Charleston Division         |                               |
  3110 MacCorkle Avenue S.E.  |                               |
  zip: 25304                  |                               |
  Phone: (800) 642-3625 (West |                               |
                Virginia only)|                               |
        (304) 348-4211        |                               |
==============================|                               |
  WISCONSIN                   |                               |
==============================|                               |
                              |                               |
  Green Bay                   |                               |
  Green Bay Poison Control Center |                           |
  St. Vincent Hospital        |                               |
  P.O. Box 13508              |                               |
  zip: 54307-3508             |                               |
  Phone: (414) 433-8100       |                               |
                              |                               |
  Madison                     |                               |
  University of Wisconsin Hospital |                          |
  Regional Poison Control Center |                            |
  600 Highland Avenue         |                               |
  zip: 53792                  |                               |
  Phone: (608) 262-3702       |                               |
                              |                               |
  Milwaukee                   |                               |
  The Poison Center           |                               |
  Children's Hospital of Wisconsin |                          |
  9000 West Wisconsin Avenue  |                               |
  P.O. Box 1997               |                               |
  zip: 53201                  |                               |
  Phone: (414) 266-2222       |                               |
==============================|                               |
  WYOMING                     |                               |
==============================|                               |
                              |                               |
  Interstate Centers          |                               |
  Rocky Mountain Poison Center |                              |
  Denver, CO                  |                               |
  Phone: (800) 442-2702       |                               |
                              |                               |
  St. Luke's Midland Regional |                               |
  Medical Center Poison Control Ctr|                          |
  Aberdeen, SD                |                               |
  Phone: (800) 592-1889       |                               |
                              |                               |
                              |                               |
                              |                               |
                              |                               |
                              |                               |
```

Product List

PRODUCT	MANUFACTURER	PHONE NUMBER
9-1-1 ANTI-INFECTION OINTMENT	S.C. JOHNSON & SON	1-800-228-5635
9-1-1 BURN CREAM	S.C. JOHNSON & SON	1-800-228-5635
ADVANAGE #3260	AUSTIN DIVERSIFIED PRODUCTS	1-800-323-6444
AGREE CONDITIONER - EXTRA BODY FORMULA	S.C. JOHNSON & SON	1-800-228-5635
AGREE CONDITIONER - MOISTURE RICH FORMULA	S.C. JOHNSON & SON	1-800-228-5635
AGREE CONDITIONER - PROTEIN ENRICHED	S.C. JOHNSON & SON	1-800-228-5635
AGREE CONDITIONER - REGULAR FORMULA	S.C. JOHNSON & SON	1-800-228-5635
AGREE SHAMPOO - REGULAR	S.C. JOHNSON & SON	1-800-228-5635
AGREE SHAMPOO - EXTRA CLEANSING	S.C. JOHNSON & SON	1-800-228-5635
AGREE SHAMPOO - EXTRA BODY	S.C. JOHNSON & SON	1-800-228-5635
"ALL" AUTOMATIC DISH WASHING DETERGENT, LIQUID	LEVER BROTHERS CO.	1-301-688-6000
"ALL" CLEAR LAUNDRY DETERGENT, LIQUID	LEVER BROTHERS CO.	1-301-688-6000
"ALL" LAUNDRY DETERGENT, LIQUID	LEVER BROTHERS CO.	1-301-688-6000
AVEENO BATH - REGULAR	S.C. JOHNSON & SON	1-800-228-5635
AVEENO BATH - OILATED	S.C. JOHNSON & SON	1-800-228-5635
AVEENO LOTION	S.C. JOHNSON & SON	1-800-228-5635
AVEENO SHOWER AND BATH OIL	S.C. JOHNSON & SON	1-800-228-5635
AVEENO SUNSTICK	S.C. JOHNSON & SON	1-800-228-5635
AVEENO BAR - MEDICATED	S.C. JOHNSON & SON	1-800-228-5635
AVEENO BAR - OILATED	S.C. JOHNSON & SON	1-800-228-5635
AVEENO BAR - REGULAR	S.C. JOHNSON & SON	1-800-228-5635
BEHOLD LEMON SCENT FURNITURE POLISH	DRACKETT PRODUCTS	1-513-632-1500
BEHOLD LIGHT SCENT FURNITURE POLISH	DRACKETT PRODUCTS	1-513-632-1500
BIZ (POWDER BLEACH)	PROCTER & GAMBLE	1-800-543-1745
BOLD (LAUNDRY GRANULES)	PROCTER & GAMBLE	1-800-543-1745
BOLD, LIQUID (LAUNDRY DETERGENT)	PROCTER & GAMBLE	1-800-543-1745

PRODUCT	MANUFACTURER	PHONE NUMBER
BOLD, ULTRA (LAUNDRY GRANULES)	PROCTER & GAMBLE	1-800-543-1745
BOUNCE PERFUMED - HOUSEHOLD FABRIC SOFTENER SHEETS	PROCTER & GAMBLE	1-800-543-1745
BOUNCE UNSCENTED - HOUSEHOLD FABRIC SOFTENER SHEETS	PROCTER & GAMBLE	1-800-543-1745
BOUNCE WITH STAIN GUARD - LIQUID FABRIC SOFTENER	PROCTER & GAMBLE	1-800-543-1745
BRITE	PROCTER & GAMBLE	1-800-543-1745
CAMAY TOILET SOAP	PROCTER & GAMBLE	1-800-543-1745
CASCADE AUTOMATIC DISHWASHER DETERGENT	PROCTER & GAMBLE	1-800-543-1745
CASCADE, LEMON LIQUID (AUTOMATIC DISH WASHING DETERGENT)	PROCTER & GAMBLE	1-800-543-1745
CASCADE, LIQUID (AUTOMATIC DISH WASHING DETERGENT)	PROCTER & GAMBLE	1-800-543-1745
CASCADE LIQUIGEL, LEMON (DISH WASHER DETERGENT)	PROCTER & GAMBLE	1-800-543-1745
CHECK-UP GEL TOOTH PASTE (TUBE OR PUMP) ALL SIZES	S.C. JOHNSON & SON	1-800-228-5635
CHECK-UP GINGIVAL TOOTH PASTE	S.C. JOHNSON & SON	1-800-228-5635
CHECK-UP TOOTH PASTE (TUBE & PUMP) ALL SIZES	S.C. JOHNSON & SON	1-800-228-5635
CHEER (LAUNDRY GRANULES)	PROCTER & GAMBLE	1-800-543-1745
CHEER, LIQUID (LAUNDRY DETERGENT)	PROCTER & GAMBLE	1-800-543-1745
CHEER, ULTRA WITH COLOR GUARD (LAUNDRY GRANULES)	PROCTER & GAMBLE	1-800-543-1745
CLEAN & CLEAR	S.C. JOHNSON & SON	1-800-228-5635
CLOROX 2 (POWDER)	CLOROX COMPANY	1-800-446-1014
CLOROX 2, LIQUID	CLOROX CO.	1-800-446-1014
CLOROX, FRESH SCENT	CLOROX COMPANY	1-800-446-1014
CLOROX, LEMON FRESH	CLOROX COMPANY	1-800-446-1014
CLOROX BLEACH, INSTITUTIONAL	CLOROX COMPANY	1-800-446-1014
CLOROX BLEACH, REGULAR	CLOROX COMPANY	1-800-446-1014
CLOROX CLEAN-UP	CLOROX COMPANY	1-800-446-1014
CLOROX PLUS	CLOROX COMPANY	1-800-446-1014
CLOROX PRE-WASH (AEROSOL)	CLOROX COMPANY	1-800-446-1014
CLOROX THICKENED PRE-WASH	CLOROX COMPANY	1-800-446-1014

PRODUCT	MANUFACTURER	PHONE NUMBER
COAST DEODORANT TOILET BAR	PROCTER & GAMBLE	1-800-543-1745
COMBAT ANT CONTROL SYSTEM	CLOROX COMPANY	1-800-446-1014
COMBAT ANT & ROACH INSTANT KILLER	CLOROX COMPANY	1-800-446-1014
COMET, LEMON FRESH- CHLORINATED CLEANSER (NON-PHOSPHATE)	PROCTER & GAMBLE	1-800-543-1745
COMET, LEMON FRESH- CHLORINATED CLEANSER (PHOSPHATE)	PROCTER & GAMBLE	1-800-543-1745
COMET, LIQUID - CHLORINATED CLEANSER	PROCTER & GAMBLE	1-800-543-1745
COMET, SAFE FOR SURFACES- CHLORINATED CLEANSER (NON-PHOSPHATE)	PROCTER & GAMBLE	1-800-543-1745
COMET, SAFE FOR SURFACES- CHLORINATED CLEANSER (PHOSPHATE)	PROCTER & GAMBLE	1-800-543-1745
COMET TOILET BOWL CLEANSER, LIQUID	PROCTER & GAMBLE	1-800-543-1745
COMPLETE FOR FURNITURE	S.C. JOHNSON & SON	1-800-228-5635
CUREL	S.C. JOHNSON & SON	1-800-228-5635
CUREL, FRAGRANCE FREE	S.C. JOHNSON & SON	1-800-228-5635
DASH, LEMON FRESH	PROCTER & GAMBLE	1-800-543-1745
DASH, LIQUID (LAUNDRY DETERGENT)	PROCTER & GAMBLE	1-800-543-1745
DAWN (LIQUID DISH WASHING DETERGENT)	PROCTER & GAMBLE	1-800-543-1745
DAWN, MOUNTAIN SPRING (LIQUID DISH WASHING DETERGENT)	PROCTER & GAMBLE	1-800-543-1745
DOWNY LIQUID FABRIC SOFTENER (REGULAR CONCENTRATION)	PROCTER & GAMBLE	1-800-543-1745
DOWNY LIQUID FABRIC SOFTENER (TRIPLE CONCENTRATION)	PROCTER & GAMBLE	1-800-543-1745
DOWNY PERFUMED HOUSEHOLD FABRIC SOFTENER	PROCTER & GAMBLE	1-800-543-1745
DRANO CRYSTAL ALL PURPOSE DRAIN OPENER	DRACKETT PRODUCTS	1-513-632-1500
DRANO INSTANT PLUNGER	DRACKETT PRODUCTS	1-513-632-1500
DRANO LIQUID DRAIN OPENER	DRACKETT PRODUCTS	1-513-632-1500
DRANO LIQUID PROFESSIONAL STRENGTH	DRACKETT PRODUCTS	1-513-632-1500
DRANO LIQUID PROFESSIONAL STRENGTH PLUS DRAIN OPENER	DRACKETT PRODUCTS	1-513-632-1500
DREFT (LAUNDRY GRANULES)	PROCTER & GAMBLE	1-800-543-1745
DREFT, LIQUID (LAUNDRY DETERGENT)	PROCTER & GAMBLE	1-800-543-1745

PRODUCT	MANUFACTURER	PHONE NUMBER
DREFT, ULTRA (LAUNDRY GRANULES)	PROCTER & GAMBLE	1-800-543-1745
DRY BRITE	PROCTER & GAMBLE	1-800-543-1745
DUSTER PLUS - FRESH SCENT	S.C. JOHNSON & SON	1-800-228-5635
DUSTER PLUS - LEMON	S.C. JOHNSON & SON	1-800-228-5635
EDGE, LIME	S.C. JOHNSON & SON	1-800-228-5635
EDGE, MEDICATED/ MENTHOL	S.C. JOHNSON & SON	1-800-228-5635
EDGE FOR SENSITIVE SKIN - ALOE FORMULA	S.C. JOHNSON & SON	1-800-228-5635
EDGE SHAVING CREME (REGULAR)	S.C. JOHNSON & SON	1-800-228-5635
EDGE, SKIN CONDITIONING - LANOLIN FORMULA	S.C. JOHNSON & SON	1-800-228-5635
EDGE FOR TOUGH BEARDS	S.C. JOHNSON & SON	1-800-228-5635
ENDUST DUSTING AND CLEANING SPRAY	DRACKETT PRODUCTS	1-513-632-1500
ENDUST LEMON DUSTING AND CLEANING SPRAY	DRACKETT PRODUCTS	1-513-632-1500
ERA (LIQUID LAUNDRY DETERGENT)	PROCTER & GAMBLE	1-800-543-1745
FAVOR	S.C. JOHNSON & SON	1-800-228-5635
FAVOR, LIQUID LEMON	S.C. JOHNSON & SON	1-800-228-5635
FIBER ALL	S.C. JOHNSON & SON	1-800-228-5635
FIBER ALL WAFERS	S.C. JOHNSON & SON	1-800-228-5635
FIBER ALL BULK LAXATIVE TABLET	S.C. JOHNSON & SON	1-800-228-5635
FORMULA 409 ALL PURPOSE CLEANER	CLOROX COMPANY	1-800-446-1014
FRESH STEP (CAT LITTER)	CLOROX COMPANY	1-800-446-1014
FUTURE	S.C. JOHNSON & SON	1-800-228-5635
GAIN (LAUNDRY GRANULES)	PROCTER AND GAMBLE	1-800-543-1745
GAIN, ULTRA (LAUNDRY GRANULES)	PROCTER & GAMBLE	1-800-543-1745
GLADE - APRIL MEADOWS WITH CHLOROPHYLL	S.C. JOHNSON & SON	1-800-228-5635
GLADE - DILUTED WATER INTERMEDIATE	S.C. JOHNSON & SON	1-800-228-5635
GLADE - EARLY SPRING WITH CHLOROPHYLL	S.C. JOHNSON & SON	1-800-228-5635
GLADE - FABRIC FRESH	S.C. JOHNSON & SON	1-800-228-5635
GLADE - GENTLE FRESH WITH CHLOROPHYLL	S.C. JOHNSON & SON	1-800-228-5635

PRODUCT	MANUFACTURER	PHONE NUMBER
GLADE - LITTER FRESH	S.C. JOHNSON & SON	1-800-228-5635
GLADE - MORNING FRESH WITH CHLOROPHYLL	S.C. JOHNSON & SON	1-800-228-5635
GLADE - POWDER FRESH WITH CHLOROPHYLL	S.C. JOHNSON & SON	1-800-228-5635
GLADE - RAIN SHOWER FRESH WITH CHLOROPHYLL	S.C. JOHNSON & SON	1-800-228-5635
GLADE - SUMMER BREEZE WITH CHLOROPHYLL	S.C. JOHNSON & SON	1-800-228-5635
GLADE - SUPER FRESH WITH CHLOROPHYLL	S.C. JOHNSON & SON	1-800-228-5635
GLADE II - COUNTRY FRESH	S.C. JOHNSON & SON	1-800-228-5635
GLADE II - POWDER FRESH	S.C. JOHNSON & SON	1-800-228-5635
GLADE II - SUPER FRESH	S.C. JOHNSON & SON	1-800-228-5635
GLADE COUNTRY - COUNTRY BERRY	S.C. JOHNSON & SON	1-800-228-5635
GLADE COUNTRY - COUNTRY PEACH	S.C. JOHNSON & SON	1-800-228-5635
GLADE COUNTRY - COUNTRY SPICE	S.C. JOHNSON & SON	1-800-228-5635
GLADE COUNTRY - FRENCH VANILLA	S.C. JOHNSON & SON	1-800-228-5635
GLADE COUNTRY - SPICED APPLE	S.C. JOHNSON & SON	1-800-228-5635
GLADE LIGHT (ALL SCENTS)	S.C. JOHNSON & SON	1-800-228-5635
GLADE LIGHT - FLORAL WHISPER	S.C. JOHNSON & SON	1-800-228-5635
GLADE LIGHT - HINT OF POWDER WITH CHLOROPHYLL	S.C. JOHNSON & SON	1-800-228-5635
GLADE SPIN FRESH (ALL FRAGRANCES)	S.C. JOHNSON & SON	1-800-228-5635
GLO-COAT	S.C. JOHNSON & SON	1-800-228-5635
GLORY RUG CLEANER (AEROSOL)	S.C. JOHNSON & SON	1-800-228-5635
GLORY RUG SHAMPOO (LIQUID)	S.C. JOHNSON & SON	1-800-228-5635
GODDARD'S ALMOND FURNITURE OIL	S.C. JOHNSON & SON	1-800-228-5635
GODDARD'S BRASS AND COPPER CLEANER	S.C. JOHNSON & SON	1-800-228-5635
GODDARD'S CLEANING AND DUSTING SPRAY	S.C. JOHNSON & SON	1-800-228-5635
GODDARD'S DRY CLEAN	S.C. JOHNSON & SON	1-800-228-5635
GODDARD'S FINE FURNITURE CLEANER	S.C. JOHNSON & SON	1-800-228-5635
GODDARD'S FINE FURNITURE CREME WITH ALMOND OIL	S.C. JOHNSON & SON	1-800-228-5635

PRODUCT	MANUFACTURER	PHONE NUMBER
GODDARD'S FINE FURNITURE CREME WITH LEMON BEES WAX	S.C. JOHNSON & SON	1-800-228-5635
GODDARD'S FINE FURNITURE POLISH WITH ALMOND OIL	S.C. JOHNSON & SON	1-800-228-5635
GODDARD'S FINE FURNITURE POLISH WITH LEMON BEES WAX	S.C. JOHNSON & SON	1-800-228-5635
GODDARD'S GLOW BRASS AND COPPER POLISH	S.C. JOHNSON & SON	1-800-228-5635
GODDARD'S LEMON FURNITURE OIL	S.C. JOHNSON & SON	1-800=228-5635
GODDARD'S LONG SHINE BRASS AND COPPER POLISH	S.C. JOHNSON & SON	1-800-228-5635
GODDARD'S LONG SHINE SILVER FOAM	S.C. JOHNSON & SON	1-800-228-5635
GODDARD'S LONG SHINE SILVER POLISH	S.C. JOHNSON & SON	1-800-228-5635
GODDARD'S PEWTER CLEANER	S.C. JOHNSON & SON	1-800-228-5635
GODDARD'S STAINLESS STEEL CLEANER	S.C. JOHNSON & SON	1-800-228-5635
HALSA CONDITIONER - CHAMOMILE	S.C. JOHNSON & SON	1-800-228-5635
HALSA CONDITIONER - CORNFLOWER	S.C. JOHNSON & SON	1-800-228-5635
HALSA CONDITIONER - GINGER ROOT	S.C. JOHNSON & SON	1-800-228-5635
HALSA CONDITIONER - MARIGOLD	S.C. JOHNSON & SON	1-800-228-5635
HALSA DANDRUFF SHAMPOO FOR PREMED HAIR	S.C. JOHNSON & SON	1-800-228-5635
HALSA DANDRUFF SHAMPOO CONCENTRATE - REGULAR	S.C. JOHNSON & SON	1-800-228-5635
HALSA DANDRUFF SHAMPOO-REGULAR	S.C. JOHNSON & SON	1-800-228-5635
HALSA HAIR SPRAY - CHAMOMILE	S.C. JOHNSON & SON	1-800-228-5635
HALSA HAIR SPRAY - GINGER ROOT	S.C. JOHNSON & SON	1-800-228-5635
HALSA HAIR SPRAY - MARIGOLD	S.C. JOHNSON & SON	1-800-228-5635
HALSA HAIR SPRAY - WALNUT LEAVES	S.C. JOHNSON & SON	1-800-228-5635
HALSA MOUSSE - CHAMOMILE	S.C. JOHNSON & SON	1-800-228-5635
HALSA MOUSSE - GINGER ROOT	S.C. JOHNSON & SON	1-800-228-5635
HALSA MOUSSE - MARIGOLD	S.C. JOHNSON & SON	1-800-228-5635
HALSA MOUSSE - WALNUT LEAVES	S.C. JOHNSON & SON	1-800-228-5635
HALSA SHAMPOO - CHAMOMILE	S.C. JOHNSON & SON	1-800-228-5635
HALSA SHAMPOO - CORNFLOWER	S.C. JOHNSON & SON	1-800-228-5635
HALSA SHAMPOO - GINGER ROOT	S.C. JOHNSON & SON	1-800-228-5635

PRODUCT	MANUFACTURER	PHONE NUMBER
HALSA SHAMPOO - MARIGOLD	S.C. JOHNSON & SON	1-800-228-5635
HALSA SHAMPOO - WALNUT LEAVES	S.C. JOHNSON & SON	1-800-228-5635
HOT SHOTS	S.C. JOHNSON & SON	1-800-228-5635
IVORY, LIQUID - TOILET SOAP	PROCTER & GAMBLE	1-800-543-1745
IVORY BAR SOAP - TOILET SOAP	PROCTER & GAMBLE	1-800-543-1745
IVORY LIQUID (DISH WASHING LIQUID)	PROCTER & GAMBLE	1-800-543-1745
IVORY SNOW (SOAP GRANULES)	PROCTER & GAMBLE	1-800-543-1745
IVORY SNOW, LIQUID (LAUNDRY DETERGENT)	PROCTER & GAMBLE	1-800-543-1745
J/WAX AUTO CARPET CLEANER	S.C. JOHNSON & SON	1-800-228-5635
J/WAX AUTO FRESHENER	S.C. JOHNSON & SON	1-800-228-5635
J/WAX BASIC BLACK	S.C. JOHNSON & SON	1-800-228-5635
J/WAX CARNU	S.C. JOHNSON & SON	1-800-228-5635
J/WAX CAR WASH	S.C. JOHNSON & SON	1-800-228-5635
J/WAX CHROME CLEANER	S.C. JOHNSON & SON	1-800-228-5635
J/WAX FINE FABRIC CLEANER	S.C. JOHNSON & SON	1-800-228-5635
J/WAX KIT PASTE WAX	S.C. JOHNSON & SON	1-800-228-5635
J/WAX LIQUID KIT WAX	S.C. JOHNSON & SON	1-800-228-5635
J/WAX SPRINT WAX	S.C. JOHNSON & SON	1-800-228-5635
J/WAX VINYL TOP AND INTERIOR CLEANER	S.C. JOHNSON & SON	1-800-228-5635
J/WAX WHITE SIDE WALL CLEANER	S.C. JOHNSON & SON	1-800-228-5635
JOY	PROCTER & GAMBLE	1-800-543-1745
JUBILEE LIQUID	S.C. JOHNSON & SON	1-800-228-5635
JUBILEE SPRAY	S.C. JOHNSON & SON	1-800-228-5635
KINGSFORD BRIQUETS	CLOROX COMPANY	1-800-446-1014
KINGSFORD WITH MESQUITE, CHARCOAL BRIQUETS	CLOROX COMPANY	1-800-446-1014
KINGSFORD BBQ BAG	CLOROX COMPANY	1-800-446-1014
KINGSFORD FIRE RINGS - CHARCOAL	CLOROX COMPANY	1-800-446-1014
KINGSFORD ODORLESS CHARCOAL LIGHTER	CLOROX COMPANY	1-800-446-1014

PRODUCT	MANUFACTURER	PHONE NUMBER
KIRK'S CoCo HARDWATER CASTILE SOAP	PROCTER & GAMBLE	1-800-543-1745
KLEAN 'N SHINE	S.C. JOHNSON & SON	1-800-228-5635
KLEAR	S.C. JOHNSON & SON	1-800-228-5635
LAVA TOILET SOAP	PROCTER & GAMBLE	1-800-543-1745
LAVA SOAP, LIQUID - TOILET SOAP	PROCTER & GAMBLE	1-800-543-1745
L'ENVIE CONDITIONER - CAPTURE	S.C. JOHNSON & SON	1-800-228-5635
L'ENVIE CONDITIONER - CYPRESS	S.C. JOHNSON & SON	1-800-228-5635
L'ENVIE CONDITIONER - EMBER MUSK	S.C. JOHNSON & SON	1-800-228-5635
L'ENVIE CONDITIONER - LEGACE'	S.C. JOHNSON & SON	1-800-228-5635
L'ENVIE CONDITIONER - MILANO	S.C. JOHNSON & SON	1-800-228-5635
L'ENVIE CONDITIONER - SIAM	S.C. JOHNSON & SON	1-800-228-5635
L'ENVIE SHAMPOO - CAPTURE	S.C. JOHNSON & SON	1-800-228-5635
L'ENVIE SHAMPOO - CYPRESS	S.C. JOHNSON & SON	1-800-228-5635
L'ENVIE SHAMPOO - EMBER MUSK	S.C. JOHNSON & SON	1-800-228-5635
L'ENVIE SHAMPOO - LEGACE'	S.C. JOHNSON & SON	1-800-228-5635
L'ENVIE SHAMPOO - MILANO	S.C. JOHNSON & SON	1-800-228-5635
L'ENVIE SHAMPOO - SIAM	S.C. JOHNSON & SON	1-800-228-5635
LESTOIL DEODORIZING RUG SHAMPOO	NOXEL CORPORATION	1-301-785-4425
LESTOIL FLOOR CLEANER	NOXEL CORPORATION	1-301-785-4425
LESTOIL HEAVY DUTY CLEANER	NOXEL CORPORATION	1-301-785-4425
LIFE BUOY DEODORANT SOAP	LEVER BROTHERS CO.	1-212-688-6000
LIQUID PLUMBER AND PROFESSIONAL STRENGTH LIQUID PLUMBER	CLOROX CO.	1-800-446-1014
LITTER GREEN	CLOROX CO.	1-800-446-1014
LUX BEAUTY SOAP	LEVER BROTHERS CO.	1-212-688-6000
LUX LIGHT DUTY LIQUID DISH WASHING DETERGENT	LEVER BROTHERS CO.	1-212-688-6000
LYSOL BRAND BATHROOM TOUCH-UPS (DISINFECTANT CLEANING WIPES)	L & F PRODUCTS	1-201-573-5700

PRODUCT	MANUFACTURER	PHONE NUMBER
LYSOL BRAND CLING THICK LIQUID TOILET BOWL CLEANER - FRESH & PINE	L & F PRODUCTS	1-201-573-5700
LYSOL BRAND DISINFESTANT	L & F PRODUCTS	1-201-573-5700
LYSOL BRAND DISINFECTANT BASIN, TUB, AND TILE CLEANER (PUMP SPRAY)	L & F PRODUCTS	1-201-573-5700
LYSOL BRAND DISINFECTANT - COUNTRY	L & F PRODUCTS	1-201-573-5700
LYSOL BRAND DISINFECTANT DEODORIZING CLEANER	L & F PRODUCTS	1-201-573-5700
LYSOL BRAND DISINFECTANT DIRECT MULTI PURPOSE CLEANER	L & F PRODUCTS	1-201-573-5700
LYSOL BRAND DISINFECTANT - PINE ACTION	L & F PRODUCTS	1-201-573-5700
LYSOL BRAND DISINFECTANT - PINE SCENT	L & F PRODUCTS	1-201-573-5700
LYSOL BRAND DISINFECTANT SPRAY - FRESH	L & F PRODUCTS	1-201-573-5700
LYSOL BRAND DISINFECTANT SPRAY - LIGHT	L & F PRODUCTS	1-201-573-5700
LYSOL BRAND DISINFECTANT SPRAY - ORIGINAL	L & F PRODUCTS	1-201-573-5700
LYSOL BRAND FOAMING DISINFECTANT BASIN, TILE, TUB AND TILE CLEANER II	L & F PRODUCTS	1-201-573-5700
MATCH LIGHT (CHARCOAL)	CLOROX CO.	1-800-446-1014
MAXFORCE PHARAOH ANT KILLER	CLOROX CO.	1-800-446-1014
MAXFORCE ROACH CONTROL SYSTEM	CLOROX CO.	1-800-466-1014
MONCHEL TOILET SOAP	PROCTER & GAMBLE	1-800-543-1745
MR. CLEAN - LIQUID SYNTHETIC DETERGENT (NON PHOSPHATE)	PROCTER & GAMBLE	1-800-543-1745
MR. CLEAN - SOFT CLEANER (LIQUID ABRASIVE CLEANSER)	PROCTER & GAMBLE	1-800-543-1745
MR. CLEAN SPRAY - ALL PURPOSE CLEANER	PROCTER & GAMBLE	1-800-543-1745
MR. MUSCLE OVEN CLEANER	DRACKETT PRODUCTS	1-513-632-1500
MURPHY LAUNDRY AID	MURPHY - PHOENIX CO.	1-216-831-0404
MURPHY OIL SOAP LIQUID	MURPHY - PHOENIX CO.	1-216-831-0404
MURPHY OIL SOAP PASTE	MURPHY - PHOENIX CO.	1-216-831-0404
MURPHY OIL SOAP SPRAY	MURPHY - PHOENIX CO.	1-216-831-0404
OFF (AEROSOL)	S.C. JOHNSON & SON	1-800-228-5635

PRODUCT	MANUFACTURER	PHONE NUMBER
OFF (TOWELETTE)	S.C. JOHNSON & SON	1-800-228-5635
OFF, PUMP SPRAY	S.C. JOHNSON & SON	1-800-228-5635
OFF, DEEP WOODS (AEROSOL)	S.C. JOHNSON & SON	1-800-228-5635
OFF, DEEP WOODS (LOTION)	S.C. JOHNSON & SON	1-800-228-5635
OFF, DEEP WOODS (PUMP SPRAY)	S.C. JOHNSON & SON	1-800-228-5635
OFF, DEEP WOODS (TOWELETTE)	S.C. JOHNSON & SON	1-800-228-5635
OFF, MAXIMUM PROTECTION	S.C. JOHNSON & SON	1-800-228-5635
ONE STEP FINE WOOD FLOOR CARE WAX	S.C. JOHNSON & SON	1-800-228-5635
OXYDOL	PROCTER & GAMBLE	1-800-543-1745
OXYDOL, ULTRA (LAUNDRY GRANULES)	PROCTER & GAMBLE	1-800-543-1745
PASTE WAX	S.C. JOHNSON & SON	1-800-228-5635
PINE-SOL BROAD SPECTRUM FORMULA	CLOROX COMPANY	1-800-446-1014
PINE-SOL MULTI-ACTION SPRAY CLEANER	CLOROX COMPANY	1-800-446-1014
PINE-SOL, SPRING PINE - LIGHT SCENT CLEANER	CLOROX CO.	1-800-446-1014
PINE-SOL SPRUCE UPS	CLOROX COMPANY	1-800-446-1014
PLEDGE	S.C. JOHNSON & SON	1-800-228-5635
PLEDGE, LEMON	S.C. JOHNSON & SON	1-800-228-5635
PLEDGE, LEMON (PUMP SPRAY)	S.C. JOHNSON & SON	1-800-228-5635
PLEDGE, SPRING FRESH (PRESSURIZED UNIT)	S.C. JOHNSON & SON	1-800-228-5635
PLEDGE, WOOD SCENT	S.C. JOHNSON & SON	1-800-228-5635
PLEDGE LEMON OIL	S.C. JOHNSON & SON	1-800-228-5635
PLUNGE LIQUID DRAIN OPENER	DRACKETT PRODUCTS	1-513-632-1500
RAID ANT TRAPS	S.C. JOHNSON & SON	1-800-228-5635
RAID ANT TRAP II	S.C. JOHNSON & SON	1-800-228-5635
RAID ANT & ROACH KILLER - LIQUID FORMULA II	S.C. JOHNSON & SON	1-800-228-5635
RAID ANT & ROACH KILLER (PRESSURIZED)	S.C. JOHNSON & SON	1-800-228-5635
RAID CRACK & CREVICE	S.C. JOHNSON & SON	1-800-228-5635

PRODUCT	MANUFACTURER	PHONE NUMBER
RAID FLYING INSECT KILLER - FORMULA V	S.C. JOHNSON & SON	1-800-228-5635
RAID FIRE ANT KILLER - FORMULA 2	S.C. JOHNSON & SON	1-800-228-5635
RAID FLEA FUMIGATION CARTRIDGE	S.C. JOHNSON & SON	1-800-228-5635
RAID FLEA KILLER	S.C. JOHNSON & SON	1-800-228-5635
RAID FLEA KILLER PLUS	S.C. JOHNSON & SON	1-800-228-5635
RAID FUMIGATOR	S.C. JOHNSON & SON	1-800-228-5635
RAID GYPSY MOTH & JAPANESE BEETLE KILLER	S.C. JOHNSON & SON	1-800-228-5635
RAID HOUSE & GARDEN BUG KILLER	S.C. JOHNSON & SON	1-800-228-5635
RAID INDOOR FOGGER PLUS	S.C. JOHNSON & SON	1-800-228-5635
RAID INDOOR FOGGER PLUS	S.C. JOHNSON & SON	1-800-228-5635
RAID MOSQUITO REPELLENT COIL	S.C. JOHNSON & SON	1-800-228-5635
RAID OUTDOOR FLEA KILLER	S.C. JOHNSON & SON	1-800-228-5635
RAID PROFESSIONAL STRENGTH ANT & ROACH KILLER (LIQUID)	S.C. JOHNSON & SON	1-800-228-5635
RAID PROFESSIONAL STRENGTH ANT & ROACH KILLER (PRESSURIZED)	S.C. JOHNSON & SON	1-800-228-5635
RAID PROFESSIONAL STRENGTH FLYING INSECT KILLER	S.C. JOHNSON & SON	1-800-228-5635
RAID ROACH CONTROLLER (BAIT)	S.C. JOHNSON & SON	1-800-228-5635
RAID ROACH & FLEA KILLER	S.C. JOHNSON & SON	1-800-228-5635
RAID ROACH FUMIGATION CARTRIDGE	S.C. JOHNSON & SON	1-800-228-5635
RAID ROACH TRAPS	S.C. JOHNSON & SON	1-800-228-5635
RAID STRIP FLYING INSECT KILLER	S.C. JOHNSON & SON	1-800-228-5635
RAID TOMATO & VEGETABLE FOGGER	S.C. JOHNSON & SON	1-800-228-5635
RAID WASP & HORNET KILLER	S.C. JOHNSON & SON	1-800-228-5635
RAID YARD GUARD	S.C. JOHNSON & SON	1-800-228-5635
RAIN BARREL	S.C. JOHNSON & SON	1-800-228-5635
RENUZIT ADJUSTABLE AIR FRESHENER (SOLID)	DRACKETT PRODUCTS CO.	1-513-632-1500
RENUZIT AIR DEODORIZER (AEROSOL)	DRACKETT PRODUCTS CO.	1-513-632-1500
RENUZIT FRAGRANCE JAR - BAYBERRY	DRACKETT PRODUCTS CO.	1-513-632-1500

PRODUCT	MANUFACTURER	PHONE NUMBER
RENUZIT FRAGRANCE JAR - CINNAMON	DRACKETT PRODUCTS CO.	1-513-632-1500
RENUZIT FRAGRANCE JAR - FLORAL POTPOURRI	DRACKETT PRODUCTS CO.	1-513-632-1500
RENUZIT FRAGRANCE JAR - FLORAL POTPOURRI (27%)	DRACKETT PRODUCTS CO.	1-513-632-1500
RENUZIT FRAGRANCE JAR - FRESH CUT FLOWERS	DRACKETT PRODUCTS CO.	1-513-632-1500
RENUZIT FRAGRANCE JAR - FRESH POTPOURRI	DRACKETT PRODUCTS CO.	1-513-632-1500
RENUZIT FRAGRANCE JAR - JADE BREEZES	DRACKETT PRODUCTS CO.	1-513-632-1500
RENUZIT FRAGRANCE JAR - ORCHARD POTPOURRI	DRACKETT PRODUCTS CO.	1-513-632-1500
RENUZIT FRAGRANCE JAR - POWDER POTPOURRI	DRACKETT PRODUCTS CO.	1-513-632-1500
RENUZIT FRESH 'n DRY SPRAY (AEROSOL)	DRACKETT PRODUCTS CO.	1-513-632-1500
RENUZIT FRESHELL LONG LASTING AIR FRESHENER - MULTIPLE FRAGRANCES	DRACKETT PRODUCTS CO.	1-513-632-1500
RENUZIT ROOMMATE LIQUID AIR FRESHENER - MULTIPLE FRAGRANCES	DRACKETT PRODUCTS CO.	1-513-632-1500
RINSO NON-PHOSPHATE POWDER DETERGENT	LEVER BROTHERS	1-212-688-6000
RHULI CREAM	S.C. JOHNSON & SON	1-800-228-5635
RHULI GEL	S.C. JOHNSON & SON	1-800-228-5635
RHULI SPRAY	S.C. JOHNSON & SON	1-800-228-5635
SAFEGUARD DEODORANT TOILET BAR SOAP	PROCTER & GAMBLE	1-800-543-1745
SAFEGUARD DS FOR DRY SKIN PROTECTION - DEODORANT TOILET BAR SOAP	PROCTER & GAMBLE	1-800-543-1745
SHIELD DEODORANT BAR	LEVER BROTHERS	1-212-688-6000
SHOUT AEROSOL	S.C. JOHNSON & SON	1-800-228-5635
SHOUT LIQUID	S.C. JOHNSON & SON	1-800-228-5635
SKINTASTIC (ALL FRAGRANCES)	S.C. JOHNSON & SON	1-800-228-5635
SLEEK LIQUID CAR WAX	S.C. JOHNSON & SON	1-800-228-5635
SLEEK SOFT PASTE	S.C. JOHNSON & SON	1-800-228-5635
SNUGGLE DRYER FABRIC SOFTENER SHEETS	LEVER BROTHERS	1-212-688-6000
SNUGGLE LIQUID FABRIC SOFTENER	LEVER BROTHERS	1-212-688-6000
SNUGGLE LIQUID FABRIC SOFTENER, YELLOW	LEVER BROTHERS	1-212-688-6000

PRODUCT	MANUFACTURER	PHONE NUMBER
SOFT SCRUB	CLOROX CO.	1-800-446-1014
SOFT SCRUB WITH BLEACH	CLOROX CO.	1-800-446-1014
SOFT SENSE ALOE FORMULA	S.C. JOHNSON & SON	1-800-228-5635
SOFT SENSE ALOE SHAVE GEL	S.C. JOHNSON & SON	1-800-228-5635
SOFT SENSE BODY MOUSSE	S.C. JOHNSON & SON	1-800-228-5635
SOFT SENSE LANOLIN SHAVE GEL	S.C. JOHNSON & SON	1-800-228-5635
SOFT SENSE SKIN LOTION	S.C. JOHNSON & SON	1-800-228-5635
SOILOVE SOIL AND STAIN REMOVER	AMERICA'S FINEST PRODUCTS CORP.	1-213-450-6555
SOLO (LIQUID LAUNDRY DETERGENT)	PROCTER & GAMBLE	1-800-543-1745
SOL-ZOL HAND CLEANER	CRESSET	1-800-367-2020
SPIC AND SPAN - BUILT SYNTHETIC DETERGENT (PHOSPHATE)	PROCTER & GAMBLE	1-800-543-1745
SPIC AND SPAN - BUILT SYNTHETIC DETERGENT (NON-PHOSPHATE)	PROCTER & GAMBLE	1-800-543-1745
SPIC AND SPAN, LIQUID (PHOSPHATE)	PROCTER & GAMBLE	1-800-543-1745
SPIC AND SPAN, LIQUID (NON-PHOSPHATE)	PROCTER & GAMBLE	1-800-543-1745
SPIC AND SPAN, PINE CLEANER - LIQUID HOUSEHOLD CLEANER	PROCTER & GAMBLE	1-800-543-1745
SPIC AND SPAN, BATHROOM CLEANER (SPRAY)	PROCTER & GAMBLE	1-800-543-1745
STEP SAVER	S.C. JOHNSON & SON	1-800-228-5635
SUNLIGHT AUTOMATIC DISH WASHING DETERGENT	LEVER BROTHERS	1-212-688-6000
SUNLIGHT AUTOMATIC DISH WASHING DETERGENT, LIQUID	LEVER BROTHERS CO.	1-212-688-6000
SUNLIGHT DISH WASHING DETERGENT (LIQUID)	LEVER BROTHERS	1-212-688-6000
SUNLIGHT GEL AUTOMATIC DISH WASHING DETERGENT	LEVER BROTHERS	1-212-688-6000
SURF LAUNDRY DETERGENT, ULTRA PHOSPHATE	LEVER BROTHERS	1-212-688-6000
SURF LAUNDRY DETERGENT, ULTRA ZEOLITE POWDER	LEVER BROTHERS	1-212-688-6000
SURF LIQUID LAUNDRY DETERGENT, NON-PHOSPHATE	LEVER BROTHERS CO.	1-212-688-6000
SURF POWDER LAUNDRY DETERGENT, NON-PHOSPHATE	LEVER BROTHERS CO.	1-212-688-6000
TACKLE	CLOROX CO.	1-800-446-1014
TIDE (LAUNDRY GRANULES)	PROCTER & GAMBLE	1-800-543-1745

PRODUCT	MANUFACTURER	PHONE NUMBER
TIDE, LIQUID (LAUNDRY DETERGENT)	PROCTER & GAMBLE	1-800-543-1745
TIDE, ULTRA (LAUNDRY GRANULES)	PROCTER & GAMBLE	1-800-543-1745
TIDE, ULTRA (O - P) (LAUNDRY GRANULES)	PROCTER & GAMBLE	1-800-543-1745
TIDE WITH BLEACH ALTERNATIVE, LIQUID (LAUNDRY DETERGENT)	PROCTER & GAMBLE	1-800-543-1745
TIDE WITH BLEACH (LAUNDRY GRANULES)	PROCTER & GAMBLE	1-800-543-1745
TIDE WITH BLEACH, ULTRA (LAUNDRY GRANULES)	PROCTER & GAMBLE	1-800-543-1745
TILEX	CLOROX CO.	1-800-446-1014
TOP JOB - LIQUID HOUSEHOLD CLEANER (NON-PHOSPHATE)	PROCTER & GAMBLE	1-800-543-1745
TOP JOB SPRAY - ALL PURPOSE CLEANER	PROCTER & GAMBLE	1-800-543-1745
TSP POWDER CLEANER	AMERICA'S FINEST PRODUCTS	1-213-450-6555
TWINKLE SILVER POLISH	DRACKETT PRODUCTS CO.	1-513-632-1500
ULTRAGESIC LOTION	S.C. JOHNSON & SON	1-800-228-5635
VANISH BLUE AUTOMATIC TOILET BOWL CLEANER	DRACKETT PRODUCTS	1-513-632-1500
VANISH BOWL FRESHENER	DRACKETT PRODUCTS	1-513-632-1500
VANISH CLEAR DROP-INS - TANK AUTOMATIC BOWL CLEANER	DRACKETT PRODUCTS	1-513-632-1500
VANISH CRYSTAL TOILET BOWL CLEANER	DRACKETT PRODUCTS	1-513-632-1500
VANISH DISINFECTANT THICK HEAVY DUTY TOILET BOWL CLEANER	DRACKETT PRODUCTS	1-513-632-1500
VANISH DROP-INS - SOLID AUTOMATIC TOILET BOWL CLEANER (BLUE)	DRACKETT PRODUCTS	1-513-632-1500
VANISH DROP-INS - SOLID AUTOMATIC BOWL CLEANER (GREEN)	DRACKETT PRODUCTS	1-513-632-1500
VANISH DROP-INS - SOLID AUTOMATIC TOILET BOWL CLEANER (GREEN) (CANADIAN)	DRACKETT PRODUCTS	1-513-632-1500
VANISH EXTRA STRENGTH FRAGRANCE BOWL FRESHENER & DEODORIZER	DRACKETT PRODUCTS	1-513-632-1500
VANISH FOAMIN' TOILET BOWL CLEANER	DRACKETT PRODUCTS	1-513-632-1500
VANISH GREEN AUTOMATIC TOILET CLEANER	DRACKETT PRODUCTS	1-513-632-1500
VANISH SOLID AUTOMATIC TOILET BOWL CLEANER	DRACKETT PRODUCTS	1-513-632-1500
VIDAL SASSOON AEROSOL PROTEIN ENRICHED HAIR SPRAY	PROCTER & GAMBLE	1-800-543-1745

PRODUCT	MANUFACTURER	PHONE NUMBER
VIDAL SASSOON COLORIFIC GELS	PROCTER & GAMBLE	1-800-543-1745
VIDAL SASSOON COLORIFIC MOUSSES	PROCTER & GAMBLE	1-800-543-1745
VIDAL SASSOON EXTRA BODY MOUSSE	PROCTER & GAMBLE	1-800-543-1745
WINDEX ENVIRO - REFILL GLASS CLEANER BLUE	DRACKETT PRODUCTS	1-513-632-1500
WINDEX GLASS CLEANER (AEROSOL)	DRACKETT PRODUCTS	1-513-632-1500
WINDEX GLASS CLEANER (BLUE)	DRACKETT PRODUCTS	1-513-632-1500
WINDEX LEMON GLASS CLEANER (YELLOW)	DRACKETT PRODUCTS	1-513-632-1500
WINDEX PROFESSIONAL STRENGTH MULTI - SURFACE AND GLASS CLEANER	DRACKETT PRODUCTS	1-513-632-1500
WINDEX PROFESSIONAL STRENGTH MULTI - SURFACE AND GLASS CLEANER	DRACKETT PRODUCTS	1-513-632-1500
WINDEX VINEGAR GLASS CLEANER (GREEN)	DRACKETT PRODUCTS	1-513-632-1500
WISK NON-PHOSPHATE POWER SCOOP POWDER LAUNDRY DETERGENT	LEVER BROTHERS	1-212-688-6000
ZACT TOOTH PASTE	S.C. JOHNSON & SON	1-800-228-5635
ZEST DEODORANT BEAUTY BAR - TOILET SOAP	PROCTER & GAMBLE	1-800-543-1745

PRODUCT: 9-1-1 ANTI-INFECTION OINTMENT

INCOMPATIBILITY: STRONG ACIDS; STRONG ALKALIES; STRONG OXIDIZERS; BASES/CAUSTICS

INGREDIENTS: MINERAL OIL   TLV: CAS#: 8042-47-5
     PETROLATUM   TLV: CAS#: 8009-03-8

IGNI TEMP: NA FP: NA LEL: NA UEL: NA VP: NA VD: NA SG: NA PS: SEMI-SOLID APPEAR: WHITE ODOR: NONE
 PH FACTOR: NA

HAZARD RATINGS: H: F: R: 0 1 0

NEUTRALIZING AGENTS:

FIRE FIGHTING:
 HIGH EXPANSION FOAM, LOW EXPANSION FOAM, ALCOHOL FOAM; DRY CHEMICAL, CARBON DIOXIDE, WATER FOG.
 PROTECTIVE CLOTHING, RUBBER GLOVES, AND BREATHING APPARATUS.

WARNING: 1] STRUCTURAL PROTECTIVE CLOTHING IS PERMEABLE, REMAIN CLEAR OF SMOKE, WATER FALL OUT AND WATER RUN OFF.
   2] KEEP OUT OF THE REACH OF CHILDREN.
   3] DO NOT MIX WITH: SWIMMING POOL ACID, DRAIN CLEANERS, LIQUID CHLORINE.
   4] FOR EXTERNAL USE ONLY. DO NOT APPLY OVER A LARGE PART OF THE BODY.
   5] MOVE CONTAINERS FROM AREA IF WITHOUT RISK, COOL EXPOSED CONTAINERS.
   6] DIKE AREA FOR CONTROL AND CONTAINMENT TO PREVENT ENTRY INTO SEWERS, DRAIN, AND WATER WAYS.

LARGE SPILL/NO FIRE/RESCUE: WEAR RUBBER OR NEOPRENE BOOTS, GLOVES.

SPILL CONTROL AND CONTAINMENT:
 HOUSEHOLD SPILL: WIPE UP MATERIAL AND PUT IN A DISPOSAL CONTAINER. WASH AREA WITH SOAP AND WATER, PICK UP WITH A
 PAPER TOWEL, PUT IN DISPOSAL CONTAINER.
 LARGE SPILL: SCOOP UP MATERIAL AND PUT IN A DISPOSAL CONTAINER. WASH SPILL AREA WITH SOAP AND WATER, PICK UP
 WITH ABSORBENT, PUT IN DISPOSAL CONTAINER.

HEALTH HAZARD INFORMATION:
 WHEN USING THIS PRODUCT WEAR: NO SPECIAL CLOTHING REQUIRED. Contact with the eyes will cause irritation.
 Ingestion may cause gastrointestinal irritation, with nausea, possible vomiting and diarrhea.

 A physician should be contacted if anyone develops any signs or symptoms and suspects that they are caused by
 exposure to this product.

FIRST AID:
 Eye Exposure:
 Flush eyes with water for at least 15 minutes while lifting the upper and lower eye lids. Contact lens should
 not be worn when working with this product, get medical attention.

 Skin Exposure:
 If irritation appears get medical attention.

 Breathing:
 Not expected to be any problem

 Swallowing:
 Immediately contact the local poison control center for advice. Keep the victim warm and at rest, get medical
 attention.

PRODUCT: 9-1-1 BURN CREAM

INCOMPATIBILITY: NONE KNOWN

INGREDIENTS: BENZOCAINE   11% TLV: CAS#: 94-09-7
             GLYCERINE   20% TLV: CAS#: 56-81-5
             REMAINING PRODUCTS   79%
             PETROLATUM   DISTERYLDIMOMIUM CHLORIDE
             ISOPROPYL PALMITATE   EMULSIFIERS
             WATER

IGNI TEMP: NA FP: NA LEL: NA UEL: NA VP: NA VD: NA SG: 1.0 PS: THICK CREAM APPEAR: WHITE ODOR: ALMOND
   PH FACTOR:

HAZARD RATINGS: H: F: R: 0 0 0

NEUTRALIZING AGENTS:

FIRE FIGHTING:
   HIGH EXPANSION FOAM, LOW EXPANSION FOAM, ALCOHOL FOAM; DRY CHEMICAL, CARBON DIOXIDE, WATER FOG.
   PROTECTIVE CLOTHING, RUBBER GLOVES, AND BREATHING APPARATUS.
WARNING: 1] STRUCTURAL PROTECTIVE CLOTHING IS PERMEABLE, REMAIN CLEAR OF SMOKE, WATER FALL OUT AND WATER RUN OFF.
         2] KEEP OUT OF THE REACH OF CHILDREN.

LARGE SPILL/NO FIRE/RESCUE: WEAR RUBBER OR NEOPRENE BOOTS, GLOVES.

SPILL CONTROL AND CONTAINMENT:
   HOUSEHOLD SPILL: WIPE UP THE MATERIAL WITH A PAPER TOWEL, PUT IN DISPOSAL CONTAINER. WASH SPILL AREA WITH SOAP
   AND WATER, PICK UP WITH PAPER TOWELS AND PUT IN DISPOSAL CONTAINER.
   LARGE SPILL: SCOOP UP MATERIAL AND PUT IN A DISPOSAL CONTAINER. SCRUB SPILL AREA WITH SOAP AND WATER. PICK UP
   WITH ABSORBENT AND PUT IN DISPOSAL CONTAINER.

HEALTH HAZARD INFORMATION:
   WHEN USING THIS PRODUCT WEAR: NO SPECIAL CLOTHING REQUIRED. Direct contact with the eyes may cause irritation.
   Prolonged or repeated skin contact may cause irritation.

   A physician should be contacted if anyone develops any signs or symptoms and suspects that they are caused by
   exposure to this product.

FIRST AID:
   Eye Exposure:
   Flush the eyes with water for at least 15 minutes while lifting the upper and lower eye lids. Contact lenses
   should not be worn when working with this product, get medical attention.

   Skin Exposure:
   Wash exposed skin using plenty of soap and water. If irritation appears get medical attention.

   Breathing:
   Not expected to be an inhalation problem.

   Swallowing:
   Immediately contact the local poison control center for advice. Keep the victim warm and at rest, get medical
   attention.

PRODUCT: ADVANAGE #3260

INCOMPATIBILITY: STRONG ACIDS; OXIDIZING AGENTS.

INGREDIENTS: META SILICATE   3.5%
         SULFINATED ANIONIC DETERGENT

IGNI TEMP: NA FP: NA LEL: NA UEL: NA VP: NA VD: NA SG: 1.036 PS: LIQUID APPEAR: DARK PURPLE ODOR: NONE
  PH FACTOR: 12.3

HAZARD RATINGS: H: F: R:

NEUTRALIZING AGENTS:

FIRE FIGHTING:
  HIGH EXPANSION FOAM, LOW EXPANSION FOAM, ALCOHOL FOAM; DRY CHEMICAL, CARBON DIOXIDE, WATER FOG.
  PROTECTIVE CLOTHING, RUBBER GLOVES, AND BREATHING APPARATUS.
WARNING: 1] STRUCTURAL PROTECTIVE CLOTHING IS PERMEABLE, REMAIN CLEAR OF SMOKE, WATER FALL OUT AND WATER RUN OFF.
      2] KEEP OUT OF THE REACH OF CHILDREN.
      3] PRODUCT IS A CORROSIVE MATERIAL.
      4] MOVE CONTAINERS FROM AREA IF WITHOUT RISK, COOL EXPOSED CONTAINERS.
      5] DIKE AREA FOR CONTROL AND CONTAINMENT TO PREVENT ENTRY INTO SEWERS, DRAIN, AND WATER WAYS.

LARGE SPILL/NO FIRE/RESCUE: WEAR RUBBER OR NEOPRENE BOOTS, GLOVES.

SPILL CONTROL AND CONTAINMENT:
  HOUSEHOLD SPILL: WEAR RUBBER GLOVES, COVER SPILL AREA WITH A 50% WATER 50% VINEGAR SOLUTION. WIPE UP MATERIAL.
  WIPE UP WITH PAPER TOWELS, PUT IN A DISPOSAL CONTAINER.
  LARGE SPILL: PICK UP LIQUID WITH ABSORBENT. PUT IN DISPOSAL CONTAINER. RINSE AREA WITH WATER, PICK UP WITH
  ABSORBENT. PUT IN DISPOSAL CONTAINER. SEAL AND REMOVE CONTAINER FROM WORK AREA.

HEALTH HAZARD INFORMATION:
  WHEN USING THIS PRODUCT WEAR: RUBBER GLOVES AND EYE PROTECTION. Contact with the eyes will cause burning.
  Contact with the skin can cause burns or irritation. Ingestion can cause burns to the mouth, throat, and
  stomach.

  A physician should be contacted if anyone develops any signs or symptoms and suspects that they are caused by
  exposure to this product.

FIRST AID:
  Eye Exposure:
  Flush eyes with water for at least 15 minutes while lifting the upper and lower eye lids. Contact lenses should
  not be worn when working with this product, get medical attention.

  Skin Exposure:
  Flush skin with water, wash with soap and water. Get medical attention.

  Breathing:
  No problem if diluted properly with good ventilation. May cause upper respiratory problems. Move the victim to
  fresh air. Get medical attention.

  Swallowing:
  If the victim is conscious, give 2-3 glasses of water to drink. Immediately contact the local poison control
  center for advice. Keep victim warm and at rest. Get medical attention

PRODUCT: AGREE CONDITIONER - EXTRA BODY FORMULA

INCOMPATIBILITY: NONE KNOWN

INGREDIENTS: ALL ARE NON-HAZARDOUS BY OSHA 1910.1200 STANDARD

IGNI TEMP: NA FP: NA LEL: NA UEL: NA VP: NA VD: NA SG: 1.0 PS: LIQUID APPEAR: OPAQUE ODOR: FRAGRANT
   PH FACTOR: 3.1 TO 3.5

HAZARD RATINGS: H: F: R: 0 0 0

FIRE FIGHTING:
   HIGH EXPANSION FOAM, LOW EXPANSION FOAM, ALCOHOL FOAM; DRY CHEMICAL, CARBON DIOXIDE, WATER FOG.
   PROTECTIVE CLOTHING, RUBBER GLOVES, AND BREATHING APPARATUS.
WARNING: 1] STRUCTURAL PROTECTIVE CLOTHING IS PERMEABLE, REMAIN CLEAR OF SMOKE, WATER FALL OUT AND WATER RUN OFF.
         2] KEEP OUT OF THE REACH OF CHILDREN.
         3] THERMAL DECOMPOSITION MAY YIELD CARBON DIOXIDE, CARBON MONOXIDE.
         4] MOVE CONTAINERS FROM AREA IF WITHOUT RISK, COOL EXPOSED CONTAINERS.
         5] DIKE AREA FOR CONTROL AND CONTAINMENT TO PREVENT ENTRY INTO SEWERS, DRAIN, AND WATER WAYS.

LARGE SPILL/NO FIRE/RESCUE: WEAR RUBBER OR NEOPRENE BOOTS, GLOVES.

SPILL CONTROL AND CONTAINMENT:
   HOUSEHOLD SPILL: WEAR RUBBER GLOVES. WIPE UP SPILLED MATERIAL AND PUT IN A DISPOSAL CONTAINER. RINSE SPILL AREA
   WITH WATER. WIPE UP AND PUT IN A DISPOSAL CONTAINER.
   LARGE SPILL: COVER SPILL AREA WITH ABSORBENT. SWEEP UP AND PUT IN A DISPOSAL CONTAINER. RINSE SPILL AREA WITH
   WATER, PICK UP WITH ABSORBENT. PUT IN DISPOSAL CONTAINER. SEAL AND REMOVE FROM WORK AREA.

HEALTH HAZARD INFORMATION:
   WHEN USING THIS PRODUCT WEAR: NO SPECIAL REQUIREMENTS. Contact with the eyes may cause transient eye irritation.
   A skin rash may appear if allergic to this product. Ingestion may cause gastrointestinal distress, nausea,
   vomiting, and diarrhea.

   A physician should be contacted if anyone develops any signs or symptoms and suspects that they are caused by
   exposure to this product.

FIRST AID:
   Eye Exposure:
   Flush eyes with water for at least 15 minutes while lifting the upper and lower eye lids. Contact lenses should
   not be worn when working with this product. Get medical attention.

   Skin Exposure:
   Rinse exposed skin with water and wash with soap and water. If rash appears, get medical attention.

   Breathing:
   Not expected to be an inhalation problem.

   Swallowing:
   Give 2 to 3 glasses of milk or water to dilute the product. Immediately contact the local poison control center
   for advice. Keep victim warm and at rest. Get medical attention.

PRODUCT: AGREE CONDITIONER - MOISTURE RICH FORMULA

INCOMPATIBILITY: NONE KNOWN

INGREDIENTS: ALL ARE NON-HAZARDOUS BY OSHA 1910.1200 STANDARD

IGNI TEMP: NA FP: NA LEL: NA UEL: NA VP: NA VD: NA SG: 1.0 PS: LIQUID APPEAR: OPAQUE ODOR: FRAGRANT
  PH FACTOR: 3.1 TO 3.5

HAZARD RATINGS: H: F: R: 0 0 0

FIRE FIGHTING:
  HIGH EXPANSION FOAM, LOW EXPANSION FOAM, ALCOHOL FOAM; DRY CHEMICAL, CARBON DIOXIDE, WATER FOG.
  PROTECTIVE CLOTHING, RUBBER GLOVES, AND BREATHING APPARATUS.
WARNING: 1] STRUCTURAL PROTECTIVE CLOTHING IS PERMEABLE, REMAIN CLEAR OF SMOKE, WATER FALL OUT AND WATER RUN OFF.
         2] KEEP OUT OF THE REACH OF CHILDREN.
         3] THERMAL DECOMPOSITION MAY YIELD CARBON DIOXIDE, CARBON MONOXIDE.
         4] MOVE CONTAINERS FROM AREA IF WITHOUT RISK, COOL EXPOSED CONTAINERS.
         5] DIKE AREA FOR CONTROL AND CONTAINMENT TO PREVENT ENTRY INTO SEWERS, DRAIN, AND WATER WAYS.

LARGE SPILL/NO FIRE/RESCUE: WEAR RUBBER OR NEOPRENE BOOTS, GLOVES.

SPILL CONTROL AND CONTAINMENT:
  HOUSEHOLD SPILL: WEAR LATEX RUBBER GLOVES. WIPE UP MATERIAL WITH PAPER TOWELS AND PUT IN A DISPOSAL CONTAINER.
  RINSE SPILL AREA WITH WATER, WIPE UP WITH PAPER TOWELS, PUT IN DISPOSAL CONTAINER.
  LARGE SPILL: COVER SPILL WITH ABSORBENT. SWEEP UP AND PLACE IN A DISPOSAL CONTAINER. RINSE SPILL AREA WITH
  WATER, PICK UP WITH ABSORBENT, PUT IN A DISPOSAL CONTAINER. SEAL AND REMOVE FROM THE WORK AREA.

HEALTH HAZARD INFORMATION:
  WHEN USING THIS PRODUCT WEAR: NO SPECIAL REQUIREMENTS. Contact with the eyes may cause slight transient
  irritation. Rash may appear on skin if allergic to product. Ingestion may cause gastrointestinal distress,
  nausea, vomiting, and diarrhea.

  A physician should be contacted if anyone develops any signs or symptoms and suspects that they are caused by
  exposure to this product.

FIRST AID:
  Eye Exposure:
  Flush the eyes with water for at least 15 minutes while lifting the upper and lower eye lids. Contact lenses
  should not be worn when working with this product. Get medical attention.

  Skin Exposure:
  Rinse material off of exposed skin. If rash appears get medical attention.

  Breathing:
  Not expected to be a problem.

  Swallowing:
  If victim is conscious, give 2 to 3 glasses of milk or water to drink. Immediately contact the local poison
  control center for advice. Keep victim warm and at rest. Get medical attention.

PRODUCT: AGREE CONDITIONER - PROTEIN ENRICHED

INCOMPATIBILITY: NONE KNOWN

INGREDIENTS: CETYL ALCOHOL   CAS#: 36653-82-4 HYDROLIZED ANIMAL PROTEIN
             CETRIMONIUM CHLORIDE CAS#: 112-02-7 HYDROXYETHYL CELLULOSE
             DIMETHYL STEARAMINE CAS#:   STEARALKONIUM CHLORIDE
             PROPYLENE GLYCOL   CAS#: 4254-15-3 DEIONIZED WATER

IGNI TEMP: NA FP: NA LEL: NA UEL: NA VP: NA VD: NA SG: 1.0 PS: LIQUID APPEAR: ODOR: FRAGRANT
    PH FACTOR: 3.2

HAZARD RATINGS: H: F: R: 1 0 0

FIRE FIGHTING:
  HIGH EXPANSION FOAM, LOW EXPANSION FOAM, ALCOHOL FOAM; DRY CHEMICAL, CARBON DIOXIDE, WATER FOG.
  PROTECTIVE CLOTHING, RUBBER GLOVES, AND BREATHING APPARATUS.
WARNING: 1] STRUCTURAL PROTECTIVE CLOTHING IS PERMEABLE, REMAIN CLEAR OF SMOKE, WATER FALL OUT AND WATER RUN OFF.
         2] KEEP OUT OF THE REACH OF CHILDREN.
         3] THERMAL DECOMPOSITION MAY YIELD CARBON DIOXIDE, CARBON MONOXIDE.
         4] MOVE CONTAINERS FROM AREA IF WITHOUT RISK, COOL EXPOSED CONTAINERS.
         5] DIKE AREA FOR CONTROL AND CONTAINMENT TO PREVENT ENTRY INTO SEWERS, DRAIN, AND WATER WAYS.

LARGE SPILL/NO FIRE/RESCUE: WEAR RUBBER OR NEOPRENE BOOTS, GLOVES.

SPILL CONTROL AND CONTAINMENT:
  HOUSEHOLD SPILL: WIPE UP MATERIAL WITH AN ABSORBENT CLOTH, RINSE OUT IN THE SINK. RINSE AREA WITH WATER, PICK UP
  WITH AN ABSORBENT CLOTH, RINSE OUT IN THE SINK. PICK ANY RESIDUE UP WITH A DAMP CLOTH.
  LARGE SPILL: COVER SPILL AREA WITH ABSORBENT. SCOOP UP AND PUT IN A DISPOSAL CONTAINER. RINSE AREA WITH WATER,
  PICK UP WITH ABSORBENT, PUT IN A DISPOSAL CONTAINER. SEAL AND REMOVE FROM THE WORK AREA.

HEALTH HAZARD INFORMATION:
  WHEN USING THIS PRODUCT WEAR: NO SPECIAL REQUIREMENTS. Contact may cause eye irritation. May cause skin
  irritation if allergic to product. Ingestion may cause gastrointestinal distress, nausea, vomiting, and
  diarrhea.

  A physician should be contacted if anyone develops any signs or symptoms and suspects that they are caused by
  exposure to this product.

FIRST AID:
  Eye Exposure:
  Flush eyes with water for at least 15 minutes while lifting the upper lower eye lids. Contact lenses should not
  be worn when working with this product. Get medical attention.

  Skin Exposure:
  Rinse with water, wash with soap and water. If irritation appears get medical attention.

  Breathing:
  Not expected to be a problem.

  Swallowing:
  If victim is conscious, give 2 to 3 glasses of milk or water to drink. Immediately contact the local poison
  control center for advice. Keep victim warm and at rest. Get medical attention.

PRODUCT: AGREE CONDITIONER - REGULAR FORMULA

INCOMPATIBILITY: NONE KNOWN

INGREDIENTS: INGREDIENTS ARE NON HAZARDOUS BY OSHA 1910.1200 STANDARDS

IGNI TEMP: NA FP: NA LEL: NA UEL: NA VP: NA VD: NA SG: 1.0 PS: LIQUID APPEAR: OPAQUE ODOR: FRAGRANT
    PH FACTOR: 3.1 TO 3.5

HAZARD RATINGS: H: F: R: 0 0 0

FIRE FIGHTING:
    HIGH EXPANSION FOAM, LOW EXPANSION FOAM, ALCOHOL FOAM; DRY CHEMICAL, CARBON DIOXIDE, WATER FOG.
    PROTECTIVE CLOTHING, RUBBER GLOVES, AND BREATHING APPARATUS.
WARNING: 1] STRUCTURAL PROTECTIVE CLOTHING IS PERMEABLE, REMAIN CLEAR OF SMOKE, WATER FALL OUT AND WATER RUN OFF.
         2] KEEP OUT OF THE REACH OF CHILDREN.
         3] THERMAL DECOMPOSITION YIELDS TOXIC CARBON DIOXIDE, CARBON MONOXIDE.
         4] MOVE CONTAINERS FROM AREA IF WITHOUT RISK, COOL EXPOSED CONTAINERS.
         5] DIKE AREA FOR CONTROL AND CONTAINMENT TO PREVENT ENTRY INTO SEWERS, DRAIN, AND WATER WAYS.

LARGE SPILL/NO FIRE/RESCUE: WEAR RUBBER OR NEOPRENE BOOTS, GLOVES.

SPILL CONTROL AND CONTAINMENT:
    HOUSEHOLD SPILL: WIPE UP MATERIAL WITH ABSORBENT MATERIAL. PUT INTO A DISPOSAL CONTAINER. RINSE SPILL AREA WITH
    WATER, PICK UP WITH ABSORBENT MATERIAL AND PLACE IN A DISPOSAL CONTAINER.
    LARGE SPILL: COVER SPILL WITH ABSORBENT. SCOOP UP AND PUT INTO A DISPOSAL CONTAINER. RINSE SPILL AREA WITH
    WATER, PICK WITH ABSORBENT, PUT INTO A DISPOSAL CONTAINER. SEAL AND REMOVE FROM WORK AREA.

HEALTH HAZARD INFORMATION:
    WHEN USING THIS PRODUCT WEAR: NO SPECIAL REQUIREMENTS. Contact with the eyes may cause slight irritation. May
    cause skin rash if allergic to the product. Ingestion may cause gastrointestinal distress, nausea, vomiting, and
    diarrhea.

    A physician should be contacted if anyone develops any signs or symptoms and suspects that they are caused by
    exposure to this product.

FIRST AID:
    Eye Exposure:
    Flush the eyes with water for 15 minutes while lifting the upper and lower eye lids. Contact lenses should not
    be worn when working with this product. If irritation persists, get medical attention.

    Skin Exposure:
    Remove contaminated clothing. Wash skin with soap and water. If irritation appears, get medical attention.

    Breathing:
    Not expected to be a problem.

    Swallowing:
    If the victim is conscious, give 2-3 glasses of milk or water to drink. Immediately contact the local poison
    control center for advice. Keep victim warm and at rest. Get medical attention.

PRODUCT: AGREE SHAMPOO - REGULAR

INCOMPATIBILITY: NONE KNOWN

INGREDIENTS: AMMONIUM LAURYL SULFATE   TLV: CAS#:
             AMMONIUM LAURETH SULFATE   TLV: CAS#:
             AMMONIUM CHLORIDE   TLV: CAS#:   12125-02-9
             CITRIC ACID   TLV: CAS#:   77-92-9
             METHYLCHLOROISOTHIAZOLINONE TLV: CAS#:
             METHYLISOTHIAZOLINONE   TLV: CAS#:

IGNI TEMP: NA FP: NA LEL: NA UEL: NA VP: NA VD: NA SG: 1.0 PS: LIQUID APPEAR: OPAQUE ODOR: FRAGRANT
    PH FACTOR: 3.2

HAZARD RATINGS: H: F: R: 1   0   0

FIRE FIGHTING:
    HIGH EXPANSION FOAM, LOW EXPANSION FOAM, ALCOHOL FOAM; DRY CHEMICAL, CARBON DIOXIDE, WATER FOG.
    PROTECTIVE CLOTHING, RUBBER GLOVES, AND BREATHING APPARATUS.
WARNING: 1] STRUCTURAL PROTECTIVE CLOTHING IS PERMEABLE, REMAIN CLEAR OF SMOKE, WATER FALL OUT AND WATER RUN OFF.
         2] KEEP OUT OF THE REACH OF CHILDREN.
         3] THERMAL DECOMPOSITION YIELDS TOXIC CARBON DIOXIDE, CARBON MONOXIDE.
         4] MOVE CONTAINERS FROM AREA IF WITHOUT RISK, COOL EXPOSED CONTAINERS.
         5] DIKE AREA FOR CONTROL AND CONTAINMENT TO PREVENT ENTRY INTO SEWERS, DRAIN, AND WATER WAYS.

LARGE SPILL/NO FIRE/RESCUE: WEAR RUBBER OR NEOPRENE BOOTS, GLOVES.

SPILL CONTROL AND CONTAINMENT:
    HOUSEHOLD SPILL: PICK UP WITH PAPER TOWELS. WASH SPILL AREA WITH WATER, PICK UP WITH PAPER TOWELS AND PUT IN
    DISPOSAL CONTAINER.
    LARGE SPILL: COVER SPILL AREA WITH ABSORBENT, SCOOP UP AND PUT INTO A DISPOSAL CONTAINER. RINSE AREA WITH WATER,
    PICK UP WITH ABSORBENT, PUT INTO A DISPOSAL CONTAINER. SEAL AND REMOVE FROM WORK AREA.

HEALTH HAZARD INFORMATION:
    WHEN USING THIS PRODUCT WEAR: NO SPECIAL REQUIREMENTS. Contact with the eyes may cause slight irritation. May
    cause skin rash if allergic to the product. May cause gastrointestinal irritation and diarrhea.

    A physician should be contacted if anyone develops any signs or symptoms and suspects that they are caused by
    exposure to this product.

FIRST AID:
    Eye Exposure:
    Flush the eyes with water for 15 minutes while lifting the upper and lower eye lids. Contact lenses should not
    be worn when working with this product. If irritation persists, get medical attention.

    Skin Exposure:
    Wash with soap and water. If rash appears, get medical attention.

    Breathing:
    Not expected to be a problem.

    Swallowing:
    If the victim is conscious, give 2-3 glasses of milk or water to drink. Immediately contact the local poison
    control center for advice. Keep the victim warm and at rest. Get medical attention.

PRODUCT: AGREE SHAMPOO - EXTRA CLEANSING

INCOMPATIBILITY: NONE KNOWN

INGREDIENTS: AMMONIUM LAURYL SULFATE	TLV:	CAS#:	
AMMONIUM LAURETH SULFATE	TLV:	CAS#:	
AMMONIUM CHLORIDE	TLV:	CAS#:	12125-02-9
CITRIC ACID	TLV:	CAS#:	77-92-9
4-BENZOPHENONE	TLV:	CAS#:	119-61-9
METHYLCHLOROISOTHIAZOLINONE	TLV:	CAS#:	
METHYLISOTHIAZOLINONE	TLV:	CAS#:	

IGNI TEMP: NA FP: NA LEL: NA UEL: NA VP: NA VD: NA SG: 1.0 PS: LIQUID APPEAR: CLEAR ODOR: FRAGRANT
  PH FACTOR: 5.0

HAZARD RATINGS: H: F: R: 1 0 0

FIRE FIGHTING:
  HIGH EXPANSION FOAM, LOW EXPANSION FOAM, ALCOHOL FOAM; DRY CHEMICAL, CARBON DIOXIDE, WATER FOG.
  PROTECTIVE CLOTHING, RUBBER GLOVES, AND BREATHING APPARATUS.
WARNING: 1] STRUCTURAL PROTECTIVE CLOTHING IS PERMEABLE, REMAIN CLEAR OF SMOKE, WATER FALL OUT AND WATER RUN OFF.
         2] KEEP OUT OF THE REACH OF CHILDREN.
         3] THERMAL DECOMPOSITION YIELDS TOXIC CARBON DIOXIDE, CARBON MONOXIDE.
         4] MOVE CONTAINERS FROM AREA IF WITHOUT RISK, COOL EXPOSED CONTAINERS.
         5] DIKE AREA FOR CONTROL AND CONTAINMENT TO PREVENT ENTRY INTO SEWERS, DRAIN, AND WATER WAYS.

LARGE SPILL/NO FIRE/RESCUE: WEAR RUBBER OR NEOPRENE BOOTS, GLOVES.

SPILL CONTROL AND CONTAINMENT:
  HOUSEHOLD SPILL: WIPE UP SPILLED MATERIAL AND PUT INTO DISPOSAL CONTAINER. RINSE AREA WITH WATER, WIPE UP AND
  PUT INTO DISPOSAL CONTAINER.
  LARGE SPILL: COVER SPILL AREA WITH ABSORBENT. SWEEP UP AND PUT INTO DISPOSAL CONTAINER. RINSE SPILL AREA WITH
  WATER, PICK UP WITH ABSORBENT, PUT INTO DISPOSAL CONTAINER. SEAL AND REMOVE FROM WORK AREA.

HEALTH HAZARD INFORMATION:
  WHEN USING THIS PRODUCT WEAR: NO SPECIAL REQUIREMENTS. Contact causes eye irritation. Contact with the skin may
  cause a rash if allergic to product. Ingestion will cause gastrointestinal distress and diarrhea.

  A physician should be contacted if anyone develops any signs or symptoms and suspects that they are caused by
  exposure to this product.

FIRST AID:
  Eye Exposure:
  Flush eyes with water for 15 minutes while lifting the upper and lower eye lids. Contact lenses should not be
  worn when working with this product, get medical attention.

  Skin Exposure:
  Wash skin with soap and water. If rash appears, get medical attention.

  Breathing:
  Not expected to be a problem.

  Swallowing:
  If the victim is conscious, give 2-3 glasses of milk or water to drink. Immediately contact the local poison
  control center for advice. Keep the victim warm and at rest. Get medical attention.

PRODUCT: AGREE SHAMPOO - EXTRA BODY

INCOMPATIBILITY: NONE KNOWN

INGREDIENTS: AMMONIUM LAURETH SULFATE   TLV: CAS#:
             AMMONIUM CHLORIDE   TLV: CAS#: 12125-02-9
             BALSAM FRAGRANCE   TLV: CAS#:
             CITRIC ACID   TLV: CAS#:   77-92-9
             HYDROLIZED ANIMAL PROTEIN   TLV: CAS#:

IGNI TEMP: NA FP: NA LEL: NA UEL: NA VP: NA VD: NA SG: 1.0 PS: LIQUID APPEAR: CLEAR THICK ODOR: FRAGRANT
  PH FACTOR: 5.0

HAZARD RATINGS: H: F: R:

FIRE FIGHTING:
  HIGH EXPANSION FOAM, LOW EXPANSION FOAM, ALCOHOL FOAM; DRY CHEMICAL, CARBON DIOXIDE, WATER FOG.
  PROTECTIVE CLOTHING, RUBBER GLOVES, AND BREATHING APPARATUS.
WARNING: 1] STRUCTURAL PROTECTIVE CLOTHING IS PERMEABLE, REMAIN CLEAR OF SMOKE, WATER FALL OUT AND WATER RUN OFF.
         2] KEEP OUT OF THE REACH OF CHILDREN.
         3] THERMAL DECOMPOSITION YIELDS TOXIC CARBON DIOXIDE, CARBON MONOXIDE.
         4] MOVE CONTAINERS FROM AREA IF WITHOUT RISK, COOL EXPOSED CONTAINERS.
         5] DIKE AREA FOR CONTROL AND CONTAINMENT TO PREVENT ENTRY INTO SEWERS, DRAIN, AND WATER WAYS.

LARGE SPILL/NO FIRE/RESCUE: WEAR RUBBER OR NEOPRENE BOOTS, GLOVES.

SPILL CONTROL AND CONTAINMENT:
  HOUSEHOLD SPILL: WIPE UP SPILLED MATERIAL AND FLUSH DOWN SINK DRAIN. WIPE UP RESIDUE WITH A DAMP CLOTH.
  LARGE SPILL: COVER SPILL AREA WITH ABSORBENT, SWEEP UP AND PUT IN A DISPOSAL CONTAINER. RINSE AREA WITH WATER,
  PICK UP WITH ABSORBENT, PUT IN A DISPOSAL CONTAINER.

HEALTH HAZARD INFORMATION:
  WHEN USING THIS PRODUCT WEAR: NO SPECIAL REQUIREMENTS. May cause irritation of the eyes. May cause skin rash if
  allergic to the product. Ingestion may cause gastrointestinal distress and diarrhea.

  A physician should be contacted if anyone develops any signs or symptoms and suspects that they are caused by
  exposure to this product.

FIRST AID:
  Eye Exposure:
  Flush with water for 15 minutes while lifting the upper and lower eye lids. Contact lens should not be worn when
  working with this product. Get medical attention.

  Skin Exposure:
  Wash with soap and water. If rash appears, get medical attention.

  Breathing:
  Not expected to be a problem.

  Swallowing:
  If the victim is conscious, give 2-3 glasses of milk or water to drink. Immediately contact the local poison
  control center for advice. Keep the victim warm and at rest. Get medical attention.

PRODUCT: "ALL" AUTOMATIC DISH WASHER DETERGENT, LIQUID

INCOMPATIBILITY: NONE KNOWN

INGREDIENTS: SODIUM HYDROXIDE   1% TLV: 2 mg/m3   CAS#: 1310-73-2
        SODIUM CARBONATE   1% TLV: CAS#: 497-19-8
        SODIUM SILICATE   1% TLV: CAS#: 1344-09-8
        MINERAL OXIDE   1% TLV: 4.6 mg/m3 CAS#: 1302-78-9
                               (non-resp dust)

IGNI TEMP: NA FP: NA LEL: NA UEL: NA VP: NA VD: NA SG: 1.31 PS: LIQUID APPEAR: CLEAR ODOR: PERFUMED
  PH FACTOR: 11.8-12.5

HAZARD RATINGS: H: F: R: 2 0 0

FIRE FIGHTING:
  HIGH EXPANSION FOAM, LOW EXPANSION FOAM, ALCOHOL FOAM; DRY CHEMICAL, CARBON DIOXIDE, WATER FOG.
  PROTECTIVE CLOTHING, RUBBER GLOVES, AND BREATHING APPARATUS.
WARNING: 1] STRUCTURAL PROTECTIVE CLOTHING IS PERMEABLE, REMAIN CLEAR OF SMOKE, WATER FALL OUT AND WATER RUN OFF.
       2] KEEP OUT OF THE REACH OF CHILDREN.
       3] THERMAL DECOMPOSITION MAY YIELD TOXIC CHLORINE GAS, OXIDES OF SULFUR.
       4] MOVE CONTAINERS FROM AREA IF WITHOUT RISK, COOL EXPOSED CONTAINERS.
       5] DIKE AREA FOR CONTROL AND CONTAINMENT TO PREVENT ENTRY INTO SEWERS, DRAIN, AND WATER WAYS.

LARGE SPILL/NO FIRE/RESCUE: WEAR RUBBER OR NEOPRENE BOOTS, GLOVES.

SPILL CONTROL AND CONTAINMENT:
  HOUSEHOLD SPILL: WIPE UP LIQUID WITH AN ABSORBENT CLOTH, RINSE OUT IN THE SINK. WIPE UP RESIDUE WITH A DAMP
  CLOTH, RINSE OUT IN THE SINK.
  LARGE SPILL: PICK UP LIQUID WITH ABSORBENT, PUT IN A DISPOSAL CONTAINER. RINSE AREA WITH WATER, PICK UP WITH
  ABSORBENT AND PUT IN A DISPOSAL CONTAINER. SEAL AND REMOVE FROM THE WORK AREA.

HEALTH HAZARD INFORMATION:
  WHEN USING THIS PRODUCT WEAR: Eye contact will cause severe stinging, tearing, itching, swelling, and redness.
  Prolonged skin contact will cause drying of the skin and redness. Ingestion will cause abdominal pain, nausea,
  vomiting, and diarrhea. Inhalation may cause irritation of the nasal passage.

  A physician should be contacted if anyone develops any signs or symptoms and suspects that they are caused by
  exposure to this product.

FIRST AID:
  Eye Exposure:
  Flush with water for 15 minutes while lifting the upper and lower eye lids. Contact lenses should not be worn
  when working with this product. Get medical attention.

  Skin Exposure:
  Remove contaminated clothing, flush skin with water. If irritation appears, get medical attention.

  Breathing:
  If any symptoms develop, move victim to fresh air.

  Swallowing:
  If victim is conscious, give 1-2 glasses of milk or water to drink. Immediately contact the local poison control
  center for advice. Keep the victim warm and at rest. Get medical attention.

PRODUCT: "ALL" CLEAR LAUNDRY DETERGENT, LIQUID

INCOMPATIBILITY: STRONG ACIDS

INGREDIENTS: SODIUM SILICATE   1% TLV: CAS#: 1344-09-8

IGNI TEMP: NA FP: NA LEL: NA UEL: NA VP: NA VD: NA SG: 1.06-1.08 PS: LIQUID   APPEAR: CLEAR   ODOR:
   PH FACTOR: 11.8 - 12.2

HAZARD RATINGS: H: F: R:

FIRE FIGHTING:
   HIGH EXPANSION FOAM, LOW EXPANSION FOAM, ALCOHOL FOAM; DRY CHEMICAL, CARBON DIOXIDE, WATER FOG.
   PROTECTIVE CLOTHING, RUBBER GLOVES, AND BREATHING APPARATUS.
WARNING: 1] STRUCTURAL PROTECTIVE CLOTHING IS PERMEABLE, REMAIN CLEAR OF SMOKE, WATER FALL OUT AND WATER RUN OFF.
         2] KEEP OUT OF THE REACH OF CHILDREN.
         3] THERMAL DECOMPOSITION MAY YIELD TOXIC SULFUR DIOXIDE.
         4] MOVE CONTAINERS FROM AREA IF WITHOUT RISK, COOL EXPOSED CONTAINERS.
         5] DIKE AREA FOR CONTROL AND CONTAINMENT TO PREVENT ENTRY INTO SEWERS, DRAIN, AND WATER WAYS.

LARGE SPILL/NO FIRE/RESCUE: WEAR RUBBER OR NEOPRENE BOOTS, GLOVES.

SPILL CONTROL AND CONTAINMENT:
   HOUSEHOLD SPILL: WIPE UP LIQUID WITH AN ABSORBENT CLOTH. RINSE OUT IN THE SINK. WIPE UP RESIDUE WITH A DAMP
   CLOTH, RINSE OUT IN THE SINK.
   LARGE SPILL: PICK UP LIQUID WITH ABSORBENT. PUT IN A DISPOSAL CONTAINER. RINSE AREA WITH WATER, PICK UP WITH
   ABSORBENT. PUT IN A DISPOSAL CONTAINER. SEAL AND REMOVE FROM THE WORK AREA.

HEALTH HAZARD INFORMATION:
   WHEN USING THIS PRODUCT WEAR: NO SPECIAL REQUIREMENT. May cause burning sensation, tearing, itching, swelling,
   and redness. Prolonged skin contact may cause drying of the skin and irritation. Ingestion may result in
   abdominal pain, nausea, vomiting, and diarrhea. Inhalation may cause irritation of the respiratory tract.

   A physician should be contacted if anyone develops any signs or symptoms and suspects that they are caused by
   exposure to this product.

FIRST AID:
   Eye Exposure:
   Flush with water for 15 minutes while lifting the upper and lower eye lids. Contact lenses should not be worn
   when working with this product. Get medical attention.

   Skin Exposure:
   Wash off with water. If irritation appears, get medical attention.

   Breathing:
   If symptoms develop, move the victim to fresh air.

   Swallowing:
   If the victim is conscious, give 2-3 glasses of milk or water to drink. Immediately contact the local poison
   control center for advice. Keep the victim warm and at rest. Get medical attention.

PRODUCT: "ALL" LAUNDRY DETERGENT, LIQUID

INCOMPATIBILITY: STRONG ACIDS.

INGREDIENTS: SODIUM SILICATE   1%   TLV: CAS#: 1344-09-8

IGNI TEMP: NA FP: NA LEL: NA UEL: NA VP: NA VD: NA SG: 1.06-1.08 PS: LIQUID   APPEAR: ODOR:
   PH FACTOR: 11.8-12.2

HAZARD RATINGS: H: F: R:

FIRE FIGHTING:
   HIGH EXPANSION FOAM, LOW EXPANSION FOAM, ALCOHOL FOAM; DRY CHEMICAL, CARBON DIOXIDE, WATER FOG.
   PROTECTIVE CLOTHING, RUBBER GLOVES, AND BREATHING APPARATUS.
WARNING: 1] STRUCTURAL PROTECTIVE CLOTHING IS PERMEABLE, REMAIN CLEAR OF SMOKE, WATER FALL OUT AND WATER RUN OFF.
         2] KEEP OUT OF THE REACH OF CHILDREN.
         3] THERMAL DECOMPOSITION MAY YIELDS TOXIC SULFUR DIOXIDE.
         4] MOVE CONTAINERS FROM AREA IF WITHOUT RISK, COOL EXPOSED CONTAINERS.
         5] DIKE AREA FOR CONTROL AND CONTAINMENT TO PREVENT ENTRY INTO SEWERS, DRAIN, AND WATER WAYS.

LARGE SPILL/NO FIRE/RESCUE: WEAR RUBBER OR NEOPRENE BOOTS, GLOVES.

SPILL CONTROL AND CONTAINMENT:
   HOUSEHOLD SPILL: WIPE UP LIQUID WITH AN ABSORBENT CLOTH, RINSE OUT IN THE SINK. WIPE UP THE RESIDUE WITH A DAMP
   CLOTH, RINSE OUT IN THE SINK.
   LARGE SPILL: PICK UP LIQUID WITH ABSORBENT, PUT IN A DISPOSAL CONTAINER. RINSE AREA WITH WATER, PICK UP WITH
   ABSORBENT, PUT IN A DISPOSAL CONTAINER. SEAL AND REMOVE FROM THE WORK AREA.

HEALTH HAZARD INFORMATION:
   WHEN USING THIS PRODUCT WEAR: NO SPECIAL REQUIREMENT. Eye contact may cause burning sensation, tearing, itching,
   swelling, and redness. Prolonged contact may cause drying of the skin and irritation. Ingestion may result in
   abdominal pain, nausea, vomiting, and diarrhea. Inhalation may result in irritation of the respiratory tract.

   A physician should be contacted if anyone develops any signs or symptoms and suspects that they are caused by
   exposure to this product.

FIRST AID:
   Eye Exposure:
   Flush with water for 15 minutes while lifting the upper and lower eye lids. Contact lenses should not be worn
   when working with this product. Get medical attention.

   Skin Exposure:
   If irritation appears, get medical attention.

   Breathing:
   If symptoms develop, move the victim to fresh air.

   Swallowing:
   If the victim is conscious, give 2-3 glasses of milk or water to drink. Immediately contact the local poison
   control center for advice. Keep the victim warm and at rest. Get medical attention.

PRODUCT: AVEENO BATH - REGULAR

INCOMPATIBILITY: NONE KNOWN

INGREDIENTS: COLLOIDAL OATMEAL   TLV: 10 10 mg/m3 (DUST)

IGNI TEMP: NA FP: NA LEL: NA UEL: NA VP: NA VD: NA SG: NA PS: POWDER APPEAR: BUFF   ODOR:
    PH FACTOR: NA

HAZARD RATINGS: H: F: R: 0 0 0

FIRE FIGHTING:
    HIGH EXPANSION FOAM, LOW EXPANSION FOAM, ALCOHOL FOAM; DRY CHEMICAL, CARBON DIOXIDE, WATER FOG.
    PROTECTIVE CLOTHING, RUBBER GLOVES, AND BREATHING APPARATUS.
WARNING: 1] STRUCTURAL PROTECTIVE CLOTHING IS PERMEABLE, REMAIN CLEAR OF SMOKE, WATER FALL OUT AND WATER RUN OFF.
         2] KEEP OUT OF THE REACH OF CHILDREN.
         3] DUST IS CAPABLE OF FORMING AN EXPLOSIVE MIXTURE IN AIR.
         4] THERMAL DECOMPOSITION YIELDS TOXIC CARBON DIOXIDE, CARBON MONOXIDE.
         5] AVOID BREATHING EXCESSIVE AMOUNTS OF DUST.
         6] WHEN WET, PRODUCTS CREATES A SLIPPING HAZARD.
         7] MOVE CONTAINERS FROM AREA IF WITHOUT RISK, COOL EXPOSED CONTAINERS.
         8] DIKE AREA FOR CONTROL AND CONTAINMENT TO PREVENT ENTRY INTO SEWERS, DRAIN, AND WATER WAYS.

LARGE SPILL/NO FIRE/RESCUE: WEAR RUBBER OR NEOPRENE BOOTS, GLOVES.

SPILL CONTROL AND CONTAINMENT:
    HOUSEHOLD SPILL: AVOID CREATING A DUST. GENTLY SWEEP UP AND PUT INTO A DISPOSAL CONTAINER.
    LARGE SPILL: AVOID CREATING A DUST. GENTLY SWEEP UP MATERIAL AND PUT INTO A DISPOSAL CONTAINER. RINSE SPILL AREA
WITH WATER, PICK UP WITH ABSORBENT, PUT INTO A DISPOSAL CONTAINER. SEAL AND REMOVE FROM THE WORK PLACE.

HEALTH HAZARD INFORMATION:
    WHEN USING THIS PRODUCT WEAR: NO SPECIAL REQUIREMENTS. Inhalation of dust may cause irritation of the nose and
    throat. May cause skin rash if allergic to the product. Ingestion may cause gastrointestinal irritation and
    diarrhea.

    A physician should be contacted if anyone develops any signs or symptoms and suspects that they are caused by
    exposure to this product.

FIRST AID:
    Eye Exposure:
    Flush eyes with water for 15 minutes. If irritation is persistent, get medical attention. Contact lenses should
    not be worn when working with this product.

    Skin Exposure:
    Wash with soap and water. If rash appears, get medical attention.

    Breathing:
    Move the victim to fresh air. Treat any symptoms as they appear. Get medical attention.

    Swallowing:
    Immediately contact the local poison control center for advice. Keep the victim warm and at rest. Get medical
    attention.

PRODUCT: AVEENO BATH - OILATED

INCOMPATIBILITY: NONE KNOWN

INGREDIENTS: COLLOIDAL OATMEAL	TLV: 10 mg/m3(DUST)	CAS#:
CALCIUM SILICATE	TLV:	CAS#:
MINERAL OIL	TLV:	CAS#: 8042-47-5
LAURETH-4	TLV:	CAS#:
LANOLIN ALCOHOL	TLV:	CAS#:

IGNI TEMP: NA FP: NA LEL: NA UEL: NA VP: NA VD: NA SG: NA PS: POWDER APPEAR: GUM LIKE ODOR:
 PH FACTOR: NA

HAZARD RATINGS: H: F: R: 0 1 0

FIRE FIGHTING:
  HIGH EXPANSION FOAM, LOW EXPANSION FOAM, ALCOHOL FOAM; DRY CHEMICAL, CARBON DIOXIDE, WATER FOG.
  PROTECTIVE CLOTHING, RUBBER GLOVES, AND BREATHING APPARATUS.
WARNING: 1] STRUCTURAL PROTECTIVE CLOTHING IS PERMEABLE, REMAIN CLEAR OF SMOKE, WATER FALL OUT AND WATER RUN OFF.
         2] KEEP OUT OF THE REACH OF CHILDREN.
         3] SPILL MAY CREATE A SLIPPING HAZARD.
         4] THERMAL DECOMPOSITION YIELDS TOXIC CARBON DIOXIDE, CARBON MONOXIDE.
         5] MOVE CONTAINERS FROM AREA IF WITHOUT RISK, COOL EXPOSED CONTAINERS.
         6] DIKE AREA FOR CONTROL AND CONTAINMENT TO PREVENT ENTRY INTO SEWERS, DRAIN, AND WATER WAYS.

LARGE SPILL/NO FIRE/RESCUE: WEAR RUBBER OR NEOPRENE BOOTS, GLOVES.

SPILL CONTROL AND CONTAINMENT:
  HOUSEHOLD SPILL: WIPE UP LIQUID WITH ABSORBENT MATERIAL. RINSE OUT IN SINK. WIPE UP RESIDUE WITH A DAMP CLOTH
  AND RINSE OUT IN THE SINK.
  LARGE SPILL: COVER SPILL AREA WITH ABSORBENT. SCOOP UP AND PUT IN A DISPOSAL CONTAINER. RINSE SPILL AREA WITH
  WATER, PICK UP WITH ABSORBENT, PUT IN A DISPOSAL CONTAINER. SEAL AND REMOVE FROM WORK AREA.

HEALTH HAZARD INFORMATION:
  WHEN USING THIS PRODUCT WEAR: NO SPECIAL REQUIREMENTS. Contact may cause eye irritation. May cause a skin rash
  if allergic to product. Ingestion may cause gastrointestinal irritation and diarrhea.

  A physician should be contacted if anyone develops any signs or symptoms and suspects that they are caused by
  exposure to this product.

FIRST AID:
  Eye Exposure:
  Flush eyes with water for 15 minutes. If irritation persists get medical attention. Contact lenses should not be
  worn when working with this product.

  Skin Exposure:
  Wash skin with water. If rash appears, get medical attention.

  Breathing:
  Not expected to be a problem.

  Swallowing:
  Immediately contact the local poison control center for advice. Keep the victim warm and at rest. Get medical
  attention.

PRODUCT: AVEENO LOTION

INCOMPATIBILITY: NONE KNOWN

INGREDIENTS: PETROLATUM   TLV: CAS#:   8009-03-8
             ISOPROPYLPALMITATE   TLV: CAS#:
             CETYL ALCOHOL   TLV: CAS#:   36653-82-4
             GLYCERINE   TLV: CAS#:   56-81-5
             SODIUM CHLORIDE   TLV: CAS#:   7647-14-5
             BENZYL ALCOHOL   TLV: CAS#:   100-51-6
             DIMETHICONE   TLV: CAS#:

IGNI TEMP: NA FP: NA LEL: NA UEL: NA VP: 760 mmHg @ 212F VD: 1.0 SG: 1.0 PS: LIQUID   APPEAR: CREAMY OFF
    WHITE   ODOR: PH FACTOR: 4.5 - 6.5

HAZARD RATINGS: H: F: R: 0 0 0

FIRE FIGHTING:
  HIGH EXPANSION FOAM, LOW EXPANSION FOAM, ALCOHOL FOAM; DRY CHEMICAL, CARBON DIOXIDE, WATER FOG.
  PROTECTIVE CLOTHING, RUBBER GLOVES, AND BREATHING APPARATUS.
WARNING: 1] STRUCTURAL PROTECTIVE CLOTHING IS PERMEABLE, REMAIN CLEAR OF SMOKE, WATER FALL OUT AND WATER RUN OFF.
         2] KEEP OUT OF THE REACH OF CHILDREN.
         3] THERMAL DECOMPOSITION YIELDS TOXIC CARBON DIOXIDE, CARBON MONOXIDE.
         4] SPILLED PRODUCT CREATES A SLIPPING HAZARD.
         5] MOVE CONTAINERS FROM AREA IF WITHOUT RISK, COOL EXPOSED CONTAINERS.
         6] DIKE AREA FOR CONTROL AND CONTAINMENT TO PREVENT ENTRY INTO SEWERS, DRAIN, AND WATER WAYS.

LARGE SPILL/NO FIRE/RESCUE: WEAR RUBBER OR NEOPRENE BOOTS, GLOVES.

SPILL CONTROL AND CONTAINMENT:
  HOUSEHOLD SPILL: WIPE UP SPILL WITH ABSORBENT CLOTH, RINSE IN SINK.
  LARGE SPILL: COVER SPILL WITH ABSORBENT. SWEEP UP AND PUT IN A DISPOSAL CONTAINER. RINSE SPILL AREA WITH WATER,
  PICK UP WITH ABSORBENT. PUT IN A DISPOSAL CONTAINER.

HEALTH HAZARD INFORMATION:
  WHEN USING THIS PRODUCT WEAR: NO SPECIAL REQUIREMENT. Contact will cause eye irritation. May cause skin rash if
  allergic to product. Ingestion will cause gastrointestinal distress and diarrhea.

  A physician should be contacted if anyone develops any signs or symptoms and suspects that they are caused by
  exposure to this product.

FIRST AID:
  Eye Exposure:
  Flush eyes with water for 15 minutes while lifting the upper and lower eye lids. Contact lenses should not be
  worn when working with this product, If irritation persists, get medical attention.

  Skin Exposure:
  Not expected to be a problem. If rash appears, get medical attention.

  Breathing:
  Not expected to be a problem.

  Swallowing:
  Immediately contact the local poison control center for advice. Keep the victim warm and at rest. Get medical
  attention.

PRODUCT: AVEENO SHOWER AND BATH OIL

INCOMPATIBILITY: NONE KNOWN

INGREDIENTS:	TLV:	CAS#:
COLLOIDAL OATMEAL	TLV:	CAS#:
MINERAL OIL	TLV:	CAS#: 8042-47-5
GLYCERYL STEARATE	TLV:	CAS#:
PEG 100 STEARATE	TLV:	CAS#: 25322-68-3
BENZYL ALCOHOL	TLV:	CAS#: 100-51-6
BENZALDEHYDE	TLV:	CAS#: 100-52-7
SILICA	TLV: 10 mg/m3(DUST)	CAS#: 63231-67-4

IGNI TEMP: NA FP: NA LEL: NA UEL: NA VP: NA VD: NA SG: .9 PS: LIQUID APPEAR: PALE YELLOW ODOR: ALMOND
   PH FACTOR: 5.8-6.5

HAZARD RATINGS: H: F: R: 0 0 0

FIRE FIGHTING:
  HIGH EXPANSION FOAM, LOW EXPANSION FOAM, ALCOHOL FOAM; DRY CHEMICAL, CARBON DIOXIDE, WATER FOG.
  PROTECTIVE CLOTHING, RUBBER GLOVES, AND BREATHING APPARATUS.
WARNING: 1] STRUCTURAL PROTECTIVE CLOTHING IS PERMEABLE, REMAIN CLEAR OF SMOKE, WATER FALL OUT AND WATER RUN OFF.
     2] KEEP OUT OF THE REACH OF CHILDREN.
     3] THERMAL DECOMPOSITION YIELDS TOXIC CARBON DIOXIDE, CARBON MONOXIDE.
     4] SPILL CREATES A SLIPPING HAZARD.
     5] MOVE CONTAINERS FROM AREA IF WITHOUT RISK, COOL EXPOSED CONTAINERS.
     6] DIKE AREA FOR CONTROL AND CONTAINMENT TO PREVENT ENTRY INTO SEWERS, DRAIN, AND WATER WAYS.

LARGE SPILL/NO FIRE/RESCUE: WEAR RUBBER OR NEOPRENE BOOTS, GLOVES.

SPILL CONTROL AND CONTAINMENT:
  HOUSEHOLD SPILL: WIPE UP MATERIAL WITH AN ABSORBENT CLOTH. RINSE IN SINK. WASH WITH SOAP AND WATER, WIPE UP DRY.
  LARGE SPILL: COVER SPILL WITH ABSORBENT. PICK UP AND PUT INTO A DISPOSAL CONTAINER. WASH AREA WITH SOAP AND
  WATER PICK UP WITH ABSORBENT, PUT INTO A DISPOSAL CONTAINER. SEAL AND REMOVE FROM WORK AREA.

HEALTH HAZARD INFORMATION:
  WHEN USING THIS PRODUCT WEAR: NO SPECIAL REQUIREMENTS. Contact may cause eye irritation. May cause skin rash if
  allergic to product. Ingestion may cause gastrointestinal irritation and diarrhea.

  A physician should be contacted if anyone develops any signs or symptoms and suspects that they are caused by
  exposure to this product.

FIRST AID:
  Eye Exposure:
  Flush eyes with water for 15 minutes while lifting the upper and lower eye lids. Contact lenses should not be
  worn when working with this product. If irritation persists, get medical attention.

  Skin Exposure:
  Wash with soap and water. If rash appears, get medical attention.

  Breathing:
  Not expected to be a problem.

  Swallowing:
  Immediately contact the local poison control center for advice. Keep the victim warm and at rest. Get medical
  attention.

PRODUCT: AVEENO SUNSTICK

INCOMPATIBILITY: NONE KNOWN

INGREDIENTS: CARNUBA WAX   5-8 %   TLV: CAS#:   8015-86-9
             BEES WAX   10-15%   TLV: CAS#:   36653-82-4
             CETYL ALCOHOL   1-3 %   TLV: CAS#:   8012-89-3
             LANOLIN OIL   7-10%   TLV: CAS#:   8006-54-0
             MINERAL OIL   7-10%   TLV: CAS#:   8042-47-5
             CASTOR OIL   50 50-58%   TLV: CAS#:   8001-79-4
             HOMOSALATE   3.5%   TLV: CAS#:   118-56-9

IGNI TEMP: NA FP: NA LEL: NA UEL: NA VP: NA VD: NA SG: NA PS: SOLID STICK APPEAR: PALE YELLOW ODOR:
  PH FACTOR: NA

HAZARD RATINGS: H: F: R: 0   0   0

FIRE FIGHTING:
  HIGH EXPANSION FOAM, LOW EXPANSION FOAM, ALCOHOL FOAM; DRY CHEMICAL, CARBON DIOXIDE, WATER FOG.
  PROTECTIVE CLOTHING, RUBBER GLOVES, AND BREATHING APPARATUS.
WARNING: 1] STRUCTURAL PROTECTIVE CLOTHING IS PERMEABLE, REMAIN CLEAR OF SMOKE, WATER FALL OUT AND WATER RUN OFF.
         2] KEEP OUT OF THE REACH OF CHILDREN.
         3] LIQUID RUN-OFF INTO SEWERS MAY CAUSE A FIRE/EXPLOSION HAZARD.
         4] PRODUCT WILL BURN YIELDING TOXIC CARBON DIOXIDE, CARBON MONOXIDE.
         5] MOVE CONTAINERS FROM AREA IF WITHOUT RISK, COOL EXPOSED CONTAINERS.
         6] DIKE AREA FOR CONTROL AND CONTAINMENT TO PREVENT ENTRY INTO SEWERS, DRAIN, AND WATER WAYS.

LARGE SPILL/NO FIRE/RESCUE: WEAR RUBBER OR NEOPRENE BOOTS, GLOVES.

SPILL CONTROL AND CONTAINMENT:
  HOUSEHOLD SPILL: PICK UP OR SCRAPE UP MATERIAL AND PUT IN DISPOSAL CONTAINER. SCRUB AREA WITH SOAP AND WATER,
  PICK UP WITH ABSORBENT CLOTH. RINSE IN SINK.
  LARGE SPILL: SCRAPE UP MATERIAL FOR RECYCLE OR DISPOSAL. SCRUB AREA WITH SOAP AND WATER. PICK UP WITH ABSORBENT.
  PUT IN A DISPOSAL CONTAINER.

HEALTH HAZARD INFORMATION:
  WHEN USING THIS PRODUCT WEAR: NO SPECIAL REQUIREMENTS. May cause eye irritation. May cause a skin rash if
  allergic to product. Ingestion will cause gastrointestinal irritation and diarrhea.

  A physician should be contacted if anyone develops any signs or symptoms and suspects that they are caused by
  exposure to this product.

FIRST AID:
Eye Exposure:
Flush eyes with water for 15 minutes while lifting the upper and lower eye lids. Contact lenses should not be
worn when working with this product. If irritation persists, get medical attention.

Skin Exposure:
Not expected to be a problem. If rash appears, get medical attention.

Breathing:
Not expected to be a problem.

Swallowing:
Immediately contact the local poison control center for advice. Keep the victim warm and at rest. Get medical
attention.

PRODUCT: AVEENOBAR, MEDICATED

INCOMPATIBILITY: NONE KNOWN

INGREDIENTS: COLLOIDAL OATMEAL   50%   TLV: CAS#:
             SULFUR   2%   TLV: CAS#: 7704-34-9
             SALICYLIC ACID   2%   TLV: CAS#:   69-72-7
             SURFACTANTS   46%   TLV: CAS#:

IGNI TEMP: NA FP: NA LEL: NA UEL: NA VP: NA VD: NA SG: NA PS: SOLID APPEAR: TAN COLOR ODOR:
    PH FACTOR: 4.8-6.0

HAZARD RATINGS: H: F: R: 0 0 0

NEUTRALIZING AGENTS:

FIRE FIGHTING:
    HIGH EXPANSION FOAM, LOW EXPANSION FOAM, ALCOHOL FOAM; DRY CHEMICAL, CARBON DIOXIDE, WATER FOG.
    PROTECTIVE CLOTHING, RUBBER GLOVES, AND BREATHING APPARATUS.
WARNING: 1] STRUCTURAL PROTECTIVE CLOTHING IS PERMEABLE, REMAIN CLEAR OF SMOKE, WATER FALL OUT AND WATER RUN OFF.
         2] KEEP OUT OF THE REACH OF CHILDREN.
         3] THERMAL DECOMPOSITION YIELDS TOXIC CARBON DIOXIDE, CARBON MONOXIDE.
         4] MAY CREATE A SLIPPING HAZARD.
         5] MOVE CONTAINERS FROM AREA IF WITHOUT RISK, COOL EXPOSED CONTAINERS.
         6] DIKE AREA FOR CONTROL AND CONTAINMENT TO PREVENT ENTRY INTO SEWERS, DRAIN, AND WATER WAYS.

LARGE SPILL/NO FIRE/RESCUE: WEAR RUBBER OR NEOPRENE BOOTS, GLOVES.

SPILL CONTROL AND CONTAINMENT:
    HOUSEHOLD SPILL: SWEEP OR SCRAPE UP MATERIAL AND PUT INTO DISPOSAL CONTAINER.
    LARGE SPILL: SWEEP OR SCOOP UP MATERIAL AND PUT INTO A DISPOSAL CONTAINER. SEAL AND REMOVE FROM WORK AREA.

HEALTH HAZARD INFORMATION:
    WHEN USING THIS PRODUCT WEAR: NO SPECIAL REQUIREMENTS. Contact may cause eye irritation. May cause skin rash if
    allergic to product. Ingestion may cause gastrointestinal distress and diarrhea.

    A physician should be contacted if anyone develops any signs or symptoms and suspects that they are caused by
    exposure to this product.

FIRST AID:
    Eye Exposure:
    Flush eyes with water for 15 minutes while lifting the upper and lower eye lids. Contact lenses should not be
    worn when working with this product. If irritation persists, get medical attention.

    Skin Exposure:
    Not expected to be a problem. If rash appears, get medical attention.

    Breathing:
    Not expected to be a problem.

    Swallowing:
    Immediately contact the local poison control center for advice. Keep the victim warm and at rest. Get medical
    attention.

PRODUCT: AVEENO BAR - OILATED

INCOMPATIBILITY: NONE KNOWN

INGREDIENTS: COLLOIDAL OATMEAL   30% TLV: CAS#:
             VEGETABLE OIL   TLV: CAS#:
             LANOLIN   TLV: CAS#: 8006-54-0
             GLYCERIN   TLV: CAS#:   56-81-5
             FRAGRANCE   TLV: CAS#:
             SURFACTANTS   TLV: CAS#:

IGNI TEMP: NA FP: NA LEL: NA UEL: NA VP: NA VD: NA SG: NA PS: SOLID APPEAR: TAN COLOR ODOR: FRAGRANCE
    PH FACTOR: 5.0-6.0

HAZARD RATINGS: H: F: R: 0 0 0

NEUTRALIZING AGENTS:
FIRE FIGHTING:
    HIGH EXPANSION FOAM, LOW EXPANSION FOAM, ALCOHOL FOAM; DRY CHEMICAL, CARBON DIOXIDE, WATER FOG.
    PROTECTIVE CLOTHING, RUBBER GLOVES, AND BREATHING APPARATUS.
WARNING: 1] STRUCTURAL PROTECTIVE CLOTHING IS PERMEABLE, REMAIN CLEAR OF SMOKE, WATER FALL OUT AND WATER RUN OFF.
         2] KEEP OUT OF THE REACH OF CHILDREN.
         3] THERMAL DECOMPOSITION YIELDS TOXIC CARBON DIOXIDE, CARBON MONOXIDE.
         4] SPILL MAY CREATE A SLIPPING HAZARD.
         5] MOVE CONTAINERS FROM AREA IF WITHOUT RISK, COOL EXPOSED CONTAINERS.
         6] DIKE AREA FOR CONTROL AND CONTAINMENT TO PREVENT ENTRY INTO SEWERS, DRAIN, AND WATER WAYS.

LARGE SPILL/NO FIRE/RESCUE: WEAR RUBBER OR NEOPRENE BOOTS, GLOVES.

SPILL CONTROL AND CONTAINMENT:
    HOUSEHOLD SPILL: PICK UP BAR, WIPE ANY RESIDUE WITH A DRY CLOTH.
    LARGE SPILL: PICK UP OR SCOOP UP MATERIAL AND PUT INTO A DISPOSAL CONTAINER. RINSE AREA WITH WATER, PICK UP WITH
    ABSORBENT, PUT IN A DISPOSAL CONTAINER. SEAL AND REMOVE FROM WORK AREA.

HEALTH HAZARD INFORMATION:
    WHEN USING THIS PRODUCT WEAR: NO SPECIAL REQUIREMENTS. Contact may cause eye irritation. Skin rash may appear if
    allergic to the product. Ingestion may cause gastrointestinal distress and diarrhea.

    A physician should be contacted if anyone develops any signs or symptoms and suspects that they are caused by
    exposure to this product.

FIRST AID:
    Eye Exposure:
    Flush eyes with water for 15 minutes while lifting the upper and lower eye lids. Contact lenses should not be
    worn when working with this product. If irritation persists, get medical attention.

    Skin Exposure:
    Not expected to be a problem. If rash appears, get medical attention.

    Breathing:
    Not expected to be a problem.

    Swallowing:
    Immediately contact the local poison control center for advice. Keep the victim warm and at rest. Get medical
    attention.

PRODUCT: AVEENO BAR - REGULAR

INCOMPATIBILITY: NONE KNOWN

INGREDIENTS: COLLOIDAL OATMEAL   50% TLV: CAS#:
             LANOLIN   TLV: CAS#: 8006-54-0
             SURFACTANTS   TLV: CAS#:

IGNI TEMP: NA FP: NA LEL: NA UEL: NA VP: NA VD: NA SG: NA PS: SOLID APPEAR: TAN COLOR ODOR:
   PH FACTOR: 4.5-6.3

HAZARD RATINGS: H: F: R: 0 0 0

NEUTRALIZING AGENTS:

FIRE FIGHTING:
   HIGH EXPANSION FOAM, LOW EXPANSION FOAM, ALCOHOL FOAM; DRY CHEMICAL, CARBON DIOXIDE, WATER FOG.
   PROTECTIVE CLOTHING, RUBBER GLOVES, AND BREATHING APPARATUS.
WARNING: 1] STRUCTURAL PROTECTIVE CLOTHING IS PERMEABLE, REMAIN CLEAR OF SMOKE, WATER FALL OUT AND WATER RUN OFF.
         2] KEEP OUT OF THE REACH OF CHILDREN.
         3] THERMAL DECOMPOSITION YIELDS TOXIC CARBON DIOXIDE, CARBON MONOXIDE.
         4] MAY CAUSE A SLIPPING HAZARD.
         5] MOVE CONTAINERS FROM AREA IF WITHOUT RISK, COOL EXPOSED CONTAINERS.
         6] DIKE AREA FOR CONTROL AND CONTAINMENT TO PREVENT ENTRY INTO SEWERS, DRAIN, AND WATER WAYS.

LARGE SPILL/NO FIRE/RESCUE: WEAR RUBBER OR NEOPRENE BOOTS, GLOVES.

SPILL CONTROL AND CONTAINMENT:
   HOUSEHOLD SPILL: PICK UP MATERIAL, WIPE UP RESIDUE WITH A DRY CLOTH.
   LARGE SPILL: SWEEP OR SCOOP UP MATERIAL AND PUT INTO A DISPOSAL CONTAINER. SEAL AND REMOVE FROM WORK AREA.

HEALTH HAZARD INFORMATION:
   WHEN USING THIS PRODUCT WEAR: NO SPECIAL REQUIREMENTS. Contact with the eyes may cause irritation. May cause
   skin rash if allergic to the product. Ingestion may cause gastroirointestinal distress and diarrhea.

   A physician should be contacted if anyone develops any signs or symptoms and suspects that they are caused by
   exposure to this product.

FIRST AID:
   Eye Exposure:
   Flush eyes with water for 15 minutes while lifting the upper and lower eye lids. Contact lenses should not be
   worn when working with this product. If irritation persists, get medical attention.

   Skin Exposure:
   Not expected to be a problem. If rash appears, get medical attention.

   Breathing:
   Not expected to be a problem.

   Swallowing:
   Immediately contact the local poison control center for advice. Keep the victim warm and at rest. Get medical
   attention.

PRODUCT: BEHOLD LEMON SCENT FURNITURE POLISH

INCOMPATIBILITY: NONE KNOWN

INGREDIENTS: HYDROCARBON SOLVENT   under 20% TLV: 500 ppm   CAS#: 64742-48-9
             HYDROCARBON PROPELLANT under 10% TLV: CAS#:

IGNI TEMP: NA FP: 100F LEL: .9 UEL: 7.0 VP: 30mmHg VD: 4.1 SG:.93 PS: LIQUID APPEAR: WHITE MILKY SPRAY
    ODOR: LEMON   PH FACTOR: NA

HAZARD RATINGS: H: F: R: 0 3 0

FIRE FIGHTING:
  HIGH EXPANSION FOAM, LOW EXPANSION FOAM, ALCOHOL FOAM; DRY CHEMICAL, CARBON DIOXIDE, WATER FOG.
  PROTECTIVE CLOTHING, RUBBER GLOVES, AND BREATHING APPARATUS.
WARNING: 1] STRUCTURAL PROTECTIVE CLOTHING IS PERMEABLE, REMAIN CLEAR OF SMOKE, WATER FALL OUT AND WATER RUN OFF.
         2] KEEP OUT OF THE REACH OF CHILDREN.
         3] REMOVE ALL IGNITION SOURCES IF WITHOUT RISK.
         4] CONTAINERS WILL EXPLODE IN A FIRE OR IF HEATED ABOVE 120F.
         5] MOVE CONTAINERS FROM AREA IF WITHOUT RISK, COOL EXPOSED CONTAINERS.
         6] DIKE AREA FOR CONTROL AND CONTAINMENT TO PREVENT ENTRY INTO SEWERS, DRAIN, AND WATER WAYS.

LARGE SPILL/NO FIRE/RESCUE: WEAR RUBBER OR NEOPRENE BOOTS, GLOVES.

SPILL CONTROL AND CONTAINMENT:
  HOUSEHOLD SPILL: WIPE DRY WITH AN ABSORBENT CLOTH. PUT IN A DISPOSAL CONTAINER.
  LARGE SPILL: PICK UP WITH ABSORBENT. RINSE AREA WITH WATER. PICK UP WITH ABSORBENT. PUT ALL ABSORBENT IN A
  DISPOSAL CONTAINER. SEAL AND REMOVE FROM WORK AREA.

HEALTH HAZARD INFORMATION:
  WHEN USING THIS PRODUCT WEAR: NO SPECIAL REQUIREMENT. Eye contact will cause stinging, tearing, itching,
  swelling, and redness. Ingestion may cause nausea, and vomiting.

  A physician should be contacted if anyone develops any signs or symptoms and suspects that they are caused by
  exposure to this product.

FIRST AID:
  Eye Exposure:
  Flush with water for 15 minutes while lifting the upper and lower eye lids. Contact lenses should not be worn
  when working with this product. Get medical attention.

  Skin Exposure:
  Wash with soap and water. If irritation develops, get medical attention.

  Breathing:
  Not expected to be a problem.

  Swallowing:
  Do not induce vomiting. Immediately contact the local poison control center for advice. Keep the victim warm and
  at rest. Get medical attention.

PRODUCT: BEHOLD LIGHT SCENT FURNITURE POLISH

INCOMPATIBILITY: NONE KNOWN

INGREDIENTS: HYDROCARBON SOLVENT   under 20% TLV: 500 ppm   CAS#: 64742-48-9
             HYDROCARBON PROPELLANT under 10% TLV: CAS#:

IGNI TEMP: NA FP: 100F LEL: .9 UEL: 7.0 VP: 30mmHg VD: 4.1 SG:.93 PS: LIQUID   APPEAR: WHITE MILKY SPRAY
   ODOR: WOODY SCENT   PH FACTOR: NA

HAZARD RATINGS: H: F: R: 0 3 0

FIRE FIGHTING:
  HIGH EXPANSION FOAM, LOW EXPANSION FOAM, ALCOHOL FOAM; DRY CHEMICAL, CARBON DIOXIDE, WATER FOG.
  PROTECTIVE CLOTHING, RUBBER GLOVES, AND BREATHING APPARATUS.
WARNING: 1] STRUCTURAL PROTECTIVE CLOTHING IS PERMEABLE, REMAIN CLEAR OF SMOKE, WATER FALL OUT AND WATER RUN OFF.
         2] KEEP OUT OF THE REACH OF CHILDREN.
         3] REMOVE ALL IGNITION SOURCES IF WITHOUT RISK.
         4] CONTAINERS WILL EXPLODE IN A FIRE OR IF HEATED ABOVE 120F.
         5] MOVE CONTAINERS FROM AREA IF WITHOUT RISK, COOL EXPOSED CONTAINERS.
         6] DIKE AREA FOR CONTROL AND CONTAINMENT TO PREVENT ENTRY INTO SEWERS, DRAIN, AND WATER WAYS.

LARGE SPILL/NO FIRE/RESCUE: WEAR RUBBER OR NEOPRENE BOOTS, GLOVES.

SPILL CONTROL AND CONTAINMENT:
  HOUSEHOLD SPILL: WIPE DRY WITH AN ABSORBENT CLOTH. PUT IN A DISPOSAL CONTAINER.
  LARGE SPILL: PICK UP WITH ABSORBENT. RINSE AREA WITH WATER. PICK UP WITH ABSORBENT. PUT ALL ABSORBENT IN A
  DISPOSAL CONTAINER. SEAL AND REMOVE FROM WORK AREA.

HEALTH HAZARD INFORMATION:
  WHEN USING THIS PRODUCT WEAR: NO SPECIAL REQUIREMENT. Eye contact will cause stinging, tearing, itching,
  swelling, and redness. Ingestion may cause nausea, vomiting.

  A physician should be contacted if anyone develops any signs or symptoms and suspects that they are caused by
  exposure to this product.

FIRST AID:
  Eye Exposure:
  Flush with water for 15 minutes while lifting the upper and lower eye lids. Contact lenses should not be worn
  when working with this product. Get medical attention.

  Skin Exposure:
  Wash with soap and water. If irritation develops, get medical attention.

  Breathing:
  Not expected to be a problem.

  Swallowing:
  Do not induce vomiting. Immediately contact the local poison control center for advice. Keep the victim warm and
  at rest. Get medical attention.

PRODUCT: BIZ (POWDER BLEACH)

INCOMPATIBILITY: ORGANIC MATERIALS; STRONG OXIDIZERS.

INGREDIENTS: SODIUM TRIPOLYPHOSPHATE   TLV: CAS#:   7758-29-4
             SODIUM SESQUICARBONATE   TLV: CAS#:   6106-20-3
             SODIUM PERBORATE   TLV: CAS#:   10486-00-7

IGNI TEMP: NA FP: NA LEL: NA UEL: NA VP: NA VD: NA SG: NA PS: POWDER APPEAR: WHITE GRANULAR BLUE SPECKLES
    ODOR: PERFUME PH FACTOR:

HAZARD RATINGS: H: F: R:

FIRE FIGHTING:
    HIGH EXPANSION FOAM, LOW EXPANSION FOAM, ALCOHOL FOAM; DRY CHEMICAL, CARBON DIOXIDE, WATER FOG.
    PROTECTIVE CLOTHING, RUBBER GLOVES, AND BREATHING APPARATUS.
WARNING: 1] STRUCTURAL PROTECTIVE CLOTHING IS PERMEABLE, REMAIN CLEAR OF SMOKE, WATER FALL OUT AND WATER RUN OFF.
         2] KEEP OUT OF THE REACH OF CHILDREN.
         3] MIXING WITH PETROLEUM BASED PRODUCTS MAY CAUSE A FIRE OR CHLORINE GAS TO BE GENERATED.
         4] MOVE CONTAINERS FROM AREA IF WITHOUT RISK, COOL EXPOSED CONTAINERS.
         5] DIKE AREA FOR CONTROL AND CONTAINMENT TO PREVENT ENTRY INTO SEWERS, DRAIN, AND WATER WAYS.

LARGE SPILL/NO FIRE/RESCUE: WEAR RUBBER/NEOPRENE BOOTS, GLOVES AND BREATHING APPARATUS.

SPILL CONTROL AND CONTAINMENT:
    HOUSEHOLD SPILL: AVOID CREATING A DUST. GENTLY SWEEP UP MATERIAL AND PLACE IN A DISPOSAL CONTAINER.
    LARGE SPILL: AVOID CREATING A DUST. GENTLY SWEEP UP MATERIAL AND PLACE IN A DISPOSAL CONTAINER, RINSE AREA WITH
    WATER, PICK UP WITH ABSORBENT, PUT IN DISPOSAL CONTAINER. SEAL AND REMOVE FROM WORK AREA.

HEALTH HAZARD INFORMATION:
    WHEN USING THIS PRODUCT WEAR: NO SPECIAL REQUIREMENTS. Contact may cause mild eye irritation, stinging, tearing,
    itching, swelling, and redness. Dust may cause respiratory tract irritation, coughing, sore throat, wheezing and
    shortness of breath. Ingestion may cause gastrointestinal irritation, nausea, vomiting, and diarrhea.

    A physician should be contacted if anyone develops any signs or symptoms and suspects that they are caused by
    exposure to this product.

FIRST AID:
Eye Exposure:
Flush eyes with water for 15 minutes while lifting the upper and lower eye lids. Contact lenses should not be
worn when working with this product. If irritation persists, get medical attention.

Skin Exposure:
Not expected to be a problem. If rash appears, get medical attention.

Breathing:
Not expected to be a problem.

Swallowing:
If the victim is conscious, give 2 to 3 glasses of milk or water to drink. Immediately contact the local poison
control center for advice. Keep the victim warm and at rest. Get medical attention.

PRODUCT: BOLD   (LAUNDRY GRANULES)

INCOMPATIBILITY: NONE KNOWN

INGREDIENTS: ALUMINOSILICATES   TLV: CAS#:
             SODIUM PHOSPHATES   TLV: CAS#: 7775-27-1
             SODIUM CARBONATE   TLV: CAS#:   497-19-8
             SODIUM SULFATE   TLV: CAS#: 7757-83-7
             SODIUM SILICATES   TLV: CAS#: 1344-09-8

IGNI TEMP: NA FP: NA LEL: NA UEL: NA VP: NA VD: NA SG: NA PS: POWDER
   APPEAR: WHITE GRANULES BLUE/GREEN SPECKLES ODOR: PERFUMED   PH FACTOR:

HAZARD RATINGS: H: F: R:

FIRE FIGHTING:
   HIGH EXPANSION FOAM, LOW EXPANSION FOAM, ALCOHOL FOAM; DRY CHEMICAL, CARBON DIOXIDE, WATER FOG.
   PROTECTIVE CLOTHING, RUBBER GLOVES, AND BREATHING APPARATUS.
WARNING: 1] STRUCTURAL PROTECTIVE CLOTHING IS PERMEABLE, REMAIN CLEAR OF SMOKE, WATER FALL OUT AND WATER RUN OFF.
         2] KEEP OUT OF THE REACH OF CHILDREN.
         3] CHECK TOP OR BOTTOM FOR WHITE RECTANGLE BOX. IF "L" PROCEEDS NUMBER, PRODUCT CONTAINS SODIUM
            PYROPHOSPHATE CAS#: 7722-85-5 TLV: 5mg/m3.
         4] MOVE CONTAINERS FROM AREA IF WITHOUT RISK, COOL EXPOSED CONTAINERS.
         5] DIKE AREA FOR CONTROL AND CONTAINMENT TO PREVENT ENTRY INTO SEWERS, DRAIN, AND WATER WAYS.

LARGE SPILL/NO FIRE/RESCUE: WEAR RUBBER OR NEOPRENE BOOTS, GLOVES.

SPILL CONTROL AND CONTAINMENT:
   HOUSEHOLD SPILL: AVOID CREATING A DUST. GENTLY SWEEP UP MATERIAL AND PUT IN A DISPOSAL CONTAINER.
   LARGE SPILL: AVOID CREATING A DUST. GENTLY SWEEP UP MATERIAL AND PUT IN A DISPOSAL CONTAINER. RINSE AREA WITH
   WATER, PICK UP WITH ABSORBENT. PUT IN A DISPOSAL CONTAINER. SEAL AND REMOVE FROM WORK AREA.

HEALTH HAZARD INFORMATION:
   WHEN USING THIS PRODUCT WEAR: NO SPECIAL REQUIREMENTS. Inhalation of dust may cause coughing, sore throat,
   wheezing, and shortness of breath. Eye contact may cause stinging, tearing, itching, swelling, and redness.
   Ingestion may cause abdominal cramps, nausea, vomiting, and diarrhea.

   A physician should be contacted if anyone develops any signs or symptoms and suspects that they are caused by
   exposure to this product.

FIRST AID:
   Eye Exposure:
   Flush eyes with water for at least 15 minutes while lifting the upper and lower eye lids. Contact lenses should
   not be worn when working with this product. If irritation persists, get medical attention.

   Skin Exposure:
   Not expected to be a problem. If rash appears, get medical attention.

   Breathing:
   Not expected to be a problem.

   Swallowing:
   If the victim is conscious, give 2-3 glasses of water to drink. Immediately contact the local poison control
   center for advice. Keep the victim warm and at rest. Get medical attention.

PRODUCT: BOLD, LIQUID (LAUNDRY DETERGENT)

INCOMPATIBILITY: NONE KNOWN

INGREDIENTS: ETHYL ALCOHOL   TLV: 1900 mg/m3   CAS#:   64-17-5
             ALKYL ETHOXYLATE   TLV: CAS#:
             NONIONIC SURFACTANTS   TLV: CAS#:
             WA WATER   TLV: CAS#: 7732-18-5
             COLORANT; PERFUME

IGNI TEMP: NA FP: NA LEL: NA UEL: NA VP: NA VD: NA SG: .99 PS: LIQUID APPEAR: CLEAR TO HAZY BLUE
   ODOR: PERFUMED   PH FACTOR: NA

HAZARD RATINGS: H: F: R:

FIRE FIGHTING:
   HIGH EXPANSION FOAM, LOW EXPANSION FOAM, ALCOHOL FOAM; DRY CHEMICAL, CARBON DIOXIDE, WATER FOG.
   PROTECTIVE CLOTHING, RUBBER GLOVES, AND BREATHING APPARATUS.
WARNING: 1] STRUCTURAL PROTECTIVE CLOTHING IS PERMEABLE, REMAIN CLEAR OF SMOKE, WATER FALL OUT AND WATER RUN OFF.
         2] KEEP OUT OF THE REACH OF CHILDREN.
         3] THERMAL DECOMPOSITION MAY YIELD CARBON DIOXIDE, CARBON MONOXIDE.
         4] MOVE CONTAINERS FROM AREA IF WITHOUT RISK, COOL EXPOSED CONTAINERS.
         5] DIKE AREA FOR CONTROL AND CONTAINMENT TO PREVENT ENTRY INTO SEWERS, DRAIN, AND WATER WAYS.

LARGE SPILL/NO FIRE/RESCUE: WEAR RUBBER OR NEOPRENE BOOTS, GLOVES.

SPILL CONTROL AND CONTAINMENT:
   HOUSEHOLD SPILL: WIPE UP LIQUID WITH AN ABSORBENT CLOTH, RINSE OUT IN THE SINK. WIPE UP RESIDUE WITH A DAMP
   CLOTH, RINSE OUT IN THE SINK.
   LARGE SPILL: PICK UP LIQUID WITH ABSORBENT, PUT IN A DISPOSAL CONTAINER. RINSE AREA WITH WATER, PICK UP WITH
   ABSORBENT. PUT IN A DISPOSAL CONTAINER. SEAL AND REMOVE FROM THE WORK AREA.

HEALTH HAZARD INFORMATION:
   WHEN USING THIS PRODUCT WEAR: NO SPECIAL REQUIREMENT. Eye contact will cause stinging, tearing, itching,
   swelling, and redness. Prolonged contact with concentrated liquid may cause drying of the skin and irritation.
   Ingestion may result in abdominal pain, nausea, vomiting, and diarrhea. Large volume ingestion may cause
   transient nervous system effects.

   A physician should be contacted if anyone develops any signs or symptoms and suspects that they are caused by
   exposure to this product.

FIRST AID:
   Eye Exposure:
   Flush with water for 15 minutes while lifting the upper and lower eye lids. Contact lenses should not be worn
   when working with this product. Get medical attention.

   Skin Exposure:
   Wash off with water. If irritation develops, get medical attention.

   Breathing:
   Not expected to be a problem.

   Swallowing:
   If the victim is conscious, give 1-2 glasses of water to drink. Immediately contact the local poison control
   center for advice. Keep the victim warm and at rest. Get medical attention.

PRODUCT: BOLD, ULTRA (LAUNDRY GRANULES)

INCOMPATIBILITY: NONE KNOWN

INGREDIENTS: SODIUM CARBONATE   TLV: CAS#: 497-19-8
            SODIUM SULFATE   TLV: CAS#: 7757-82-6
            SODIUM SILICATE   TLV: CAS#: 1344-09-8
            ANIONIC SURFACTANT   TLV: CAS#:
            ALUMINOSILICATE   TLV: CAS#:
            CATIONIC ANTISTATIC, ENZYMES, PERFUME

IGNI TEMP: NA FP: NA LEL: NA UEL: NA VP: NA VD: NA SG: NA PS: POWDER APPEAR: WHITE - BLUE SPECKS
   ODOR: PERFUMED   PH FACTOR: NA

HAZARD RATINGS: H: F: R:

FIRE FIGHTING:
  HIGH EXPANSION FOAM, LOW EXPANSION FOAM, ALCOHOL FOAM; DRY CHEMICAL, CARBON DIOXIDE, WATER FOG.
  PROTECTIVE CLOTHING, RUBBER GLOVES, AND BREATHING APPARATUS.
WARNING: 1] STRUCTURAL PROTECTIVE CLOTHING IS PERMEABLE, REMAIN CLEAR OF SMOKE, WATER FALL OUT AND WATER RUN OFF.
       2] KEEP OUT OF THE REACH OF CHILDREN.
       3] THERMAL DECOMPOSITION MAY YIELD CARBON DIOXIDE, CARBON MONOXIDE.
       4] MOVE CONTAINERS FROM AREA IF WITHOUT RISK, COOL EXPOSED CONTAINERS.
       5] DIKE AREA FOR CONTROL AND CONTAINMENT TO PREVENT ENTRY INTO SEWERS, DRAIN, AND WATER WAYS.

LARGE SPILL/NO FIRE/RESCUE: WEAR RUBBER OR NEOPRENE BOOTS, GLOVES.

SPILL CONTROL AND CONTAINMENT:
  HOUSEHOLD SPILL: GENTLY SWEEP UP POWDER AND PUT IN A DISPOSAL CONTAINER. PICK UP RESIDUE WITH A DAMP MOP. RINSE
  OUT IN THE SINK.
  LARGE SPILL: AVOID CREATING A DUST. SCOOP UP POWDER. PUT IN A DISPOSAL CONTAINER. RINSE AREA WITH WATER. PICK UP
  WITH ABSORBENT. PUT IN A DISPOSAL CONTAINER. SEAL AND REMOVE FROM THE WORK AREA.

HEALTH HAZARD INFORMATION:
  WHEN USING THIS PRODUCT WEAR: NO SPECIAL REQUIREMENT. Inhalation of dust may cause coughing, sore throat,
  wheezing, and shortness of breath. Eye contact may cause stinging, tearing, itching, swelling, and redness.
  Ingestion may result in nausea, vomiting, and diarrhea.

  A physician should be contacted if anyone develops any signs or symptoms and suspects that they are caused by
  exposure to this product.

FIRST AID:
  Eye Exposure:
  Flush with water for 15 minutes while lifting the upper and lower eye lids. Contact lenses should not be worn
  when working with this product. Get medical attention.

  Skin Exposure:
  Wash with water. If irritation develops, get medical attention.

  Breathing:
  If irritation develops, move victim to fresh air.

  Swallowing:
  If the victim is conscious, give 2-3 glasses of milk or water to drink. Immediately contact the local poison
  control center for advice. Keep the victim warm and at rest. Get medical attention.

PRODUCT: BOUNCE PERFUMED - HOUSEHOLD FABRIC SOFTENER SHEETS

INCOMPATIBILITY: NONE KNOWN

INGREDIENTS: ALL ARE NON-HAZARDOUS BY OSHA 1910.1200 STANDARD

IGNI TEMP: NA FP: NA LEL: NA UEL: NA VP: NA VD: NA SG: NA PS: SHEETS/SOLID APPEAR: OPAQUE WHITE
    ODOR: WOODY PERFUME PH FACTOR: NA

HAZARD RATINGS: H: F: R:

NEUTRALIZING AGENTS:

FIRE FIGHTING:
    HIGH EXPANSION FOAM, LOW EXPANSION FOAM, ALCOHOL FOAM; DRY CHEMICAL, CARBON DIOXIDE, WATER FOG.
    PROTECTIVE CLOTHING, RUBBER GLOVES, AND BREATHING APPARATUS.
WARNING: 1] STRUCTURAL PROTECTIVE CLOTHING IS PERMEABLE, REMAIN CLEAR OF SMOKE, WATER FALL OUT AND WATER RUN OFF.
         2] KEEP OUT OF THE REACH OF CHILDREN.
         3] THERMAL DECOMPOSITION YIELDS TOXIC CARBON DIOXIDE, CARBON MONOXIDE.
         4] MOVE CONTAINERS FROM AREA IF WITHOUT RISK, COOL EXPOSED CONTAINERS.
         5] DIKE AREA FOR CONTROL AND CONTAINMENT TO PREVENT ENTRY INTO SEWERS, DRAIN, AND WATER WAYS. .

LARGE SPILL/NO FIRE/RESCUE: WEAR RUBBER OR NEOPRENE BOOTS, GLOVES.

SPILL CONTROL AND CONTAINMENT:
    HOUSEHOLD SPILL: PICK UP ROLL AND PUT BACK INTO CONTAINER.
    LARGE SPILL: PICK UP MATERIAL FOR SALVAGE. FOR DISPOSAL, PLACE IN CONTAINER. SEAL AND REMOVE FROM WORK AREA.

HEALTH HAZARD INFORMATION:
    WHEN USING THIS PRODUCT WEAR: NO SPECIAL REQUIREMENTS. Eye contact may cause irritation. Prolonged or repeated
    skin contact may cause irritation. Ingestion of product from sheets may cause gastrointestinal irritation,
    nausea, vomiting, and diarrhea.

    A physician should be contacted if anyone develops any signs or symptoms and suspects that they are caused by
    exposure to this product.

FIRST AID:
Eye Exposure:
Flush eyes with water for 15 minutes while lifting the upper and lower eye lids. Contact lenses should not be
worn when working with this product. If irritation persists, get medical attention.

Skin Exposure:
Not expected to be a problem. If irritation appears, get medical attention.

Breathing:
Not expected to be a problem.

Swallowing:
Immediately contact the local poison control center for advice. Keep the victim warm and at rest. Get medical
attention.

PRODUCT: BOUNCE UNSCENTED - HOUSEHOLD FABRIC SOFTENER SHEETS

INCOMPATIBILITY: NONE KNOWN

INGREDIENTS: ALL ARE NON-HAZARDOUS BY OSHA 1910.1200 STANDARD

IGNI TEMP: NA FP: NA LEL: NA UEL: NA VP: NA VD: NA SG: NA PS: SHEETS/SOLID APPEAR: OPAQUE WHITE   ODOR:
  PH FACTOR: NA

HAZARD RATINGS: H: F: R:

FIRE FIGHTING:
  HIGH EXPANSION FOAM, LOW EXPANSION FOAM, ALCOHOL FOAM; DRY CHEMICAL, CARBON DIOXIDE, WATER FOG.
  PROTECTIVE CLOTHING, RUBBER GLOVES, AND BREATHING APPARATUS.
WARNING: 1] STRUCTURAL PROTECTIVE CLOTHING IS PERMEABLE, REMAIN CLEAR OF SMOKE, WATER FALL OUT AND WATER RUN OFF.
         2] KEEP OUT OF THE REACH OF CHILDREN.
         3] THERMAL DECOMPOSITION YIELDS TOXIC CARBON DIOXIDE, CARBON MONOXIDE.
         4] MOVE CONTAINERS FROM AREA IF WITHOUT RISK, COOL EXPOSED CONTAINERS.
         5] DIKE AREA FOR CONTROL AND CONTAINMENT TO PREVENT ENTRY INTO SEWERS, DRAIN, AND WATER WAYS.

LARGE SPILL/NO FIRE/RESCUE: WEAR RUBBER OR NEOPRENE BOOTS, GLOVES.

SPILL CONTROL AND CONTAINMENT:
  HOUSEHOLD SPILL: PICK UP ROLL AND PUT BACK INTO CONTAINER.
  LARGE SPILL: PICK UP BOXES FOR SALVAGE. FOR DISPOSAL, PUT INTO A CONTAINER FOR DISPOSAL.

HEALTH HAZARD INFORMATION:
  WHEN USING THIS PRODUCT WEAR: NO SPECIAL REQUIREMENTS. Contact with the eyes may cause irritation. Prolonged or
  repeated skin contact may cause irritation. Ingestion of product from sheets may cause gastrointestinal
  irritation, nausea, vomiting, and diarrhea.

  A physician should be contacted if anyone develops any signs or symptoms and suspects that they are caused by
  exposure to this product.

FIRST AID:
  Eye Exposure:
  Flush eyes with water for 15 minutes while lifting the upper and lower eye lids. Contact lenses should not be
  worn when working with this product. If irritation persists, get medical attention.

  Skin Exposure:
  Not expected to be a problem. If irritation appears, get medical attention.

  Breathing:
  Not expected to be a problem.

  Swallowing:
  Immediately contact the local poison control center for advice. Keep the victim warm and at rest. Get medical
  attention.

PRODUCT: BOUNCE WITH STAIN GUARD - LIQUID FABRIC SOFTENER

INCOMPATIBILITY: NONE KNOWN

INGREDIENTS: ALL ARE NON-HAZARDOUS BY OSHA 1910.1200 STANDARD

IGNI TEMP: NA FP: 200F LEL: NA UEL: NA VP: NA VD: NA SG: 1.0 PS: LIQUID APPEAR: WHITE OPAQUE
    ODOR: SOFT FLORAL PH FACTOR:

HAZARD RATINGS: H: F: R:

FIRE FIGHTING:
    HIGH EXPANSION FOAM, LOW EXPANSION FOAM, ALCOHOL FOAM; DRY CHEMICAL, CARBON DIOXIDE, WATER FOG.
    PROTECTIVE CLOTHING, RUBBER GLOVES, AND BREATHING APPARATUS.
WARNING: 1] STRUCTURAL PROTECTIVE CLOTHING IS PERMEABLE, REMAIN CLEAR OF SMOKE, WATER FALL OUT AND WATER RUN OFF.
         2] KEEP OUT OF THE REACH OF CHILDREN.
         3] PRODUCT MUST BE PRE-HEATED CONSIDERABLY FOR IGNITION TO OCCUR.
         4] DO NOT DISPOSE OF PRODUCT IN A LAND FILL.
         5] THERMAL DECOMPOSITION PRODUCTS ARE NOT KNOWN.
         6] MOVE CONTAINERS FROM AREA IF WITHOUT RISK, COOL EXPOSED CONTAINERS.
         7] DIKE AREA FOR CONTROL AND CONTAINMENT TO PREVENT ENTRY INTO SEWERS, DRAIN, AND WATER WAYS.

LARGE SPILL/NO FIRE/RESCUE: WEAR RUBBER OR NEOPRENE BOOTS, GLOVES.

SPILL CONTROL AND CONTAINMENT:
    HOUSEHOLD SPILL: WIPE UP MATERIAL WITH AN ABSORBENT CLOTH. RINSE AREA WITH WATER AND WIPE DRY. WEAR LATEX RUBBER
    GLOVES DURING PICK UP OR WASH HANDS WITH SOAP AND WATER AFTER COMPLETION.
    LARGE SPILL: COVER SPILL WITH ABSORBENT. PICK UP AND PLACE IN A DISPOSAL CONTAINER. RINSE AREA WITH WATER, PICK
    UP WITH ABSORBENT, PUT IN DISPOSAL CONTAINER. SEAL AND REMOVE FROM WORK AREA.

HEALTH HAZARD INFORMATION:
    WHEN USING THIS PRODUCT WEAR: LATEX OR RUBBER GLOVES DURING CLEANUP OF SPILL. Health hazards are unknown.

    A physician should be contacted if anyone develops any signs or symptoms and suspects that they are caused by
    exposure to this product.

FIRST AID:
Eye Exposure:
Immediately flush eyes with water for at least 15 minutes while lifting the upper and lower eye lids. Contact
lenses should not be worn when working with this product, get medical attention.

Skin Exposure:
If product gets on skin immediately remove contaminated clothing and wash off with soap and water. If irritation
appear, get medical attention.

Breathing:
Not expected to be a problem.

Swallowing:
If the victim is conscious, give 2 to 3 glasses of milk or water to drink and immediately contact the local
poison control center for advice. Keep the victim warm and at rest. Get medical attention.

PRODUCT: BRITE

INCOMPATIBILITY: NONE KNOWN

INGREDIENTS: MODIFIED ACRYLIC POLYMERS   10-20% TLV: CAS#:
             DIETHYLENE GLYCOL MONOETHYL ETHER 1-3% TLV: CAS#: 111-90-0
             AMMONIUM HYDROXIDE   1-4% TLV: 25 ppm CAS#: 1336-21-6
             WATER   80-90% TLV: CAS#: 7732-18-5

IGNI TEMP: NA FP: NA LEL: NA UEL: NA VP: NA VD: NA SG: 1.02 PS: LIQUID APPEAR: TAN TO GRAY
    ODOR: AMMONICAL   PH FACTOR: 9.5-9.9

HAZARD RATINGS: H: F: R: 1 0 0

FIRE FIGHTING:
    HIGH EXPANSION FOAM, LOW EXPANSION FOAM, ALCOHOL FOAM; DRY CHEMICAL, CARBON DIOXIDE, WATER FOG.
    PROTECTIVE CLOTHING, RUBBER GLOVES, AND BREATHING APPARATUS.
WARNING: 1] STRUCTURAL PROTECTIVE CLOTHING IS PERMEABLE, REMAIN CLEAR OF SMOKE, WATER FALL OUT AND WATER RUN OFF.
         2] KEEP OUT OF THE REACH OF CHILDREN.
         3] THERMAL DECOMPOSITION YIELDS TOXIC CARBON DIOXIDE, CARBON MONOXIDE.
         4] MOVE CONTAINERS FROM AREA IF WITHOUT RISK, COOL EXPOSED CONTAINERS.
         5] DIKE AREA FOR CONTROL AND CONTAINMENT TO PREVENT ENTRY INTO SEWERS, DRAIN, AND WATER WAYS.

LARGE SPILL/NO FIRE/RESCUE: WEAR RUBBER OR NEOPRENE BOOTS, GLOVES.

SPILL CONTROL AND CONTAINMENT:
    HOUSEHOLD SPILL: WIPE UP MATERIAL WITH ABSORBENT CLOTH AND PUT IN A DISPOSAL CONTAINER. RINSE AREA WITH WATER
    AND PICK UP WITH ABSORBENT CLOTH AND PUT IN A DISPOSAL CONTAINER. WASH HANDS WHEN COMPLETED.
    LARGE SPILL: COVER WITH ABSORBENT MATERIAL, PICK UP AND PUT INTO A DISPOSAL CONTAINER. RINSE WITH WATER, PICK UP
    WITH ABSORBENT AND PUT IN A DISPOSAL CONTAINER. SEAL AND REMOVE FROM WORK AREA.

HEALTH HAZARD INFORMATION:
    WHEN USING THIS PRODUCT WEAR: LATEX RUBBER GLOVES. Contact with the eyes may cause irritation. Prolonged or
    repeated skin contact may cause irritation.

    A physician should be contacted if anyone develops any signs or symptoms and suspects that they are caused by
    exposure to this product.

FIRST AID:
    Eye Exposure:
    Flush eyes with water for 15 minutes while lifting the upper and lower eye lids. Contact lenses should not be
    worn when working with this product. If irritation persists, get medical attention.

    Skin Exposure:
    Not expected to be a problem. If irritation appears, get medical attention.

    Breathing:
    Not expected to be a problem.

    Swallowing:
    Immediately contact the local poison control center for advice. Keep the victim warm and at rest. Get medical
    attention.

PRODUCT: CAMAY TOILET SOAP

INCOMPATIBILITY: NONE KNOWN

INGREDIENTS: SODIUM COCOATE   TLV: CAS#:
             SODIUM TALLOWATE   TLV: CAS#:
             COCONUT ACID   TLV: CAS#:
             SODIUM CHLORIDE   TLV: CAS#:   7647-14-5
             TETRASODIUM EDTA   TLV: CAS#:
             TITANIUM DIOXIDE   TLV: CAS#:   13463-67-7

IGNI TEMP: NA FP: NA LEL: NA UEL: NA VP: NA VD: NA SG: 1.04 PS: SOLID APPEAR: PINK BAR   ODOR: FLORAL
    PH FACTOR: NA

HAZARD RATINGS: H: F: R:

FIRE FIGHTING:
   HIGH EXPANSION FOAM, LOW EXPANSION FOAM, ALCOHOL FOAM; DRY CHEMICAL, CARBON DIOXIDE, WATER FOG.
   PROTECTIVE CLOTHING, RUBBER GLOVES, AND BREATHING APPARATUS.
WARNING: 1] STRUCTURAL PROTECTIVE CLOTHING IS PERMEABLE, REMAIN CLEAR OF SMOKE, WATER FALL OUT AND WATER RUN OFF.
         2] KEEP OUT OF THE REACH OF CHILDREN.
         3] MAY CAUSE A SLIPPING HAZARD.
         4] MOVE CONTAINERS FROM AREA IF WITHOUT RISK, COOL EXPOSED CONTAINERS.
         5] DIKE AREA FOR CONTROL AND CONTAINMENT TO PREVENT ENTRY INTO SEWERS, DRAIN, AND WATER WAYS.

LARGE SPILL/NO FIRE/RESCUE: WEAR RUBBER OR NEOPRENE BOOTS, GLOVES.

SPILL CONTROL AND CONTAINMENT:
   HOUSEHOLD SPILL: PICK UP BAR AND WIPE UP ANY LIQUID WHERE IT LANDED.
   LARGE SPILL: PICK UP BARS FOR RECYCLE. SWEEP UP ANY RESIDUE, RINSE WITH WATER AND PICK UP WITH ABSORBENT. PUT
   INTO A DISPOSAL CONTAINER. SEAL AND REMOVE FROM WORK AREA.

HEALTH HAZARD INFORMATION:
   WHEN USING THIS PRODUCT WEAR: NO SPECIAL REQUIREMENTS. Contact with the eyes may cause irritation. Contact with
   skin may cause a rash if allergic to this product.

   A physician should be contacted if anyone develops any signs or symptoms and suspects that they are caused by
   exposure to this product.

FIRST AID:
Eye Exposure:
Flush eyes with water for 15 minutes while lifting the upper and lower eye lids. Contact lenses should not be
worn when working with this product. If irritation persists, get medical attention.

Skin Exposure:
Not expected to be a problem. If irritation appears, change to another soap. If irritation persists, get medical
attention.

Breathing:
Not expected to be a problem.

Swallowing:
If the victim is conscious, give 2-3 glasses of water to drink. Immediately contact the local poison control
center for advice. Keep the victim warm and at rest. Get medical attention.

```
PRODUCT: CASCADE   AUTOMATIC DISHWASHER DETERGENT

INCOMPATIBILITY: NONE KNOWN

INGREDIENTS: COMPLEX SODIUM PHOSPHATES     TLV:          CAS#:
             SODIUM CARBONATE              TLV:          CAS#:  497-19-8
             CHLORINE BLEACH               TLV:          CAS#: 7681-52-9
             SODIUM SILICATE               TLV:          CAS#: 1344-09-8
             SODIUM SULFATE                TLV:          CAS#: 7757-82-6

IGNI TEMP: NA  FP: NA  LEL: NA  UEL: NA  VP: NA  VD: NA  SG: NA  PS: GRANULAR  APPEAR: BLUE & WHITE
      ODOR: PERFUMED    PH FACTOR: NA

HAZARD RATINGS:    H:  F:  R:
```

FIRE FIGHTING:
   HIGH EXPANSION FOAM, LOW EXPANSION FOAM, ALCOHOL FOAM; DRY CHEMICAL, CARBON DIOXIDE, WATER FOG.
   PROTECTIVE CLOTHING, RUBBER GLOVES, AND BREATHING APPARATUS.
WARNING: 1] STRUCTURAL PROTECTIVE CLOTHING IS PERMEABLE, REMAIN CLEAR OF SMOKE, WATER FALL OUT AND WATER RUN OFF.
         2] KEEP OUT OF THE REACH OF CHILDREN.
         3] THERMAL DECOMPOSITION MAY YIELD TOXIC FUMES.
         4] MOVE CONTAINERS FROM AREA IF WITHOUT RISK, COOL EXPOSED CONTAINERS.
         5] DIKE AREA FOR CONTROL AND CONTAINMENT TO PREVENT ENTRY INTO SEWERS, DRAIN, AND WATER WAYS.

LARGE SPILL/NO FIRE/RESCUE: WEAR RUBBER OR NEOPRENE BOOTS, GLOVES.

SPILL CONTROL AND CONTAINMENT:
   HOUSEHOLD SPILL: AVOID CREATING A DUST. SWEEP UP MATERIAL AND PUT IN A DISPOSAL CONTAINER. WIPE UP RESIDUE WITH
   A DAMP CLOTH. RINSE IN SINK.
   LARGE SPILL: AVOID CREATING A DUST. GENTLY SWEEP UP MATERIAL AND PUT INTO A DISPOSAL CONTAINER. RINSE AREA WITH
   WATER, PICK UP WITH ABSORBENT, PUT INTO A DISPOSAL CONTAINER.

HEALTH HAZARD INFORMATION:
   WHEN USING THIS PRODUCT WEAR: NO SPECIAL REQUIREMENTS. Contact with the eyes may cause irritation. Prolonged
   skin contact may cause drying of skin. Ingestion may cause gastrointestinal irritation, nausea, vomiting, and
   diarrhea.

   A physician should be contacted if anyone develops any signs or symptoms and suspects that they are caused by
   exposure to this product.

FIRST AID:
   Eye Exposure:
   Flush eyes with water for 15 minutes while lifting the upper and lower eye lids. Contact lenses should not be
   worn when working with this product. If irritation persists, get medical attention.

   Skin Exposure:
   Rinse off with water, rub in a hand lotion. If irritation appears, get medical attention.

   Breathing:
   Not expected to be a problem.

   Swallowing:
   If the victim is conscious, give 3-4 glasses of milk, water or citrus fruit juice (orange juice etc.) to drink.
   Immediately contact the local poison control center for advice. Keep the victim warm and at rest. Get medical
   attention.

PRODUCT: CASCADE, LEMON LIQUID   (AUTOMATIC DISH WASHING DETERGENT)

INCOMPATIBILITY: TOILET BOWL CLEANERS; RUST REMOVERS; ACIDS; VINEGAR; AMMONIA OR PRODUCT WITH AMMONIA.

INGREDIENTS: SODIUM PHOSPHATE   TLV: CAS#:
             SODIUM CARBONATE   TLV: CAS#: 497-19-8
             SODIUM HYPOCHLORITE   TLV: CAS#: 7681-52-9
             SODIUM SILICATES   TLV: CAS#: 1344-09-8

IGNI TEMP: NA FP: NA LEL: NA UEL: NA VP: NA VD: NA SG: NA PS: LIQUID APPEAR: THICK YELLOW
    ODOR: LEMON   PH FACTOR: NA

HAZARD RATINGS: H: F: R:

FIRE FIGHTING:
  HIGH EXPANSION FOAM, LOW EXPANSION FOAM, ALCOHOL FOAM; DRY CHEMICAL, CARBON DIOXIDE, WATER FOG.
     PROTECTIVE CLOTHING, RUBBER GLOVES, AND BREATHING APPARATUS.
WARNING: 1] STRUCTURAL PROTECTIVE CLOTHING IS PERMEABLE, REMAIN CLEAR OF SMOKE, WATER FALL OUT AND WATER RUN OFF.
         2] KEEP OUT OF THE REACH OF CHILDREN.
         3] THERMAL DECOMPOSITION YIELDS TOXIC CHLORINE GAS.
         4] DO NOT MIX WITH INCOMPATIBLE PRODUCT LISTED ABOVE, YIELDS TOXIC CHLORINE OR AMMONIA GAS.
         5] MOVE CONTAINERS FROM AREA IF WITHOUT RISK, COOL EXPOSED CONTAINERS.
         6] DIKE AREA FOR CONTROL AND CONTAINMENT TO PREVENT ENTRY INTO SEWERS, DRAIN, AND WATER WAYS.

LARGE SPILL/NO FIRE/RESCUE: WEAR RUBBER OR NEOPRENE BOOTS, GLOVES.

SPILL CONTROL AND CONTAINMENT:
  HOUSEHOLD SPILL: WIPE UP LIQUID WITH AN ABSORBENT CLOTH, RINSE OUT IN THE SINK. WIPE UP RESIDUE WITH A DAMP
  CLOTH. RINSE OUT IN THE SINK.
  LARGE SPILL: COVER SPILL WITH ABSORBENT, SCOOP UP AND PUT IN A DISPOSAL CONTAINER. RINSE AREA WITH WATER, PICK
  UP WITH ABSORBENT, PUT IN A DISPOSAL CONTAINER. SEAL AND REMOVE FROM THE WORK AREA.

HEALTH HAZARD INFORMATION:
  WHEN USING THIS PRODUCT WEAR: NO SPECIAL REQUIREMENT. Ingestion will cause burning sensation of the mouth, and
  throat with nausea and vomiting. Eye contact will cause stinging, tearing, itching, swelling, and redness. Skin
  contact may cause skin irritation, redness, and swelling.

  A physician should be contacted if anyone develops any signs or symptoms and suspects that they are caused by
  exposure to this product.

FIRST AID:
  Eye Exposure:
  Flush with water for 15 minutes while lifting the upper and lower eye lids. Contact lenses should not be worn
  when working with this product. If irritation persists, get medical attention.

  Skin Exposure:
  Rinse off with water.

  Breathing:
  Not expected to be a problem.

  Swallowing:
  Do not induce vomiting. If victim is conscious, give 2-3 glasses of milk or water to drink. Immediately contact
  the local poison control center for advice. Keep the victim warm and at rest. Get medical attention.

PRODUCT: CASCADE, LIQUID (AUTOMATIC DISH WASHING DETERGENT)

INCOMPATIBILITY: STRONG ACIDS; AMMONIA.

INGREDIENTS: COMPLEX SODIUM PHOSPHATE   TLV: CAS#:
             SODIUM CARBONATE   TLV: CAS#: 497-19-8
             CHLORINE BLEACH   TLV: CAS#: 7681-52-9
             SODIUM SILICATE   TLV: CAS#:
             COLORANT; PERFUME

IGNI TEMP: NA FP: NA LEL: NA UEL: NA VP: NA VD: NA SG: 1.2-1.3 PS: LIQUID APPEAR: LIGHT GREEN
    ODOR: PERFUMED   PH FACTOR: 12.2

HAZARD RATINGS: H: F: R:

FIRE FIGHTING:
    HIGH EXPANSION FOAM, LOW EXPANSION FOAM, ALCOHOL FOAM; DRY CHEMICAL, CARBON DIOXIDE, WATER FOG.
    PROTECTIVE CLOTHING, RUBBER GLOVES, AND BREATHING APPARATUS.
WARNING: 1] STRUCTURAL PROTECTIVE CLOTHING IS PERMEABLE, REMAIN CLEAR OF SMOKE, WATER FALL OUT AND WATER RUN OFF.
         2] KEEP OUT OF THE REACH OF CHILDREN.
         3] THERMAL DECOMPOSITION MAY YIELD TOXIC CHLORINE GAS.
         4] MOVE CONTAINERS FROM AREA IF WITHOUT RISK, COOL EXPOSED CONTAINERS.
         5] DIKE AREA FOR CONTROL AND CONTAINMENT TO PREVENT ENTRY INTO SEWERS, DRAIN, AND WATER WAYS.

LARGE SPILL/NO FIRE/RESCUE: WEAR RUBBER OR NEOPRENE BOOTS, GLOVES.

SPILL CONTROL AND CONTAINMENT:
    HOUHOUSEHOLD SPILL: WIPE UP LIQUID WITH AN ABSORBENT CLOTH, RINSE OUT IN THE SINK. WIPE UP RESIDUE WITH AN DAMP
    CLOTH.
    LARGE SPILL: PICK UP WITH ABSORBENT, PUT IN A DISPOSAL CONTAINER. RINSE AREA WITH WATER, PICK UP WITH ABSORBENT.
    PUT IN A DISPOSAL CONTAINER. SEAL AND REMOVE FROM THE WORK AREA.

HEALTH HAZARD INFORMATION:
    WHEN USING THIS PRODUCT WEAR: NO SPECIAL REQUIREMENT. Ingestion may result in a burning sensation to the mouth
    and throat with nausea, vomiting, and diarrhea. Eye contact will cause stinging, tearing, itching, swelling, and
    redness. Prolonged contact with concentrated liquid may cause skin redness and swelling with irritation.

    A physician should be contacted if anyone develops any signs or symptoms and suspects that they are caused by
    exposure to this product.

FIRST AID:
    Eye Exposure:
    Flush with water for 15 minutes while lifting the upper and lower eye lids. Contact lenses should not be worn
    when working with this product. Get medical attention.

    Skin Exposure:
    Wash off of the skin. If redness or swelling develop, get medical attention.

    Breathing:
    Not expected to be a problem.

    Swallowing:
    If the victim is conscious, give 1-2 glasses of milk or water to drink. Immediately contact the local poison
    control center for advice. Keep the victim warm and at rest. Get medical attention.

PRODUCT: CASCADE LIQUIGEL, LEMON (AUTOMATIC DISH WASHING DETERGENT)

INCOMPATIBILITY: STRONG ACIDS; TOILET BOWL CLEANERS; RUST REMOVERS; VINEGAR; AMMONIA; PRODUCTS WITH AMMONIA.

INGREDIENTS: POTASSIUM CARBONATE   TLV: CAS#: 584-08-7
             SODIUM HYPOCHLORITE   TLV: CAS#: 7681-52-9
             WATER   TLV: CAS#: 7732-18-5
             COMPLEX POTASSIUM   SODIUM PHOSPHATES

IGNI TEMP: NA FP: NA LEL: NA UEL: NA VP: NA VD: NA SG: 1.4 PS: LIQUID APPEAR: THICK YELLOW GEL
    ODOR: LEMON   PH FACTOR: 12.2

HAZARD RATINGS: H: F: R:

FIRE FIGHTING:
   HIGH EXPANSION FOAM, LOW EXPANSION FOAM, ALCOHOL FOAM; DRY CHEMICAL, CARBON DIOXIDE, WATER FOG.
   PROTECTIVE CLOTHING, RUBBER GLOVES, AND BREATHING APPARATUS.
WARNING: 1] STRUCTURAL PROTECTIVE CLOTHING IS PERMEABLE, REMAIN CLEAR OF SMOKE, WATER FALL OUT AND WATER RUN OFF.
         2] KEEP OUT OF THE REACH OF CHILDREN.
         3] HAZARDOUS DECOMPOSITION YIELDS TOXIC CHLORINE.
         4] DO NOT MIX WITH INCOMPATIBLE PRODUCTS LISTED ABOVE, YIELDS TOXIC CHLORINE OR AMMONIA GAS.
         5] MOVE CONTAINERS FROM AREA IF WITHOUT RISK, COOL EXPOSED CONTAINERS.
         6] DIKE AREA FOR CONTROL AND CONTAINMENT TO PREVENT ENTRY INTO SEWERS, DRAIN, AND WATER WAYS.

LARGE SPILL/NO FIRE/RESCUE: WEAR RUBBER OR NEOPRENE BOOTS, GLOVES.

SPILL CONTROL AND CONTAINMENT:
   HOUSEHOLD SPILL: WIPE UP LIQUID WITH AN ABSORBENT CLOTH. RINSE OUT IN THE SINK. WIPE UP RESIDUE WITH A DAMP
   CLOTH, RINSE OUT IN THE SINK.
   LARGE SPILL: COVER SPILL WITH ABSORBENT. SCOOP UP AND PUT IN A DISPOSAL CONTAINER. RINSE AREA WITH WATER, PICK
   UP WITH ABSORBENT. PUT IN A DISPOSAL CONTAINER. SEAL AND REMOVE FROM THE WORK AREA.

HEALTH HAZARD INFORMATION:
   WHEN USING THIS PRODUCT WEAR: NO SPECIAL REQUIREMENTS. Product is a skin and eye irritant. Eye contact causes
   stinging, tearing, itching, swelling, and redness. Skin contact may cause redness and swelling. Ingestion may
   cause nausea and vomiting, moderate to severe gastrointestinal irritation. Inhalation of dust may cause
   coughing, sore throat, wheezing, and shortness of breath.

   A physician should be contacted if anyone develops any signs or symptoms and suspects that they are caused by
   exposure to this product.

FIRST AID:
   Eye Exposure:
   Flush with water for 15 minutes while lifting the upper and lower eye lids. Contact lenses should not be worn
   when working with this product. If irritation persists, get medical attention.

   Skin Exposure:
   Rinse skin with water.

   Breathing:
   If problems develop, move victim to fresh air.

   Swallowing:
   Do not induce vomiting. If victim is conscious, give 2-3 glasses of milk or water to drink. Immediately contact
   the local poison control center for advice. Keep the victim warm and at rest. Get medical attention.

PRODUCT: CHECK-UP GEL TOOTH PASTE (TUBE OR PUMP) ALL SIZES

INCOMPATIBILITY: NONE KNOWN

INGREDIENTS: ALL ARE NON-HAZARDOUS BY OSHA 1910.1200 STANDARD

IGNI TEMP: NA FP: NA LEL: NA UEL: NA VP: NA VD: NA SG: 1.31-1.34 PS: GEL APPEAR: THICK GREEN ODOR:
   PH FACTOR: 6.5

HAZARD RATINGS: H: F: R: 0 0 0

FIRE FIGHTING:
   HIGH EXPANSION FOAM, LOW EXPANSION FOAM, ALCOHOL FOAM; DRY CHEMICAL, CARBON DIOXIDE, WATER FOG.
   PROTECTIVE CLOTHING, RUBBER GLOVES, AND BREATHING APPARATUS.
WARNING: 1] STRUCTURAL PROTECTIVE CLOTHING IS PERMEABLE, REMAIN CLEAR OF SMOKE, WATER FALL OUT AND WATER RUN OFF.
         2] KEEP OUT OF THE REACH OF CHILDREN.
         3] WILL CREATE A SLIPPING HAZARD.
         4] MOVE CONTAINERS FROM AREA IF WITHOUT RISK, COOL EXPOSED CONTAINERS.
         5] DIKE AREA FOR CONTROL AND CONTAINMENT TO PREVENT ENTRY INTO SEWERS, DRAIN, AND WATER WAYS.

LARGE SPILL/NO FIRE/RESCUE: WEAR RUBBER OR NEOPRENE BOOTS, GLOVES.

SPILL CONTROL AND CONTAINMENT:
   HOUSEHOLD SPILL: WIPE UP WITH ABSORBENT CLOTH. PICK UP RESIDUE WITH DAMP CLOTH.
   LARGE SPILL: COVER WITH ABSORBENT. PICK UP AND PUT INTO A DISPOSAL CONTAINER. RINSE AREA WITH WATER, PICK UP
   WITH ABSORBENT, PUT INTO A DISPOSAL CONTAINER. SEAL AND REMOVE FROM THE WORK AREA.

HEALTH HAZARD INFORMATION:
   WHEN USING THIS PRODUCT WEAR: NO SPECIAL REQUIREMENTS. Contact with the eyes may cause irritation. Ingestion of
   a large amount may cause gastrointestinal irritation with diarrhea. Possible nausea and vomiting.

   A physician should be contacted if anyone develops any signs or symptoms and suspects that they are caused by
   exposure to this product.

FIRST AID:
   Eye Exposure:
   Flush eyes with water for 15 minutes while lifting the upper and lower eye lids. Contact lenses should not be
   worn when working with this product. If irritation persists, get medical attention.

   Skin Exposure:
   Not expected to be a problem.

   Breathing:
   Not expected to be a problem.

   Swallowing:
   Immediately contact the local poison control center for advice. Keep the victim warm and at rest. Get medical
   attention.

PRODUCT: CHECK-UP GINGIVAL TOOTH PASTE

INCOMPATIBILITY: NONE KNOWN

INGREDIENTS: ALL ARE NON-HAZARDOUS BY OSHA 1910.1200 STANDARD

IGNI TEMP: NA FP: NA LEL: NA UEL: NA VP: NA VD: NA SG: 1.3-1.4 PS: PASTE APPEAR: OFF WHITE
    ODOR: SPEARMINT   PH FACTOR: 5.0 TO 5.8

HAZARD RATINGS: H: F: R:

FIRE FIGHTING:
    HIGH EXPANSION FOAM, LOW EXPANSION FOAM, ALCOHOL FOAM; DRY CHEMICAL, CARBON DIOXIDE, WATER FOG.
    PROTECTIVE CLOTHING, RUBBER GLOVES, AND BREATHING APPARATUS.
WARNING: 1] STRUCTURAL PROTECTIVE CLOTHING IS PERMEABLE, REMAIN CLEAR OF SMOKE, WATER FALL OUT AND WATER RUN OFF.
         2] KEEP OUT OF THE REACH OF CHILDREN.
         3] SPILL WILL CREATE A SLIPPING HAZARD.
         4] MOVE CONTAINERS FROM AREA IF WITHOUT RISK, COOL EXPOSED CONTAINERS.
         5] DIKE AREA FOR CONTROL AND CONTAINMENT TO PREVENT ENTRY INTO SEWERS, DRAIN, AND WATER WAYS.

LARGE SPILL/NO FIRE/RESCUE: WEAR RUBBER OR NEOPRENE BOOTS, GLOVES.

SPILL CONTROL AND CONTAINMENT:
    HOUSEHOLD SPILL: WIPE UP WITH ABSORBENT CLOTH AND RINSE IN THE SINK. WIPE UP RESIDUE WITH A DAMP CLOTH AND RINSE
    IN THE SINK.
    LARGE SPILL: COVER SPILL AREA WITH ABSORBENT. PICK UP AND PUT INTO A DISPOSAL CONTAINER. RINSE AREA WITH WATER,
    PICK UP WITH ABSORBENT, PUT INTO A DISPOSAL CONTAINER. SEAL AND REMOVE FROM THE WORK AREA.

HEALTH HAZARD INFORMATION:
    WHEN USING THIS PRODUCT WEAR: NO SPECIAL REQUIREMENTS. Contact with the eyes may cause irritation. Ingestion of
    a large amount may cause gastrointestinal irritation. Possible nausea, vomiting, and diarrhea.

    A physician should be contacted if anyone develops any signs or symptoms and suspects that they are caused by
    exposure to this product.

FIRST AID:
Eye Exposure:
Flush the eyes with water for 15 minutes while lifting the upper and lower eye lids. Contact lenses should not
be worn when working with this product. If irritation persists, get medical attention.

Skin Exposure:
Not expected to be a problem.

Breathing:
Not expected to be a problem.

Swallowing:
Immediately contact the local poison control center for advice. Keep the victim warm and at rest. Get medical
attention.

PRODUCT: CHECK-UP TOOTH PASTE (TUBE OR PUMP) ALL SIZES

INCOMPATIBILITY: NONE KNOWN

INGREDIENTS: ALL ARE NON-HAZARDOUS BY OSHA 1910.1200 STANDARD

IGNI TEMP:NA   FP: NA   LEL: NA   UEL: NA   VP: NA   VD: NA   SG: 1.31-1.34   PS: PASTE   APPEAR: WHITE   ODOR:
   PH FACTOR: 6.4 TO 6.8

HAZARD RATINGS: H: F: R:

FIRE FIGHTING:
   HIGH EXPANSION FOAM, LOW EXPANSION FOAM, ALCOHOL FOAM; DRY CHEMICAL, CARBON DIOXIDE, WATER FOG.
   PROTECTIVE CLOTHING, RUBBER GLOVES, AND BREATHING APPARATUS.
WARNING: 1] STRUCTURAL PROTECTIVE CLOTHING IS PERMEABLE, REMAIN CLEAR OF SMOKE, WATER FALL OUT AND WATER RUN OFF.
         2] KEEP OUT OF THE REACH OF CHILDREN.
         3] SPILL WILL CREATE A SLIPPING HAZARD.
         4] MOVE CONTAINERS FROM AREA IF WITHOUT RISK, COOL EXPOSED CONTAINERS.
         5] DIKE AREA FOR CONTROL AND CONTAINMENT TO PREVENT ENTRY INTO SEWERS, DRAIN, AND WATER WAYS.

LARGE SPILL/NO FIRE/RESCUE: WEAR RUBBER OR NEOPRENE BOOTS, GLOVES.

SPILL CONTROL AND CONTAINMENT:
   HOUSEHOLD SPILL: WIPE UP WITH ABSORBENT CLOTH AND RINSE IN THE SINK. PICK UP RESIDUE WITH A DAMP CLOTH. RINSE IN
   THE SINK.
   LARGE SPILL: SCOOP UP MATERIAL AND PUT INTO A DISPOSAL CONTAINER. RINSE AREA WITH WATER, PICK UP WITH ABSORBENT,
   PUT INTO A DISPOSAL CONTAINER.

HEALTH HAZARD INFORMATION:
   WHEN USING THIS PRODUCT WEAR: NO SPECIAL REQUIREMENT. Contact with the eyes may cause irritation. Ingestion of a
   large amount may cause gastrointestinal irritation with nausea, vomiting, and diarrhea.

   A physician should be contacted if anyone develops any signs or symptoms and suspects that they are caused by
   exposure to this product.

FIRST AID:
   Eye Exposure:
   Flush eyes with water for 15 minutes while lifting the upper and lower eye lids. Contact lenses should not be
   worn when working with this product. If irritation persists, get medical attention.

   Skin Exposure:
   Not expected to be a problem.

   Breathing:
   Not expected to be a problem.

   Swallowing:
   Immediately contact the local poison control center for advice. Keep the victim warm and at rest. Get medical
   attention.

PRODUCT: CHEER   (LAUNDRY GRANULES)

INCOMPATIBILITY: NONE KNOWN

INGREDIENTS: COMPLEX SODIUM PHOSPHATE OR
              ALUMINOSILICATES   TLV: CAS#:
              SODIUM CARBONATE   TLV: CAS#:   497-19-8
              SODIUM SULFATE   TLV: CAS#:   7757-82-6
              SODIUM SILICATE   TLV: CAS#:   1344-09-8

IGNI TEMP: NA FP: NA LEL: NA UEL: NA VP: NA VD: NA SG: NA PS: GRANULAR APPEAR: BLUE   ODOR: PERFUMED
     PH FACTOR: NA

HAZARD RATINGS: H: F: R:
FIRE FIGHTING:
    HIGH EXPANSION FOAM, LOW EXPANSION FOAM, ALCOHOL FOAM; DRY CHEMICAL, CARBON DIOXIDE, WATER FOG.
    PROTECTIVE CLOTHING, RUBBER GLOVES, AND BREATHING APPARATUS.
WARNING: 1] STRUCTURAL PROTECTIVE CLOTHING IS PERMEABLE, REMAIN CLEAR OF SMOKE, WATER FALL OUT AND WATER RUN OFF.
         2] KEEP OUT OF THE REACH OF CHILDREN.
         3] CHECK WHITE RECTANGLE ON THE TOP OR BOTTOM. IF NUMBER BEGINS WITH "L" PRODUCT HAS SODIUM PYROPHOSPHATE
            IN IT. TLV: 5 mg/m3 CAS: 7722-88-5.
         4] MOVE CONTAINERS FROM AREA IF WITHOUT RISK, COOL EXPOSED CONTAINERS.
         5] DIKE AREA FOR CONTROL AND CONTAINMENT TO PREVENT ENTRY INTO SEWERS, DRAIN, AND WATER WAYS.

LARGE SPILL/NO FIRE/RESCUE: WEAR RUBBER OR NEOPRENE BOOTS, GLOVES.

SPILL CONTROL AND CONTAINMENT:
    HOUSEHOLD SPILL: AVOID CREATING A DUST. SWEEP UP MATERIAL AND PUT IN A DISPOSAL CONTAINER. WIPE UP RESIDUE WITH
    A DAMP CLOTH.
    LARGE SPILL: AVOID CREATING A DUST. GENTLY SWEEP UP MATERIAL AND PUT IN A DISPOSAL CONTAINER. RINSE AREA WITH
    WATER, PICK UP WITH ABSORBENT, PUT INTO A DISPOSAL CONTAINER.

HEALTH HAZARD INFORMATION:
    WHEN USING THIS PRODUCT WEAR: NO SPECIAL REQUIREMENTS. Dust may create eye irritation problem and irritation of
    the respiratory tract. Ingestion may cause gastrointestinal irritation with nausea, vomiting, and diarrhea.

    A physician should be contacted if anyone develops any signs or symptoms and suspects that they are caused by
    exposure to this product.

FIRST AID:
Eye Exposure:
Flush eyes with water for 15 minutes while lifting the upper and lower eye lids. Contact lenses should not be
worn when working with this product. If irritation persists, get medical attention.

Skin Exposure:
Not expected to be a problem. Rinse off with water.

Breathing:
Move the victim to fresh air. Treat any symptoms as they appear. If required get medical attention.

Swallowing:
If the victim is conscious, give 2-3 glasses of milk or water to drink. Immediately contact the local poison
control center for advice. Keep the victim warm and at rest. Get medical attention.

PRODUCT: CHEER, LIQUID (LAUNDRY DETERGENT)

INCOMPATIBILITY: ACIDS AND ACID FUMES

INGREDIENTS: ETHYL ALCOHOL   TLV: 1900 mg/m3   CAS#:   64-17-5
           ANIONIC SURFACTANTS   TLV: CAS#:
           NONIONIC SURFACTANTS   TLV: CAS#:
           COLORANT   TLV: CAS#:
           PERFUME   TLV: CAS#:

IGNI TEMP: NA FP: NA LEL: NA UEL: NA VP: NA VD: NA SG: 1.08 PS: LIQUID APPEAR: DARK BLUE ODOR: PERFUMED
  PH FACTOR:

HAZARD RATINGS: H: F: R:

FIRE FIGHTING:
  HIGH EXPANSION FOAM, LOW EXPANSION FOAM, ALCOHOL FOAM; DRY CHEMICAL, CARBON DIOXIDE, WATER FOG.
  PROTECTIVE CLOTHING, RUBBER GLOVES, AND BREATHING APPARATUS.
WARNING: 1] STRUCTURAL PROTECTIVE CLOTHING IS PERMEABLE, REMAIN CLEAR OF SMOKE, WATER FALL OUT AND WATER RUN OFF.
       2] KEEP OUT OF THE REACH OF CHILDREN.
       3] THERMAL DECOMPOSITION MAY YIELD CARBON DIOXIDE, CARBON MONOXIDE.
       4] MOVE CONTAINERS FROM AREA IF WITHOUT RISK, COOL EXPOSED CONTAINERS.
       5] DIKE AREA FOR CONTROL AND CONTAINMENT TO PREVENT ENTRY INTO SEWERS, DRAIN, AND WATER WAYS.

LARGE SPILL/NO FIRE/RESCUE: WEAR RUBBER OR NEOPRENE BOOTS, GLOVES.

SPILL CONTROL AND CONTAINMENT:
  HOUSEHOLD SPILL: WIPE UP WITH AN ABSORBENT CLOTH, RINSE OUT IN THE SINK. WIPE UP RESIDUE WITH A DAMP CLOTH,
  RINSE OUT IN THE SINK.
  LARGE SPILL: PICK UP WITH ABSORBENT, PUT IN A DISPOSAL CONTAINER. RINSE AREA WITH WATER, PICK UP WITH ABSORBENT.
  PUT IN A DISPOSAL CONTAINER. SEAL AND REMOVE FROM THE WORK AREA.

HEALTH HAZARD INFORMATION:
  WHEN USING THIS PRODUCT WEAR: NO SPECIAL REQUIREMENT. Eye contact may cause stinging, tearing, itching,
  swelling, and redness. Prolonged contact with the concentrated liquid may cause skin drying and irritation.
  Ingestion may result in abdominal pain, nausea, vomiting, and diarrhea.

  A physician should be contacted if anyone develops any signs or symptoms and suspects that they are caused by
  exposure to this product.

FIRST AID:
  Eye Exposure:
  Flush with water for 15 minutes while lifting the upper and lower eye lids. Contact lenses should not be worn
  when working with thishis product. Get medical attention.

  Skin Exposure:
  Wash off with water. If irritation develops, get medical attention.

  Breathing:
  Not expected to be a problem.

  Swallowing:
  If the victim is conscious, give 1-2 glasses of water to drink. Immediately contact the local poison control
  center for advice. Keep the victim warm and at rest. Get medical attention.

PRODUCT: CHEER, ULTRA WITH COLOR GUARD (LAUNDRY GRANULES)

INCOMPATIBILITY: NONE KNOWN

INGREDIENTS: SODIUM SILICATES   TLV: CAS#: 1344-09-8
             SODIUM SULFATE   TLV: CAS#: 7757-82-6
             SODIUM CARBONATE   TLV: CAS#: 497-19-8
             CITRIC ACID   TLV: CAS#:   77-92-9
             ALUMINOSILICATE, ANIONIC SURFACTANTS, SODIUM PHOSPHATE, ENZYMES, PERFUME

IGNI TEMP: NA FP: NA LEL: NA UEL: NA VP: NA VD: NA SG: NA PS: POWDER APPEAR: WHITE - BLUE SPECKS
    ODOR: PERFUMED   PH FACTOR: NA

HAZARD RATINGS: H: F: R:

FIRE FIGHTING:
   HIGH EXPANSION FOAM, LOW EXPANSION FOAM, ALCOHOL FOAM; DRY CHEMICAL, CARBON DIOXIDE, WATER FOG.
   PROTECTIVE CLOTHING, RUBBER GLOVES, AND BREATHING APPARATUS.
WARNING: 1] STRUCTURAL PROTECTIVE CLOTHING IS PERMEABLE, REMAIN CLEAR OF SMOKE, WATER FALL OUT AND WATER RUN OFF.
         2] KEEP OUT OF THE REACH OF CHILDREN.
         3] THERMAL DECOMPOSITION MAY YIELD CARBON DIOXIDE, CARBON MONOXIDE.
         4] MOVE CONTAINERS FROM AREA IF WITHOUT RISK, COOL EXPOSED CONTAINERS.
         5] DIKE AREA FOR CONTROL AND CONTAINMENT TO PREVENT ENTRY INTO SEWERS, DRAIN, AND WATER WAYS.

LARGE SPILL/NO FIRE/RESCUE: WEAR RUBBER OR NEOPRENE BOOTS, GLOVES.

SPILL CONTROL AND CONTAINMENT:
   HOUSEHOLD SPILL: SWEEP UP AND PUT IN A DISPOSAL CONTAINER. PICK UP RESIDUE WITH A DAMP MOP. RINSE OUT IN THE
   SINK.
   LARGE SPILL: AVOID CREATING A DUST. SWEEP UP POWDER. PUT IN A DISPOSAL CONTAINER. RINSE AREA WITH WATER. PICK UP
   WITH ABSORBENT. PUT IN A DISPOSAL CONTAINER. SEAL AND REMOVE FROM THE WORK AREA.

HEALTH HAZARD INFORMATION:
   WHEN USING THIS PRODUCT WEAR: NO SPECIAL REQUIREMENT. Inhalation may cause coughing, sore throat, wheezing, and
   shortness of breath. Eye contact may cause stinging, tearing, itching, swelling, and redness. Ingestion may
   result in nausea, vomiting, and diarrhea.

   A physician should be contacted if anyone develops any signs or symptoms and suspects that they are caused by
   exposure to this product.

FIRST AID:
Eye Exposure:
Flush with water for 15 minutes while lifting the upper and lower eye lids. Contact lenses should not be worn
when working with this product. Get medical attention.

Skin Exposure:
Wash with water. If irritation develops, get medical attention.

Breathing:
If irritation develops, move victim to fresh air.

Swallowing:
If the victim is conscious, give 2-3 glasses of milk or water to drink. Immediately contact the local poison
control center for advice. Keep the victim warm and at rest. Get medical attention.

PRODUCT: CLEAN & CLEAR

INCOMPATIBILITY: NONE KNOWN

INGREDIENTS: MODIFIED ACRYLIC POLYMERS 5-10% TLV: CAS#:
           SURFACTANTS   1-4 % TLV: CAS#:
           WATER   80-90% TLV: CAS#: 7732-18-5

IGNI TEMP: NA FP: NA LEL: NA UEL: NA VP: NA VD: NA SG: NA PS: LIQUID APPEAR: CLEAR BLUE ODOR: EVERGREEN
   PH FACTOR: 8.7

HAZARD RATINGS: H: F: R:

FIRE FIGHTING:
  HIGH EXPANSION FOAM, LOW EXPANSION FOAM, ALCOHOL FOAM; DRY CHEMICAL, CARBON DIOXIDE, WATER FOG.
  PROTECTIVE CLOTHING, RUBBER GLOVES, AND BREATHING APPARATUS.
WARNING: 1] STRUCTURAL PROTECTIVE CLOTHING IS PERMEABLE, REMAIN CLEAR OF SMOKE, WATER FALL OUT AND WATER RUN OFF.
        2] KEEP OUT OF THE REACH OF CHILDREN.
        3] THERMAL DECOMPOSITION YIELDS TOXIC CARBON DIOXIDE, CARBON MONOXIDE.
        4] SPILL MAY CREATE A SLIPPING HAZARD.
        5] MOVE CONTAINERS FROM AREA IF WITHOUT RISK, COOL EXPOSED CONTAINERS.
        6] DIKE AREA FOR CONTROL AND CONTAINMENT TO PREVENT ENTRY INTO SEWERS, DRAIN, AND WATER WAYS.

LARGE SPILL/NO FIRE/RESCUE: WEAR RUBBER OR NEOPRENE BOOTS, GLOVES.

SPILL CONTROL AND CONTAINMENT:
  HOUSEHOLD SPILL: WIPE UP WITH ABSORBENT CLOTH, RINSE IN SINK. WIPE UP RESIDUE WITH A DAMP CLOTH. RINSE IN SINK.
  LARGE SPILL: COVER SPILL WITH ABSORBENT. PICK UP AND PUT INTO A DISPOSAL CONTAINER. SCRUB AREA WITH DETERGENT
  AND WATER. PICK UP WITH ABSORBENT, PUT INTO DISPOSAL CONTAINER. SEAL AND REMOVE FROM WORK AREA.

HEALTH HAZARD INFORMATION:
  WHEN USING THIS PRODUCT WEAR: LATEX RUBBER GLOVES. Contact may cause eye irritation. Prolonged or repeated skin
  contact may cause irritation. Ingestion will cause gastrointestinal irritation with nausea, vomiting, and
  diarrhea.

  A physician should be contacted if anyone develops any signs or symptoms and suspects that they are caused by
  exposure to this product.

FIRST AID:
  Eye Exposure:
  Flush eyes with water for 15 minutes while lifting the upper and lower eye lids. Contact lenses should not be
  worn when working with this product. If irritation persists, get medical attention.

  Skin Exposure:
  If liquid get onto the skin, flush with water and wash with soap and water. If irritation appears, get medical
  attention.

  Breathing:
  Not expected to be a problem.

  Swallowing:
  Immediately contact the local poison control center for advice. Keep the victim warm and at rest. Get medical
  attention.

PRODUCT: CLOROX 2 (POWDER)

INCOMPATIBILITY: STRONG ACIDS.

ug = micrograms/m3

INGREDIENTS: SODIUM PERBORATE   5-25%   TLV: CAS#: 10486-00-7
             SUBTILISINS PROTEOLYTIC .5-2% TLV: .06ug/m3   CAS#: 1395-21-7
             ESPERASE   TLV: 3.9ug/m3   CAS#: 9072-77-2
             SAVINASE   TLV: 5.6ug/m3   CAS#: 9014-01-1
             AMYLASE   LESS THAN .5% TLV: 1.5ug/m3   CAS#: 9000-90-2

IGNI TEMP: NA FP: NA LEL: NA UEL: NA VP: NA VD: NA SG: NA PS: POWDER APPEAR: WHITE ODOR: CHLORINE
    PH FACTOR: 11.1

HAZARD RATINGS: H: F: R: 3   0   0

FIRE FIGHTING:
   HIGH EXPANSION FOAM, LOW EXPANSION FOAM, ALCOHOL FOAM; DRY CHEMICAL, CARBON DIOXIDE, WATER FOG.
   PROTECTIVE CLOTHING, RUBBER GLOVES, AND BREATHING APPARATUS.
WARNING: 1] STRUCTURAL PROTECTIVE CLOTHING IS PERMEABLE, REMAIN CLEAR OF SMOKE, WATER FALL OUT AND WATER RUN OFF.
         2] KEEP OUT OF THE REACH OF CHILDREN.
         3] CONTACT WITH ACIDS CAUSES EXOTHERMIC REACTION GENERATING HEAT AND TOXIC FUMES AND GASES.
         4] MOVE CONTAINERS FROM AREA IF WITHOUT RISK, COOL EXPOSED CONTAINERS.
         5] DIKE AREA FOR CONTROL AND CONTAINMENT TO PREVENT ENTRY INTO SEWERS, DRAIN, AND WATER WAYS.

LARGE SPILL/NO FIRE/RESCUE: WEAR RUBBER OR NEOPRENE BOOTS, GLOVES.

SPILL CONTROL AND CONTAINMENT:
   HOUSEHOLD SPILL: AVOID CREATING A DUST. SWEEP UP POWDER. PUT IN DISPOSAL CONTAINER. WIPE UP RESIDUE WITH A DAMP
   CLOTH AND RINSE IN SINK.
   LARGE SPILL: AVOID CREATING A DUST. GENTLY SWEEP UP MATERIAL AND PUT IN A DISPOSAL CONTAINER. RINSE AREA WITH
   WATER, PICK UP WITH ABSORBENT. PUT IN A DISPOSAL CONTAINER. SEAL AND REMOVE FROM WORK AREA.

HEALTH HAZARD INFORMATION:
   WHEN USING THIS PRODUCT WEAR: NO SPECIAL REQUIREMENTS. Product will cause moderate eye irritation, tearing,
   burning sensation, itching, swelling, and redness. Ingestion is harmful. Inhalation of large quantities of dust
   may cause sensitization; symptoms range from hay fever to asthma.

   A physician should be contacted if anyone develops any signs or symptoms and suspects that they are caused by
   exposure to this product.

FIRST AID:
Eye Exposure:
Immediately flush eyes with water for at least 15 minutes while lifting the upper and lower eye lids. Contact
lenses should not be worn when working with this product, get medical attention.

Skin Exposure:
If powder gets on the skin, rinse with water. If irritation appears, get medical attention.

Breathing:
If breathing becomes labored, move the victim to fresh air. Treat symptoms as they appear. Get medical
attention.

Swallowing:
If the victim is conscious, give 2 to 3 glasses of water to drink. Immediately contact the local poison control
center for advice. Keep the victim warm and at rest. Get medical attention.

PRODUCT: CLOROX 2, LIQUID

INCOMPATIBILITY: NONE LISTED

INGREDIENTS: HYDROGEN PEROXIDE   .5-5%   TLV: CAS#: 7722-84-1

IGNI TEMP: NA FP: NA LEL: NA UEL: NA VP: P: NA VD: NA SG: 1.02 PS: LIQUID APPEAR: BLUE   ODOR: FRAGRANCE
    PH FACTOR: 3.8

HAZARD RATINGS: H: F: R:

FIRE FIGHTING:
    HIGH EXPANSION FOAM, LOW EXPANSION FOAM, ALCOHOL FOAM; DRY CHEMICAL, CARBON DIOXIDE, WATER FOG.
    PROTECTIVE CLOTHING, RUBBER GLOVES, AND BREATHING APPARATUS.
WARNING: 1] STRUCTURAL PROTECTIVE CLOTHING IS PERMEABLE, REMAIN CLEAR OF SMOKE, WATER FALL OUT AND WATER RUN OFF.
         2] KEEP OUT OF THE REACH OF CHILDREN.
         3] MOVE CONTAINERS FROM AREA IF WITHOUT RISK, COOL EXPOSED CONTAINERS.
         4] DIKE AREA FOR CONTROL AND CONTAINMENT TO PREVENT ENTRY INTO SEWERS, DRAIN, AND WATER WAYS.

LARGE SPILL/NO FIRE/RESCUE: WEAR RUBBER OR NEOPRENE BOOTS, GLOVES.

SPILL CONTROL AND CONTAINMENT:
    HOUSEHOLD SPILL: WIPE UP LIQUID WITH AN ABSORBENT CLOTH, RINSE OUT IN THE SINK. WIPE UP RESIDUE WITH A DAMP
    CLOTH, RINSE OUT IN THE SINK.
    LARGE SPILL: PICK UP WITH NON-ORGANIC ABSORBENT. PUT IN A DISPOSAL CONTAINER. RINSE AREA WITH WATER, PICK UP
    WITH ABSORBENT. PUT IN A DISPOSAL CONTAINER. SEAL AND REMOVE FROM THE WORK AREA.

HEALTH HAZARD INFORMATION:
    WHEN USING THIS PRODUCT WEAR: NO SPECIAL REQUIREMENT. Eye contact will cause stinging, tearing, itching,
    swelling, and redness. Prolonged skin contact may cause irritation.

    A physician should be contacted if anyone develops any signs or symptoms and suspects that they are caused by
    exposure to this product.

FIRST AID:
    Eye Exposure:
    Flush with water for 15 minutes while lifting the upper and lower eye lids. Contact lenses should not be worn
    when working with this product. Get medical attention.

    Skin Exposure:
    Wash off with water. If irritation develops, get medical attention.

    Breathing:
    Not expected to be a problem.

    Swallowing:
    Expected to be a low hazard. If victim is conscious, give 1-2 glass of water to drink. Immediately contact the
    local poison control center for advice. Keep the victim warm and at rest. Get medical attention.

PRODUCT: CLOROX, FRESH SCENT

INCOMPATIBILITY: TOILET BOWL CLEANERS; RUST REMOVERS; VINEGAR; ACIDS; AMMONIA; OR PRODUCTS CONTAINING AMMONIA.

INGREDIENTS: SODIUM HYPOCHLORITE   5.25%   TLV: CAS#: 7681-52-9
             SODIUM CHLORATE   TLV: CAS#: 7775-09-9

IGNI TEMP: NA FP: NA LEL: NA UEL: NA VP: NA VD: NA SG: 1.09 PS: LIQUID APPEAR: CLEAR ODOR: CHLORINE
   PH FACTOR: 12.5

HAZARD RATINGS: H: F: R: 2   0   1   OXY

FIRE FIGHTING:
   HIGH EXPANSION FOAM, LOW EXPANSION FOAM, ALCOHOL FOAM; DRY CHEMICAL, CARBON DIOXIDE, WATER FOG.
   PROTECTIVE CLOTHING, RUBBER GLOVES, AND BREATHING APPARATUS.
WARNING: 1] STRUCTURAL PROTECTIVE CLOTHING IS PERMEABLE, REMAIN CLEAR OF SMOKE, WATER FALL OUT AND WATER RUN OFF.
         2] KEEP OUT OF THE REACH OF CHILDREN.
         3] THERMAL DECOMPOSITION AT 212F YIELDS CHLORINE VAPORS.
         4] DO NOT MIX WITH ANY OF THE PRODUCTS LISTED AS INCOMPATIBLE ABOVE.
         5] MOVE CONTAINERS FROM AREA IF WITHOUT RISK. COOL EXPOSED CONTAINERS.
         6] DIKE AREA FOR CONTROL AND CONTAINMENT TO PREVENT ENTRY INTO SEWERS, DRAIN, AND WATER WAYS.

LARGE SPILL/NO FIRE/RESCUE: WEAR NON-SEALED CHEMICAL PROTECTIVE CLOTHING RUBBER OR NEOPRENE BOOTS, GLOVES AND
                            BREATHING APPARATUS.

SPILL CONTROL AND CONTAINMENT:
   HOUSEHOLD SPILL: WEAR LATEX GLOVES, WIPE UP LIQUID WITH ABSORBENT CLOTH. RINSE OUT IN THE SINK. WIPE UP RESIDUE
   WITH A DAMP CLOTH UNTIL CHLORINE ODOR IS GONE. RINSE SPILL AREA WITH WATER AND WIPE UP WITH A CLOTH. RINSE OUT
   IN THE SINK.
   LARGE SPILL: COVER SPILL AREA WITH A NON-ORGANIC ABSORBENT. PICK UP AND PUT INTO A DISPOSAL CONTAINER. RINSE
   AREA WITH WATER, PICK UP WITH ABSORBENT, PUT IN A DISPOSAL CONTAINER. SEAL AND REMOVE FROM THE WORK AREA.

HEALTH HAZARD INFORMATION:
   WHEN USING THIS PRODUCT WEAR: LATEX GLOVES. Skin contact may cause an irritating rash. Contact with the eyes
   causes severe injury (but temporary). Inhalation will cause irritation of the nose, throat, and lungs. Ingestion
   will cause abdominal pain, nausea, vomiting, and burns of the mouth, throat, and stomach.

   A physician should be contacted if anyone develops any signs or symptoms and suspects that they are caused by
   exposure to this product.

FIRST AID:
   Eye Exposure:
   Flush with water for 15 minutes while lifting the upper and lower eye lids. Contact lenses should not be worn
   when working with this product. Get medical attention.

   Skin Exposure:
   Wash with soap and water. If irritation appears, get medical attention.

   Breathing:
   Move victim to fresh air. Treat symptoms as they appear. If needed, get medical attention.

   Swallowing:
   If victim is conscious, give 3-4 glasses of water to drink. Immediately contact the local poison control center
   for advice. Keep the victim warm and at rest. Get medical attention.

PRODUCT: CLOROX, LEMON FRESH

INCOMPATIBILITY: TOILET BOWL CLEANERS; RUST REMOVERS; VINEGAR; ACIDS;AMMONIA; OR PRODUCT WITH AMMONIA.

INGREDIENTS: SODIUM HYPOCHLORITE   TLV: CAS#: 7681-52-9

IGNI TEMP: NA FP: NA LEL: NA UEL: NA VP: NA VD: NA SG: 1.085 PS: LIQUID APPEAR: CLEAR ODOR: CHLORINE
    PH FACTOR: 12.5

HAZARD RATINGS: H: F: R: 2 0 1   OXY   HAZARD CLASS   5.1

FIRE FIGHTING:
  HIGH EXPANSION FOAM, LOW EXPANSION FOAM, ALCOHOL FOAM; DRY CHEMICAL, CARBON DIOXIDE, WATER FOG.
  PROTECTIVE CLOTHING, RUBBER GLOVES, AND BREATHING APPARATUS.
WARNING: 1] STRUCTURAL PROTECTIVE CLOTHING IS PERMEABLE, REMAIN CLEAR OF SMOKE, WATER FALL OUT AND WATER RUN OFF.
         2] KEEP OUT OF THE REACH OF CHILDREN.
         3] THERMAL DECOMPOSITION AT 212F YIELDS SODIUM CHLORATE.
         4] IF A FIRE CONTAINERS WILL RUPTURE RELEASING SODIUM CHLORATE.
         5] DO NOT MIX WITH INCOMPATIBLE PRODUCTS LIST ABOVE, YIELDS TOXIC CHLORINE OR AMMONIA GAS.
         6] MOVE CONTAINERS FROM AREA IF WITHOUT RISK, COOL EXPOSED CONTAINERS.
         7] DIKE AREA FOR CONTROL AND CONTAINMENT TO PREVENT ENTRY INTO SEWERS, DRAIN, AND WATER WAYS.

LARGE SPILL/NO FIRE/RESCUE: WEAR RUBBER OR NEOPRENE BOOTS, GLOVES.

SPILL CONTROL AND CONTAINMENT:
  HOUSEHOLD SPILL: DILUTE WITH WATER. WIPE UP WITH AN ABSORBENT CLOTH, RINSE OUT IN THE SINK. RINSE AREA WITH
  WATER AND WIPE UP AT LEAST TWO TIMES. DRY WITH ABSORBENT CLOTH.
  LARGE SPILL: DIKE AND DILUTE WITH WATER. APPLY NON-ORGANIC ABSORBENT. PICK UP AND PUT IN A DISPOSAL CONTAINER.
  RINSE AREA WITH WATER, PICK UP WITH ABSORBENT, PUT IN A DISPOSAL CONTAINER. SEAL AND REMOVE FROM THE WORK AREA.

HEALTH HAZARD INFORMATION:
  WHEN USING THIS PRODUCT WEAR: NO SPECIAL REQUIREMENT. Contact with the eyes causes severe but temporary eye
  injury. May cause irritation of the skin. Ingestion will cause nausea, and vomiting. Inhalation of vapor or mist
  will cause irrirritation of the nose, throat, and lungs.

  A physician should be contacted if anyone develops any signs or symptoms and suspects that they are caused by
  exposure to this product.

FIRST AID:
  Eye Exposure:
  Flush with water for 15 minutes while lifting the upper and lower eye lids. Contact lenses should not be worn
  when working with this product. If irritation persists, get medical attention.

  Skin Exposure:
  Flush with water, wash with soap and water.

  Breathing:
  If problems develop, move victim to fresh air.

  Swallowing:
  If the victim is conscious, give 2-3 glasses of water to drink. Immediately contact the local poison control
  center for advice. Keep the victim warm and at rest. Get medical attention.

PRODUCT: CLOROX BLEACH, INSTITUTIONAL

INCOMPATIBILITY: TOILET BOWL CLEANERS; RUST REMOVERS; VINEGAR; AMMONIA OR PRODUCTS CONTAINING AMMONIA.

INGREDIENTS: SODIUM HYPOCHLORITE   5.25%   TLV: CAS#: 7681-52-9

IGNI TEMP: NA FP: NA LEL: NA UEL: NA VP: NA VD: NA SG: 1.09 PS: LIQUID   APPEAR: CLEAR, LIGHT YELLOW
    ODOR: CHLORINE   PH FACTOR: 11.4

HAZARD RATINGS: H: F: R: 2 0 1 OXY   HAZARD CLASS   5.1

FIRE FIGHTING:
    HIGH EXPANSION FOAM, LOW EXPANSION FOAM, ALCOHOL FOAM; DRY CHEMICAL, CARBON DIOXIDE, WATER FOG.
    PROTECTIVE CLOTHING, RUBBER GLOVES, AND BREATHING APPARATUS.
WARNING: 1] STRUCTURAL PROTECTIVE CLOTHING IS PERMEABLE, REMAIN CLEAR OF SMOKE, WATER FALL OUT AND WATER RUN OFF.
         2] KEEP OUT OF THE REACH OF CHILDREN.
         3] DO NOT MIX WITH PRODUCTS LISTED AS INCOMPATIBLE ABOVE, YIELDS TOXIC CHLORINE GAS AND OTHER CHLORINATED
            COMPOUNDS.
         4] PRODUCT DECOMPOSES AT 212F YIELDING SODIUM CHLORATE.
         5] MOVE CONTAINERS FROM AREA IF WITHOUT RISK, COOL EXPOSED CONTAINERS.
         6] DIKE AREA FOR CONTROL AND CONTAINMENT TO PREVENT ENTRY INTO SEWERS, DRAIN, AND WATER WAYS.

LARGE SPILL/NO FIRE/RESCUE: WEAR NON-SEALED CHEMICAL PROTECTIVE CLOTHING, RUBBER BOOTS, GLOVES AND BREATHING
                           APPARATUS.

SPILL CONTROL AND CONTAINMENT:
    HOUSEHOLD SPILL: WEAR LATEX RUBBER GLOVES FOR SPILL CLEAN-UP. WIPE UP THE LIQUID WITH AN ABSORBENT CLOTH, RINSE
    OUT IN THE SINK. RINSE AREA WITH WATER, WIPE OR MOP UP LIQUID, RINSE OUT IN THE SINK.
    LARGE SPILL: COVER SPILL AREA WITH NON-ORGANIC ABSORBENT. SWEEP UP AND PUT IN A PLASTIC LINED DISPOSAL
    CONTAINER. RINSE AREA WITH WATER, PICK UP WITH ABSORBENT, PUT IN A DISPOSAL CONTAINER. SEAL AND REMOVE FROM THE
    WORK AREA.

HEALTH HAZARD INFORMATION:
    WHEN USING THIS PRODUCT WEAR: NO SPECIAL REQUIREMENT. Eye contact causes severe but temporary injury. May cause
    skin irritation. Ingestion can cause abdominal pain, nausea, vomiting, and damage to the mouth, throat, and
    stomach. Inhalation may cause irritation of the nose, throat, and lungs. **Under normal household usage the
    likelihood of any adverse health effects is low.**

    A physician should be contacted if anyone develops any signs or symptoms and suspects that they are caused by
    exposure to this product.

FIRST AID:
Eye Exposure:
Flush with water for at least 15 minutes while lifting the upper and lower eye lids. Contact lenses should not
be worn when working with this product. Get medical attention.

Skin Exposure:
RemRemove contaminated clothing. Wash skin with soap and water. If irritation appears, get medical attention.

Breathing:
If breathing problems develop, move the victim to fresh air. Treat symptoms as they appear.

Swallowing:
If the victim is conscious, give 2-3 glasses of water to drink. Immediately contact the local poison control
center for advice. Keep the victim warm and at rest. Get medical attention.

PRODUCT: CLOROX BLEACH, REGULAR

INCOMPATIBILITY: TOILET BOWL CLEANERS; RUST REMOVERS; VINEGAR; ACIDS; AMMONIA AND PRODUCTS CONTAINING AMMONIA.

INGREDIENTS: SODIUM HYPOCHLORITE   5.25% TLV: CAS#: 7681-52-9

IGNI TEMP: NA FP: NA LEL: NA UEL: NA VP: NA VD: NA SG: 1.085 PS: LIQUID APPEAR: CLEAR YELLOW
     ODOR: CHLORINE   PH FACTOR: 11.4

HAZARD RATINGS: H: F: R: 2 0 1   OXY   HAZARD CLASS   8

FIRE FIGHTING:
   HIGH EXPANSION FOAM, LOW EXPANSION FOAM, DRY CHEMICAL, CARBON DIOXIDE, WATER.
   PROTECTIVE CLOTHING, RUBBER GLOVES, AND BREATHING APPARATUS.
WARNING: 1] STRUCTURAL PROTECTIVE CLOTHING IS PERMEABLE, REMAIN CLEAR OF SMOKE, WATER FALL OUT AND WATER RUN OFF.
         2] KEEP OUT OF THE REACH OF CHILDREN.
         3] THERMAL DECOMPOSITION AT 212F RELEASING CHLORINE GAS.
         4] DO NOT MIX WITH INCOMPATIBLE PRODUCTS LISTED ABOVE. RELEASES CHLORINE GAS AND OTHER CHLORINE BY
            PRODUCTS.
         5] MOVE CONTAINERS FROM AREA IF WITHOUT RISK, COOL EXPOSED CONTAINERS.
         6] DIKE AREA FOR CONTROL AND CONTAINMENT TO PREVENT ENTRY INTO SEWERS, DRAIN, AND WATER WAYS.

LARGE SPILL/NO FIRE/RESCUE: EVACUATE NON-ESSENTIAL PERSONNEL 500 FEET FROM HAZARD AREA. WEAR NON-SEALED CHEMICAL
                            PROTECTIVE CLOTHING, RUBBER OR NEOPRENE BOOTS, GLOVES, BREATHING APPARATUS.

SPILL CONTROL AND CONTAINMENT:
   HOUSEHOLD SPILL: VENTILATE AREA. WEAR RUBBER GLOVES. WIPE UP DRY WITH AN ABSORBENT CLOTH. RINSE OUT IN THE
   SINK. SCRUB AREA WITH DETERGENT AND WATER. WIPE UP WITH AN ABSORBENT CLOTH. RINSE OUT IN THE SINK.
   LARGE SPILL: PICK UP WITH ABSORBENT. RINSE AREA WITH WATER. PICK UP WITH ABSORBENT. PUT ALL ABSORBENT IN A
   DISPOSAL CONTAINER. SEAL AND REMOVE FROM THE WORK AREA.

HEALTH HAZARD INFORMATION:
   WHEN USING THIS PRODUCT WEAR: WEAR RUBBER GLOVES. Eye contact will cause severe injury. Skin contact may cause
   irritation. Inhalation of vapors or mist may cause severe irritation of the nose, throat, and lungs. Ingestion
   will cause burning of the mouth, throat, and stomach.

   A physician should be contacted if anyone develops any signs or symptoms and suspects that they are caused by
   exposure to this product.

FIRST AID:
   Eye Exposure:
   Flush with water for 15 minutes while lifting the upper and lower eye lids. Contact lenses should not be worn
   when working with this product. Get medical attention.

   Skin Exposure:
   Remove contaminated clothing, flush with water. If irritation develops, get medical attention.

   Breathing:
   If irritation develops, move victim to fresh air.

   Swallowing:
   If conscious, give little sips of milk to ease burning sensation. Immediately contact the local poison control
   center for advice. Keep the victim warm and at rest. Get medical attention.

PRODUCT: CLOROX CLEAN-UP

INCOMPATIBILITY: TOILET BOWL CLEANERS; RUST REMOVERS; VINEGAR; ACIDS; AMMONIA AND PRODUCTS WITH AMMONIA IN THEM.

INGREDIENTS: SODIUM HYDROXIDE   .5-2.0% TLV: 2 mg/m3   CAS#: 1310-73-2
             SODIUM HYPOCHLORITE   2-5%   TLV: CAS#: 7681-52-9
             SODIUM CHLORATE   TLV: CAS#: 7775-09-9

IGNI TEMP: NA FP: NA LEL: NA UEL: NA VP: NA VD: NA SG: 1.034 PS: LIQUID APPEAR: YELLOW   ODOR: CHLORINE
    PH FACTOR: 12.4 TO 12.8

HAZARD RATINGS: H: F: R: 2 0 0   OXY

FIRE FIGHTING:
    HIGH EXPANSION FOAM, LOW EXPANSION FOAM, ALCOHOL FOAM; DRY CHEMICAL, CARBON DIOXIDE, WATER FOG.
    PROTECTIVE CLOTHING, RUBBER GLOVES, AND BREATHING APPARATUS.
WARNING: 1] STRUCTURAL PROTECTIVE CLOTHING IS PERMEABLE, REMAIN CLEAR OF SMOKE, WATER FALL OUT AND WATER RUN OFF.
         2] KEEP OUT OF THE REACH OF CHILDREN.
         3] DO NOT MIX WITH INCOMPATIBLE MATERIALS LISTED ABOVE. YIELDS TOXIC FUMES AND GASES OF CHLORINE AND
            OTHER CHLORINATED BY-PRODUCTS.
         4] MOVE CONTAINERS FROM AREA IF WITHOUT RISK, COOL EXPOSED CONTAINERS.
         5] DIKE AREA FOR CONTROL AND CONTAINMENT TO PREVENT ENTRY INTO SEWERS, DRAIN, AND WATER WAYS.

LARGE SPILL/NO FIRE/RESCUE: WEAR NON-SEALED RUBBER CHEMICAL PROTECTIVE CLOTHING, BOOTS, GLOVES AND BREATHING
                           APPARATUS.

SPILL CONTROL AND CONTAINMENT:
    HOUSEHOLD SPILL: WIPE UP LIQUID WITH ABSORBENT CLOTH, RINSE OUT IN SINK. WIPE UP RESIDUE WITH A DAMP CLOTH.
    RINSE OUT IN SINK. APPLY A LITTLE WATER ON SPILL AREA. WIPE UP WITH A DRY ABSORBENT CLOTH. RINSE IN SINK.
    LARGE SPILL: COVER SPILL AREA WITH ABSORBENT. PICK UP AND PUT INTO A DISPOSAL CONTAINER. RINSE AREA WITH WATER,
    PICK UP WITH ABSORBENT. PUT IN A DISPOSAL CONTAINER. SEAL AND REMOVE FROM WORK AREA.

HEALTH HAZARD INFORMATION:
    WHEN USING THIS PRODUCT WEAR: LATEX RUBBER GLOVES: Contact with the eyes will cause moderate irritation with
    tearing, burning sensation, itching, swelling, and redness. Prolonged or repeated skin contact may cause a
    rash-like irritation. If ingested may cause abdominal pain, nausea, vomiting, and diarrhea. Inhalation of the
    vapors may cause irritation of the nose, throat, and lungs.

    A physician should be contacted if anyone develops any signs or symptoms and suspects that they are caused by
    exposure to this product.

FIRST AID:
Eye Exposure:
Flush eyes with water for at least 15 minutes while lifting the upper and lower eye lids. Contact lenses should
not be worn when working with this product. If irritation persists, get medical attention.

Skin Exposure:
Rinse skin under flowing water, then wash with soap and water. If irritation appears, get medical attention.

Breathing:
Move the victim to fresh air. Treat symptoms as they appear. If needed, get medical attention.

Swallowing:
Immediately contact the local poison control center for advice. Keep the victim warm and at rest. Get medical
attention.

PRODUCT: CLOROX PLUS

INCOMPATIBILITY: TOILET BOWL CLEANERS; RUST REMOVERS; VINEGAR; AMMONIA AND PRODUCTS CONTAINING AMMONIA.

INGREDIENTS: SODIUM HYDROXIDE   0.5-2%   TLV: 2 mg/m3   CAS#: 1310-73-2
             SODIUM HYPOCHLORITE   2-5%   TLV: CAS#: 7681-52-9
             SODIUM CHLORATE   TLV: CAS#: 7775-09-9

IGNI TEMP: NA FP: NA LEL: NA UEL: NA VP: NA VD: NA SG: 1.034 PS: LIQUID APPEAR: CLEAR YELLOW
    ODOR: CHLORINE   PH FACTOR: 12.4 TO 12.8

HAZARD RATINGS: H: F: R: 2 0 1   OXY
FIRE FIGHTING:
  HIGH EXPANSION FOAM, LOW EXPANSION FOAM, ALCOHOL FOAM; DRY CHEMICAL, CARBON DIOXIDE, WATER FOG.
  PROTECTIVE CLOTHING, RUBBER GLOVES, AND BREATHING APPARATUS.
WARNING: 1] STRUCTURAL PROTECTIVE CLOTHING IS PERMEABLE, REMAIN CLEAR OF SMOKE, WATER FALL OUT AND WATER RUN OFF.
         2] KEEP OUT OF THE REACH OF CHILDREN.
         3] DO NOT MIX WITH INCOMPATIBLE PRODUCTS LISTED ABOVE, YIELDS TOXIC FUMES AND GASES OF CHLORINE AND OTHER
            CHLORINATED BY-PRODUCTS.
         4] MOVE CONTAINERS FROM AREA IF WITHOUT RISK, COOL EXPOSED CONTAINERS.
         5] DIKE AREA FOR CONTROL AND CONTAINMENT TO PREVENT ENTRY INTO SEWERS, DRAIN, AND WATER WAYS.

LARGE SPILL/NO FIRE/RESCUE: WEAR NON-SEALED RUBBER CHEMICAL PROTECTIVE CLOTHING, BOOTS, GLOVES AND BREATHING
                           APPARATUS.

SPILL CONTROL AND CONTAINMENT:
  HOUSEHOLD SPILL: WIPE UP LIQUID WITH ABSORBENT CLOTH, RINSE OUT IN SINK. WIPE UP RESIDUE WITH A DAMP CLOTH.
  APPLY A LITTLE WATER ON AREA, WIPE UP WITH A DRY ABSORBENT CLOTH, RINSE OUT IN SINK.
  LARGE SPILL: COVER SPILL WITH ABSORBENT, PICK UP AND PUT INTO A DISPOSAL CONTAINER. RINSE AREA WITH WATER, PICK
  UP WITH ABSORBENT. PUT INTO A DISPOSAL CONTAINER. SEAL AND REMOVE FROM WORK AREA.

HEALTH HAZARD INFORMATION:
  WHEN USING THIS PRODUCT WEAR: LATEX RUBBER GLOVES. Contact with eyes may cause moderate eye irritation with
  stinging, burning sensation, tearing, itching, swelling, and redness. May cause rash-like skin irritation.
  Ingestion may cause abdominal pain with nausea, vomiting, and diarrhea. Inhalation of vapors may cause
  irritation to the nose, throat and lungs.

  A physician should be contacted if anyone develops any signs or symptoms and suspects that they are caused by
  exposure to this product.

FIRST AID:
  Eye Exposure:
  Flush eyes with water for at least 15 minutes while lifting the upper and lower eye lids. Contact lenses should
  not be worn when working with this product. If irritation persists, get medical attention.

  Skin Exposure:
  Wash with soap and water. If irritation appears, get medical attention.

  Breathing:
  Move victim to fresh air. Treat symptoms as they appear. If needed, get medical attention.

  Swallowing:
  If victim is conscious, give 2-3 glasses of water to drink. Immediately contact the local poison control center
  for advice. Keep the victim warm and at rest. Get medical attention.

PRODUCT: CLOROX   PRE-WASH (AEROSOL)

INCOMPATIBILITY: PEROXIDES; CHLORINE BLEACH; CLOROX 2.

INGREDIENTS: MINERAL SPIRITS   50-75% TLV: 50 ppm   CAS#: 64741-65-7
             MINERAL SPIRITS   TLV: 100 ppm   CAS#: 64742-48-9
             LIQUID PETROLEUM GAS   5-25% TLV: 1000 ppm   CAS#: 68476-85-7
             ETHYLENE GLYCOL   2-5%   TLV: 50 ppm   CAS#: 107-21-1

IGNI TEMP: NA FP: -2F LEL: NA UEL: NA VP: NA VD: NA SG: NA PS: LIQUID APPEAR: CLEAR MIST ODOR: FRAGRANCE
  PH FACTOR: NA

HAZARD RATINGS: H: F: R: 2 4 0

FIRE FIGHTING:
  HIGH EXPANSION FOAM, LOW EXPANSION FOAM, ALCOHOL FOAM; DRY CHEMICAL, CARBON DIOXIDE, WATER FOG.
  PROTECTIVE CLOTHING, RUBBER GLOVES, AND BREATHING APPARATUS.
WARNING: 1] STRUCTURAL PROTECTIVE CLOTHING IS PERMEABLE, REMAIN CLEAR OF SMOKE, WATER FALL OUT AND WATER RUN OFF.
         2] KEEP OUT OF THE REACH OF CHILDREN.
         3] DO NOT MIX WITH INCOMPATIBLE PRODUCTS LISTED ABOVE.
         4] CONTAINERS MAY EXPLODE IN A FIRE OR IF HEATED ABOVE 120F.
         5] REMOVE ALL SOURCES OF IGNITION. DO NOT SPRAY AROUND OPEN FLAME.
         6] MOVE CONTAINERS FROM AREA IF WITHOUT RISK, COOL EXPOSED CONTAINERS.
         7] DIKE AREA FOR CONTROL AND CONTAINMENT TO PREVENT ENTRY INTO SEWERS, DRAIN, AND WATER WAYS.

LARGE SPILL/NO FIRE/RESCUE: WEAR NON-SEALED TYVEK SUIT, RUBBER OR NEOPRENE BOOTS, GLOVES AND BREATHING APPARATUS.

SPILL CONTROL AND CONTAINMENT:
  HOUSEHOLD SPILL: WIPE UP LIQUID WITH ABSORBENT CLOTH. RINSE IN SINK. WIPE UP RESIDUE WITH DAMP CLOTH, RINSE OUT
  IN SINK.
  LARGE SPILL: COVER SPILL WITH ABSORBENT, PICK UP AND PUT IN A DISPOSAL CONTAINER. RINSE AREA WITH WATER, PICK UP
  WITH ABSORBENT, PUT IN DISPOSAL CONTAINER. SEAL AND REMOVE FROM WORK AREA.

HEALTH HAZARD INFORMATION:
  WHEN USING THIS PRODUCT WEAR: NO SPECIAL REQUIREMENTS. Product will cause eye and skin irritation. Prolonged
  inhalation will cause dizziness, headache, and unconsciousness. Overexposure to ethylene glycol may cause
  central nervous system damage, heart failure, kidney damage, and birth defects.

  A physician should be contacted if anyone develops any signs or symptoms and suspects that they are caused by
  exposure to this product.

FIRST AID:
  Eye Exposure:
  Flush eyes with water for 15 minutes while lifting the upper and lower eye lids. Contact lenses should not be
  worn when working with this product. If irritation persists, get medical attention.

  Skin Exposure:
  Wash skin with soap and water. If irritation appears, get medical attention.

  Breathing:
  Move victim to fresh air. Treat symptoms as they appear. If needed, get medical attention.

  Swallowing:
  Do not induce vomiting. Immediately contact the local poison control center for advice. Keep the victim warm and
  at rest. Get medical attention.

PRODUCT: CLOROX THICKENED PRE-WASH

INCOMPATIBILITY: NONE KNOWN

INGREDIENTS: SURFACTANTS   5-25% TLV: CAS#: 68412-54-4

IGNI TEMP: NA FP: NA LEL: NA UEL: NA VP: NA VD: NA SG: 1.03 PS: LIQUID APPEAR: CLEAR THICK
    ODOR: FRAGRANCE   PH FACTOR: 6.5-7.0

HAZARD RATINGS: H: F: R: 1 0 0

FIRE FIGHTING:
    HIGH EXPANSION FOAM, LOW EXPANSION FOAM, ALCOHOL FOAM; DRY CHEMICAL, CARBON DIOXIDE, WATER FOG.
    PROTECTIVE CLOTHING, RUBBER GLOVES, AND BREATHING APPARATUS.
WARNING: 1] STRUCTURAL PROTECTIVE CLOTHING IS PERMEABLE, REMAIN CLEAR OF SMOKE, WATER FALL OUT AND WATER RUN OFF.
         2] KEEP OUT OF THE REACH OF CHILDREN.
         3] MOVE CONTAINERS FROM AREA IF WITHOUT RISK, COOL EXPOSED CONTAINERS.
         4] DIKE AREA FOR CONTROL AND CONTAINMENT TO PREVENT ENTRY INTO SEWERS, DRAIN, AND WATER WAYS.

LARGE SPILL/NO FIRE/RESCUE: WEAR RUBBER OR NEOPRENE BOOTS, GLOVES.

SPILL CONTROL AND CONTAINMENT:
    HOUSEHOLD SPILL: WIPE UP LIQUID WITH ABSORBENT CLOTH, RINSE IN SINK. WIPE UP RESIDUE WITH A DAMP CLOTH, RINSE IN
    SINK.
    LARGE SPILL: COVER SPILL WITH ABSORBENT, PICK UP AND PUT INTO A DISPOSAL CONTAINER. RINSE AREA WITH WATER, PICK
    UP WITH ABSORBENT, PUT INTO DISPOSAL CONTAINER. SEAL AND REMOVE FROM WORK AREA.

HEALTH HAZARD INFORMATION:
    WHEN USING THIS PRODUCT WEAR: Contact will cause mild eye irritation.

    A physician should be contacted if anyone develops any signs or symptoms and suspects that they are caused by
    exposure to this product.

FIRST AID:
    Eye Exposure:
    Flush eyes with water for 15 minutes while lifting the upper and lower eye lids. Contact lenses should not be
    worn when working with this product. If irritation persists, get medical attention.

    Skin Exposure:
    Not expected to be a problem. If irritation appears, get medical attention.

    Breathing:
    Not expected to be a problem. If breathing becomes labored, move victim to fresh air.

    Swallowing:
    Not expected to be a problem but contact the local poison control center for advice.

PRODUCT: COAST DEODORANT TOILET BAR

INCOMPATIBILITY: NONE KNOWN

INGREDIENTS: NON-HAZARDOUS BY OSHA 1910.1200 STANDARDS
             SODIUM TALLOWATE   COCOATE OR PALM KERNELATE
             SODIUM CHLORIDE   COCONUT OR PALM KERNEL ACID
             TITANIUM DIOXIDE   TETRASODIUM EDTA
             2,6-DI-TERT-BUTYL-PARA-CRESOL

IGNI TEMP: NA FP: NA LEL: NA UEL: NA VP: NA VD: NA SG: 1.03 PS: SOLID APPEAR: STRIPED BAR
     ODOR: SCENTED   PH FACTOR: NA

HAZARD RATINGS: H: F: R:

FIRE FIGHTING:
    HIGH EXPANSION FOAM, LOW EXPANSION FOAM, ALCOHOL FOAM; DRY CHEMICAL, CARBON DIOXIDE, WATER FOG.
    PROTECTIVE CLOTHING, RUBBER GLOVES, AND BREATHING APPARATUS.
WARNING: 1] STRUCTURAL PROTECTIVE CLOTHING IS PERMEABLE, REMAIN CLEAR OF SMOKE, WATER FALL OUT AND WATER RUN OFF.
         2] KEEP OUT OF THE REACH OF CHILDREN.
         3] RESIDUE MAY CREATE A SLIPPING HAZARD.
         4] MOVE CONTAINERS FROM AREA IF WITHOUT RISK, COOL EXPOSED CONTAINERS.
         5] DIKE AREA FOR CONTROL AND CONTAINMENT TO PREVENT ENTRY INTO SEWERS, DRAIN, AND WATER WAYS.

LARGE SPILL/NO FIRE/RESCUE: WEAR RUBBER OR NEOPRENE BOOTS, GLOVES.

SPILL CONTROL AND CONTAINMENT:
    HOUSEHOLD SPILL: PICK UP BAR AND WIPE UP RESIDUE WITH ABSORBENT CLOTH.
    LARGE SPILL: LIQUID: COVER WITH ABSORBENT, SCOOP UP, PUT INTO A DISPOSAL CONTAINER. SCRUB AREA WITH WATER, PICK
    UP WITH ABSORBENT. PUT INTO DISPOSAL CONTAINER. SEAL AND REMOVE FROM WORK AREA.

HEALTH HAZARD INFORMATION:
    WHEN USING THIS PRODUCT WEAR: NO SPECIAL REQUIREMENTS. Eye contact may cause irritation with tearing, burning
    sensation, stinging, itching, swelling, and redness. Ingestion may cause abdominal pain, nausea, vomiting, and
    diarrhea.

    A physician should be contacted if anyone develops any signs or symptoms and suspects that they are caused by
    exposure to this product.

FIRST AID:
    Eye Exposure:
    Flush eyes with water for 15 minutes while lifting the upper and lower eye lids. Contact lenses should not be
    worn when working with this product. If irritation persists, get medical attention.

    Skin Exposure:
    Not expected to be a problem.

    Breathing:
    Not expected to be a problem.

    Swallowing:
    Contact the local poison control center for advice. Keep the victim warm and at rest. Get medical attention.

PRODUCT: COMBAT ANT CONTROL SYSTEM

INCOMPATIBILITY: NONE KNOWN

INGREDIENTS: HYDRAMETHLNON   1%   TLV: 1.4 mg/m3   CAS#: 67485-29-4

IGNI TEMP:NA FP: 200F LEL:NA UEL: NA VP: NA VD: NA SG: 1.44 PS: WAXY MASS APPEAR: YELLOWISH BROWN
    ODOR: ODORLESS   PH FACTOR: NA

HAZARD RATINGS: H: F: R:

FIRE FIGHTING:
  HIGH EXPANSION FOAM, LOW EXPANSION FOAM, ALCOHOL FOAM; DRY CHEMICAL, CARBON DIOXIDE, WATER FOG.
  PROTECTIVE CLOTHING, RUBBER GLOVES, AND BREATHING APPARATUS.
WARNING: 1] STRUCTURAL PROTECTIVE CLOTHING IS PERMEABLE, REMAIN CLEAR OF SMOKE, WATER FALL OUT AND WATER RUN OFF.
         2] KEEP OUT OF THE REACH OF CHILDREN.
         3] THERMAL DECOMPOSITION YIELDS TOXIC CARBON DIOXIDE, CARBON MONOXIDE, METHANE, OXIDES OF NITROGEN,
            OXIDES OF HYDROGEN, OXIDES OF FLUORIDE.
         4] MOVE CONTAINERS FROM AREA IF WITHOUT RISK, COOL EXPOSED CONTAINERS.
         5] DIKE AREA FOR CONTROL AND CONTAINMENT TO PREVENT ENTRY INTO SEWERS, DRAIN, AND WATER WAYS.

LARGE SPILL/NO FIRE/RESCUE: WEAR RUBBER OR NEOPRENE BOOTS, GLOVES.

SPILL CONTROL AND CONTAINMENT:
  HOUSEHOLD SPILL: USE CAUTION NOT TO TOUCH PRODUCT. PICK UP WITH PAPER TOWEL OR RAG AND PUT EVERYTHING IN A
  DISPOSAL CONTAINER.
  LARGE SPILL: SCOOP UP CONTAINERS AND MATERIAL, PUT INTO A DISPOSAL CONTAINER. SCRUB AREA WITH DETERGENT AND
  WATER, PICK UP WITH ABSORBENT, PUT INTO DISPOSAL CONTAINER. SEAL AND REMOVE FROM WORK AREA.

HEALTH HAZARD INFORMATION:
  WHEN USING THIS PRODUCT WEAR: Use care not to touch waxy material. Product is practically non-toxic by
  ingestion. May cause eye irritation.

  A physician should be contacted if anyone develops any signs or symptoms and suspects that they are caused by
  exposure to this product.

FIRST AID:
  Eye Exposure:
  Flush eyes with water for 15 minutes while lifting the upper and lower eye lids. Contact lenses should not be
  worn when working with this product. If irritation persists, get medical attention.

  Skin Exposure:
  Wash with soap and water after handling. Not expected to be a problem.

  Breathing:
  Not expected to be a problem.

  Swallowing:
  Immediately contact the local poison control center for advice. Keep the victim warm and at rest. Get medical
  attention if advised.

PRODUCT: COMBAT   ANT & ROACH INSTANT KILLER

INCOMPATIBILITY: NONE KNOWN

INGREDIENTS: d-trans-ALLETHEIN   .06% TLV: CAS#: 28434-00-6
             PIPERONYL BUTOXIDE   .12% TLV: CAS#:   51-03-6
             n-OCTYL BICYCLOHEPTANE DICARBOXIMIDE   .2% TLV: CAS#: 113-48-4
             FENVALERATE   .2% TLV: CAS#: 51630-58-1
             PETROLEUM DISTILLATE   70-95% TLV: 500 ppm   CAS#: 64742-47-8
             CARBON DIOXIDE   3% TLV: 5000 ppm   CAS#: 124-38-9
                (or)
             1,1,1-TRICHLOROETHANE   10% TLV: 350 ppm   CAS#:   71-55-4
             PROPELLANT-A46   TLV: 800 ppm
                (ISOBUTANE   85%   CAS#:   75-28-5
                 PROPANE   15%   CAS#:   74-98-6)

IGNI TEMP:NA FP: 140F LEL:NA UEL:NA VP: 70-90 PSIG @ 70F VD: NA SG: .80 PS: LIQUID CONCENTRATE
   APPEAR: YELLOW   ODOR: KEROSENE PH FACTOR: NA

HAZARD RATINGS: H: F: R:
FIRE FIGHTING:
   HIGH EXPANSION FOAM, LOW EXPANSION FOAM, ALCOHOL FOAM; DRY CHEMICAL, CARBON DIOXIDE, WATER FOG.
   PROTECTIVE CLOTHING, RUBBER GLOVES, AND BREATHING APPARATUS.
WARNING: 1] STRUCTURAL PROTECTIVE CLOTHING IS PERMEABLE, REMAIN CLEAR OF SMOKE, WATER FALL OUT AND WATER RUN OFF.
         2] KEEP OUT OF THE REACH OF CHILDREN.
         3] CONTAINERS MAY EXPLODE IN A FIRE OR IF HEATED TO ABOVE 130F.
         4] THERMAL DECOMPOSITION YIELDS TOXIC CARBON DIOXIDE, CARBON MONOXIDE.
         5] DO NOT SPRAY TOWARD FACE, DO NOT GET ON SKIN, DO NOT INHALE VAPORS.
         6] MOVE CONTAINERS FROM AREA IF WITHOUT RISK, COOL EXPOSED CONTAINERS.
         7] DIKE AREA FOR CONTROL AND CONTAINMENT TO PREVENT ENTRY INTO SEWERS, DRAIN, AND WATER WAYS.

LARGE SPILL/NO FIRE/RESCUE: WEAR NON-SEALED CHEMICAL PROTECTIVE CLOTHING, BOOTS GLOVES AND BREATHING APPARATUS.

SPILL CONTROL AND CONTAINMENT:
   HOUSEHOLD SPILL: VENTILATE AREA. WEAR NEOPRENE GLOVES, WIPE UP LIQUID WITH ABSORBENT CLOTH. PUT INTO DISPOSAL
   CONTAINER. SCRUB AREA WITH SOAP AND WATER. PICK UP WITH ABSORBENT CLOTH. PUT ALL MATERIAL INCLUDING GLOVES IN A
   DISPOSAL CONTAINER.
      LARGE SPILL: COVER WITH ABSORBENT. PICK UP AND PUT INTO A DISPOSAL CONTAINER. SCRUB AREA WITH DETERGENT AND
   WATER, PICK UP WITH ABSORBENT. PUT IN A DISPOSAL CONTAINER. SEAL AND REMOVE FROM WORK AREA.

HEALTH HAZARD INFORMATION:
   Product may be harmful by inhalation, skin absorption, ingestion. May cause eye irritation.

   A physician should be contacted if anyone develops any signs or symptoms and suspects that they are caused by
   exposure to this product.

FIRST AID:
   Eye Exposure:
   Flush eyes with water for 15 minutes while lifting the upper and lower eye lids. Contact lenses should not be
   worn when working with this product. If irritation persists, get medical attention.

   Skin Exposure:
   Wash with soap and water. If irritation appears, get medical attention.

   Breathing:
   Move victim to fresh air. Treat symptoms as they appear

   Swallowing:
   Do not induce vomiting. Immediately contact the local poison control center for advice. Keep the victim warm and
   at rest. Get medical attention.

PRODUCT: COMET, LEMON FRESH - CHLORINATED CLEANSER (NON-PHOSPHATES)

INCOMPATIBILITY: AMMONIA; ACIDS.

INGREDIENTS: CALCIUM CARBONATE   TLV: CAS#: 471-34-1
             SODIUM CARBONATE   TLV: CAS#: 497-19-8

IGNI TEMP: NA FP: NA LEL: NA UEL: NA VP: NA VD: NA SG: NA PS: POWDER APPEAR: GREEN ODOR: LEMON
   PH FACTOR: NA

HAZARD RATINGS: H: F: R:

FIRE FIGHTING:
   HIGH EXPANSION FOAM, LOW EXPANSION FOAM, ALCOHOL FOAM; DRY CHEMICAL, CARBON DIOXIDE, WATER FOG.
   PROTECTIVE CLOTHING, RUBBER GLOVES, AND BREATHING APPARATUS.
WARNING: 1] STRUCTURAL PROTECTIVE CLOTHING IS PERMEABLE, REMAIN CLEAR OF SMOKE, WATER FALL OUT AND WATER RUN OFF.
         2] KEEP OUT OF THE REACH OF CHILDREN.
         3] HAZARDOUS DECOMPOSITION YIELDS TOXIC CHLORINE GAS.
         4] MOVE CONTAINERS FROM AREA IF WITHOUT RISK, COOL EXPOSED CONTAINERS.
         5] DIKE AREA FOR CONTROL AND CONTAINMENT TO PREVENT ENTRY INTO SEWERS, DRAIN, AND WATER WAYS.

LARGE SPILL/NO FIRE/RESCUE: WEAR NON-SEALED CHEMICAL PROTECTIVE CLOTHING RUBBER OR NEOPRENE BOOTS, GLOVES AND
                           BREATHING APPARATUS.

SPILL CONTROL AND CONTAINMENT:
   HOUSEHOLD SPILL: AVOID CREATING OR BREATHING THE DUST. WIPE UP WITH A DAMP CLOTH. RINSE OUT IN THE SINK. WIPE UP
   RESIDUE WITH DAMP CLOTH. RINSE OUT IN THE SINK.
   LARGE SPILL: AVOID CREATING A DUST. GENTLY SWEEP UP MATERIAL AND PUT IN A DISPOSAL CONTAINER. RINSE AREA WITH
   WATER, PICK UP WITH ABSORBENT. PUT IN A DISPOSAL CONTAINER. SEAL AND REMOVE FROM WORK AREA.

HEALTH HAZARD INFORMATION:
   WHEN USING THIS PRODUCT WEAR: NO SPECIAL REQUIREMENTS. Product is a mild eye irritant, with stinging, burning
   sensation, tearing, itching, swelling and redness. May cause skin irritating rash. Inhalation will cause burning
   sensation of the nasal passage. Ingestion will cause gastrointestinal irritation with nausea, vomiting, and
   diarrhea.

   A physician should be contacted if anyone develops any signs or symptoms and suspects that they are caused by
   exposure to this product.

FIRST AID:
   Eye Exposure:
   Flush eyes with water for 15 minutes while lifting the upper and lower eye lids. Contact lenses should not be
   worn when working with this product. Get medical attention.

   Skin Exposure:
   Wash with soap and water. If irritation appears, get medical attention.

   Breathing:
   Move the victim to fresh air. Treat symptoms as they appear. If needed get medical attention.

   Swallowing:
   If the victim is conscious, give 2-3 glasses of milk or water to drink. Immediately contact the local poison
   control center for advice. Keep the victim warm and at rest. Get medical attention.

PRODUCT: COMET, LEMON FRESH CHLORINATED CLEANSER (PHOSPHATE)

INCOMPATIBILITY: AMMONIA; ACIDS.

INGREDIENTS: TETRA SODIUM PYROPHOSPHATES   TLV: 5 mg/m3   CAS#: 7722-88-5
             CALCIUM CARBONATE   TLV: CAS#: 471-34-1
             SODIUM CARBONATE   TLV: CAS#: 497-19-8

IGNI TEMP: NA FP: NA LEL: NA UEL: NA VP: NA VD: NA SG: NA PS: POWDER APPEAR: GREEN ODOR: LEMON
   PH FACTOR: NA

HAZARD RATINGS: H: F: R:

FIRE FIGHTING:
  HIGH EXPANSION FOAM, LOW EXPANSION FOAM, ALCOHOL FOAM; DRY CHEMICAL, CARBON DIOXIDE, WATER FOG.
  PROTECTIVE CLOTHING, RUBBER GLOVES, AND BREATHING APPARATUS.
WARNING: 1] STRUCTURAL PROTECTIVE CLOTHING IS PERMEABLE, REMAIN CLEAR OF SMOKE, WATER FALL OUT AND WATER RUN OFF.
         2] KEEP OUT OF THE REACH OF CHILDREN.
         3] HAZARDOUS DECOMPOSITION YIELDS TOXIC CHLORINE GAS.
         4] MOVE CONTAINERS FROM AREA IF WITHOUT RISK, COOL EXPOSED CONTAINERS.
         5] DIKE AREA FOR CONTROL AND CONTAINMENT TO PREVENT ENTRY INTO SEWERS, DRAIN, AND WATER WAYS.

LARGE SPILL/NO FIRE/RESCUE: WEAR NON-SEALED CHEMICAL PROTECTIVE CLOTHING RUBBER OR NEOPRENE BOOTS, GLOVES AND
                           BREATHING APPARATUS.

SPILL CONTROL AND CONTAINMENT:
  HOUSEHOLD SPILL: AVOID CREATING OR BREATHING THE DUST. WIPE UP WITH A DAMP CLOTH. RINSE OUT IN SINK. WIPE UP
  RESIDUE WITH A DAMP CLOTH. RINSE OUT IN SINK.
  LARGE SPILL: AVOID CREATING A DUST. GENTLY SWEEP UP MATERIAL AND PUT IN A DISPOSAL CONTAINER. RINSE AREA WITH
  WATER, PICK UP WITH ABSORBENT, PUT IN A DISPOSAL CONTAINER. SEAL AND REMOVE FROM WORK AREA.

HEALTH HAZARD INFORMATION:
  WHEN USING THIS PRODUCT WEAR: NO SPECIAL REQUIREMENTS. Product is a mild eye irritant with stinging, burning
  sensation, tearing, swelling, and redness. Mild skin irritant, may cause a rash. Inhalation will cause a burning
  sensation in the nasal passage. Ingestion will cause gastrointestinal irritation with nausea, vomiting, and
  diarrhea.

  A physician should be contacted if anyone develops any signs or symptoms and suspects that they are caused by
  exposure to this product.

FIRST AID:
Eye Exposure:
Flush eyes with water for 15 minutes while lifting the upper and lower eye lids. Contact lenses should not be
worn when working with this product. Get medical attention.

Skin Exposure:
Wash with soap and water. If irritation appears, get medical attention.

Breathing:
Move the victim to fresh air. Treat symptoms as they appear. If needed, get medical attention.

Swallowing:
If the victim is conscious, give 2-3 glasses of milk or water to drink. Immediately contact the local poison
control center for advice. Keep the victim warm and at rest. Get medical attention.

PRODUCT: COMET, LIQUID - CHLORINATED CLEANSER

INCOMPATIBILITY: AMMONIA; ACIDS.

INGREDIENTS: SODIUM HYPOCHLORITE   TLV: CAS#: 7681-52-9

IGNI TEMP: NA FP: 200 LEL: NA UEL: NA VP: NA VD: NA SG: 1.08 PS: LIQUID APPEAR: GREEN OPAQUE ODOR: CEDAR
    PINE PH FACTOR: NA

HAZARD RATINGS: H: F: R:

NEUTRALIZING AGENTS:

FIRE FIGHTING:
    HIGH EXPANSION FOAM, LOW EXPANSION FOAM, ALCOHOL FOAM; DRY CHEMICAL, CARBON DIOXIDE, WATER FOG.
    PROTECTIVE CLOTHING, RUBBER GLOVES, AND BREATHING APPARATUS.
WARNING: 1] STRUCTURAL PROTECTIVE CLOTHING IS PERMEABLE, REMAIN CLEAR OF SMOKE, WATER FALL OUT AND WATER RUN OFF.
         2] KEEP OUT OF THE REACH OF CHILDREN.
         3] HAZARDOUS DECOMPOSITION YIELDS TOXIC CHLORINE GAS.
         4] MOVE CONTAINERS FROM AREA IF WITHOUT RISK, COOL EXPOSED CONTAINERS.
         5] DIKE AREA FOR CONTROL AND CONTAINMENT TO PREVENT ENTRY INTO SEWERS, DRAIN, AND WATER WAYS.

LARGE SPILL/NO FIRE/RESCUE: WEAR NON-SEALED CHEMICAL PROTECTIVE CLOTHING RUBBER OR NEOPRENE BOOTS, GLOVES AND
                           BREATHING APPARATUS.

SPILL CONTROL AND CONTAINMENT:
    HOUSEHOLD SPILL: WIPE UP LIQUID WITH AN ABSORBENT CLOTH, RINSE OUT IN THE SINK. WIPE UP RESIDUE WITH A DAMP
    CLOTH, RINSE OUT IN THE SINK.
    LARGE SPILL: COVER SPILL AREA WITH ABSORBENT. PICK UP AND PUT IN A DISPOSAL CONTAINER. RINSE AREA WITH WATER,
    PICK UP WITH ABSORBENT. PUT IN A DISPOSAL CONTAINER. SEAL AND REMOVE FROM THE WORK AREA.

HEALTH HAZARD INFORMATION:
    WHEN USING THIS PRODUCT WEAR: NO SPECIAL REQUIREMENTS. Product is a mild eye irritant with stinging, burning
    sensation, tearing, itching, swelling, and redness. Skin contact may cause irritating rash. Inhalation will
    cause nasal burning sensation. Ingestion will cause gastrointestinal irritation with nausea, vomiting, and
    diarrhea.

    A physician should be contacted if anyone develops any signs or symptoms and suspects that they are caused by
    exposure to this product.

FIRST AID:
Eye Exposure:
Flush eyes with water for 15 minutes while lifting the upper and lower eye lids. Contact lenses should not be
worn when working with this product. Get medical attention.

Skin Exposure:
Wash with soap and water. If irritation appears, get medical attnetion.

Breathing:
Move the victim to fresh air. Treat any symptoms as they appear. If needed, get medical attention.

Swallowing:
If the victim is conscious, give 2-3 glasses of milk or water to drink. Immediately contact the local poison
control center for advice. Keep the victim warm and at rest. Get medical attention.

PRODUCT: COMET,   SAFE FOR SURFACES - CHLORINATED CLEANSER (NON-PHOSPHATE)

INCOMPATIBILITY: AMMONIA; ACIDS.

INGREDIENTS: SODIUM DICHLORO-s-TRIAZINE
            TRIONE DIHYDRATE   TLV: CAS#:
            SODIUM SULFATE   TLV: CAS#: 7757-82-6
            SODIUM CARBONATE   TLV: CAS#: 497-19-8
            CALCIUM CARBONATE   TLV: CAS#: 471-34-1
            SODIUM ALKYL BENZENE SULFONATE TLV: CAS#:

IGNI TEMP: NA FP: NA LEL: NA UEL: NA VP: NA VD: NA SG: NA PS: POWDER APPEAR: GREEN ODOR: CEDAR PINE
    PH FACTOR: NA

HAZARD RATINGS: H: F: R:

FIRE FIGHTING:
    HIGH EXPANSION FOAM, LOW EXPANSION FOAM, ALCOHOL FOAM; DRY CHEMICAL, CARBON DIOXIDE, WATER FOG.
    PROTECTIVE CLOTHING, RUBBER GLOVES, AND BREATHING APPARATUS.
WARNING: 1] STRUCTURAL PROTECTIVE CLOTHING IS PERMEABLE, REMAIN CLEAR OF SMOKE, WATER FALL OUT AND RUN OFF.
         2] KEEP OUT OF THE REACH OF CHILDREN.
         3] HAZARDOUS DECOMPOSITION YIELDS TOXIC CHLORINE GAS.
         4] MOVE CONTAINERS FROM AREA IF WITHOUT RISK, COOL EXPOSED CONTAINERS.
         5] DIKE AREA FOR CONTROL AND CONTAINMENT TO PREVENT ENTRY INTO SEWERS, DRAIN, AND WATER WAYS.

LARGE SPILL/NO FIRE/RESCUE: WEAR NON-SEALED CHEMICAL PROTECTIVE CLOTHING RUBBER OR NEOPRENE BOOTS, GLOVES AND
                           BREATHING APPARATUS.

SPILL CONTROL AND CONTAINMENT:
    HOUSEHOLD SPILL: AVOID CREATING OR BREATHING DUST. PICK UP WITH A DAMP CLOTH AND RINSE OUT IN SINK. WIPE UP
    RESIDUE WITH A DAMP CLOTH, RINSE OUT IN SINK.
    LARGE SPILL: AVOID CREATING A DUST. GENTLY SWEEP UP MATERIAL AND PUT INTO A DISPOSAL CONTAINER. RINSE AREA WITH
    WATER, PICK UP WITH ABSORBENT, PUT IN A DISPOSAL CONTAINER. SEAL AND REMOVE FROM THE WORK AREA.

HEALTH HAZARD INFORMATION:
    WHEN USING THIS PRODUCT WEAR: NO SPECIAL REQUIREMENTS. Product is a mild eye irritant will cause stinging,
    tearing, burning sensation, itching, swelling, and redness. Contact with the skin may cause an irritating rash.
    Inhalation causes irritation to the upper respiratory tract. Ingestion will cause irritation of the mouth,
    throat, and stomach with nausea, vomiting, and diarrhea.

    A physician should be contacted if anyone develops any signs or symptoms and suspects that they are caused by
    exposure to this product.

FIRST AID:
Eye Exposure:
Flush with water for 15 minutes while lifting the upper and lower eye lids. Contact lenses should not be worn
when working with this product. Get medical attention.

Skin Exposure:
Wash skin with soap and water. If irritation appears, get medical attention.

Breathing:
Move the victim to fresh air. Treat symptoms as they appear. If needed, get medical attention.

Swallowing:
If the victim is conscious, give 2-3 glasses of milk or water to drink. Immediately contact the local poison
control center for advice. Keep the victim warm and at rest. Get medical attention.

PRODUCT: COMET, SAFE FOR SURFACES - CHLORINATED CLEANSER (PHOSPHATE)

INCOMPATIBILITY: AMMONIA; ACIDS.

INGREDIENTS: TETRA SODIUM PYROPHOSPHATE   TLV: 5 mg/m3   CAS#: 7722-88-5
             SODIUM CARBONATE   TLV: CAS#: 497-19-8
             CALCIUM CARBONATE   TLV: CAS#: 471-34-1
             SODIUM ALKYLBENZENE SULFONATE   TLV: CAS#:

IGNI TEMP: NA FP: NA LEL: NA UEL: NA VP: NA VD: NA SG: NA PS: POWDER APPEAR: GREEN ODOR: CEDAR PINE
    PH FACTOR: NA

HAZARD RATINGS: H: F: R:

FIRE FIGHTING:
    HIGH EXPANSION FOAM, LOW EXPANSION FOAM, ALCOHOL FOAM; DRY CHEMICAL, CARBON DIOXIDE, WATER FOG.
    PROTECTIVE CLOTHING, RUBBER GLOVES, AND BREATHING APPARATUS.
WARNING: 1] STRUCTURAL PROTECTIVE CLOTHING IS PERMEABLE, REMAIN CLEAR OF SMOKE, WATER FALL OUT AND WATER RUN OFF.
         2] KEEP OUT OF THE REACH OF CHILDREN.
         3] HAZARDOUS DECOMPOSITION YIELDS TOXIC CHLORINE GAS.
         4] MOVE CONTAINERS FROM AREA IF WITHOUT RISK, COOL EXPOSED CONTAINERS.
         5] DIKE AREA FOR CONTROL AND CONTAINMENT TO PREVENT ENTRY INTO SEWERS, DRAIN, AND WATER WAYS.

LARGE SPILL/NO FIRE/RESCUE: WEAR NON-SEALED CHEMICAL PROTECTIVE CLOTHING RUBBER OR NEOPRENE BOOTS, GLOVES AND
                           BREATHING APPARATUS.

SPILL CONTROL AND CONTAINMENT:
    HOUSEHOLD SPILL: AVOID CREATING OR BREATHING THE DUST. WIPE UP WITH A DAMP ABSORBENT CLOTH, RINSE IN SINK. WIPE
    UP RESIDUE AND RINSE IN SINK.
    LARGE SPILL: AVOID CREATING A DUST. GENTLY SWEEP UP MATERIAL AND PUT INTO A DISPOSAL CONTAINER. RINSE AREA WITH
    WATER, PICK UP WITH ABSORBENT, PUT IN A DISPOSAL CONTAINER. SEAL AND REMOVE FROM WORK AREA.

HEALTH HAZARD INFORMATION:
    WHEN USING THIS PRODUCT WEAR: NO SPECIAL REQUIREMENTS. Product is a mild skin and eye irritant. Dust irritates
    the mucous membranes. Causes stinging, burning sensation, tearing, itching, swelling, and redness of the eyes.
    Skin contact may cause irritating rash. Ingestion will cause irritation of the gastrointestinal tract with
    nausea, vomiting, and diarrhea.

    A physician should be contacted if anyone develops any signs or symptoms and suspects that they are caused by
    exposure to this product.

FIRST AID:
    Eye Exposure:
    Flush eyes with water for 15 minutes while lifting the upper and lower eye lids. Contact lenses should not be
    worn when working with this product. Get medical attention.

    Skin Exposure:
    Wash skin with soap and water. If irritation appears, get medical attention.

    Breathing:
    Move the victim to fresh air. Treat symptoms as they appear. If needed get medical attention.

    Swallowing:
    If the victim is conscious, give 2-3 glasses of milk or water to drink. Immediately contact the local poison
    control center for advice. Keep the victim warm and at rest. Get medical attention.

PRODUCT: COMET TOILET BOWL CLEANSER, LIQUID

INCOMPATIBILITY: DRAIN CLEANERS; RUST REMOVERS; ACIDS; AMMONIA; PRODUCTS WITH AMMONIA

INGREDIENTS: SODIUM HYPOCHLORITE   TLV: CAS#: 7681-52-9
             PERFUME   TLV: CAS#:
             COLORANTS   TLV: CAS#:

IGNI TEMP: NA FP: 200F LEL: NA UEL: NA VP: NA VD: NA SG: 1.08 PS: LIQUID APPEAR: GREEN ODOR: CEDAR PINE
PH FACTOR:

HAZARD RATINGS: H: F: R:

FIRE FIGHTING:
   HIGH EXPANSION FOAM, LOW EXPANSION FOAM, ALCOHOL FOAM; DRY CHEMICAL, CARBON DIOXIDE, WATER FOG.
   PROTECTIVE CLOTHING, RUBBER GLOVES, AND BREATHING APPARATUS.
WARNING: 1] STRUCTURAL PROTECTIVE CLOTHING IS PERMEABLE, REMAIN CLEAR OF SMOKE, WATER FALL OUT AND WATER RUN OFF.
         2] KEEP OUT OF THE REACH OF CHILDREN.
         3] DO NOT MIX WITH INCOMPATIBLE PRODUCTS LISTED ABOVE, YIELDS CHLORINE GAS.
         4] THERMAL DECOMPOSITION MAY YIELD TOXIC CHLORINE GAS.
         5] AVOID EXCESSIVE HEAT ABOVE 110F.
         6] MOVE CONTAINERS FROM AREA IF WITHOUT RISK, COOL EXPOSED CONTAINERS.
         7] DIKE AREA FOR CONTROL AND CONTAINMENT TO PREVENT ENTRY INTO SEWERS, DRAIN, AND WATER WAYS.

LARGE SPILL/NO FIRE/RESCUE: WEAR RUBBER OR NEOPRENE BOOTS, GLOVES AND BREATHING APPARATUS.

SPILL CONTROL AND CONTAINMENT:
   HOUSEHOLD SPILL: WIPE UP WITH AN ABSORBENT CLOTH, RINSE OUT IN THE SINK. WIPE UP RESIDUE WITH A DAMP CLOTH,
   RINSE OUT IN THE SINK.
   LARGE SPILL: PICK UP WITH ABSORBENT, PUT IN A DISPOSAL CONTAINER. RINSE AREA WITH WATER, PICK UP WITH ABSORBENT.
   PUT IN A DISPOSAL CONTAINER. SEAL AND REMOVE FROM THE WORK AREA.

HEALTH HAZARD INFORMATION:
   WHEN USING THIS PRODUCT WEAR: LATEX OR NEOPRENE GLOVES. Liquid will cause skin irritation. Eye contact will
   cause stinging, tearing, itching, swelling, and redness. Ingestion will cause abdominal pain, nausea, vomiting,
   and diarrhea.

   A physician should be contacted if anyone develops any signs or symptoms and suspects that they are caused by
   exposure to this product.

FIRST AID:
   Eye Exposure:
   Flush with water for 15 minutes while lifting the upper and lower eye lids. Contact lenses should not be worn
   when working with this product. Get medical attention.

   Skin Exposure:
   Wash with plenty of soap and water. If irritation develops, get medical attention.

   Breathing:
   Not expected to be a problem on small spill. Large spill may need organic vapor mask or breathing apparatus.

   Swallowing:
   If the victim is conscious, give 1-2 glasses of water to drink. Immediately contact the local poison control
   center for advice. Keep the victim warm and at rest. Get medical attention.

PRODUCT: COMPLETE FOR FURNITURE

INCOMPATIBILITY: NONE KNOWN

INGREDIENTS: SILICONE, SURFACTANTS 2-4% TLV: CAS#:
    ISOPARAFFINIC HYDROCARBON   mfg recommended
      SOLVENT   1-3% TLV: 400 ppm   CAS#: 64742-48-9
    FRAGRANCE   0.5% TLV: CAS#:
    WATER   92-96% TLV: CAS#: 7732-18-5

IGNI TEMP: NA FP: NA LEL: NA UEL: NA VP: NA VD: NA SG: 1.0 PS: LIQUID APPEAR: WHITE OPAQUE ODOR: LEMON
  PH FACTOR: 6.6

HAZARD RATINGS: H: F: R:

FIRE FIGHTING:
  HIGH EXPANSION FOAM, LOW EXPANSION FOAM, ALCOHOL FOAM; DRY CHEMICAL, CARBON DIOXIDE, WATER FOG.
  PROTECTIVE CLOTHING, RUBBER GLOVES, AND BREATHING APPARATUS.
WARNING: 1] STRUCTURAL PROTECTIVE CLOTHING IS PERMEABLE, REMAIN CLEAR OF SMOKE, WATER FALL OUT AND WATER RUN OFF.
     2] KEEP OUT OF THE REACH OF CHILDREN.
     3] SPILL CAN CREATE A SLIPPING HAZARD.
     4] MOVE CONTAINERS FROM AREA IF WITHOUT RISK, COOL EXPOSED CONTAINERS.
     5] DIKE AREA FOR CONTROL AND CONTAINMENT TO PREVENT ENTRY INTO SEWERS, DRAIN, AND WATER WAYS.

LARGE SPILL/NO FIRE/RESCUE: WEAR RUBBER OR NEOPRENE BOOTS, GLOVES.

SPILL CONTROL AND CONTAINMENT:
  HOUSEHOLD SPILL: WIPE UP LIQUID WITH ABSORBENT CLOTH, PUT IN A DISPOSAL CONTAINER. WASH SPILL AREA WITH SOAP AND
  WATER, PICK UP WITH ABSORBENT CLOTH.
  LARGE SPILL: COVER SPILL WITH ABSORBENT. PICK UP AND PUT INTO A DISPOSAL CONTAINER. SCRUB SPILL AREA WITH
  DETERGENT AND WATER, PICK UP WITH ABSORBENT, PUT INTO A DISPOSAL CONTAINER. SEAL AND REMOVE FROM WORK AREA.

HEALTH HAZARD INFORMATION:
  WHEN USING THIS PRODUCT WEAR: NO SPECIAL REQUIREMENTS. Contact may cause eye irritation, stinging, itching,
  tearing, swelling, and redness. Skin contact may cause skin irritation if allergic to the product. Ingestion
  will cause gastrointestinal irritation with nausea, vomiting, and diarrhea.

  A physician should be contacted if anyone develops any signs or symptoms and suspects that they are caused by
  exposure to this product.

FIRST AID:
  Eye Exposure:
  Flush the eyes with water for 15 minutes while lifting the upper and lower eye lids. Contact lenses should not
  be worn when working with this product. If irritation persists, get medical attention.

  Skin Exposure:
  Remove contaminated clothing, wash with soap and water. If rash appears, get medical attention.

  Breathing:
  Not expected to be a problem.

  Swallowing:
  Immediately contact the local poison control center for advice. Keep the victim warm and at rest. Get medical
  attention.

PRODUCT: CUREL

INCOMPATIBILITY: NONE KNOWN

```
INGREDIENTS: GLYCERIN             TLV:        CAS#:    56-85-5
             PETROLATUM           TLV:        CAS#:  8008-03-8
             ISOPROPYL PALMITATE  TLV:        CAS#:
             1-HEXADECANOL        TLV:        CAS#: 36653-82-4
             METHYLPARABEN        TLV:        CAS#:    99-76-3
             PROPYLPARABEN        TLV:        CAS#:    94-13-3
```

IGNI TEMP: NA FP: NA LEL: NA UEL: NA VP: NA VD: NA SG: 1.0 PS: LIQUID APPEAR: LIGHT CREAMY ODOR: FRAGRANT
    PH FACTOR: 5.3

HAZARD RATINGS: H: F: R: 0   0   0

FIRE FIGHTING:
    HIGH EXPANSION FOAM, LOW EXPANSION FOAM, ALCOHOL FOAM; DRY CHEMICAL, CARBON DIOXIDE, WATER FOG.
    PROTECTIVE CLOTHING, RUBBER GLOVES, AND BREATHING APPARATUS.
WARNING: 1] STRUCTURAL PROTECTIVE CLOTHING IS PERMEABLE, REMAIN CLEAR OF SMOKE, WATER FALL OUT AND WATER RUN OFF.
         2] KEEP OUT OF THE REACH OF CHILDREN.
         3] THERMAL DECOMPOSITION YIELDS TOXIC CARBON DIOXIDE, CARBON MONOXIDE.
         4] SPILL WILL CREATE A SLIPPING HAZARD.
         5] MOVE CONTAINERS FROM AREA IF WITHOUT RISK, COOL EXPOSED CONTAINERS.
         6] DIKE AREA FOR CONTROL AND CONTAINMENT TO PREVENT ENTRY INTO SEWERS, DRAIN, AND WATER WAYS.

LARGE SPILL/NO FIRE/RESCUE: WEAR RUBBER OR NEOPRENE BOOTS, GLOVES.

SPILL CONTROL AND CONTAINMENT:
    HOUSEHOLD SPILL: WIPE UP LIQUID WITH AN ABSORBENT CLOTH. PUT INTO A DISPOSAL CONTAINER. WIPE UP RESIDUE WITH A
    DAMP CLOTH, RINSE OUT IN THE SINK.
    LARGE SPILL: COVER SPILL WITH ABSORBENT. PICK UP AND PUT INTO A DISPOSAL CONTAINER. WASH SPILL AREA WITH
    DETERGENT AND WATER, PICK UP WITH ABSORBENT, PUT IN A DISPOSAL CONTAINER. SEAL AND REMOVE FROM WORK AREA.

HEALTH HAZARD INFORMATION:
    WHEN USING THIS PRODUCT WEAR: NO SPECIAL REQUIREMENTS. Contact with the eyes may cause irritation. May cause
    skin irritation if allergic to product. Ingestion will cause gastrointestinal irritation, possibly with nausea,
    vomiting, and diarrhea.

    A physician should be contacted if anyone develops any signs or symptoms and suspects that they are caused by
    exposure to this product.

FIRST AID:
    Eye Exposure:
    Flush eyes with water for 15 minutes while lifting the upper and lower eye lids. Contact lenses should not be
    worn when working with this product. If irritation persists, get medical attention.

    Skin Exposure:
    Wash skin with soap and water. If irritation appears, get medical attention.

    Breathing:
    Not expected to be a problem.

    Swallowing:
    Immediately contact the local poison control center for advice. Keep the victim warm and at rest. Get medical
    attention.

PRODUCT: CUREL,   FRAGRANCE FREE

INCOMPATIBILITY: NONE KNOWN

INGREDIENTS: GLYCERIN   TLV: CAS#:   56-85-5
             PETROLATUM   TLV: CAS#: 8008-03-8
             ISOPROPYL PALMITATE   TLV: CAS#:
             1-HEXADECANOL   TLV: CAS#: 36653-82-4
             METHYLPARABEN   TLV: CAS#:   99-76-3
             PROPYLPARABEN   TLV: CAS#:   94-13-3

IGNI TEMP: NA FP: NA LEL: NA UEL: NA VP: NA VD: NA SG: 1.0 PS: LIQUID APPEAR: LIGHT CREAMY ODOR: FRAGRANT
   PH FACTOR: 5.3

HAZARD RATINGS: H: F: R: 0   0   0

FIRE FIGHTING:
   HIGH EXPANSION FOAM, LOW EXPANSION FOAM, ALCOHOL FOAM; DRY CHEMICAL, CARBON DIOXIDE, WATER FOG.
   PROTECTIVE CLOTHING, RUBBER GLOVES, AND BREATHING APPARATUS.
WARNING: 1] STRUCTURAL PROTECTIVE CLOTHING IS PERMEABLE, REMAIN CLEAR OF SMOKE, WATER FALL OUT AND WATER RUN OFF.
         2] KEEP OUT OF THE REACH OF CHILDREN.
         3] THERMAL DECOMPOSITION YIELDS TOXIC CARBON DIOXIDE, CARBON MONOXIDE.
         4] SPILL WILL CREATE A SLIPPING HAZARD.
         5] MOVE CONTAINERS FROM AREA IF WITHOUT RISK, COOL EXPOSED CONTAINERS.
         6] DIKE AREA FOR CONTROL AND CONTAINMENT TO PREVENT ENTRY INTO SEWERS,   DRAIN, AND WATER WAYS.

LARGE SPILL/NO FIRE/RESCUE: WEAR RUBBER OR NEOPRENE BOOTOOTS, GLOVES.

SPILL CONTROL AND CONTAINMENT:
   HOUSEHOLD SPILL: WIPE UP LIQUID WITH AN ABSORBENT CLOTH. PUT INTO A DISPOSAL CONTAINER. WIPE UP RESIDUE WITH A
   DAMP CLOTH, RINSE OUT IN THE SINK.
   LARGE SPILL: COVER SPILL WITH ABSORBENT. PICK UP AND PUT INTO A DISPOSAL CONTAINER. WASH SPILL AREA WITH
   DETERGENT AND WATER, PICK UP WITH ABSORBENT, PUT IN A DISPOSAL CONTAINER. SEAL AND REMOVE FROM WORK AREA.

HEALTH HAZARD INFORMATION:
   WHEN USING THIS PRODUCT WEAR: NO SPECIAL REQUIREMENTS. Contact with the eyes may cause irritation. May cause
   skin irritation if allergic to product. Ingestion will cause gastrointestinal irritation, possibly with nausea,
   vomiting, and diarrhea.

   A physician should be contacted if anyone develops any signs or symptoms and suspects that they are caused by
   exposure to this product.

FIRST AID:
   Eye Exposure:
   Flush eyes with water for 15 minutes while lifting the upper and lower eye lids. Contact lenses should not be
   worn when working with this product. If irritation persists, get medical attention.

   Skin Exposure:
   Wash skin with soap and water. If irritation appears, get medical attention.

   Breathing:
   Not expected to be a problem.

   Swallowing:
   Immediately contact the local poison control center for advice. Keep the victim warm and at rest. Get medical
   attention.

PRODUCT: DASH, LEMON FRESH

INCOMPATIBILITY: NONE KNOWN

INGREDIENTS: ANIONIC SURFACTANTS   TLV: CAS#:
           SODIUM CARBONATE   TLV: CAS#: 497-19-8
           SODIUM SULFATE   TLV: CAS#: 7757-82-6
           SODIUM SILICATES   TLV: CAS#: 1344-09-8

IGNI TEMP: NA FP: NA LEL: NA UEL: NA VP: NA VD: NA SG: NA PS: POWDER APPEAR: WHITE GRANULAR
    ODOR: LEMON   PH FACTOR: NA

HAZARD RATINGS: H: F: R:

FIRE FIGHTING:
    HIGH EXPANSION FOAM, LOW EXPANSION FOAM, ALCOHOL FOAM; DRY CHEMICAL, CARBON DIOXIDE, WATER FOG.
    PROTECTIVE CLOTHING, RUBBER GLOVES, AND BREATHING APPARATUS.
WARNING: 1] STRUCTURAL PROTECTIVE CLOTHING IS PERMEABLE, REMAIN CLEAR OF SMOKE, WATER FALL OUT AND WATER RUN OFF.
       2] KEEP OUT OF THE REACH OF CHILDREN.
       3] THERMAL DECOMPOSITION YIELDS TOXIC CARBON DIOXIDE, CARBON MONOXIDE.
       4] MOVE CONTAINERS FROM AREA IF WITHOUT RISK, COOL EXPOSED CONTAINERS.
       5] DIKE AREA FOR CONTROL AND CONTAINMENT TO PREVENT ENTRY INTO SEWERS, DRAIN, AND WD WATER WAYS.

LARGE SPILL/NO FIRE/RESCUE: WEAR RUBBER OR NEOPRENE BOOTS, GLOVES.

SPILL CONTROL AND CONTAINMENT:
    HOUSEHOLD SPILL: SWEEP UP POWDER AND PUT IN A DISPOSAL CONTAINER.
    LARGE SPILL: AVOID CREATING A DUST. GENTLY SWEEP UP MATERIAL AND PUT IN A DISPOSAL CONTAINER. RINSE AREA WITH
    WATER, PICK UP WITH ABSORBENT. PUT IN A DISPOSAL CONTAINER. SEAL AND REMOVE FROM THE WORK AREA.

HEALTH HAZARD INFORMATION:
    WHEN USING THIS PRODUCT WEAR: NO SPECIAL REQUIREMENTS. Inhalation of dust will cause coughing, sore throat,
    wheezing, and shortness of breath. Contact with the eyes will cause stinging, tearing, itching, swelling, and
    redness. Ingestion will cause nausea, vomiting, and diarrhea.

    A physician should be contacted if anyone develops any signs or symptoms and suspects that they are caused by
    exposure to this product.

FIRST AID:
    Eye Exposure:
    Flush with water for 15 minutes while lifting the upper and lower eye lids. Contact lenses should not be worn
    when working with this product. If irritation persists, get medical attention.

    Skin Exposure:
    Rinse off with water.

    Breathing:
    If problem develop move victim to fresh air.

    Swallowing:
    If victim is conscious, give 2-3 glasses of water to drink. Immediately contact the local poison control center
    for advice. Keep the victim warm and at rest. Get medical attention.

PRODUCT: DASH, LIQUID (LAUNDRY DETERGENT)

INCOMPATIBILITY: ACIDS AND ACID FUMES

INGREDIENTS: ETHYL ALCOHOL   TLV: 1900 mg/m3   CAS#:   64-17-5
           PROPYLENE GLYCOL   TLV: CAS#: 4254-15-3
           ANIONIC SURFACTANTS   TLV: CAS#:
           NONIONIC SURFACTANTS TLV: CAS#:
           LAURATE; CITRATE; PERFUME

IGNI TEMP: NA FP: NA LEL: NA UEL: NA VP: NA VD: NA SG: 1.06 PS: LIQUID APPEAR: DARK AMBER   ODOR: LEMON
  PH FACTOR:

HAZARD RATINGS: H: F: R:

FIRE FIGHTING:
  HIGH EXPANSION FOAM, LOW EXPANSION FOAM, ALCOHOL FOAM; DRY CHEMICAL, CARBON DIOXIDE, WATER FOG.
  PROTECTIVE CLOTHING, RUBBER GLOVES, AND BREATHING APPARATUS.
WARNING: 1] STRUCTURAL PROTECTIVE CLOTHING IS PERMEABLE, REMAIN CLEAR OF SMOKE, WATER FALL OUT AND WATER RUN OFF.
        2] KEEP OUT OF THE REACH OF CHILDREN.
        3] THERMAL DECOMPOSITION MAY YIELD CARBON DIOXIDE, CARBON MONOXIDE.
        4] MOVE CONTAINERS FROM AREA IF WITHOUT RISK, COOL EXPOSED CONTAINERS.
        5] DIKE AREA FOR CONTROL AND CONTAINMENT TO PREVENT ENTRY INTO SEWERS, DRAIN, AND WATER WAYS.

LARGE SPILL/NO FIRE/RESCUE: WEAR RUBBER OR NEOPRENE BOOTS, GLOVES.

SPILL CONTROL AND CONTAINMENT:
  HOUSEHOLD SPILL: WIPE UP WITH AN ABSORBENT CLOTH, RINSE OUT IN THE SINK. WIPE UP RESIDUE WITH A DAMP CLOTH,
  RINSE OUT IN THE SINK.
  LARGE SPILL: PICK UP WITH ABSORBENT, PUT IN A DISPOSAL CONTAINER. RINSE AREA WITH WATER, PICK UP WITH ABSORBENT.
  PUT IN A DISPOSAL CONTAINER. SEAL AND REMOVE FROM THE WORK AREA.

HEALTH HAZARD INFORMATION:
  WHEN USING THIS PRODUCT WEAR: NO SPECIAL REQUIREMENT. Eye contact may cause stinging, tearing, itching,
  swelling, and redness. Prolonged contact with concentrated liquid may cause drying of the skin and irritation.
  Ingestion may result in abdominal pain, nausea, vomiting, and diarrhea.

  A physician should be contacted if anyone develops any signs or symptoms and suspects that they are caused by
  exposure to this product.

FIRST AID:
Eye Exposure:
Flush with water for 15 minutes while lifting the upper and lower eye lids. Contact lenses should not be worn
when working with this product. Get medical attention.

Skin Exposure:
Wash with water. If irritation develops, get medical attention.

Breathing:
Not expected to be a problem.

Swallowing:
If the victim is conscious, give 1-2 glasses of water to drink. Immediately contact the local poison control
center for advice. Keep the victim warm and at rest. Get medical attention.

PRODUCT: DAWN (LIQUID DISH WASHING DETERGENT)

INCOMPATIBILITY: CHLORINE BLEACH

INGREDIENTS: ANIONIC SURFACTANT   TLV: CAS#:
             NONIONIC SURFACTANT   TLV: CAS#:
             ETHYL ALCOHOL   TLV: 1900 mg/m3 CAS#:   64-17-5

IGNI TEMP: NA FP: 116F LEL: NA UEL: NA VP: NA VD: NA SG: 1.03 PS: LIQUID APPEAR: CLEAR BLUE ODOR: PERFUME
   PH FACTOR: NA

HAZARD RATINGS: H: F: R:

FIRE FIGHTING:
  HIGH EXPANSION FOAM, LOW EXPANSION FOAM, ALCOHOL FOAM; DRY CHEMICAL, CARBON DIOXIDE, WATER FOG.
  PROTECTIVE CLOTHING, RUBBER GLOVES, AND BREATHING APPARATUS.
WARNING: 1] STRUCTURAL PROTECTIVE CLOTHING IS PERMEABLE, REMAIN CLEAR OF SMOKE, WATER FALL OUT AND WATER RUN OFF.
         2] KEEP OUT OF THE REACH OF CHILDREN.
         3] ANIONIC AND NONIONIC SURFACTANTS ARE MODERATELY TOXIC MATERIALS.
         4] SPILL WILL CREATE A SLIPPING HAZARD.
         5] MOVE CONTAINERS FROM AREA IF WITHOUT RISK, COOL EXPOSED CONTAINERS.
         6] DIKE AREA FOR CONTROL AND CONTAINMENT TO PREVENT ENTRY INTO SEWERS, DRAIN, AND WATER WAYS.

LARGE SPILL/NO FIRE/RESCUE: WEAR RUBBER OR NEOPRENE BOOTS, GLOVES.

SPILL CONTROL AND CONTAINMENT:
  HOUSEHOLD SPILL: WIPE UP LIQUID WITH ABSORBENT CLOTH. RINSE OUT IN SINK. WIPE UP RESIDUE WITH DAMP CLOTHS UNTIL
  NO SUDS APPEAR. RINSE IN SINK.
  LARGE SPILL: COVER SPILL WITH ABSORBENT, PICK UP AND PUT INTO A DISPOSAL CONTAINER. RINSE AREA WITH WATER, PICK
  UP WITH ABSORBENT, PUT INTO DISPOSAL CONTAINER. SEAL AND REMOVE FROM WORK AREA.

HEALTH HAZARD INFORMATION:
  WHEN USING THIS PRODUCT WEAR: NO SPECIAL REQUIREMENTS. Eye contact will cause stinging, burning sensation,
  tearing, itching, swelling, and redness. Prolonged skin contact with concentrate may cause skin to dry and
  scale. Ingestion will cause abdominal pain, nausea, vomiting, and diarrhea.

  A physician should be contacted if anyone develops any signs or symptoms and suspects that they are caused by
  exposure to this product.

FIRST AID:
Eye Exposure:
Flush with water for 15 minutes while lifting the upper and lower eye lids. Contact lenses should not be worn
when working with this product. If irritation persists, get medical attention.

Skin Exposure:
Wash off with water and apply a skin lotion.

Breathing:
Not expected to be a problem.

Swallowing:
If victim is conscious, give 2-3 glasses of milk or water to drink. Immediately contact the local poison control
center for advice. Keep the victim warm and at rest. Get medical attention as soon as possible.

PRODUCT: DAWN, MOUNTAIN SPRING (LIQUID DISH WASHING DETERGENT)

INCOMPATIBILITY: CHLORINE BLEACH.

INGREDIENTS: ANIONIC SURFACTANTS   TLV: CAS#:
             NONIONIC SURFACTANTS   TLV: CAS#:
             ETHYL ALCOHOL   TLV: 1900 mg/m3   CAS#:   64-17-5
             WATER   TLV: CAS#: 7732-18-5
             COLORANT, PERFUME.

IGNI TEMP: NA FP: 116F LEL: NA UEL: NA VP: NA VD: NA SG: 1.03 PS: LIQUID APPEAR: CLEAR GREEN
    ODOR: PERFUMED   PH FACTOR: NA

HAZARD RATINGS: H: F: R:

FIRE FIGHTING:
    HIGH EXPANSION FOAM, LOW EXPANSION FOAM, ALCOHOL FOAM; DRY CHEMICAL, CARBON DIOXIDE, WATER FOG.
    PROTECTIVE CLOTHING, RUBBER GLOVES, AND BREATHING APPARATUS.
WARNING: 1] STRUCTURAL PROTECTIVE CLOTHING IS PERMEABLE, REMAIN CLEAR OF SMOKE, WATER FALL OUT AND WATER RUN OFF.
         2] KEEP OUT OF THE REACH OF CHILDREN.
         3] THERMAL DECOMPOSITION MAY YIELD TOXIC FUMES AND VAPORS.
         4] MOVE CONTAINERS FROM AREA IF WITHOUT RISK, COOL EXPOSED CONTAINERS.
         5] DIKE AREA FOR CONTROL AND CONTAINMENT TO PREVENT ENTRY INTO SEWERS, DRAIN, AND WATER WAYS.

LARGE SPILL/NO FIRE/RESCUE: WEAR RUBBER OR NEOPRENE BOOTS, GLOVES.

SPILL CONTROL AND CONTAINMENT:
    HOUSEHOLD SPILL: WIPE UP WITH AN ABSORBENT CLOTH. RINSE OUT IN THE SINK. WIPE UP RESIDUE WITH A DAMP CLOTH.
    RINSE OUT IN THE SINK.
    LARGE SPILL: PICK UP WITH ABSORBENT. PUT IN A DISPOSAL CONTAINER. RINSE AREA WITH WATER. PICK UP WITH ABSORBENT.
    PUT IN A DISPOSAL CONTAINER. SEAL AND REMOVE FROM THE WORK AREA.

HEALTH HAZARD INFORMATION:
    WHEN USING THIS PRODUCT WEAR: NO SPECIAL REQUIREMENT. Eye contact will cause stinging, tearing, itching,
    swelling, and redness. Prolonged contact may cause drying of the skin and irritation. Ingestion will result in
    abdominal pain, nausea, vomiting, and diarrhea.

    A physician should be contacted if anyone develops any signs or symptoms and suspects that they are caused by
    exposure to this product.

FIRST AID:
Eye Exposure:
Flush with water for 15 minutes while lifting the upper and lower eye lids. Contact lenses should not be worn
when working with this product. Get medical attention.

Skin Exposure:
Wash with water. If irritation develops, get medical attention.

Breathing:
Not expected to be a problem.

Swallowing:
If the victim is conscious, give 2-3 glasses of water to drink. Immediately contact the local poison control
center for advice. Keep the victim warm and at rest. Get medical attention.

PRODUCT: DOWNY LIQUID FABRIC SOFTENER (REGULAR CONCENTRATION)

INCOMPATIBILITY: NONE KNOWN

INGREDIENTS: NONIONIC SURFACTANTS   TLV: CAS#:
             CATIONIC SURFACTANTS   7-9% TLV: CAS#:

IGNI TEMP: NA FP: 200F LEL: NA UEL: NA VP: NA VD: NA SG: 1 PS: LIQUID APPEAR: BLUE OR YELLOW
     ODOR: FLORAL   PH FACTOR: NA

HAZARD RATINGS: H: F: R:

FIRE FIGHTING:
   HIGH EXPANSION FOAM, LOW EXPANSION FOAM, ALCOHOL FOAM; DRY CHEMICAL, CARBON DIOXIDE, WATER FOG.
   PROTECTIVE CLOTHING, RUBBER GLOVES, AND BREATHING APPARATUS.
WARNING: 1] STRUCTURAL PROTECTIVE CLOTHING IS PERMEABLE, REMAIN CLEAR OF SMOKE, WATER FALL OUT AND WATER RUN OFF.
         2] KEEP OUT OF THE REACH OF CHILDREN.
         3] THESE SURFACTANTS ARE MODERATELY TOXIC BY INGESTION.
         4] PRODUCT MUST BE PREHEATED FOR IGNITION.
         5] THERMAL DECOMPOSITION YIELDS TOXIC CARBON DIOXIDE, CARBON MONOXIDE.
         6] MOVE CONTAINERS FROM AREA IF WITHOUT RISK, COOL EXPOSED CONTAINERS.
         7] DIKE AREA FOR CONTROL AND CONTAINMENT TO PREVENT ENTRY INTO SEWERS, DRAIN, AND WATER WAYS.

LARGE SPILL/NO FIRE/RESCUE: WEAR RUBBER OR NEOPRENE BOOTS, GLOVES.

SPILL CONTROL AND CONTAINMENT:
   HOUSEHOLD SPILL: PICK UP WITH AN ABSORBENT CLOTH AND PUT IN A DISPOSAL CONTAINER. WIPE UP RESIDUE WITH A DAMP
   CLOTH AND RINSE OUT IN THE SINK.
   LARGE SPILL: COVER SPILL WITH ABSORBENT. PICK UP AND PUT IN A DISPOSAL CONTAINER. RINSE AREA WITH WATER, PICK UP
   WITH ABSORBENT, PUT IN DISPOSAL CONTAINER. SEAL AND REMOVE FROM WORK AREA.

HEALTH HAZARD INFORMATION:
   WHEN USING THIS PRODUCT WEAR: NO SPECIAL REQUIREMENTS. Eye contact causes irritation. Ingestion of product will
   cause abdominal pain, nausea, vomiting, and diarrhea. Skin contact may cause mild irritation.

   A physician should be contacted if anyone develops any signs or symptoms and suspects that they are caused by
   exposure to this product.

FIRST AID:
   Eye Exposure:
   Flush eyes with water for 15 minutes while lifting the upper and lower eye lids. Contact lenses should not be
   worn when working with this product. If irritation persists, get medical attention.

   Skin Exposure:
   Wash with soap and water. If irritation persists, get medical attention.

   Breathing:
   Not expected to be a problem.

   Swallowing:
   If victim is conscious, give 2-3 glasses of milk or water to drink. Immediately contact the local poison control
   center for advice. Keep the victim warm and at rest. Get medical attention.

PRODUCT: DOWNY LIQUID FABRIC SOFTENER (TRIPLE CONCENTRATION)

INCOMPATIBILITY: NONE KNOWN

INGREDIENTS: NONIONIC SURFACTANTS   TLV: CAS#:
            CATIONIC SURFACTANTS   TLV: CAS#:
            ALUMINOSILICATE   TLV: CAS#:

IGNI TEMP: NA FP: 200F LEL: NA UEL: NA VP: NA VD: NA SG: 1 PS: LIQUID APPEAR: BLUE
    ODOR: ORIENTAL FLORAL PH FACTOR: NA

HAZARD RATINGS: H: F: R:

FIRE FIGHTING:
   HIGH EXPANSION FOAM, LOW EXPANSION FOAM, ALCOHOL FOAM; DRY CHEMICAL, CARBON DIOXIDE, WATER FOG.
   PROTECTIVE CLOTHING, RUBBER GLOVES, AND BREATHING APPARATUS.
WARNING: 1] STRUCTURAL PROTECTIVE CLOTHING IS PERMEABLE, REMAIN CLEAR OF SMOKE, WATER FALL OUT AND WATER RUN OFF.
        2] KEEP OUT OF THE REACH OF CHILDREN.
        3] THESE SURFACTANTS ARE MODERATELY TOXIC BY INGESTION.
        4] PRODUCT MUST BE PREHEATED FOR IGNITION.
        5] THERMAL DECOMPOSITION YIELDS TOXIC CARBON DIOXIDE, CARBON MONOXIDE.
        6] MOVE CONTAINERS FROM AREA IF WITHOUT RISK, COOL EXPOSED CONTAINERS.
        7] DIKE AREA FOR CONTROL AND CONTAINMENT TO PREVENT ENTRY INTO SEWERS, DRAIN, AND WATER WAYS.

LARGE SPILL/NO FIRE/RESCUE: WEAR RUBBER OR NEOPRENE BOOTS, GLOVES. DUST MASK OR BREATHING APPARATUS.

SPILL CONTROL AND CONTAINMENT:
   HOUSEHOLD SPILL: WIPE UP LIQUID WITH A ABSORBENT CLOTH,PICK UP AND PUT IN A DISPOSAL CONTAINER. WIPE UP RESIDUE
   WITH A DAMP CLOTH AND RINSE OUT IN THE SINK.
   LARGE SPILL: COVER SPILL WITH ABSORBENT, PICK UP AND PUT IN A DISPOSAL CONTAINER. RINSE AREA WITH WATER, PICK UP
   WITH ABSORBENT, PUT IN DISPOSAL CONTAINER. SEAL AND REMOVE FROM WORK AREA.

HEALTH HAZARD INFORMATION:
   WHEN USING THIS PRODUCT WEAR: NO SPECIAL REQUIREMENTS. Eye contact causes irritation. Ingestion of product will
   cause abdominal pain, nausea, vomiting, and diarrhea. Skin contact may cause mild irritation.

   A physician should be contacted if anyone develops any signs or symptoms and suspects that they are caused by
   exposure to this product.

FIRST AID:
   Eye Exposure:
   Flush eyes with water for 15 minutes while lifting the upper and lower eye lids. Contact lenses should not be
   worn when working with this product. If irritation persists, get medical attention.

   Skin Exposure:
   Wash with soap and water. If irritation persists, get medical attention.

   Breathing:
   Not expected to be a problem.

   Swallowing:
   If victim is conscious, give 2-3 glasses of milk or water to drink. Immediately contact the local poison control
   center for advice. Keep the victim warm and at rest. Get medical attention.

PRODUCT: DOWNY   PERFUMED HOUSEHOLD FABRIC SOFTENER

INCOMPATIBILITY: NONE KNOWN

INGREDIENTS: NONIONIC SURFACTANTS   TLV: CAS#:
             CATIONIC SURFACTANTS   7-9% TLV: CAS#:
             ALUMINOSILICATE   TLV: CAS#:

IGNI TEMP: NA FP: NA LEL: NA UEL: NA VP: NA VD: NA SG: NA PS: SOLID APPEAR: BLUE SHEETS
     ODOR: FRUITY SWEET   PH FACTOR: 3-4

HAZARD RATINGS: H: F: R:

FIRE FIGHTING:
  HIGH EXPANSION FOAM, LOW EXPANSION FOAM, ALCOHOL FOAM; DRY CHEMICAL, CARBON DIOXIDE, WATER FOG.
  PROTECTIVE CLOTHING, RUBBER GLOVES, AND BREATHING APPARATUS.
WARNING: 1] STRUCTURAL PROTECTIVE CLOTHING IS PERMEABLE, REMAIN CLEAR OF SMOKE, WATER FALL OUT AND WATER RUN OFF.
         2] KEEP OUT OF THE REACH OF CHILDREN.
         3] THESE SURFACTANTS ARE MODERATELY TOXIC BY INGESTION.
         4] THERMAL DECOMPOSITION YIELDS TOXIC CARBON DIOXIDE, CARBON MONOXIDE.
         5] MOVE CONTAINERS FROM AREA IF WITHOUT RISK, COOL EXPOSED CONTAINERS.
         6] DIKE AREA FOR CONTROL AND CONTAINMENT TO PREVENT ENTRY INTO SEWERS, DRAIN, AND WATER WAYS.

LARGE SPILL/NO FIRE/RESCUE: WEAR RUBBER OR NEOPRENE BOOTS, GLOVES, DUST MASK MAY BE NEEDED ON LARGE SPILL.

SPILL CONTROL AND CONTAINMENT:
  HOUSEHOLD SPILL: PICK UP AND PUT IN A DISPOSAL CONTAINER. WIPE UP RESIDUE WITH A DAMP CLOTH.
  LARGE SPILL: AVOID CREATING A DUST. GENTLY SWEEP UP AND PUT IN A DISPOSAL CONTAINER. RINSE AREA WITH WATER, PICK
  UP WITH ABSORBENT, PUT IN DISPOSAL CONTAINER. SEAL AND REMOVE FROM WORK AREA.

HEALTH HAZARD INFORMATION:
  WHEN USING THIS PRODUCT WEAR: NO SPECIAL REQUIREMENTS. Eye contact causes irritation. Ingestion of product from
  sheet will cause abdominal pain, nausea, vomiting, and diarrhea. Skin contact may cause mild irritation.

  A physician should be contacted if anyone develops any signs or symptoms and suspects that they are caused by
  exposure to this product.

FIRST AID:
  Eye Exposure:
  Flush eyes with water for 15 minutes while lifting the upper and lower eye lids. Contact lenses should not be
  worn when working with this product. If irritation persists, get medical attention.

  Skin Exposure:
  Wash with soap and water. If irritation persists, get medical attention.

  Breathing:
  Not expected to be a problem.

  Swallowing:
  If victim is conscious, give 2-3 glasses of milk or water to drink. Immediately contact the local poison control
  center for advice. Keep the victim warm and at rest. Get medical attention.

PRODUCT: DRANO   CRYSTAL ALL PURPOSE DRAIN OPENER

INCOMPATIBILITY: WATER; HOUSEHOLD/DRAIN/TOILET BOWL CLEANERS; ACIDS; PRODUCTS CONTAINER AMMONIA; AMMONIA; VINEGAR.

INGREDIENTS: SODIUM HYDROXIDE   under 60%   TLV: 2 mg/m3 ceil   CAS#: 1310-73-2
             ALUMINUM CHIPS   under 5%   TLV: 10 mg/m3 dust   CAS#: 7429-90-5

IGNI TEMP: NA FP: NA LEL: NA UEL: NA VP: NA VD: NA SG: NA PS: CRYSTALS & ALUMINUM CHIPS   APPEAR: GREENISH
   ODOR: PH FACTOR: 14

HAZARD RATINGS: H: F: R: 3 0 2   HAZARD CLASS   8

FIRE FIGHTING:
   HIGH EXPANSION FOAM, LOW EXPANSION FOAM, ALCOHOL FOAM; DRY CHEMICAL, CARBON DIOXIDE, WATER FOG.
   PROTECTIVE CLOTHING, RUBBER GLOVES, AND BREATHING APPARATUS.
WARNING: 1] STRUCTURAL PROTECTIVE CLOTHING IS PERMEABLE, REMAIN CLEAR OF SMOKE, WATER FALL OUT AND WATER RUN OFF.
         2] KEEP OUT OF THE REACH OF CHILDREN.
         3] DO NOT MIX WITH INCOMPATIBLE PRODUCTS LISTED ABOVE. HAZARDOUS/TOXIC GASSES WILL BE PRODUCED
         4] MOVE CONTAINERS FROM AREA IF WITHOUT RISK, COOL EXPOSED CONTAINERS.
         5] DIKE AREA FOR CONTROL AND CONTAINMENT TO PREVENT ENTRY INTO SEWERS, DRAIN, AND WATER WAYS.

LARGE SPILL/NO FIRE/RESCUE: WEAR NON-SEALED CHEMICAL PROTECTIVE CLOTHING, BOOTS, GLOVES AND BREATHING APPARATUS.

SPILL CONTROL AND CONTAINMENT:
   HOUSEHOLD SPILL: WEAR RUBBER GLOVES. SWEEP UP CRYSTALS. PUT IN A DISPOSAL CONTAINER. WIPE UP RESIDUE WITH A DAMP
   CLOTH. RINSE OUT IN THE SINK.
   LARGE SPILL: AVOID CREATING A DUST. SWEEP UP AND PUT IN A DISPOSAL CONTAINER. RINSE AREA WITH WATER. PICK UP
   WITH ABSORBENT. PUT IN DISPOSAL CONTAINER. SEAL AND REMOVE FROM THE WORK AREA.

HEALTH HAZARD INFORMATION:
   WHEN USING THIS PRODUCT WEAR: NO SPECIAL REQUIREMENT. Ingestion causes burns of the mouth, throat, and stomach.
   Spontaneous vomiting may occur. Eye contact will cause severe pain and damage to the tissue. Skin contact may
   cause severe pain and tissue damage.

   A physician should be contacted if anyone develops any signs or symptoms and suspects that they are caused by
   exposure to this product.

FIRST AID:
   Eye Exposure:
   Flush with water for 15 minutes while lifting the upper and lower eye lids. Contact lenses should not be worn
   when working with this product. Get medical attention.

   Skin Exposure:
   Remove contaminated clothing, flush the skin with water for 15 minutes. If irritation develops, get medical
   attention.

   Breathing:
   Not expected to be a problem. If mixed with an incompatible product, open doors, windows and evacuate the area.

   Swallowing:
   Do not induce vomiting. If victim is conscious, rinse out the mouth. Give 2-3 glasses of milk or water to drink.
   Immediately contact the local poison control center for advice. Keep the victim warm and at rest. Get medical
   attention.

PRODUCT: DRANO INSTANT PLUNGER

INCOMPATIBILITY: NONE KNOWN

INGREDIENTS: BUTANE   under 10% TLV: 800 ppm CAS#: 106-97-8
             DICHLORODIFLUOROMETHANE under 90% TLV: 1000 ppm CAS#: 75-71-8
             1,2-DICHLORO-1,1,2,2-TETRA
             FLUOROETHANE   under 10% TLV: 1000 ppm CAS#: 76-14-2

IGNI TEMP:NA FP: 20F LEL:NA UEL: NA VP: .05mmHg VD: 4.0 SG: .9 PS: LIQUID APPEAR: CLEAR ODOR:
    PH FACTOR: 3.4

HAZARD RATINGS: H: F: R: 1 1 0

FIRE FIGHTING:
    HIGH EXPANSION FOAM, LOW EXPANSION FOAM, ALCOHOL FOAM; DRY CHEMICAL, CARBON DIOXIDE, WATER FOG.
    PROTECTIVE CLOTHING, RUBBER GLOVES, AND BREATHING APPARATUS.
WARNING: 1] STRUCTURAL PROTECTIVE CLOTHING IS PERMEABLE, REMAIN CLEAR OF SMOKE, WATER FALL OUT AND WATER RUN OFF.
         2] KEEP OUT OF THE REACH OF CHILDREN.
         3] REMOVE ALL IGNITION SOURCES IF WITHOUT RISK.
         4] CONTAINER WILL EXPLODE IN A FIRE OR IF HEATED ABOVE 120F.
         5] MOVE CONTAINERS FROM AREA IF WITHOUT RISK, COOL EXPOSED CONTAINERS.
         6] DIKE AREA FOR CONTROL AND CONTAINMENT TO PREVENT ENTRY INTO SEWERS, DRAIN, AND WATER WAYS.

LARGE SPILL/NO FIRE/RESCUE: WEAR RUBBER OR NEOPRENE BOOTS, GLOVES.

SPILL CONTROL AND CONTAINMENT:
    HOUSEHOLD SPILL: WIPE UP DRY WITH AN ABSORBENT CLOTH. RINSE AREA WITH WATER. WIPE UP WITH AN ABSORBENT CLOTH.
    PUT ALL CLOTHS IN A DISPOSAL CONTAINER.
    LARGE SPILL: PICK UP WITH ABSORBENT. RINSE AREA WITH WATER. PICK UP WITH ABSORBENT. PUT ALL ABSORBENT IN A
    DISPOSAL CONTAINER. SEAL AND REMOVE FROM THE WORK AREA.

HEALTH HAZARD INFORMATION:
    WHEN USING THIS PRODUCT WEAR: NO SPECIAL REQUIREMENT. Eye contact will cause stinging, tearing, itching,
    swelling, and redness. Ingestion may cause nausea and vomiting.

    A physician should be contacted if anyone develops any signs or symptoms and suspects that they are caused by
    exposure to this product.

FIRST AID:
    Eye Exposure:
    Flush with water for 15 minutes while lifting the upper and lower eye lids. Contact lenses should not be worn
    when working with this product. Get medical attention.

    Skin Exposure:
    Remove contaminated clothing. Flush with water. Wash with soap and water. If irritation develops, get medical
    attention.

    Breathing:
    Not expected to be a problem.

    Swallowing:
    Immediately contact the local poison control center for advice. Keep the victim warm and at rest. Get medical
    attention.

PRODUCT: DRANO LIQUID DRAIN OPENER

INCOMPATIBILITY: PRODUCTS WITH AMMONIA; HOUSEHOLD/TOILET BOWL/DRAIN CLEANERS; AMMONIA; ACIDS; VINEGAR.

INGREDIENTS: SODIUM HYPOCHLORITE   under 10% TLV: CAS#: 7681-52-9
             SODIUM HYDROXIDE   under 2% TLV: 2 mg/m3 CAS#: 1310-73-2

IGNI TEMP: NA FP: NA LEL: NA UEL: NA VP: NA VD: NA SG: 1.105 PS: LIQUID APPEAR: CLEAR ODOR: BLEACH
  PH FACTOR: 12

HAZARD RATINGS: H: F: R: 3 0 1   HAZARD CLASS   8

FIRE FIGHTING:
  HIGH EXPANSION FOAM, LOW EXPANSION FOAM, ALCOHOL FOAM; DRY CHEMICAL, CARBON DIOXIDE, WATER FOG.
  PROTECTIVE CLOTHING, RUBBER GLOVES, AND BREATHING APPARATUS.
WARNING: 1] STRUCTURAL PROTECTIVE CLOTHING IS PERMEABLE, REMAIN CLEAR OF SMOKE, WATER FALL OUT AND WATER RUN OFF.
         2] KEEP OUT OF THE REACH OF CHILDREN.
         3] DO NOT MIX WITH INCOMPATIBLE PRODUCTS LISTED ABOVE. HAZARDOUS/TOXIC GASSES WILL BE PRODUCED.
         4] MOVE CONTAINERS FROM AREA IF WITHOUT RISK, COOL EXPOSED CONTAINERS.
         5] DIKE AREA FOR CONTROL AND CONTAINMENT TO PREVENT ENTRY INTO SEWERS, DRAIN, AND WATER WAYS.

LARGE SPILL/NO FIRE/RESCUE: WEAR RUBBER OR NEOPRENE BOOTS, GLOVES.

SPILL CONTROL AND CONTAINMENT:
  HOUSEHOLD SPILL: WEAR RUBBER GLOVES. WIPE UP DRY WITH AN ABSORBENT CLOTH. RINSE AREA WITH WATER. WIPE UP WITH
  ABSORBENT CLOTH. RINSE OUT IN THE SINK. WASH HANDS WITH SOAP AND WATER.
  LARGE SPILL: PICK UP WITH ABSORBENT. RINSE AREA WITH WATER. PICK UP WITH ABSORBENT. PUT ALL ABSORBENT IN A
  DISPOSAL CONTAINER. SEAL AND REMOVE FROM THE WORK AREA.

HEALTH HAZARD INFORMATION:
  WHEN USING THIS PRODUCT WEAR: NO SPECIAL REQUIREMENT. Eye contact will cause severe burns and tissue damage.
  Ingestion will cause burns to the mouth, throat, and stomach with difficult breathing. Inhalation may cause
  irritation. Skin contact may cause burns.

  A physician should be contacted if anyone develops any signs or symptoms and suspects that they are caused by
  exposure to this product.

FIRST AID:
  Eye Exposure:
  Flush with water for 15 minutes while lifting the upper and lower eye lids. Contact lenses should not be worn
  when working with this product. Get medical attention.

  Skin Exposure:
  Remove contaminated clothing. Flush skin with water, wash with soap and water. If irritation develops, get
  medical attention.

  Breathing:
  If irritation develops, move victim to fresh air.

  Swallowing:
  Do not induce vomiting. If victim is conscious, rinse out the mouth, give 2-3 glasses of milk or water to drink.
  Immediately contact the local poison control center for advice. Keep the victim warm and at rest. Get medical
  attention.

PRODUCT: DRANO LIQUID PROFESSIONAL STRENGTH

INCOMPATIBILITY: PRODUCTS WITH AMMONIA; HOUSEHOLD/TOILET BOWL/DRAIN CLEANERS; AMMONIA; ACIDS; VINEGAR.

INGREDIENTS: SODIUM HYPOCHLORITE   under 10% TLV: CAS#: 7681-52-9
             SODIUM HYDROXIDE   under 2% TLV: 2 mg/m3 CAS#: 1310-73-2

IGNI TEMP: NA FP: NA LEL: NA UEL: NA VP: NA VD: NA SG: 1.105 PS: LIQUID APPEAR: CLEAR ODOR: BLEACH
   PH FACTOR: 12

HAZARD RATINGS: H: F: R: 3 0 1   HAZARD CLASS   8

FIRE FIGHTING:
   HIGH EXPANSION FOAM, LOW EXPANSION FOAM, ALCOHOL FOAM; DRY CHEMICAL, CARBON DIOXIDE, WATER FOG.
   PROTECTIVE CLOTHING, RUBBER GLOVES, AND BREATHING APPARATUS.
WARNING: 1] STRUCTURAL PROTECTIVE CLOTHING IS PERMEABLE, REMAIN CLEAR OF SMOKE, WATER FALL OUT AND WATER RUN OFF.
         2] KEEP OUT OF THE REACH OF CHILDREN.
         3] DO NOT MIX WITH INCOMPATIBLE PRODUCTS LISTED ABOVE. HAZARDOUS/TOXIC GASSES WILL BE PRODUCED.
         4] MOVE CONTAINERS FROM AREA IF WITHOUT RISK, COOL EXPOSED CONTAINERS.
         5] DIKE AREA FOR CONTROL AND CONTAINMENT TO PREVENT ENTRY INTO SEWERS, DRAIN, AND WATER WAYS.

LARGE SPILL/NO FIRE/RESCUE: WEAR RUBBER OR NEOPRENE BOOTS, GLOVES.

SPILL CONTROL AND CONTAINMENT:
   HOUSEHOLD SPILL: WEAR RUBBER GLOVES. WIPE UP DRY WITH AN ABSORBENT CLOTH. RINSE AREA WITH WATER. WIPE UP WITH
   ABSORBENT CLOTH. RINSE OUT IN THE SINK. WASH HANDS WITH SOAP AND WATER.
   LARGE SPILL: PICK UP WITH ABSORBENT. RINSE AREA WITH WATER. PICK UP WITH ABSORBENT. PUT ALL ABSORBENT IN A
   DISPOSAL CONTAINER. SEAL AND REMOVE FROM THE WORK AREA.

HEALTH HAZARD INFORMATION:
   WHEN USING THIS PRODUCT WEAR: NO SPECIAL REQUIREMENT. Eye contact will cause severe burns and tissue damage.
   Ingestion will cause burns to the mouth, throat, and stomach with difficult breathing. Inhalation may cause
   irritation. Skin contact may cause burns.

   A physician should be contacted if anyone develops any signs or symptoms and suspects that they are caused by
   exposure to this product.

FIRST AID:
   Eye Exposure:
   Flush with water for 15 minutes while lifting the upper and lower eye lids. Contact lenses should not be worn
   when working with this product. Get medical attention.

   Skin Exposure:
   Remove contaminated clothing. Flush skin with water, wash with soap and water. If irritation develops, get
   medical attention.

   Breathing:
   If irritation develops, move victim to fresh air.

   Swallowing:
   Do not induce vomiting. If victim is conscious, rinse out the mouth, give 2-3 glasses of milk or water to drink.
   Immediately contact the local poison control center for advice. Keep the victim warm and at rest. Get medical
   attention.

PRODUCT: DRANO LIQUID PROFESSIONAL STRENGTH PLUS DRAIN OPENER

INCOMPATIBILITY: PRODUCTS WITH AMMONIA; HOUSEHOLD/TOILET BOWL/DRAIN CLEANERS; AMMONIA; ACIDS; VINEGAR.

INGREDIENTS: SODIUM HYPOCHLORITE under 10% TLV: CAS#: 7681-52-9
             SODIUM HYDROXIDE   under 2.5% TLV: 2 mg/m3 CAS#: 1310-73-2

IGNI TEMP: NA FP: NA LEL: NA UEL: NA VP: NA VD: NA SG: 1.113 PS: LIQUID APPEAR: CLEAR ODOR: BLEACH
   PH FACTOR: 14

HAZARD RATINGS: H: F: R: 3 0 1   HAZARD CLASS   8

FIRE FIGHTING:
   HIGH EXPANSION FOAM, LOW EXPANSION FOAM, ALCOHOL FOAM; DRY CHEMICAL, CARBON DIOXIDE, WATER FOG.
   PROTECTIVE CLOTHING, RUBBER GLOVES, AND BREATHING APPARATUS.
WARNING: 1] STRUCTURAL PROTECTIVE CLOTHING IS PERMEABLE, REMAIN CLEAR OF SMOKE, WATER FALL OUT AND WATER RUN OFF.
         2] KEEP OUT OF THE REACH OF CHILDREN.
         3] DO NOT MIX WITH INCOMPATIBLE PRODUCTS LISTED ABOVE. HAZARDOUS/TOXIC GASSES WILL BE PRODUCED.
         4] MOVE CONTAINERS FROM AREA IF WITHOUT RISK, COOL EXPOSED CONTAINERS.
         5] DIKE AREA FOR CONTROL AND CONTAINMENT TO PREVENT ENTRY INTO SEWERS, DRAIN, AND WATER WAYS.

LARGE SPILL/NO FIRE/RESCUE: WEAR RUBBER OR NEOPRENE BOOTS, GLOVES.

SPILL CONTROL AND CONTAINMENT:
   HOUSEHOLD SPILL: WEAR RUBBER GLOVES. WIPE UP DRY WITH AN ABSORBENT CLOTH. RINSE AREA WITH WATER. WIPE UP WITH
   ABSORBENT CLOTH. RINSE OUT IN THE SINK. WASH HANDS WITH SOAP AND WATER.
   LARGE SPILL: PICK UP WITH ABSORBENT. RINSE AREA WITH WATER. PICK UP WITH ABSORBENT. PUT ALL ABSORBENT IN A
   DISPOSAL CONTAINER. SEAL AND REMOVE FROM THE WORK AREA.

HEALTH HAZARD INFORMATION:
   WHEN USING THIS PRODUCT WEAR: NO SPECIAL REQUIREMENT. Eye contact will cause severe burns and tissue damage.
   Ingestion will cause burns to the mouth, throat, and stomach with difficult breathing. Inhalation may cause
   irritation. Skin contact may cause burns.

   A physician should be contacted if anyone develops any signs or symptoms and suspects that they are caused by
   exposure to this product.

FIRST AID:
   Eye Exposure:
   Flush with water for 15 minutes while lifting the upper and lower eye lids. Contact lenses should not be worn
   when working with this product. Get medical attention.

   Skin Exposure:
   Remove contaminated clothing. Flush skin with water, wash with soap and water. If irritation develops, get
   medical attention.

   Breathing:
   If irritation develops, move victim to fresh air.

   Swallowing:
   Do not induce vomiting. If victim is conscious, rinse out the mouth, give 2-3 glasses of milk or water to drink.
   Immediately contact the local poison control center for advice. Keep the victim warm and at rest. Get medical
   attention.

PRODUCT: DREFT   (LAUNDRY GRANULES)

INCOMPATIBILITY: NONE KNOWN

INGREDIENTS: SODIUM CARBONATE   TLV: CAS#:   497-19-8
             SODIUM PERBORATE TETRAHYDRATE TLV: CAS#: 10486-00-7
             SODIUM SILICATE   TLV: CAS#: 1344-09-8
             SODIUM SULFATE   TLV: CAS#: 7757-82-6
             ANIONIC SURFACTANTS   COMPLEX SODIUM PHOSPHATES
             ALUMINOSILICATE

IGNI TEMP: NA FP: NA LEL: NA UEL: NA VP: NA VD: NA SG: NA PS: POWDER APPEAR: PINK   ODOR: PERFUME
   PH FACTOR:

HAZARD RATINGS: H: F: R:

FIRE FIGHTING:
   HIGH EXPANSION FOAM, LOW EXPANSION FOAM, ALCOHOL FOAM; DRY CHEMICAL, CARBON DIOXIDE, WATER FOG.
   PROTECTIVE CLOTHING, RUBBER GLOVES, AND BREATHING APPARATUS.
WARNING: 1] STRUCTURAL PROTECTIVE CLOTHING IS PERMEABLE, REMAIN CLEAR OF SMOKE, WATER FALL OUT AND WATER RUN OFF.
         2] KEEP OUT OF THE REACH OF CHILDREN.
         3] THESE SURFACTANTS ARE MODERATELY TOXIC BY INGESTION.
         4] THERMAL DECOMPOSITION MAY YIELD TOXIC FUMES AND VAPORS.
         5] MOVE CONTAINERS FROM AREA IF WITHOUT RISK, COOL EXPOSED CONTAINERS.
         6] DIKE AREA FOR CONTROL AND CONTAINMENT TO PREVENT ENTRY INTO SEWERS, DRAIN, AND WATER WAYS.

LARGE SPILL/NO FIRE/RESCUE: WEAR RUBBER OR NEOPRENE BOOTS, GLOVES. DUST MASK OR BREATHING APPARATUS.

SPILL CONTROL AND CONTAINMENT:
   HOUSEHOLD SPILL: AVOID CREATING A DUST. SCOOP UP AND PUT IN A DISPOSAL CONTAINER. WIPE UP RESIDUE WITH A DAMP
   CLOTH, RINSE OUT IN THE SINK.
   LARGE SPILL: AVOID CREATING A DUST. PICK UP MATERIAL, PUT IN A DISPOSAL CONTAINER. RINSE AREA WITH WATER, PICK
   UP WITH ABSORBENT. PUT IN DISPOSAL CONTAINER. SEAL AND REMOVE FROM WORK AREA.

HEALTH HAZARD INFORMATION:
   WHEN USING THIS PRODUCT WEAR: NO SPECIAL REQUIREMENTS. Inhalation of dust will cause coughing, sore throat,
   wheezing, and shortness of breath. Eye contact will cause stinging, burning sensation, tearing, itching,
   swelling, and redness. Ingestion will cause abdominal pain with nausea, vomiting, and diarrhea.

   A physician should be contacted if anyone develops any signs or symptoms and suspects that they are caused by
   exposure to this product.

FIRST AID:
   Eye Exposure:
   Flush with water for 15 minutes while lifting the upper and lower eye lids. Contact lenses should not be worn
   when working with this product. Get medical attention.

   Skin Exposure:
   Rinse off with water. If irritation appears, get medical attention.

   Breathing:
   Move the victim to fresh air. Treat symptoms as they appear. If needed get medical attention.

   Swallowing:
   If victim is conscious, give 2-3 glasses of milk or water to drink. Immediately contact the local poison control
   center for advice. Keep the victim warm and at rest. Get medical attention.

PRODUCT: DREFT, LIQUID (LAUNDRY DETERGENT)

INCOMPATIBILITY: ACIDS AND ACID FUMES.

INGREDIENTS: ETHYL ALCOHOL   TLV: 1900 mg/m3   CAS#:   64-17-5
             ANIONIC SURFACTANTS   TLV: CAS#:
             NONIONIC SURFACTANTS   TLV: CAS#:
             WATER   TLV: CAS#: 7732-18-5
             PERFUME   TLV: CAS#:

IGNI TEMP: NA FP: NA LEL: NA UEL: NA VP: NA VD: NA SG: 1.06 PS: LIQUID APPEAR: DARK AMBER
    ODOR: PERFUMED   PH FACTOR:

HAZARD RATINGS: H: F: R:

FIRE FIGHTING:
    HIGH EXPANSION FOAM, LOW EXPANSION FOAM, ALCOHOL FOAM; DRY CHEMICAL, CARBON DIOXIDE, WATER FOG.
    PROTECTIVE CLOTHING, RUBBER GLOVES, AND BREATHING APPARATUS.
WARNING: 1] STRUCTURAL PROTECTIVE CLOTHING IS PERMEABLE, REMAIN CLEAR OF SMOKE, WATER FALL OUT AND WATER RUN OFF.
         2] KEEP OUT OF THE REACH OF CHILDREN.
         3] THERMAL DECOMPOSITION MAY YIELD CARBON DIOXIDE, CARBON MONOXIDE.
         4] MOVE CONTAINERS FROM AREA IF WITHOUT RISK, COOL EXPOSED CONTAINERS.
         5] DIKE AREA FOR CONTROL AND CONTAINMENT TO PREVENT ENTRY INTO SEWERS, DRAIN, AND WATER WAYS.

LARGE SPILL/NO FIRE/RESCUE: WEAR RUBBER OR NEOPRENE BOOTS, GLOVES.

SPILL CONTROL AND CONTAINMENT:
    HOUSEHOLD SPILL: WIPE UP WITH AN ABSORBENT CLOTH, RINSE OUT IN THE SINK. WIPE UP RESIDUE WITH A DAMP CLOTH,
    RINSE OUT IN THE SINK.
    LARGE SPILL: PICK UP WITH ABSORBENT, PUT IN A DISPOSAL CONTAINER. RINSE AREA WITH WATER, PICK UP WITH ABSORBENT,
    PUT IN A DISPOSAL CONTAINER. SEAL AND REMOVE FROM THE WORK AREA.

HEALTH HAZARD INFORMATION:
    WHEN USING THIS PRODUCT WEAR: NO SPECIAL REQUIREMENT. Eye contact may cause stinging, tearing, itching,
    swelling, and redness. Prolonged contact with the concentrated liquid may cause drying of the skin and
    irritation. Ingestion may result in abdominal pain, nausea, vomiting, and diarrhea.

    A physician should be contacted if anyone develops any signs or symptoms and suspects that they are caused by
    exposure to this product.

FIRST AID:
Eye Exposure:
Flush with water for 15 minutes while lifting the upper and lower eye lids. Contact lenses should not be worn
when working with this product. Get medical attention.

Skin Exposure:
Wash skin with water. If irritation develops, get medical attention.

Breathing:
Not expected to be a problem.

Swallowing:
If the victim is conscious, give 1-2 glasses of water to drink. Immediately contact the local poison control
center for advice. Keep the victim warm and at rest. Get medical attention.

PRODUCT: DREFT, ULTRA (LAUNDRY GRANULES)

INCOMPATIBILITY: NONE KNOWN

INGREDIENTS: SODIUM CARBONATE   TLV: CAS#:   497-19-8
             SODIUM SULFATE   TLV: CAS#:   7757-82-6
             ANIONIC SURFACTANTS   TLV: CAS#:
             ALUMINOSILICATES   TLV: CAS#:
             PERFUME   TLV: CAS#:

IGNI TEMP: NA FP: NA LEL: NA UEL: NA VP: NA VD: NA SG: NA PS: POWDER APPEAR: WHITE - PINK SPECKS
     ODOR: PERFUMED   PH FACTOR: NA

HAZARD RATINGS: H: F: R:

FIRE FIGHTING:
  HIGH EXPANSION FOAM, LOW EXPANSION FOAM, ALCOHOL FOAM; DRY CHEMICAL, CARBON DIOXIDE, WATER FOG.
  PROTECTIVE CLOTHING, RUBBER GLOVES, AND BREATHING APPARATUS.
WARNING: 1] STRUCTURAL PROTECTIVE CLOTHING IS PERMEABLE, REMAIN CLEAR OF SMOKE, WATER FALL OUT AND WATER RUN OFF.
         2] KEEP OUT OF THE REACH OF CHILDREN.
         3] THERMAL DECOMPOSITION MAY YIELD CARBON DIOXIDE, CARBON MONOXIDE.
         4] MOVE CONTAINERS FROM AREA IF WITHOUT RISK, COOL EXPOSED CONTAINERS.
         5] DIKE AREA FOR CONTROL AND CONTAINMENT TO PREVENT ENTRY INTO SEWERS, DRAIN, AND WATER WAYS.

LARGE SPILL/NO FIRE/RESCUE: WEAR RUBBER OR NEOPRENE BOOTS, GLOVES.

SPILL CONTROL AND CONTAINMENT:
  HOUSEHOLD SPILL: SWEEP UP POWDER AND PUT IN A DISPOSAL CONTAINER. DAMP MOP UP THE RESIDUE. RINSE OUT IN THE
  SINK.
  LARGE SPILL: AVOID CREATING A DUST. SCOOP UP POWDER. PUT IN A DISPOSAL CONTAINER. RINSE THE AREA WITH WATER.
  PICK UP WITH ABSORBENT. PUT IN A DISPOSAL CONTAINER. SEAL AND REMOVE FROM THE WORK AREA.

HEALTH HAZARD INFORMATION:
  WHEN USING THIS PRODUCT WEAR: NO SPECIAL REQUIREMENT. Inhalation may cause coughing, sore throat, wheezing, and
  shortness of breath. Eye contact will cause stinging, tearing, itching, swelling, and redness. Ingestion may
  result in nausea, vomiting, and diarrhea.

  A physician should be contacted if anyone develops any signs or symptoms and suspects that they are caused by
  exposure to this product.

FIRST AID:
Eye Exposure:
Flush with water for 15 minutes while lifting the upper and lower eye lids. Contact lenses should not be worn
when working with this product. Get medical attention.

Skin Exposure:
Wash with water. If irritation develops, get medical attention.

Breathing:
If irritation develops, move victim to fresh air.

Swallowing:
If the victim is conscious, give 2-3 glasses of milk or water to drink. Immediately contact the local poison
control center for advice. Keep the victim warm and at rest. Get medical attention.

PRODUCT: DRY BRITE

INCOMPATIBILITY: NONE KNOWN

INGREDIENTS: ISOPROPYL ALCOHOL   3-6%   TLV: 400 ppm   CAS#:   67-63-0
            ISOPARAFFINIC   MFG recommended
            HYDROCARBON SOLVENT   1%   TLV: 400 ppm   CAS#:
            WAX,SILICONE SURFACTANT 1-3%   TLV: CAS#:
            WATER   90-95% TLV: CAS#: 7732-18-5

IGNI TEMP: NA FP: 103F LEL: NA UEL: NA VP: NA VD: NA SG: .98 PS: LIQUID APPEAR: MILKY ODOR: ALCOHOL
   PH FACTOR: 5.7

HAZARD RATINGS: H: F: R: 0 2 0

FIRE FIGHTING:
   HIGH EXPANSION FOAM, LOW EXPANSION FOAM, ALCOHOL FOAM; DRY CHEMICAL, CARBON DIOXIDE, WATER FOG.
   PROTECTIVE CLOTHING, RUBBER GLOVES, AND BREATHING APPARATUS.
WARNING: 1] STRUCTURAL PROTECTIVE CLOTHING IS PERMEABLE, REMAIN CLEAR OF SMOKE, WATER FALL OUT AND WATER RUN OFF.
         2] KEEP OUT OF THE REACH OF CHILDREN.
         3] THERMAL DECOMPOSITION YIELDS TOXIC CARBON DIOXIDE, CARBON MONOXIDE.
         4] REMOVE ALL IGNITION SOURCES IF WITHOUT RISK.
         5] MOVE CONTAINERS FROM AREA IF WITHOUT RISK, COOL EXPOSED CONTAINERS.
         6] DIKE AREA FOR CONTROL AND CONTAINMENT TO PREVENT ENTRY INTO SEWERS, DRAIN, AND WATER WAYS.

LARGE SPILL/NO FIRE/RESCUE: WEAR NON-SEALED CHEMICAL PROTECTIVE CLOTHING RUBBER OR NEOPRENE BOOTS, GLOVES AND
                           BREATHING APPARATUS.

SPILL CONTROL AND CONTAINMENT:
   HOUSEHOLD SPILL: VENTILATE AREA. WIPE UP LIQUID WITH ABSORBENT CLOTH, RINSE OUT IN SINK. WIPE UP RESIDUE WITH A
   DAMP CLOTH, RINSE OUT IN SINK.
   LARGE SPILL: COVER SPILL WITH ABSORBENT. PICK UP AND PUT IN A DISPOSAL CONTAINER. RINSE AREA WITH WATER, PICK UP
   WITH ABSORBENT, PUT IN DISPOSAL CONTAINER. SEAL AND REMOVE FROM WORK AREA.

HEALTH HAZARD INFORMATION:
   WHEN USING THIS PRODUCT WEAR: NO SPECIAL REQUIREMENTS. Eye contact may cause irritation. Ingestion will cause
   abdominal pain, nausea, vomiting, and diarrhea.

   A physician should be contacted if anyone develops any signs or symptoms and suspects that they are caused by
   exposure to this product.

FIRST AID:
   Eye Exposure:
   Flush with water for 15 minutes while lifting the upper and lower eye lids. Contact lenses should not be worn
   when working with this product, get medical attention.

   Skin Exposure:
   Wash with soap and water. If irritation appears, get medical attention

   Breathing:
   Not expected to be a problem.

   Swallowing:
   Immediately contact the local poison control center for advice. Keep the victim warm and at rest, get medical
   attention.

PRODUCT: DUSTER PLUS   FRESH SCENT

INCOMPATIBILITY: NONE KNOWN

INGREDIENTS: HYDROCARBON OIL   5-10% TLV: 5 mg/m3(mist) CAS#:
             ISOPARAFFINIC   MFG recommended
             HYDROCARBON SOLVENT   25-35% TLV: 400 ppm   CAS#:
             PROPANE   TLV: 1000 ppm   CAS#:   74-98-6
             ISOBUTANE   TLV: CAS#:   75-28-5
             N-BUTANE   TLV: CAS#:   106-97-8
             WATER   TLV: CAS#: 7732-18-5

IGNI TEMP: NA FP: 20F LEL: NA UEL: NA VP: NA VD: NA SG: .75 PS: LIQUID APPEAR: SPRAY MIST ODOR: PLEASANT
   PH FACTOR:

HAZARD RATINGS: H: F: R: 0 4 0

FIRE FIGHTING:
   HIGH EXPANSION FOAM, LOW EXPANSION FOAM, ALCOHOL FOAM; DRY CHEMICAL, CARBON DIOXIDE, WATER FOG.
   PROTECTIVE CLOTHING, RUBBER GLOVES, AND BREATHING APPARATUS.
WARNING: 1] STRUCTURAL PROTECTIVE CLOTHING IS PERMEABLE, REMAIN CLEAR OF SMOKE, WATER FALL OUT AND WATER RUN OFF.
         2] KEEP OUT OF THE REACH OF CHILDREN.
         3] CONTAINERS MAY EXPLODE IN A FIRE OR OF HEATED ABOVE 120F.
         4] REMOVE ALL SOURCES OF IGNITION IF WITHOUT RISK.
         5] MOVE CONTAINERS FROM AREA IF WITHOUT RISK, COOL EXPOSED CONTAINERS.
         6] DIKE AREA FOR CONTROL AND CONTAINMENT TO PREVENT ENTRY INTO SEWERS, DRAIN, AND WATER WAYS.

LARGE SPILL/NO FIRE/RESCUE: WEAR NON-SEALED CHEMICAL PROTECTIVE CLOTHING RUBBER OR NEOPRENE BOOTS, GLOVES AND
                           BREATHING APPARATUS.

SPILL CONTROL AND CONTAINMENT:
   HOUSEHOLD SPILL: WIPE UP WITH AN ABSORBENT CLOTH. RINSE OUT IN SINK. WIPE UP RESIDUE WITH A DAMP CLOTH, RINSE
   OUT IN SINK.
   LARGE SPILL: COVER SPILL WITH ABSORBENT. PICK UP AND PUT INTO A DISPOSAL CONTAINER. SCRUB AREA WITH DETERGENT
   AND WATER. PICK UP WITH ABSORBENT, PUT IN DISPOSAL CONTAINER. SEAL AND REMOVE FROM WORK AREA.

HEALTH HAZARD INFORMATION:
   WHEN USING THIS PRODUCT VENTILATE AREA. Inhalation of vapors may cause upper respiratory irritation. Skin
   contact may cause a slight irritation. Ingestion will cause abdominal pain, nausea, vomiting, and diarrhea.

   A physician should be contacted if anyone develops any signs or symptoms and suspects that they are caused by
   exposure to this product.

FIRST AID:
   Eye Exposure:
   Flush with water for 15 minutes while lifting the upper and lower eye lids. Contact lenses should not be worn
   when working with this product, get medical attention.

   Skin Exposure:
   Wash skin with soap and water. If irritation appears, get medical attention.

   Breathing:
   Move victim to fresh air. Treat any symptoms as they appear. If needed, get medical attention.

   Swallowing:
   Immediately contact the local poison control center for advice. Keep the victim warm and at rest, get medical
   attention.

PRODUCT: DUSTER PLUS - LEMON

INCOMPATIBILITY: NONE KNOWN

INGREDIENTS: HYDROCARBON OIL   5-10% TLV: 5 mg/m3(mist) CAS#:
           ISOPARAFFINIC   MFG recommended
           HYDROCARBON SOLVENT   25-35% TLV: 400 ppm   CAS#:
           PROPANE   TLV: 1000 ppm   CAS#:   74-98-6
           ISOBUTANE   TLV: CAS#:   75-28-5
           N-BUTANE   TLV: CAS#:   106-97-8
           WATER   TLV: CAS#: 7732-18-5

IGNI TEMP: NA   FP: 20F   LEL: NA   UEL: NA   VP: NA   VD: NA   SG: .75   PS: LIQUID   APPEAR: SPRAY MIST ST   ODOR: PLEASANT
   PH FACTOR:

HAZARD RATINGS: H: F: R: 0   4   0

FIRE FIGHTING:
   HIGH EXPANSION FOAM, LOW EXPANSION FOAM, ALCOHOL FOAM; DRY CHEMICAL, CARBON DIOXIDE, WATER FOG.
   PROTECTIVE CLOTHING, RUBBER GLOVES, AND BREATHING APPARATUS.
WARNING: 1] STRUCTURAL PROTECTIVE CLOTHING IS PERMEABLE, REMAIN CLEAR OF SMOKE, WATER FALL OUT AND WATER RUN OFF.
         2] KEEP OUT OF THE REACH OF CHILDREN.
         3] CONTAINERS MAY EXPLODE IN A FIRE OR IF HEATED ABOVE 120F.
         4] REMOVE ALL SOURCES OF IGNITION IF WITHOUT RISK.
         5] MOVE CONTAINERS FROM AREA IF WITHOUT RISK, COOL EXPOSED CONTAINERS.
         6] DIKE AREA FOR CONTROL AND CONTAINMENT TO PREVENT ENTRY INTO SEWERS, DRAIN, AND WATER WAYS.

LARGE SPILL/NO FIRE/RESCUE: WEAR NON-SEALED CHEMICAL PROTECTIVE CLOTHING RUBBER OR NEOPRENE BOOTS, GLOVES AND
                               BREATHING APPARATUS.   .

SPILL CONTROL AND CONTAINMENT:
   HOUSEHOLD SPILL: WIPE UP WITH AN ABSORBENT CLOTH. RINSE OUT IN SINK. WIPE UP RESIDUE WITH A DAMP CLOTH, RINSE
   OUT IN SINK.
   LARGE SPILL: COVER SPILL WITH ABSORBENT. PICK UP AND PUT INTO A DISPOSAL CONTAINER. SCRUB AREA WITH DETERGENT
   AND WATER. PICK UP WITH ABSORBENT, PUT IN DISPOSAL CONTAINER. SEAL AND REMOVE FROM WORK AREA.

HEALTH HAZARD INFORMATION:
   WHEN USING THIS PRODUCT VENTILATE AREA. Inhalation of vapors may cause upper respiratory irritation. Skin
   contact may cause a slight irritation. Ingestistion will cause abdominal pain, nausea, vomiting, and diarrhea.

   A physician should be contacted if anyone develops any signs or symptoms and suspects that they are caused by
   exposure to this product.

FIRST AID:
   Eye Exposure:
   Flush with water for 15 minutes while lifting the upper and lower eye lids. Contact lenses should not be worn
   when working with this product, get medical attention.

   Skin Exposure:
   Wash skin with soap and water. If irritation appears, get medical attention.

   Breathing:
   Move victim to fresh air. Treat any symptoms as they appear. If needed, get medical attention.

   Swallowing:
   Immediately contact the local po po poison control center for advice. Keep the victim warm and at rest, get medical
   attention.

PRODUCT: EDGE, LIME

INCOMPATIBILITY: NONE KNOWN

INGREDIENTS: PENTANE   1-4% TLV: 600 ppm   CAS#: 109-66-0
             ISOBUTANE   3-6% TLV: CAS#: 75-28-5
             NON HAZARDOUS INGREDIENTS   TLV: CAS#:

IGNI TEMP: NA FP: 20F LEL: NA UEL: NA VP: NA VD: NA SG: NA PS: LIQUID GEL APPEAR: CLEAR
     ODOR: MILD FRAGRANCE   PH FACTOR: NA

HAZARD RATINGS: H: F: R: 1 4 0

FIRE FIGHTING:
  HIGH EXPANSION FOAM, LOW EXPANSION FOAM, ALCOHOL FOAM; DRY CHEMICAL, CARBON DIOXIDE, WATER FOG.
  PROTECTIVE CLOTHING, RUBBER GLOVES, AND BREATHING APPARATUS.
WARNING: 1] STRUCTURAL PROTECTIVE CLOTHING IS PERMEABLE, REMAIN CLEAR OF SMOKE, WATER FALL OUT AND WATER RUN OFF.
         2] KEEP OUT OF THE REACH OF CHILDREN.
         3] REMOVE ALL SOURCES OF IGNITION IF WITH OUT RISK.
         4] CONTAINERS WILL EXPLODE IN A FIRE OR IF HEATED ABOVE 120F.
         5] THERMAL DECOMPOSITION MAY YIELD CARBON DIOXIDE, CARBON MONOXIDE.
         6] MOVE CONTAINERS FROM AREA IF WITHOUT RISK, COOL EXPOSED CONTAINERS.
         7] DI DIKE AREA FOR CONTROL AND CONTAINMENT TO PREVENT ENTRY INTO SEWERS, DRAIN, AND WATER WAYS.

LARGE SPILL/NO FIRE/RESCUE: WEAR RUBBER OR NEOPRENE BOOTS, GLOVES.

SPILL CONTROL AND CONTAINMENT:
  HOUSEHOLD SPILL: WIPE UP GEL WITH A DAMP CLOTH. RINSE AREA WITH WATER. WIPE UP ALMOST DRY WITH AN ABSORBENT
  CLOTH.
  LARGE SPILL: COVER WITH ABSORBENT, PICK UP AND PUT IN A DISPOSAL CONTAINER. SCRUB AREA WITH WATER. PICK UP WITH
  ABSORBENT. PUT IN A DISPOSAL CONTAINER. SEAL AND REMOVE FROM THE WORK AREA.

HEALTH HAZARD INFORMATION:
  WHEN USING THIS PRODUCT WEAR: NO SPECIAL REQUIREMENT. Eye contact can cause stinging, tearing, itching,
  swelling, and redness. Ingestion may cause nausea and vomiting.

  A physician should be contacted if anyone develops any signs or symptoms and suspects that they are caused by
  exposure to this product.

FIRST AID:
  Eye Exposure:
  Flush with water for 15 minutes while lifting the upper and lower eye lids. Contact lenses should not be worn
  when working with this product. Get medical attention.

  Skin Exposure:
  Rinse off with water.

  Breathing:
  Not expected to be a problem.

  Swallowing:
  Immediately contact the local poison control center for advice. Keep the victim warm and at rest. Get medical
  attention.

PRODUCT: EDGE, MEDICATED/MENTHOL

INCOMPATIBILITY: NONE KNOWN.

INGREDIENTS: PENTANE   1-4% TLV: 600 ppm   CAS#: 109-66-0
             ISOBUTANE   3-6% TLV: CAS#: 75-28-5
             NON-HAZARDOUS INGREDIENTS   TLV: CAS#:

IGNI TEMP: NA FP: 20F LEL: NA UEL: NA VP: NA VD: NA SG: NA PS: LIQUID GEL APPEAR: CLEAR ODOR: MILD
  PH FACTOR: NA

HAZARD RATINGS: H: F: R: 1 4 0

FIRE FIGHTING:
   HIGH EXPANSION FOAM, LOW EXPANSION FOAM, ALCOHOL FOAM; DRY CHEMICAL, CARBON DIOXIDE, WATER FOG.
   PROTECTIVE CLOTHING, RUBBER GLOVES, AND BREATHING APPARATUS.
WARNING: 1) STRUCTURAL PROTECTIVE CLOTHING IS PERMEABLE, REMAIN CLEAR OF SMOKE, WATER FALL OUT AND WATER RUN OFF.
        2) KEEP OUT OF THE REACH OF CHILDREN.
        3) REMOVE ALL SOURCES OF IGNITION IF WITHOUT RISK.
        4) CONTAINERS WILL EXPLODE IN A FIRE OR IF HEATED ABOVE 120F.
        5) THERMAL DECOMPOSITION MAY YIELD CARBON DIOXIDE, CARBON MONOXIDE.
        6) MOVE CONTAINERS FROM AREA IF WITHOUT RISK, COOL EXPOSED CONTAINERS.
        7) DIKE AREA FOR CONTROL AND CONTAINMENT TO PREVENT ENTRY INTO SEWERS, DRAIN, AND WATER WAYS.

LARGE SPILL/NO FIRE/RESCUE: WEAR RUBBER OR NEOPRENE BOOTS, GLOVES.

SPILL CONTROL AND CONTAINMENT:
   HOUSEHOLD SPILL: WIPE UP GEL WITH AN ABSORBENT CLOTH. RINSE OUT IN THE SINK. WIPE UP RESIDUE WITH A DAMP CLOTH.
   RINSE OUT IN THE SINK.
   LARGE SPILL: PICK UP WITH ABSORBENT. PUT IN A DISPOSAL CONTAINER. RINSE AREA WITH WATER. PICK UP WITH ABSORBENT.
   PUT IN A DISPOSAL CONTAINER. SEAL AND REMOVE FROM THE WORK AREA.

HEALTH HAZARD INFORMATION:
   WHEN USING THIS PRODUCT WEAR: NO SPECIAL REQUIREMENT. Eye contact may cause stinging, tearing, itching,
   swelling, and redness.

   A physician should be contacted if anyone develops any signs or symptoms and suspects that they are caused by
   exposure to this product.

FIRST AID:
   Eye Exposure:
   Flush with water for 15 minutes while lifting the upper and lower eye lids. Contact lenses should not be worn
   when working with this product. Get medical attention.

   Skin Exposure:
   Not expected to be a problem.

   Breathing:
   Not expected to be a problem.

   Swallowing:
   Immediately contact the local poison control center for advice. Keep the victim warm and at rest. Get medical
   attention.

PRODUCT: EDGE   FOR SENSITIVE SKIN - ALOE FORMULA

INCOMPATIBILITY: NONE KNOWN

INGREDIENTS: PENTANE   1-4%   TLV: 600 ppm   CAS#:   109-66-0
             ISOBUTANE   3-6%   TLV: CAS#:   75-28-5
             SORBITOL   1-4%   TLV: CAS#:   50-70-4
             TRIETHANOLAMINE   5-8%   TLV: CAS#:   102-71-6
             PALMITIC/STEARIC ACID 10-14% TLV: CAS#:   57-10-3
             MONOGLYCERDIES   1-3%   TLV: CAS#: 68990-53-4
             WATER   TLV: CAS#: 7732-18-5

IGNI TEMP: NA FP: 20F LEL: NA UEL: NA VP: NA VD: NA SG: NA PS: LIQUID GEL APPEAR: CLEAR
     ODOR: MILD FRAGRANCE   PH FACTOR: NA

HAZARD RATINGS: H: F: R: 1 4 0

FIRE FIGHTING:
   HIGH EXPANSION FOAM, LOW EXPANSION FOAM, ALCOHOL FOAM; DRY CHEMICAL, CARBON DIOXIDE, WATER FOG.
   PROTECTIVE CLOTHING, RUBBER GLOVES, AND BREATHING APPARATUS.
WARNING: 1] STRUCTURAL PROTECTIVE CLOTHING IS PERMEABLE, REMAIN CLEAR OF SMOKE, WATER FALL OUT AND WATER RUN OFF.
         2] KEEP OUT OF THE REACH OF CHILDREN.
         3] CONTAINERS MAY EXPLODE IN A FIRE OR IF HEATED ABOVE 120F.
         4] REMOVE ALL SOURCES OF IGNITION IF WITHOUT RISK.
         5] MOVE CONTAINERS FROM AREA IF WITHOUT RISK, COOL EXPOSED CONTAINERS.
         6] DIKE AREA FOR CONTROL AND CONTAINMENT TO PREVENT ENTRY INTO SEWERS, DRAIN, AND WATER WAYS.

LARGE SPILL/NO FIRE/RESCUE: WEAR RUBBER OR NEOPRENE BOOTS, GLOVES.

SPILL CONTROL AND CONTAINMENT:
   HOUSEHOLD SPILL: WIPE UP GEL WITH ABSORBENT CLOTH, RINSE OUT IN SINK. WIPE UP RESIDUE WITH A DAMP CLOTH, RINSE
   OUT IN SINK.
   LARGE SPILL: SCOOP UP GEL AND PUT IN A DISPOSAL CONTAINER. COVER RESIDUE WITH ABSORBENT. SWEEP UP AND PUT IN
   DISPOSAL CONTAINER. RINSE AREA WITH WATER AND PICK UP WITH ABSORBENT, PUT IN CONTAINER. SEAL AND REMOVE FROM
   AREA.

HEALTH HAZARD INFORMATION:
   WHEN USING THIS PRODUCT WEAR: NO SPECIAL REQUIREMENTS. Gel may cause eye irritation. Ingestion may cause
   abdominal pain, nausea, vomiting and, diarrhea.

   A physician should be contacted if anyone develops any signs or symptoms and suspects that they are caused by
   exposure to this product.

FIRST AID:
Eye Exposure:
Flush with water for 15 minutes while lifting the upper and lower eye lids. Contact lenses should not be worn
when working with this product, get medical attention.

Skin Exposure:
Wash off with soap and water. If irritation appears, get medical attention.

Breathing:
Not expected to be a problem.

Swallowing:
Immediately contact the local poison control center for advice. Keep the victim warm and at rest, get medical
attention.

PRODUCT: EDGE   SHAVING CREME   (REGULAR)

INCOMPATIBILITY: NONE KNOWN.

INGREDIENTS: PENTANE   1-4% TLV: 600 ppm   CAS#: 109-66-0
            ISOBUTANE   3-6% TLV: CAS#: 75-28-5
            BALANCE NON HAZARDOUS MATERIALS

IGNI TEMP: NA FP: 20F LEL: NA UEL: NA VP: NA VD: NA SG: NA PS: GEL APPEAR: CLEAR   ODOR: MILD FRAGRANCE
   PH FACTOR: NA

HAZARD RATINGS: H: F: R:

NEUTRALIZING AGENTS:

FIRE FIGHTING:
   HIGH EXPANSION FOAM, LOW EXPANSION FOAM, ALCOHOL FOAM; DRY CHEMICAL, CARBON DIOXIDE, WATER FOG.
   PROTECTIVE CLOTHING, RUBBER GLOVES, AND BREATHING APPARATUS.
WARNING: 1] STRUCTURAL PROTECTIVE CLOTHING IS PERMEABLE, REMAIN CLEAR OF SMOKE, WATER FALL OUT AND WATER RUN OFF.
        2] KEEP OUT OF THE REACH OF CHILDREN.
        3] CONTAINERS WILL EXPLODE IN A FIRE OR IF HEATED ABOVE 120F.
        4] SPILL WILL CREATE A SLIPPING HAZARD.
        5] MOVE CONTAINERS FROM AREA IF WITHOUT RISK, COOL EXPOSED CONTAINERS.
        6] DIKE AREA FOR CONTROL AND CONTAINMENT TO PREVENT ENTRY INTO SEWERS, DRAIN, AND WATER WAYS.

LARGE SPILL/NO FIRE/RESCUE: WEAR RUBBER OR NEOPRENE BOOTS, GLOVES.

SPILL CONTROL AND CONTAINMENT:
   HOUSEHOLD SPILL: WIPE UP GEL WITH AN ABSORBENT CLOTH, RINSE OUT IN THE SINK. WIPE UP RESIDUE WITH A DAMP CLOTH
   AND RINSE OUT IN THE SINK.
   LARGE SPILL: SCOOP UP THE GEL AND PUT IN A DISPOSAL CONTAINER. COVER RESIDUE WITH ABSORBENT. PICK UP AND PUT IN
   DISPOSAL CONTAINER. RINSE AREA WITH WATER, PICK UP WITH ABSORBENT, PUT IN DISPOSAL CONTAINER. SEAL AND REMOVE
   FROM THE WORK AREA.

HEALTH HAZARD INFORMATION:
   WHEN USING THIS PRODUCT WEAR: NO SPECIAL REQUIREMENTS. Gel may cause eye irritation. Ingestion may cause
   abdominal pain with nausea, vomiting, and diarrhea.

   A physician should be contacted if anyone develops any signs or symptoms and suspects that they are caused by
   exposure to this product.

FIRST AID:
   Eye Exposure:
   Flush with water for 15 minutes while lifting the upper and lower eye lids. Contact lenses should not be worn
   when working with this product, get medical attention.

   Skin Exposure:
   Wash with soap and water. If irritation appears, get medical attention.

   Breathing:
   Not expected to be a problem.

   Swallowing:
   Immediately contact the local poison control center for advice. Keep the victim warm and at rest, get medical
   attention.

PRODUCT: EDGE, SKIN CONDITIONING - LANOLIN FORMULA

INCOMPATIBILITY: NONE KNOWN

INGREDIENTS: PENTENE   1-4%   TLV: 600 ppm   CAS#:   109-66-0
             ISOBUTANE   3-6%   TLV: CAS#:   75-28-5
             SORBITOL   1-4%   TLV: CAS#:   50-70-4
             TRIETHANOLAMINE   5-8%   TLV: CAS#:   102-71-6
             PALMITIC/STEARIC ACID   10-14%   TLV: CAS#:   57-10-3
             MONOGLYCERIDES   1-7%   TLV: CAS#: 69990-53-4
             WATER   75-80%   TLV: CAS#: 7732-18-5

IGNI TEMP: NA FP: 20F LEL: NA UEL: NA VP: NA VD: NA SG: NA PS: GEL APPEAR: CLEAR ODOR: MILD FRAGRANCE
     PH FACTOR: NA

HAZARD RATINGS: H: F: R: 1   4   0

FIRE FIGHTING:
   HIGH EXPANSION FOAM, LOW EXPANSION FOAM, ALCOHOL FOAM; DRY CHEMICAL, CARBON DIOXIDE, WATER FOG.
   PROTECTIVE CLOTHING, RUBBER GLOVES, AND BREATHING APPARATUS.
WARNING: 1] STRUCTURAL PROTECTIVE CLOTHING IS PERMEABLE, REMAIN CLEAR OF SMOKE, WATER FALL OUT AND WATER RUN OFF.
         2] KEEP OUT OF THE REACH OF CHILDREN.
         3] REMOVE ALL IGNITION SOURCES IF WITHOUT RISK.
         4] CONTAINERS WILL EXPLODE IN A FIRE OR IF HEATED ABOVE 120F.
         5] THERMAL DECOMPOSITION MAY YIELD CARBON DIOXIDE, CARBON MONOXIDE.
         6] MOVE CONTAINERS FROM AREA IF WITHOUT RISK, COOL EXPOSED CONTAINERS.
         7] DIKE AREA FOR CONTROL AND CONTAINMENT TO PREVENT ENTRY INTO SEWERS, DRAIN, AND WATER WAYS.

LARGE SPILL/NO FIRE/RESCUE: WEAR RUBBER OR NEOPRENE BOOTS, GLOVES.

SPILL CONTROL AND CONTAINMENT:
   HOUSEHOLD SPILL: WIPE UP WITH AN ABSORBENT CLOTH. RINSE OUT IN THE SINK. WIPE UP RESIDUE WITH A DAMP CLOTH.
   RINSE OUT IN THE SINK.
   LARGE SPILL: SCOOP UP GEL AND PUT IN A DISPOSAL CONTAINER. RINSE AREA WITH WATER. PICK UP WITH ABSORBENT. PUT
   ALL ABSORBENT IN A DISPOSAL CONTAINER. SEAL AND REMOVE FROM THE WORK AREA.

HEALTH HAZARD INFORMATION:
   WHEN USING THIS PRODUCT WEAR: NO SPECIAL REQUIREMENT. Eye contact may cause stinging, tearing, itching,
   swelling, and redness. Ingestion may cause nausea and vomiting.

   A physician should be contacted if anyone develops any signs or symptoms and suspects that they are caused by
   exposure to this product.

FIRST AID:
   Eye Exposure:
   Flush with water for 15 minutes while lifting the upper and lower eye lids. Contact lenses should not be worn
   when working with this product. Get medical attention.

   Skin Exposure:
   Not expected to be a problem.

   Breathing:
   Not expected to be a problem.

   Swallowing:
   Immediately contact the local poison control center for advice. Keep the victim warm and at rest. Get medical
   attention.

PRODUCT: EDGE   FOR TOUGH BEARDS

INCOMPATIBILITY: NONE KNOWN.

INGREDIENTS: PENTANE   1-4% TLV: 600 ppm   CAS#: 109-66-0
             ISOBUTANE   3-6% TLV: CAS#: 75-28-5
             BALANCE NON HAZARDOUS MATERIALS

IGNI TEMP: NA FP: 20F LEL: NA UEL: NA VP: NA VD: NA SG: NA PS:GEL APPEAR: CLEAR   ODOR: MILD FRAGRANCE
    PH FACTOR: NA

HAZARD RATINGS: H: F: R:

NEUTRALIZING AGENTS:

FIRE FIGHTING:
   HIGH EXPANSION FOAM, LOW EXPANSION FOAM, ALCOHOL FOAM; DRY CHEMICAL, CARBON DIOXIDE, WATER FOG.
   PROTECTIVE CLOTHING, RUBBER GLOVES, AND BREATHING APPARATUS.
WARNING: 1] STRUCTURAL PROTECTIVE CLOTHING IS PERMEABLE, REMAIN CLEAR OF SMOKE, WATER FALL OUT AND WATER RUN OFF.
         2] KEEP OUT OF THE REACH OF CHILDREN.
         3] CONTAINERS WILL EXPLODE IN A FIRE OR IF HEATED ABOVE 120F.
         4] SPILL WILL CREATE A SLIPPING HAZARD.
         5] MOVE CONTAINERS FROM AREA IF WITHOUT RISK, COOL EXPOSED CONTAINERS.
         6] DIKE AREA FOR CONTROL AND CONTAINMENT TO PREVENT ENTRY INTO SEWERS, DRAIN, AND WATER WAYS.

LARGE SPILL/NO FIRE/RESCUE: WEAR RUBBER OR NEOPRENE BOOTS, GLOVES.

SPILL CONTROL AND CONTAINMENT:
   HOUSEHOLD SPILL: WIPE UP GEL WITH AN ABSORBENT CLOTH, RINSE OUT IN THE SINK. WIPE UP RESIDUE WITH A DAMP CLOTH
   AND RINSE OUT IN THE SINK.
   LARGE SPILL: SCOOP UP THE GEL AND PUT IN A DISPOSAL CONTAINER. COVER RESIDUE WITH ABSORBENT. PICK UP AND PUT IN
   DISPOSAL CONTAINER. RINSE AREA WITH WATER, PICK UP WITH ABSORBENT, PUT IN DISPOSAL CONTAINER. SEAL AND REMOVE
   FROM THE WORK AREA.

HEALTH HAZARD INFORMATION:
   WHEN USING THIS PRODUCT WEAR: NO SPECIAL REQUIREMENTS. Gel may cause eye irritation. Ingestion may cause
   abdominal pain with nausea, vomiting, and diarrhea.

   A physician should be contacted if anyone develops any signs or symptoms and suspects that they are caused by
   exposure to this product.

FIRST AID:
   Eye Exposure:
   Flush with water for 15 minutes while lifting the upper and lower eye lids. Contact lenses should not be worn
   when working with this product, get medical attention.

   Skin Exposure:
   Wash with soap and water. If irritation appears, get medical attention.

   Breathing:
   Not expected to be a problem.

   Swallowing:
   Immediately contact the local poison control center for advice. Keep the victim warm and at rest, get medical
   attention.

PRODUCT: ENDUST   DUSTING AND CLEANING SPRAY

INCOMPATIBILITY: NONE KNOWN

INGREDIENTS: ISOBUTANE   under 20% TLV: CAS#:   75-28-5
             HYDROCARBON SOLVENT under 20% TLV: CAS#: 64742-48-9
             PARAFFINIC OIL   under 20% TLV: CAS#:

IGNI TEMP: NA FP: 100F LEL: 1.8 UEL: 9.5 VP: 30 mmHg VD: 4.1 SG: .88 PS: LIQUID APPEAR: MILKY WHITE SPRAY
     ODOR: CITRUS   PH FACTOR: NA

HAZARD RATINGS: H: F: R: 0 3 0

FIRE FIGHTING:
  HIGH EXPANSION FOAM, LOW EXPANSION FOAM, ALCOHOL FOAM; DRY CHEMICAL, CARBON DIOXIDE, WATER FOG.
  PROTECTIVE CLOTHING, RUBBER GLOVES, AND BREATHING APPARATUS.
WARNING: 1] STRUCTURAL PROTECTIVE CLOTHING IS PERMEABLE, REMAIN CLEAR OF SMOKE, WATER FALL OUT AND WATER RUN OFF.
         2] KEEP OUT OF THE REACH OF CHILDREN.
         3] CONTAINERS WILL EXPLODE IN A FIRE OR IF HEATED ABOVE 120F.
         4] REMOVE ALL IGNITION SOURCES IF WITHOUT RISK.
         5] VENTILATE ENCLOSED AREA FOR A LARGE SPILL AND EVACUATE.
         6] MOVE CONTAINERS FROM AREA IF WITHOUT RISK, COOL EXPOSED CONTAINERS.
         7] DIKE AREA FOR CONTROL AND CONTAINMENT TO PREVENT ENTRY INTO SEWERS, DRAIN, AND WATER WAYS.

LARGE SPILL/NO FIRE/RESCUE: WEAR RUBBER OR NEOPRENE BOOTS, GLOVES.

SPILL CONTROL AND CONTAINMENT:
  HOUSEHOLD SPILL: WIPE UP DRY WITH AN ABSORBENT CLOTH. RINSE OUT IN THE SINK. WIPE UP RESIDUE WITH A DAMP CLOTH.
  RINSE OUT IN THE SINK.
  LARGE SPILL: PICK UP WITH ABSORBENT. RINSE AREA WITH WATER. PICK UP WITH ABSORBENT. PUT ALL ABSORBENT IN A
  DISPOSAL CONTAINER. SEAL AND REMOVE FROM THE WORK AREA.

HEALTH HAZARD INFORMATION:
  WHEN USING THIS PRODUCT WEAR: NO SPECIAL REQUIREMENT. Eye contact will cause stinging, tearing, itching,
  swelling, and redness. Ingestion may cause nausea and vomiting.

  A physician should be contacted if anyone develops any signs or symptoms and suspects that they are caused by
  exposure to this product.

FIRST AID:
  Eye Exposure:
  Flush with water for 15 minutes while lifting the upper and lower eye lids. Contact lenses should not be worn
  when working with this product. Get medical attention.

  Skin Exposure:
  Wash with soap and water.

  Breathing:
  Not expected to be a problem.

  Swallowing:
  Do not induce vomiting. If victim is conscious, give 2-3 glasses of water to drink. Immediately contact the
  local poison control center for advice. Keep the victim warm and at rest. Get medical attention.

PRODUCT: ENDUST   LEMON DUSTING AND CLEANING SPRAY

INCOMPATIBILITY: NONE KNOWN

INGREDIENTS: ISOBUTANE   under 20% TLV: CAS#:   75-28-5
             HYDROCARBON SOLVENT under 20% TLV: CAS#: 64742-48-9
             PARAFFINIC OIL   under 20% TLV: CAS#:

IGNI TEMP: NA FP: 100F LEL: 1.8 UEL: 9.5 VP: 30 mmHg VD: 4.1 SG: .88 PS: LIQUID APPEAR: MILKY WHITE SPRAY
    ODOR: LEMON   PH FACTOR: NA

HAZARD RATINGS: H: F: R: 0 3 0

FIRE FIGHTING:
  HIGH EXPANSION FOAM, LOW EXPANSION FOAM, ALCOHOL FOAM; DRY CHEMICAL, CARBON DIOXIDE, WATER FOG.
  PROTECTIVE CLOTHING, RUBBER GLOVES, AND BREATHING APPARATUS.
WARNING: 1] STRUCTURAL PROTECTIVE CLOTHING IS PERMEABLE, REMAIN CLEAR OF SMOKE, WATER FALL OUT AND WATER RUN OFF.
         2] KEEP OUT OF THE REACH OF CHILDREN.
         3] CONTAINERS WILL EXPLODE IN A FIRE OR IF HEATED ABOVE 120F.
         4] REMOVE ALL IGNITION SOURCES IF WITHOUT RISK.
         5] VENTILATE ENCLOSED AREA FOR A LARGE SPILL AND EVACUATE.
         6] MOVE CONTAINERS FROM AREA IF WITHOUT RISK, COOL EXPOSED CONTAINERS.
         7] DIKE AREA FOR CONTROL AND CONTAINMENT TO PREVENT ENTRY INTO SEWERS, DRAIN, AND WATER WAYS.

LARGE SPILL/NO FIRE/RESCUE: WEAR RUBBER OR NEOPRENE BOOTS, GLOVES.

SPILL CONTROL AND CONTAINMENT:
  HOUSEHOLD SPILL: WIPE UP DRY WITH AN ABSORBENT CLOTH. RINSE OUT IN THE SINK. WIPE UP RESIDUE WITH A DAMP CLOTH.
  RINSE OUT IN THE SINK.
  LARGE SPILL: PICK UP WITH ABSORBENT. RINSE AREA WITH WATER. PICK UP WITH ABSORBENT. PUT ALL ABSORBENT IN A
  DISPOSAL CONTAINER. SEAL AND REMOVE FROM THE WORK AREA.

HEALTH HAZARD INFORMATION:
  WHEN USING THIS PRODUCT WEAR: NO SPECIAL REQUIREMENT. Eye contact will cause stinging, tearing, itching,
  swelling, and redness. Ingestion may cause nausea and vomiting.

  A physician should be contacted if anyone develops any signs or symptoms and suspects that they are caused by
  exposure to this product.

FIRST AID:
  Eye Exposure:
  Flush with water for 15 minutes while lifting the upper and lower eye lids. Contact lenses should not be worn
  when working with this product. Get medical attention.

  Skin Exposure:
  Wash with soap and water.

  Breathing:
  Not expected to be a problem.

  Swallowing:
  Do not induce vomiting. If victim is conscious, give 2-3 glasses of water to drink. Immediately contact the
  local poison control center for advice. Keep the victim warm and at rest. Get medical attention.

PRODUCT: ERA   (LIQUID LAUNDRY DETERGENT)

INCOMPATIBILITY: NONE KNOWN

INGREDIENTS: ANIONIC SURFACTANTS   TLV: CAS#:
            NONIONIC SURFACTANTS   TLV: CAS#:
            ETHYL ALCOHOL   TLV: 1900 mg/m3   CAS#:   64-17-5
            PROPYLENE GLYCOL   TLV: CAS#: 4254-15-3

IGNI TEMP: NA FP: 142F LEL: NA UEL: NA VP: NA VD: NA NA SG: NA PS: LIQUID APPEAR: CLEAR BLUE   ODOR: PERFUME
    PH FACTOR:

HAZARD RATINGS: H: F: R:

FIRE FIGHTING:
    HIGH EXPANSION FOAM, LOW EXPANSION FOAM, ALCOHOL FOAM; DRY CHEMICAL, CARBON DIOXIDE, WATER FOG.
    PROTECTIVE CLOTHING, RUBBER GLOVES, AND BREATHING APPARATUS.
WARNING: 1] STRUCTURAL PROTECTIVE CLOTHING IS PERMEABLE, REMAIN CLEAR OF SMOKE, WATER FALL OUT AND WATER RUN OFF.
         2] KEEP OUT OF THE REACH OF CHILDREN.
         3] THESE SURFACTANTS ARE MODERATELY TOXIC BY INGESTION
         4] THERMAL DECOMPOSITION MAY YIELD CARBON DIOXIDE, CARBON MONOXIDE.
         5] MOVE CONTAINERS FROM AREA IF WITHOUT RISK, COOL EXPOSED CONTAINERS.
         6] DIKE AREA FOR CONTROL AND CONTAINMENT TO PREVENT ENTRY INTO SEWERS, DRAIN, AND WATER WAYS.

LARGE SPILL/NO FIRE/RESCUE: WEAR RUBBER OR NEOPRENE BOOTS, GLOVES, AND EYE PROTECTION.

SPILL CONTROL AND CONTAINMENT:
    HOUSEHOLD SPILL: WIPE UP LIQUID WITH AN ABSORBENT CLOTH, RINSE OUT IN THE SINK. WIPE UP RESIDUE WITH A DAMP
    CLOTH AND RINSE OUT IN THE SINK.
    LARGE SPILL: COVER SPILL AREA WITH ABSORBENT. PICK UP AND PUT IN A DISPOSAL CONTAINER. RINSE AREA WITH WATER,
    PICK UP WITH ABSORBENT. PUT IN A DISPOSAL CONTAINER. SEAL AND REMOVE FROM THE WORK AREA.

HEALTH HAZARD INFORMATION:
    WHEN USING THIS PRODUCT WEAR: NO SPECIAL REQUIREMENTS. Contact with the eyes will cause stinging, burning
    sensation, tearing, itching, swelling, and redness. Ingestion will cause abdominal pain with nausea, vomiting,
    and diarrhea.

    A physician should be contacted if anyone develops any signs or symptoms and suspects that they are caused by
    exposure to this product.

FIRST AID:
    Eye Exposure:
    Flush with water for 15 minutes while lifting the upper and lower eye lids.   Contact lenses should not be worn
    when working with this product, get medical attention.

    Skin Exposure:
    Not expected to be a problem. If irritation appears, get medical attention.

    Breathing:
    Not expected to be a problem.

    Swallowing:
    If victim is conscious, give 2-3 glasses of milk or water to drink. Immediately contact the local poison control
    center for advice. Keep the victim warm and at rest, get medical attention.

PRODUCT: FAVOR

INCOMPATIBILITY: NONE KNOWN

INGREDIENTS: SILICONE   2-10%   TLV: CAS#: 63148-62-9
             ISOPARAFFINIC   Mfg recommended
             HYDROCARBON SOLVENT   10-20% TLV: 400 ppm   CAS#: 64742-48-9
             ISOBUTANE   TLV: CAS#:   75-28-5
             PROPANE   TLV: 1000 ppm   CAS#:   74-98-6
             N-BUTANE   TLV: 800 ppm   CAS#:   106-97-8

IGNI TEMP: NA FP: 20F LEL: NA UEL: NA VP: NA VD: NA SG: .9 PS: LIQUID APPEAR: SPRY MIST   ODOR: LEMON
   PH FACTOR: NA

HAZARD RATINGS: H: F: R: 1 4 0

FIRE FIGHTING:
   HIGH EXPANSION FOAM, LOW EXPANSION FOAM, ALCOHOL FOAM; DRY CHEMICAL, CARBON DIOXIDE, WATER FOG.
   PROTECTIVE CLOTHING, RUBBER GLOVES, AND BREATHING APPARATUS.
WARNING: 1] STRUCTURAL PROTECTIVE CLOTHING IS PERMEABLE, REMAIN CLEAR OF SMOKE, WATER FALL OUT AND WATER RUN OFF.
         2] KEEP OUT OF THE REACH OF CHILDREN.
         3] REMOVE ALL SOURCES OF IGNITION IF WITHOUT RISK.
         4] CONTAINERS WILL EXPLODE IN A FIRE OR IF HEATED ABOVE 120F.
         5] THERMAL DECOMPOSITION YIELDS TOXIC CARBON DIOXIDE, CARBON MONOXIDE.
         6] PROPELLANT WILL ACT AS AN ASPHYXIANT (SUFFOCATING)
         7] MOVE CONTAINERS FROM AREA IF WITHOUT RISK, COOL EXPOSED CONTAINERS.
         8] DIKE AREA FOR CONTROL AND CONTAINMENT TO PREVENT ENTRY INTO SEWERS, DRAIN, AND WATER WAYS.

LARGE SPILL/NO FIRE/RESCUE: WEAR RUBBER OR NEOPRENE BOOTS, GLOVES.

SPILL CONTROL AND CONTAINMENT:
   HOUSEHOLD SPILL: VENTILATE THE AREA. WIPE UP LIQUID WITH AN ABSORBENT CLOTH. RINSE OUT IN THE SINK. WIPE UP THE
   RESIDUE WITH A DAMP CLOTH, RINSE OUT IN THE SINK.
   LARGE SPILL: COVER LIQUID WITH ABSORBENT. SWEEP UP AND PUT INTO A DISPOSAL CONTAINER. SCRUB AREA WITH A
   DETERGENT AND WATER, PICK UP WITH ABSORBENT, PUT INTO A DISPOSAL CONTAINER. SEAL AND REMOVE FROM THE WORK AREA.

HEALTH HAZARD INFORMATION:
   WHEN USING THIS PRODUCT WEAR: NO SPECIAL REQUIREMENTS. Contact with the eyes will cause irritation. Ingestion
   may cause abdominal pain, nausea, vomiting, and diarrhea. Prolonged inhalation of spray mist may cause upper
   respiratory irritation.

   A physician should be contacted if anyone develops any signs or symptoms and suspects that they are caused by
   exposure to this product.

FIRST AID:
   Eye Exposure:
   Flush with water for 15 minutes while lifting the upper and lower eye lids. Contact lenses should not be worn
   when working with this product. Get medical attention.

   Skin Exposure:
   Wash with soap and water. If irritation appears, get medical attention.

   Breathing:
   Not expected to be a problem. If problems occur, move victim to fresh air. if needed get medical attention.

   Swallowing:
   Do not induce vomiting. Immediately contact the local poison control center for advice. Keep the victim warm and
   at rest. Get medical attention.

PRODUCT: FAVOR, LIQUID LEMON

INCOMPATIBILITY: NONE KNOWN

INGREDIENTS: ISOPARAFFINIC   Recommended
             HYDROCARBON SOLVENT   1%   TLV: 400 ppm   CAS#: 64742-48-9
             WATER   90-97%   TLV: CAS#: 7732-18-5
             FRAGRANCE   1%   TLV: CAS#:
             SILICONE_____   TLV: CAS#:
             WAXES_____   2-5% TLV: CAS#:
             SURFACTANTS_____/   TLV: CAS#:

IGNI TEMP: NA FP: NA LEL: NA UEL: NA VP: NA VD: NA SG: 1 PS: LIQUID APPEAR: WHITE OPAQUE
      ODOR: LEMON   PH FACTOR: 6.9

HAZARD RATINGS: H: F: R:

FIRE FIGHTING:
  HIGH EXPANSION FOAM, LOW EXPANSION FOAM, ALCOHOL FOAM; DRY CHEMICAL, CARBON DIOXIDE, WATER FOG.
  PROTECTIVE CLOTHING, RUBBER GLOVES, AND BREATHING APPARATUS.
WARNING: 1] STRUCTURAL PROTECTIVE CLOTHING IS PERMEABLE, REMAIN CLEAR OF SMOKE, WATER FALL OUT AND WATER RUN OFF.
         2] KEEP OUT OF THE REACH OF CHILDREN.
         3] THERMAL DECOMPOSITION MAY YIELD CARBON DIOXIDE, CARBON MONOXIDE.
         4] MOVE CONTAINERS FROM AREA IF WITHOUT RISK, COOL EXPOSED CONTAINERS.
         5] DIKE AREA FOR CONTROL AND CONTAINMENT TO PREVENT ENTRY INTO SEWERS, DRAIN, AND WATER WAYS.

LARGE SPILL/NO FIRE/RESCUE: WEAR RUBBER OR NEOPRENE BOOTS, GLOVES.

SPILL CONTROL AND CONTAINMENT:
  HOUSEHOLD SPILL: WIPE UP WITH AN ABSORBENT CLOTH, RINSE OUT IN THE SINK. WIPE UP RESIDUE WITH W DAMP CLOTH.
  LARGE SPILL: PICK UP WITH ABSORBENT, PUT IN A DISPOSAL CONTAINER. RINSE AREA WITH WATER, PICK UP WITH ABSORBENT.
  PUT IN A DISPOSAL CONTAINER, SEAL AND REMOVE FROM THE WORK AREA.

HEALTH HAZARD INFORMATION:
  WHEN USING THIS PRODUCT WEAR: NO SPECIAL REQUIREMENT. Eye contact may cause stinging, tearing, itching,
  swelling, and redness. Ingestion may cause abdominal pain, nausea, vomiting, and diarrhea.

  A physician should be contacted if anyone develops any signs or symptoms and suspects that they are caused by
  exposure to this product.

FIRST AID:
Eye Exposure:
Flush with water for 15 minutes while lifting the upper and lower eye lids. Contact lenses should not be worn
when working with this product. Get medical attention.

  Skin Exposure:
Wash skin with water. If irritation develops, get medical attention.

  Breathing:
Not expected to be a problem.

  Swallowing:
Immediately contact the local poison control center for advice. Keep the victim warm and at rest. Get medical
attention.

PRODUCT: FIBER ALL

INCOMPATIBILITY: NONE KNOWN

INGREDIENTS: ALL ARE NON-HAZARDOUS BY OSHA 1910.1200 STANDARD

IGNI TEMP: NA FP: NA LEL: NA UEL: NA VP: NA VD: NA SG: NA PS: POWDER APPEAR: OFF-WHITE ODOR: NA
   PH FACTOR: NA

HAZARD RATINGS: H: F: R:

FIRE FIGHTING:
   HIGH EXPANSION FOAM, LOW EXPANSION FOAM, ALCOHOL FOAM; DRY CHEMICAL, CARBON DIOXIDE, WATER FOG.
   PROTECTIVE CLOTHING, RUBBER GLOVES, AND BREATHING APPARATUS.
WARNING: 1] STRUCTURAL PROTECTIVE CLOTHING IS PERMEABLE, REMAIN CLEAR OF SMOKE, WATER FALL OUT AND WATER RUN OFF.
         2] KEEP OUT OF THE REACH OF CHILDREN.
         3] THERMAL DECOMPOSITION MAY YIELD TOXIC CARBON DIOXIDE, CARBON MONOXIDE.
         4] MOVE CONTAINERS FROM AREA IF WITHOUT RISK, COOL EXPOSED CONTAINERS.
         5] DIKE AREA FOR CONTROL AND CONTAINMENT TO PREVENT ENTRY INTO SEWERS, DRAIN, AND WATER WAYS.

LARGE SPILL/NO FIRE/RESCUE: WEAR RUBBER OR NEOPRENE BOOTS, GLOVES.

SPILL CONTROL AND CONTAINMENT:
   HOUSEHOLD SPILL: WIPE UP POWDER WITH A DAMP CLOTH, RINSE OUT IN THE SINK.
   LARGE SPILL: AVOID CREATING A DUST. GENTLY SWEEP UP POWDER AND PUT INTO A DISPOSAL CONTAINER. RINSE AREA WITH
   WATER, PICK UP WITH ABSORBENT, PUT IN A DISPOSAL CONTAINER. SEAL AND REMOVE FROM THE WORK AREA.

HEALTH HAZARD INFORMATION:
   WHEN USING THIS PRODUCT WEAR: NO SPECIAL REQUIREMENTS. Powder or dust may cause eye irritation. Dust may cause
   upper respiratory tract irritation.

   A physician should be contacted if anyone develops any signs or symptoms and suspects that they are caused by
   exposure to this product.

FIRST AID:
   Eye Exposure:
   Flush with water for 15 minutes while lifting the upper and lower eye lids. Contact lenses should not be worn
   when working with this product. If irritation persists, get medical attention.

   Skin Exposure:
   Not expected to be a problem.

   Breathing:
   Not expected to be a problem.

   Swallowing:
   Immediately contact the local poison control center for advice. Keep the victim warm and at rest. Get medical
   attention.

PRODUCT: FIBER ALL WAFERS

INCOMPATIBILITY: NONE KNOWN

INGREDIENTS: ALL ARE NON-HAZARDOUS BY OSHA 1910.1200 STANDARD

IGNI TEMP: NA FP: NA LEL: NA UEL: NA VP: NA VD: NA SG: NA PS: POWDER APPEAR: OFF-WHITE ODOR: NA
   PH FACTOR: NA

HAZARD RATINGS: H: F: R:

FIRE FIGHTING:
   HIGH EXPANSION FOAM, LOW EXPANSION FOAM, ALCOHOL FOAM; DRY CHEMICAL, CARBON DIOXIDE, WATER FOG.
   PROTECTIVE CLOTHING, RUBBER GLOVES, AND BREATHING APPARATUS.
WARNING: 1] STRUCTURAL PROTECTIVE CLOTHING IS PERMEABLE, REMAIN CLEAR OF SMOKE, WATER FALL OUT AND WATER RUN OFF.
         2] KEEP OUT OF THE REACH OF CHILDREN.
         3] THERMAL DECOMPOSITION MAY YIELD TOXIC CARBON DIOXIDE, CARBON MONOXIDE.
         4] MOVE CONTAINERS FROM AREA IF WITHOUT RISK, COOL EXPOSED CONTAINERS.
         5] DIKE AREA FOR CONTROL AND CONTAINMENT TO PREVENT ENTRY INTO SEWERS, DRAIN, AND WATER WAYS.

LARGE SPILL/NO FIRE/RESCUE: WEAR RUBBER OR NEOPRENE BOOTS, GLOVES.

SPILL CONTROL AND CONTAINMENT:
   HOUSEHOLD SPILL: WIPE UP POWDER WITH A DAMP CLOTH, RINSE OUT IN THE SINK.
   LARGE SPILL: AVOID CREATING A DUST. GENTLY SWEEP UP POWDER AND PUT INTO A DISPOSAL CONTAINER. RINSE AREA WITH
   WATER, PICK UP WITH ABSORBENT, PUT IN A DISPOSAL CONTAINER. SEAL AND REMOVE FROM THE WORK AREA.

HEALTH HAZARD INFORMATION:
   WHEN USING THIS PRODUCT WEAR: NO SPECIAL REQUIREMENTS. Powder or dust may cause eye irritation. Dust may cause
   upper respiratory tract irritation.

   A physician should be contacted if anyone develops any signs or symptoms and suspects that they are caused by
   exposure to this product.

FIRST AID:
   Eye Exposure:
   Flush with water for 15 minutes while lifting the upper and lower eye lids. Contact lenses should not be worn
   when working with this product. If irritation persists, get medical attention.

   Skin Exposure:
   Not expected to be a problem.

   Breathing:
   Not expected to be a problem.

   Swallowing:
   Immediately contact the local poison control center for advice. Keep the victim warm and at rest. Get medical
   attention.

PRODUCT: FIBER ALL BULK LAXATIVE TABLET

INCOMPATIBILITY: NONE KNOWN

INGREDIENTS: ALL ARE NON-HAZARDOUS BY OSHA 1910.1200 STANDARD

IGNI TEMP: NA FP: NA LEL: NA UEL: NA VP: NA VD: NA SG: NA PS: SOLID APPEAR: TABLET   ODOR: NA
   PH FACTOR: NA

HAZARD RATINGS: H: F: R: 0 0 0

FIRE FIGHTING:
   HIGH EXPANSION FOAM, LOW EXPANSION FOAM, ALCOHOL FOAM; DRY CHEMICAL, CARBON DIOXIDE, WATER FOG.
   PROTECTIVE CLOTHING, RUBBER GLOVES, AND BREATHING APPARATUS.
WARNING: 1] STRUCTURAL PROTECTIVE CLOTHING IS PERMEABLE, REMAIN CLEAR OF SMOKE, WATER FALL OUT AND WATER RUN OFF.
         2] KEEP OUT OF THE REACH OF CHILDREN.
         3] THERMAL DECOMPOSITION MAY YIELD TOXIC CARBON DIOXIDE, CARBON MONOXIDE.
         4] MOVE CONTAINERS FROM AREA IF WITHOUT RISK, COOL EXPOSED CONTAINERS.
         5] DIKE AREA FOR CONTROL AND CONTAINMENT TO PREVENT ENTRY INTO SEWERS, DRAIN, AND WATER WAYS.

LARGE SPILL/NO FIRE/RESCUE: WEAR RUBBER OR NEOPRENE BOOTS, GLOVES.

SPILL CONTROL AND CONTAINMENT:
   HOUSEHOLD SPILL: PICK UP TABLETS, WIPE UP RESIDUE WITH A DAMP CLOTH. RINSE OUT IN THE SINK.
   LARGE SPILL: AVOID CREATING A DUST. SCOOP UP TABLETS AND PUT INTO A DISPOSAL CONTAINER. RINSE AREA WITH WATER,
   PICK UP WITH ABSORBENT, PUT INTO A DISPOSAL CONTAINER. SEAL AND REMOVE FROM THE WORK AREA.

HEALTH HAZARD INFORMATION:
   WHEN USING THIS PRODUCT WEAR: NO SPECIAL REQUIREMENTS. Ingestion may cause abdominal cramps, and diarrhea.
   Overdose may lead to excessive fluid loss and dehydration. Dust may cause eye irritation.

   A physician should be contacted if anyone develops any signs or symptoms and suspects that they are caused by
   exposure to this product.

FIRST AID:
   Eye Exposure:
   Flush with water for 15 minutes while lifting the upper and lower eye lids. Contact lenses should not be worn
   when working with this product. Get medical attention.

   Skin Exposure:
   Not expected to be a problem.

   Breathing:
   Not expected to be a problem.

   Swallowing:
   If overdose is suspected or if dehydration takes place, immediately contact the local poison control center for
   advice. Keep the victim warm and at rest. Get medical attention.

PRODUCT: FORMULA 409 ALL PURPOSE CLEANER

INCOMPATIBILITY: NONE KNOWN

INGREDIENTS: ETHYLENE GLYCOL MONOBUTYL ETHER .5-5% TLV: 25 ppm CAS#: 111-76-2

IGNI TEMP: NA FP: NA LEL: NA UEL: NA VP: NA VD: NA SG: 1.02 PS: LIQUID APPEAR: GREEN ODOR:
    PH FACTOR: 12.4

HAZARD RATINGS: H: F: R: 1 0 0

FIRE FIGHTING:
    HIGH EXPANSION FOAM, LOW EXPANSION FOAM, ALCOHOL FOAM; DRY CHEMICAL, CARBON DIOXIDE, WATER FOG.
    PROTECTIVE CLOTHING, RUBBER GLOVES, AND BREATHING APPARATUS.
WARNING: 1] STRUCTURAL PROTECTIVE CLOTHING IS PERMEABLE, REMAIN CLEAR OF SMOKE, WATER FALL OUT AND WATER RUN OFF.
         2] KEEP OUT OF THE REACH OF CHILDREN.
         3] ABSORPTION THROUGH THE SKIN OR PROLONGED INHALATION MAY CAUSE EXPOSURE LIMIT TO BE EXCEEDED.
         4] WEAR SAFETY GLASSES AND RUBBER OR NEOPRENE GLOVES FOR PROLONGED EXPOSURE.
         5] MOVE CONTAINERS FROM AREA IF WITHOUT RISK, COOL EXPOSED CONTAINERS.
         6] DIKE AREA FOR CONTROL AND CONTAINMENT TO PREVENT ENTRY INTO SEWERS, DRAIN, AND WATER WAYS.

LARGE SPILL/NO FIRE/RESCUE: WEAR RUBBER OR NEOPRENE BOOTS, GLOVES.

SPILL CONTROL AND CONTAINMENT:
    HOUSEHOLD SPILL: WIPE UP LIQUID WITH ABSORBENT CLOTH. RINSE OUT IN THE SINK.
    LARGE SPILL: COVER SPILL WITH ABSORBENT. SWEEP UP AND PUT INTO A DISPOSAL CONTAINER. RINSE AREA WITH WATER. PICK
    UP WITH ABSORBENT, PUT INTO A DISPOSAL CONTAINER. SEAL AND REMOVE FROM THE WORK AREA.

HEALTH HAZARD INFORMATION:
    WHEN USING THIS PRODUCT WEAR: Contact may cause eye irritation. Prolonged exposure may cause a skin rash as well
    as overexposure. Inhalation may cause nasal irritation. Ingestion may cause abdominal pain, nausea, vomiting,
    and diarrhea.

    A physician should be contacted if anyone develops any signs or symptoms and suspects that they are caused by
    exposure to this product.

FIRST AID:
Eye Exposure:
Flush with water for 15 minutes while lifting the upper and lower eye lids. Contact lenses should not be worn
when working with this product. If irritation persists, get medical attention.

Skin Exposure:
Wash with soap and water. If irritation appears, get medical attention.

Breathing:
If nasal irritation occurs, move to fresh air. If irritation persists, get medical attention.

Swallowing:
Immediately contact the local poison control center for advice. Keep the victim warm and at rest. Get medical
attention.

PRODUCT: FRESH STEP (CAT LITTER)

INCOMPATIBILITY: NONE KNOWN

INGREDIENTS: CLAY   75%   TLV: 10 mg/m3(DUST) CAS#: 8031-18-3
             SILICA   LESS THEN 3%   TLV: .1 mg/m3(DUST) CAS#: 14808-60-7

IGNI TEMP: NA FP: NA LEL: NA UEL: NA VP: NA VD: NA SG: NA PS: GRANULES APPEAR: WHITE ODOR:
  PH FACTOR: NA

HAZARD RATINGS: H: F: R: 1 0 0

FIRE FIGHTING:
  HIGH EXPANSION FOAM, LOW EXPANSION FOAM, ALCOHOL FOAM; DRY CHEMICAL, CARBON DIOXIDE, WATER FOG.
  PROTECTIVE CLOTHING, RUBBER GLOVES, AND BREATHING APPARATUS.
WARNING: 1] STRUCTURAL PROTECTIVE CLOTHING IS PERMEABLE, REMAIN CLEAR OF SMOKE, WATER FALL OUT AND WATER RUN OFF.
         2] KEEP OUT OF THE REACH OF CHILDREN.
         3] SILICA (QUARTZ) MAY BE CARCINOGENIC.
         4] AVOID INHALATION AND CONTACT WITH DUST.
         5] MOVE CONTAINERS FROM AREA IF WITHOUT RISK, COOL EXPOSED CONTAINERS.
         6] DIKE AREA FOR CONTROL AND CONTAINMENT TO PREVENT ENTRY INTO SEWERS, DRAIN, AND WATER WAYS.

LARGE SPILL/NO FIRE/RESCUE: WEAR RUBBER OR NEOPRENE BOOTS, GLOVES, DUST MASK.

SPILL CONTROL AND CONTAINMENT:
  HOUSEHOLD SPILL: AVOID CREATING A DUST. SWEEP UP MATERIAL AND PLACE IN A DISPOSAL CONTAINER. WIPE UP RESIDUE
  WITH A DAMP CLOTH, RINSE OUT IN THE SINK.
  LARGE SPILL: AVOID CREATING A DUST BY APPLYING A WATER SPRAY OVER SPILL AREA. SCOOP UP MATERIAL AND PUT INTO A
  DISPOSAL CONTAINER. DAMP MOP AREA TO PICK UP RESIDUE.

HEALTH HAZARD INFORMATION:
  WHEN USING THIS PRODUCT: AVOID BREATHING DUST. Inhalation of dust may cause irritation of the nose and throat.
  Prolonged or repeated exposure may cause coughing, shortness of breath, scaring of the lungs, and cancer. Under
  normal household usage and conditions, the likelihood of adverse health effects is low.

  A physician should be contacted if anyone develops any signs or symptoms and suspects that they are caused by
  exposure to this product.

FIRST AID:
  Eye Exposure:
  Flush with water for 15 minutes while lifting the upper and lower eye lids. Contact lenses should not be worn
  when working with this product. If irritation persists, get medical attention.

  Skin Exposure:
  Wash exposed skin with soap and water.

  Breathing:
  Move the victim to fresh air. Treat symptoms as they appear. If required, get medical attention.

  Swallowing:
  Not an apparent hazard. If ingested by a child, immediately contact the local poison control center for advice.
  Keep the victim warm and at rest. Get medical attention.

PRODUCT: FUTURE

INCOMPATIBILITY: NONE KNOWN.

INGREDIENTS: STYRENE/ACRYLIC COPOLYMERS   5-15% TLV: CAS#:
             DIETHYLENEGLYCOL MONOETHYLETHER 3-5%   TLV: 30 ppm CAS#: 110-90-0
             TRIBUTOXY ETHYL PHOSPHATE   1-3%   TLV: CAS#: 78-51-3
             WATER   75-85% TLV: CAS#: 7732-18-5

IGNI TEMP: NA FP: NA LEL: NA UEL: NA VP:760 mmHg @ 212F VD: NA SG: 1.02 PS: LIQUID   APPEAR: COLORLESS
     ODOR: MILD   PH FACTOR: 8.5

HAZARD RATINGS: H: F: R: 1 0 0

FIRE FIGHTING:
  HIGH EXPANSION FOAM, LOW EXPANSION FOAM, ALCOHOL FOAM; DRY CHEMICAL, CARBON DIOXIDE, WATER FOG.
  PROTECTIVE CLOTHING, RUBBER GLOVES, AND BREATHING APPARATUS.
WARNING: 1] STRUCTURAL PROTECTIVE CLOTHING IS PERMEABLE, REMAIN CLEAR OF SMOKE, WATER FALL OUT AND WATER RUN OFF.
         2] KEEP OUT OF THE REACH OF CHILDREN.
         3] THERMAL DECOMPOSITION MAY YIELD TOXIC CARBON DIOXIDE, CARBON MONOXIDE.
         4] MOVE CONTAINERS FROM AREA IF WITHOUT RISK, COOL EXPOSED CONTAINERS.
         5] DIKE AREA FOR CONTROL AND CONTAINMENT TO PREVENT ENTRY INTO SEWERS, DRAIN, AND WATER WAYS.

LARGE SPILL/NO FIRE/RESCUE: WEAR RUBBER OR NEOPRENE BOOTS, GLOVES.

SPILL CONTROL AND CONTAINMENT:
  HOUSEHOLD SPILL: WIPE UP LIQUID WITH AN ABSORBENT CLOTH. PUT IN A DISPOSAL CONTAINER. WASH SPILL AREA WITH SOAP
  AND WATER, WIPE UP WITH AN ABSORBENT CLOTH. PUT INTO A DISPOSAL CONTAINER.
  LARGE SPILL: COVER SPILL WITH ABSORBENT. PICK UP AND PUT INTO A DISPOSAL CONTAINER. SCRUB AREA WITH A DETERGENT
  AND WATER. PICK UP WITH ABSORBENT. PUT IN A DISPOSAL CONTAINER. SEAL AND REMOVE FROM THE WORK AREA.

HEALTH HAZARD INFORMATION:
  WHEN USING THIS PRODUCT WEAR: NO SPECIAL REQUIREMENTS. Eye contact will cause irritation. Ingestion may cause
  abdominal pain, nausea, vomiting, and diarrhea.

  A physician should be contacted if anyone develops any signs or symptoms and suspects that they are caused by
  exposure to this product.

FIRST AID:
  Eye Exposure:
  Flush with water for 15 minutes   while lifting the upper and lower eye lids. Contact lenses should not be worn
  when working with this product. Get medical attention.

  Skin Exposure:
  Rinse skin with water, wash with soap and water.

  Breathing:
  Not expected to be a problem.

  Swallowing:
  Immediately contact the local poison control center for advice. Keep the victim warm and at rest. Get medical
  attention.

PRODUCT: GAIN   (LAUNDRY GRANULES)

INCOMPATIBILITY: NONE KNOWN

INGREDIENTS: ANIONIC SURFACTANTS   TLV: CAS#:
             COMPLEX SODIUM PHOSPHATE   TLV: CAS#:
         OR ALUMINOSILICATES   TLV: CAS#:
             SODIUM CARBONATE   TLV: CAS#: 497-19-8
             SODIUM SILICATE   TLV: CAS#: 1344-09-8
             SODIUM SULFATE   TLV: CAS#: 7757-82-6

IGNI TEMP: NA FP: NA LEL: NA UEL: NA VP: NA VD: NA SG: NA PS: POWDER APPEAR: GREEN ODOR: PERFUME
   PH FACTOR:

HAZARD RATINGS: H: F: R:

FIRE FIGHTING:
   HIGH EXPANSION FOAM, LOW EXPANSION FOAM, ALCOHOL FOAM; DRY CHEMICAL, CARBON DIOXIDE, WATER FOG.
   PROTECTIVE CLOTHING, RUBBER GLOVES, AND BREATHING APPARATUS.
WARNING: 1] STRUCTURAL PROTECTIVE CLOTHING IS PERMEABLE, REMAIN CLEAR OF SMOKE, WATER FALL OUT AND WATER RUN OFF.
         2] KEEP OUT OF THE REACH OF CHILDREN.
         3] IF THE BAR CODE NUMBER ON TOP OR BOTTOM BEGINS WITH THE LETTER "L", PRODUCT CONTAINS SODIUM
            PYROPHOSPHATE TLV: 5 mg/m3 CAS#: 7722-88-5.
         4] ANIONIC SURFACTANTS ARE MILDLY TOXIC BY INGESTION.
         5] MOVE CONTAINERS FROM AREA IF WITHOUT RISK, COOL EXPOSED CONTAINERS.
         6] DIKE AREA FOR CONTROL AND CONTAINMENT TO PREVENT ENTRY INTO SEWERS, DRAIN, AND WATER WAYS.

LARGE SPILL/NO FIRE/RESCUE: WEAR RUBBER OR NEOPRENE BOOTS, GLOVES.

SPILL CONTROL AND CONTAINMENT:
   HOUSEHOLD SPILL: AVOID CREATING A DUST. SWEEP UP POWDER AND PUT IN A DISPOSAL CONTAINER. WIPE UP RESIDUE WITH A
   DAMP CLOTH, RINSE OUT IN THE SINK.
   LARGE SPILL: AVOID CREATING A DUST. SCOOP UP POWDER, PUT IN A DISPOSAL CONTAINER. RINSE AREA WITH WATER, PICK UP
   WITH ABSORBENT AND PUT IN A DISPOSAL CONTAINER. SEAL AND REMOVE FROM THE WORK AREA.

HEALTH HAZARD INFORMATION:
   WHEN USING THIS PRODUCT WEAR: NO SPECIAL REQUIREMENTS. Powder can cause eye stinging, burning sensation,
   tearing, itching, swelling, and redness. Inhalation of dust can cause coughing, sore throat, wheezing, and
   shortness of breath. Ingestion may cause abdominal pain, nausea, vomiting, and diarrhea.

A physician should be contacted if anyone develops any signs or symptoms and suspects that they are caused by
exposure to this product.

FIRST AID:
   Eye Exposure:
   Flush with water for 15 minutes while lifting the upper and lower eye lids. Contact lenses should not be worn
   when working with this product, get medical attention.

   Skin Exposure:
   Wash with water. If irritation appears, get medical attention.

   Breathing:
   Move the victim to fresh air. Treat any symptoms as they appear. If needed get medical attention.

   Swallowing:
   If the victim is conscious, give 2-3 glasses of milk or water to drink. Immediately contact the local poison
   control center for advice. Keep the victim warm and at rest, get medical attention.

PRODUCT: GAIN, ULTRA (LAUNDRY GRANULES)

INCOMPATIBILITY: NONE KNOWN

INGREDIENTS: SODIUM SULFATE   TLV: CAS#: 7757-82-6
             SODIUM CARBONATE   TLV: CAS#:   497-19-8
             ANIONIC SURFACTANT   TLV: CAS#:
             ALUMINOSILICATE   TLV: CAS#:
             PERFUME   TLV: CAS#:

IGNI TEMP: NA FP: NA LEL: NA UEL: NA VP: NA VD: NA SG: NA PS: POWDER APPEAR: WHITE - GREEN SPECKLES
      ODOR: PERFUMED   PH FACTOR: NA

HAZARD RATINGS: H: F: R:

FIRE FIGHTING:
   HIGH EXPANSION FOAM, LOW EXPANSION FOAM, ALCOHOL FOAM; DRY CHEMICAL, CARBON DIOXIDE, WATER FOG.
   PROTECTIVE CLOTHING, RUBBER GLOVES, AND BREATHING APPARATUS.
WARNING: 1] STRUCTURAL PROTECTIVE CLOTHING IS PERMEABLE, REMAIN CLEAR OF SMOKE, WATER FALL OUT AND WATER RUN OFF.
         2] KEEP OUT OF THE REACH OF CHILDREN.
         3] THERMAL DECOMPOSITION MAY YIELD CARBON DIOXIDE, CARBON MONOXIDE.
         4] MOVE CONTAINERS FROM AREA IF WITHOUT RISK, COOL EXPOSED CONTAINERS.
         5] DIKE AREA FOR CONTROL AND CONTAINMENT TO PREVENT ENTRY INTO SEWERS, DRAIN, AND WATER WAYS.

LARGE SPILL/NO FIRE/RESCUE: WEAR RUBBER OR NEOPRENE BOOTS, GLOVES.

SPILL CONTROL AND CONTAINMENT:
   HOUSEHOLD SPILL: SWEEP UP POWDER. PUT IN A DISPOSAL CONTAINER. DAMP MOP UP RESIDUE. RINSE OUT IN THE SINK.
   LARGE SPILL: AVOID CREATING A DUST. SCOOP UP POWDER. PUT IN A DISPOSAL CONTAINER. RINSE AREA WITH WATER. PICK UP
   WITH ABSORBENT. PUT IN A DISPOSAL CONTAINER. SEAL AND REMOVE FROM THE WORK AREA.

HEALTH HAZARD INFORMATION:
   WHEN USING THIS PRODUCT WEAR: NO SPECIAL REQUIREMENT. Inhalation may cause coughing, sore throat, wheezing, and
   shortness of breath. Eye contact will cause stinging, tearing, itching, swelling, and redness. Ingestion may
   result in nausea, vomiting, and diarrhea.

   A physician should be contacted if anyone develops any signs or symptoms and suspects that they are caused by
   exposure to this product.

FIRST AID:
Eye Exposure:
Flush with water for 15 minutes while lifting the upper and lower eye lids. Contact lenses should not be worn
when working with this product. Get medical attention.

Skin Exposure:
Wash with water. If irritation develops, get medical attention.

Breathing:
If irritation develops, move victim to fresh air.

Swallowing:
If the victim is conscious, give 2-3 glasses of milk or water to drink. Immediately contact the local poison
control center for advice. Keep the victim warm and at rest. Get medical attention.

PRODUCT: GLADE   APRIL MEADOWS WITH CHLOROPHYLL

INCOMPATIBILITY: NONE KNOWN

INGREDIENTS: PROPANE   5-10%   TLV: 1000 ppm   CAS#:   74-98-6
             ISOBUTANE   5-10%   TLV: CAS#:   75-28-5
             N-BUTANE   5-10%   TLV: 800 ppm   CAS#: 106-97-8
             WATER   65-75%   TLV: CAS#: 7732-18-5

IGNI TEMP: NA FP: 20F LEL: NA UEL: NA VP: NA VD: NA SG: .8 PS: LIQUID APPEAR: SPRAY MIST ODOR: FRAGRANCE
   PH FACTOR: NA

HAZARD RATINGS: H: F: R: 0 4 0

FIRE FIGHTING:
   HIGH EXPANSION FOAM, LOW EXPANSION FOAM, ALCOHOL FOAM; DRY CHEMICAL, CARBON DIOXIDE, WATER FOG.
   PROTECTIVE CLOTHING, RUBBER GLOVES, AND BREATHING APPARATUS.
WARNING: 1] STRUCTURAL PROTECTIVE CLOTHING IS PERMEABLE, REMAIN CLEAR OF SMOKE, WATER FALL OUT AND WATER RUN OFF.
         2] KEEP OUT OF THE REACH OF CHILDREN.
         3] CONTAINERS WILL EXPLODE IN A FIRE OR IF HEATED ABOVE 120F.
         4] REMOVE ALL SOURCES OF IGNITION IF WITHOUT RISK.
         5] PROPELLANT WILL ACT AS AN ASPHYXIANT.
         6] THERMAL DECOMPOSITION YIELDS TOXIC CARBON DIOXIDE, CARBON MONOXIDE.
         7] MOVE CONTAINERS FROM AREA IF WITHOUT RISK, COOL EXPOSED CONTAINERS.
         8] DIKE AREA FOR CONTROL AND CONTAINMENT TO PREVENT ENTRY INTO SEWERS, DRAIN, AND WATER WAYS.

LARGE SPILL/NO FIRE/RESCUE: WEAR RUBBER OR NEOPRENE BOOTS, GLOVES.

SPILL CONTROL AND CONTAINMENT:
   HOUSEHOLD SPILL: VENTILATE THE IMMEDIATE AREA. WIPE UP LIQUID WITH ABSORBENT CLOTH, RINSE OUT IN THE SINK. WIPE
   UP RESIDUE WITH A DAMP CLOTH. RINSE OUT IN THE SINK.
   LARGE SPILL: COVER LIQUID WITH ABSORBENT. SWEEP UP AND PUT INTO A DISPOSAL CONTAINER. RINSE AREA WITH WATER,
   PICK UP WITH ABSORBENT, PUT IN A DISPOSAL CONTAINER. SEAL AND REMOVE FROM THE WORK AREA.

HEALTH HAZARD INFORMATION:
   WHEN USING THIS PRODUCT WEAR: NO SPECIAL REQUIREMENT. Contact with the eyes may cause irritation. Prolonged
   inhalation of spray mist may cause upper respiratory irritation.

   A physician should be contacted if anyone develops any signs or symptoms and suspects that they are caused by
   exposure to this product.

FIRST AID:
   Eye Exposure:
   Flush with water for 15 minutes while lifting the upper and lower eye lids. Contact lenses should not be worn
   when working with this product. If irritation persists, get medical attention.

   Skin Exposure:
   Wash skin with soap and water.

   Breathing:
   Move the victim to fresh air. Treat symptoms as they appear.

   Swallowing:
   Immediately contact the local poison control center for advice. Keep the victim warm and at rest. Get medical
   attention.

PRODUCT: GLADE DILUTED WATER INTERMEDIATE

INCOMPATIBILITY: NONE KNOWN

INGREDIENTS: SODIUM HYDROXIDE LESS THEN 1% TLV: 2 mg/m3   CAS#: 1310-73-2
             POTASSIUM PHOSPHATE   1-3% TLV: 5 mg/m3(DUST) CAS#:

IGNI TEMP: NA FP: NA LEL: NA UEL: NA VP:760 mmHg @ 212F VD: NA SG: 1.0 PS: LIQUID   APPEAR: WATER WHITE
   ODOR: BLAND   PH FACTOR: 8.0 to 8.5

HAZARD RATINGS: H: F: R: 1 0 0

NEUTRALIZING AGENTS:

FIRE FIGHTING:
   HIGH EXPANSION FOAM, LOW EXPANSION FOAM, ALCOHOL FOAM; DRY CHEMICAL, CARBON DIOXIDE, WATER FOG.
   PROTECTIVE CLOTHING, RUBBER GLOVES, AND BREATHING APPARATUS.
WARNING: 1] STRUCTURAL PROTECTIVE CLOTHING IS PERMEABLE, REMAIN CLEAR OF SMOKE, WATER FALL OUT AND WATER RUN OFF.
         2] KEEP OUT OF THE REACH OF CHILDREN.
         3] THERMAL DECOMPOSITION MAY YIELD TOXIC CARBON DIOXIDE, CARBON MONOXIDE.
         4] MOVE CONTAINERS FROM AREA IF WITHOUT RISK, COOL EXPOSED CONTAINERS.
         5] DIKE AREA FOR CONTROL AND CONTAINMENT TO PREVENT ENTRY INTO SEWERS, DRAIN, AND WATER WAYS.

LARGE SPILL/NO FIRE/RESCUE: WEAR RUBBER OR NEOPRENE BOOTS, GLOVES.

SPILL CONTROL AND CONTAINMENT:
   HOUSEHOLD SPILL: WIPE UP LIQUID WITH ABSORBENT CLOTH, RINSE OUT IN THE SINK. WIPE UP RESIDUE WITH A DAMP CLOTH.
   RINSE OUT IN THE SINK.
   LARGE SPILL: COVER LIQUID WITH ABSORBENT. PICK UP AND PUT INTO A DISPOSAL CONTAINER. RINSE AREA WITH WATER, PICK
   UP WITH ABSORBENT, PUT INTO A DISPOSAL CONTAINERS. SEAL AND REMOVE FROM THE WORK AREA.

HEALTH HAZARD INFORMATION:
   WHEN USING THIS PRODUCT WEAR: NO SPECIAL REQUIREMENT. Contact with the eyes may cause irritation. Prolonged or
   repeated contact with skin may cause irritation. Ingestion may cause irritation of the mouth, throat, and
   stomach. May cause nausea.

   A physician should be contacted if anyone develops any signs or symptoms and suspects that they are caused by
   exposure to this product.

FIRST AID:
   Eye Exposure:
   Flush with water for 15 minutes while lifting the upper and lower eye lids. Contact lenses should not be worn
   when working with this product. If irritation persists, get medical attention.

   Skin Exposure:
   Flush exposed skin, wash with soap and water.

   Breathing:
   Not expected to be a problem.

   Swallowing:
   Do not induce vomiting. Immediately contact the local poison control center for advice. Keep the victim warm and
   at rest. Get medical attention.

PRODUCT: GLADE - EARLY SPRING WITH CHLOROPHYLL

INCOMPATIBILITY: NONE KNOWN

INGREDIENTS:				
PROPANE	5-10%	TLV: 1000 ppm	CAS#:	74-98-6
ISOBUTANE	5-10%	TLV:	CAS#:	75-28-5
N-BUTANE	5-10%	TLV: 800 ppm	CAS#:	106-97-8
WATER	65-75%	TLV:	CAS#:	7732-18-5

IGNI TEMP: NA FP: 20F LEL: NA UEL: NA VP: NA VD: NA SG: .8 PS: LIQUID APPEAR: SPRAY MIST ODOR: FRAGRANCE
PH FACTOR: NA

HAZARD RATINGS: H: F: R: 0 4 0

FIRE FIGHTING:
   HIGH EXPANSION FOAM, LOW EXPANSION FOAM, ALCOHOL FOAM; DRY CHEMICAL, CARBON DIOXIDE, WATER FOG.
   PROTECTIVE CLOTHING, RUBBER GLOVES, AND BREATHING APPARATUS.
WARNING: 1] STRUCTURAL PROTECTIVE CLOTHING IS PERMEABLE, REMAIN CLEAR OF SMOKE, WATER FALL OUT AND WATER RUN OFF.
         2] KEEP OUT OF THE REACH OF CHILDREN.
         3] CONTAINERS WILL EXPLODE IN A FIRE OR IF HEATED ABOVE 120F.
         4] REMOVE ALL SOURCES OF IGNITION IF WITHOUT RISK.
         5] PROPELLANT WILL ACT AS AN ASPHYXIANT.
         6] THERMAL DECOMPOSITION YIELDS TOXIC CARBON DIOXIDE, CARBON MONOXIDE.
         7] MOVE CONTAINERS FROM AREA IF WITHOUT RISK, COOL EXPOSED CONTAINERS.
         8] DIKE AREA FOR CONTROL AND CONTAINMENT TO PREVENT ENTRY INTO SEWERS, DRAIN, AND WATER WAYS.

LARGE SPILL/NO FIRE/RESCUE: WEAR RUBBER OR NEOPRENE BOOTS, GLOVES.

SPILL CONTROL AND CONTAINMENT:
   HOUSEHOLD SPILL: VENTILATE THE IMMEDIATE AREA. WIPE UP LIQUID WITH ABSORBENT CLOTH, RINSE OUT IN THE SINK. WIPE
   UP RESIDUE WITH A DAMP CLOTH. RINSE OUT IN THE SINK.
   LARGE SPILL: COVER LIQUID WITH ABSORBENT. SWEEP UP AND PUT INTO A DISPOSAL CONTAINER. RINSE AREA WITH WATER,
   PICK UP WITH ABSORBENT, PUT IN A DISPOSAL CONTAINER. SEAL AND REMOVE FROM THE WORK AREA.

HEALTH HAZARD INFORMATION:
   WHEN USING THIS PRODUCT WEAR: NO SPECIAL REQUIREMENT. Contact with the eyes may cause irritation. Prolonged
   inhalation of spray mist may cause upper respiratory irritation.

   A physician should be contacted if anyone develops any signs or symptoms and suspects that they are caused by
   exposure to this product.

FIRST AID:
Eye Exposure:
Flush with water for 15 minutes while lifting the upper and lower eye lids. Contact lenses should not be worn
when working with this product. If irritation persists, get medical attention.

Skin Exposure:
Wash skin with soap and water.

Breathing:
Move the victim to fresh air. Treat symptoms as they appear.

Swallowing:
Immediately contact the local poison control center for advice. Keep the victim warm and at rest. Get medical
attention.

PRODUCT: GLADE - FABRIC FRESH

INCOMPATIBILITY: NONE KNOWN

INGREDIENTS: PROPANE_____   TLV: 1000 ppm   CAS#: 74-98-6
             ISOBUTANE_____ 97-99%   TLV: CAS#: 75-28-5
             N-BUTANE____/   TLV: 800 ppm   CAS#: 106-97-8
             ETHYL ALCOHOL   1-2%   TLV: 1000 ppm   CAS#: 64-17-5
             FRAGRANCE   LESS THEN 1%   TLV: CAS#:

IGNI TEMP: NA   FP: 20F   LEL: 1.8   UEL: 9.5   VP: 3138 mmHg @ 70F   VD: 1.9   SG: .57   PS: LIQUID   APPEAR: SPRAY MIST
     ODOR: FRAGRANT   PH FACTOR: NA

HAZARD RATINGS: H: F: R: 0   4   0

FIRE FIGHTING:
    HIGH EXPANSION FOAM, LOW EXPANSION FOAM, ALCOHOL FOAM; DRY CHEMICAL, CARBON DIOXIDE, WATER FOG.
    PROTECTIVE CLOTHING, RUBBER GLOVES, AND BREATHING APPARATUS.
WARNING: 1] STRUCTURAL PROTECTIVE CLOTHING IS PERMEABLE, REMAIN CLEAR OF SMOKE, WATER FALL OUT AND WATER RUN OFF.
         2] KEEP OUT OF THE REACH OF CHILDREN.
         3] CONTAINERS WILL EXPLODE IN A FIRE OR IF HEATED TO ABOVE 120F.
         4] REMOVE ALL SOURCES OF IGNITION IF WITHOUT RISK.
         5] GAS FUMES ARE CAPABLE OF FORMING AN EXPLOSIVE MIXTURE IN AIR.
         6] THERMAL DECOMPOSITION YIELDS TOXIC CARBON DIOXIDE, CARBON MONOXIDE.
         7] MOVE CONTAINERS FROM AREA IF WITHOUT RISK, COOL EXPOSED CONTAINERS.
         8] DIKE AREA FOR CONTROL AND CONTAINMENT TO PREVENT ENTRY INTO SEWERS, DRAIN, AND WATER WAYS.

LARGE SPILL/NO FIRE/RESCUE: WEAR RUBBER OR NEOPRENE BOOTS, GLOVES.

SPILL CONTROL AND CONTAINMENT:
    HOUSEHOLD SPILL: VENTILATE IMMEDIATE AREA. IF POSSIBLE, REMOVE LEAKING CONTAINER OUTSIDE.
    LARGE SPILL: STOP LEAK IF POSSIBLE OR RELOCATE CONTAINER TO A WELL VENTILATED AREA.

HEALTH HAZARD INFORMATION:
    WHEN USING THIS PRODUCT WEAR: NO SPECIAL REQUIREMENT. Vapors act as an asphyxiant. Symptoms include drowsiness,
    stuporous, confused or anesthetized, unconsciousness, coma, and death by suffocation.

    A physician should be contacted if anyone develops any signs or symptoms and suspects that they are caused by
    exposure to this product.

FIRST AID:
    Eye Exposure:
    Flush with water for 15 minutes while lifting the upper and lower eye lids. Contact lenses should not be worn
    when working with this product. If irritation persists, get medical attention.

    Skin Exposure:
    Wash skin with soap and water.

    Breathing:
    Move victim to fresh air. If breathing is labored give oxygen. Keep the
    victim warm and at rest. Get Immediate medical attention.

    Swallowing:
    Immediately contact the local poison control center for advice. Keep the victim warm and at rest. Get medical
    attention.

PRODUCT: GLADE - GENTLE FRESH WITH CHLOROPHYLL

INCOMPATIBILITY: NONE KNOWN

INGREDIENTS: PROPANE   5-10%   TLV: 1000 ppm   CAS#:   74-98-6
             ISOBUTANE   5-10%   TLV: CAS#:   75-28-5
             N-BUTANE   5-10%   TLV: 800 ppm   CAS#:   106-97-8
             WATER   65-75%   TLV: CAS#:   7732-18-5

IGNI TEMP: NA FP: 20F LEL: NA UEL: NA VP: NA VD: NA SG: .8 PS: LIQUID APPEAR: SPRAY MIST ODOR: FRAGRANCE
   PH FACTOR: NA

HAZARD RATINGS: H: F: R: 0 4 0

FIRE FIGHTING:
   HIGH EXPANSION FOAM, LOW EXPANSION FOAM, ALCOHOL FOAM; DRY CHEMICAL, CARBON DIOXIDE, WATER FOG.
   PROTECTIVE CLOTHING, RUBBER GLOVES, AND BREATHING APPARATUS.
WARNING: 1] STRUCTURAL PROTECTIVE CLOTHING IS PERMEABLE, REMAIN CLEAR OF SMOKE, WATER FALL OUT AND WATER RUN OFF.
         2] KEEP OUT OF THE REACH OF CHILDREN.
         3] CONTAINERS WILL EXPLODE IN A FIRE OR IF HEATED ABOVE 120F.
         4] REMOVE ALL SOURCES OF IGNITION IF WITHOUT RISK.
         5] PROPELLANT WILL ACT AS AN ASPHYXIANT.
         6] THERMAL DECOMPOSITION YIELDS TOXIC CARBON DIOXIDE, CARBON MONOXIDE.
         7] MOVE CONTAINERS FROM AREA IF WITHOUT RISK, COOL EXPOSED CONTAINERS.
         8] DIKE AREA FOR CONTROL AND CONTAINMENT TO PREVENT ENTRY INTO SEWERS, DRAIN, AND WATER WAYS.

LARGE SPILL/NO FIRE/RESCUE: WEAR RUBBER OR NEOPRENE BOOTS, GLOVES.

SPILL CONTROL AND CONTAINMENT:
   HOUSEHOLD SPILL: VENTILATE THE IMMEDIATE AREA. WIPE UP LIQUID WITH ABSORBENT CLOTH, RINSE OUT IN THE SINK. WIPE
   UP RESIDUE WITH A DAMP CLOTH. RINSE OUT IN THE SINK.
   LARGE SPILL: COVER LIQUID WITH ABSORBENT. SWEEP UP AND PUT INTO A DISPOSAL CONTAINER. RINSE AREA WITH WATER,
   PICK UP WITH ABSORBENT, PUT IN A DISPOSAL CONTAINER. SEAL AND REMOVE FROM THE WORK AREA.

HEALTH HAZARD INFORMATION:
   WHEN USING THIS PRODUCT WEAR: NO SPECIAL REQUIREMENT. Contact with the eyes may cause irritation. Prolonged
   inhalation of spray mist may cause upper respiratory irritation.

   A physician should be contacted if anyone develops any signs or symptoms and suspects that they are caused by
   exposure to this product.

FIRST AID:
   Eye Exposure:
   Flush with water for 15 minutes while lifting the upper and lower eye lids. Contact lenses should not be worn
   when working with this product. If irritation persists, get medical attention.

   Skin Exposure:
   Wash skin with soap and water.

   Breathing:
   Move the victim to fresh air. Treat symptoms as they appear.

   Swallowing:
   Immediately contact the local poison control center for advice. Keep the victim warm and at rest. Get medical
   attention.

PRODUCT: GLADE - LITTER FRESH

INCOMPATIBILITY: STRONG ACIDS; (e.g. SWIMMING POOL ACID)

INGREDIENTS: FRAGRANCE   1-3%   TLV: CAS#:
             CALCIUM CARBONATE 97-99%   TLV: 10 mg/m3(DUST) CAS#: 131-65-3

IGNI TEMP: NA FP: NA LEL: NA UEL: NA VP: NA VD: NA SG: NA PS: SOLID APPEAR: WHITE GRANULES
     ODOR: FRAGRANT   PH FACTOR: NA

HAZARD RATINGS: H: F: R: 0 0 0

FIRE FIGHTING:
   HIGH EXPANSION FOAM, LOW EXPANSION FOAM, ALCOHOL FOAM; DRY CHEMICAL, CARBON DIOXIDE, WATER FOG.
   PROTECTIVE CLOTHING, RUBBER GLOVES, AND BREATHING APPARATUS.
WARNING: 1] STRUCTURAL PROTECTIVE CLOTHING IS PERMEABLE, REMAIN CLEAR OF SMOKE, WATER FALL OUT AND WATER RUN OFF.
         2] KEEP OUT OF THE REACH OF CHILDREN.
         3] THERMAL DECOMPOSITION ABOVE 1500°F YIELDS TOXIC CARBON DIOXIDE AND CARBON MONOXIDE.
         4] MOVE CONTAINERS FROM AREA IF WITHOUT RISK, COOL EXPOSED CONTAINERS.
         5] DIKE AREA FOR CONTROL AND CONTAINMENT TO PREVENT ENTRY INTO SEWERS, DRAIN, AND WATER WAYS.

LARGE SPILL/NO FIRE/RESCUE: WEAR RUBBER OR NEOPRENE BOOTS, GLOVES.

SPILL CONTROL AND CONTAINMENT:
   HOUSEHOLD SPILL: AVOID CREATING A DUST. SWEEP UP MATERIAL, PUT IN CONTAINER FOR USE.
   LARGE SPILL: AVOID CREATING A DUST. SWEEP UP MATERIAL AND PUT IN DISPOSAL CONTAINER. RINSE AREA WITH WATER, PICK
   UP WITH ABSORBENT, PUT INTO A DISPOSAL CONTAINER. SEAL AND REMOVE FROM THE WORK AREA.

HEALTH HAZARD INFORMATION:
   WHEN USING THIS PRODUCT WEAR: NO SPECIAL REQUIREMENT. Inhalation of dust may cause irritation of the nose,
   throat, and lungs with coughing, and sneezing. Dust may cause eye irritation.

   A physician should be contacted if anyone develops any signs or symptoms and suspects that they are caused by
   exposure to this product.

FIRST AID:
   Eye Exposure:
   Flush with water for 15 minutes while lifting the upper and lower eye lids. Contact lenses should not be worn
   when working with this product. If irritation persists, get medical attention.

   Skin Exposure:
   Not expected to be a problem.

   Breathing:
   Move the victim to fresh air. If breathing becomes labored, give oxygen. If needed, get medical attention.

   Swallowing:
   Immediately contact the local poison control center for advice. Keep the victim warm and at rest. Get medical
   attention.

PRODUCT: GLADE - MORNING FRESH WITH CHLOROPHYLL

INCOMPATIBILITY: NONE KNOWN

INGREDIENTS: PROPANE   5-10%   TLV: 1000 ppm   CAS#:   74-98-6
             ISOBUTANE   5-10%   TLV: CAS#:   75-28-5
             N-BUTANE   5-10%   TLV: 800 ppm   CAS#:   106-97-8
             WATER   65-75%   TLV: CAS#:   7732-18-5

IGNI TEMP: NA FP: 20F LEL: NA UEL: NA VP: NA VD: NA SG: 0.8 PS: LIQUID APPEAR: SPRAY MIST ODOR: FRAGRANCE
   PH FACTOR: NA

HAZARD RATINGS: H: F: R: 0 4 0

FIRE FIGHTING:
   HIGH EXPANSION FOAM, LOW EXPANSION FOAM, ALCOHOL FOAM; DRY CHEMICAL, CARBON DIOXIDE, WATER FOG.
   PROTECTIVE CLOTHING, RUBBER GLOVES, AND BREATHING APPARATUS.
WARNING: 1] STRUCTURAL PROTECTIVE CLOTHING IS PERMEABLE, REMAIN CLEAR OF SMOKE, WATER FALL OUT AND WATER RUN OFF.
         2] KEEP OUT OF THE REACH OF CHILDREN.
         3] CONTAINERS MAY EXPLODE IN A FIRE OR IF HEATED ABOVE 120F.
         4] THERMAL DECOMPOSITION YIELDS TOXIC CARBON DIOXIDE, CARBON MONOXIDE.
         5] REMOVE ALL SOURCES OF IGNITION IF WITHOUT RISK.
         6] MOVE CONTAINERS FROM AREA IF WITHOUT RISK, COOL EXPOSED CONTAINERS.
         7] DIKE AREA FOR CONTROL AND CONTAINMENT TO PREVENT ENTRY INTO SEWERS, DRAIN, AND WATER WAYS.

LARGE SPILL/ NO FIRE/ RESCUE: WEAR RUBBER OR NEOPRENE BOOTS, GLOVES.

SPILL CONTROL AND CONTAINMENT:
   HOUSEHOLD SPILL: VENTILATE AREA. WIPE UP LIQUID WITH ABSORBENT MATERIAL. PUT INTO A DISPOSAL CONTAINER. RINSE
   WITH WATER, PICK UP WITH ABSORBENT MATERIAL, PUT INTO DISPOSAL CONTAINER.
   LARGE SPILL: COVER SPILL WITH ABSORBENT, SCOOP UP AND PUT INTO A DISPOSAL CONTAINER. RINSE AREA WITH WATER, PICK
   UP WITH ABSORBENT, PUT INTO DISPOSAL CONTAINER. SEAL AND REMOVE FROM WORK PLACE.

HEALTH HAZARD INFORMATION:
   WHEN USING THIS PRODUCT WEAR: NO SPECIAL REQUIREMENTS.

   A physician should be contacted if anyone develops any signs or symptoms and suspects that they are caused by
   exposure to this product.

FIRST AID:
   Eye Exposure:
   Flush with water for 15 minutes while lifting the upper and lower eye lids. Contact lens should not be worn when
   working with this product. Get medical attention.

   Skin Exposure:
   Not expected to be a problem.

   Breathing:
   Not expected to be a problem.

   Swallowing:
   Contact the local poison control center for advice. Keep victim warm and at rest. Get medical attention.

PRODUCT: GLADE - POWDER FRESH WITH CHLOROPHYLL

INCOMPATIBILITY: NONE KNOWN

INGREDIENTS: PROPANE   5-10%   TLV: 1000 ppm   CAS#:   74-98-6
             ISOBUTANE   5-10%   TLV: CAS#:   75-28-5
             N-BUTANE   5-10%   TLV: 800 ppm   CAS#:   106-97-8
             WATER   65-75%   TLV: CAS#:   7732-18-5

IGNI TEMP: NA FP: 20F LEL: NA UEL: NA VP: NA VD: NA SG: 0.8 PS: LIQUID APPEAR: SPRAY MIST ODOR: FRAGRANCE
    PH FACTOR: NA

HAZARD RATINGS: H: F: R: 0   4   0

FIRE FIGHTING:
  HIGH EXPANSION FOAM, LOW EXPANSION FOAM, ALCOHOL FOAM; DRY CHEMICAL, CARBON DIOXIDE, WATER FOG.
  PROTECTIVE CLOTHING, RUBBER GLOVES, AND BREATHING APPARATUS.
WARNING: 1] STRUCTURAL PROTECTIVE CLOTHING IS PERMEABLE, REMAIN CLEAR OF SMOKE, WATER FALL OUT AND WATER RUN OFF.
         2] KEEP OUT OF THE REACH OF CHILDREN.
         3] CONTAINERS MAY EXPLODE IN A FIRE OR IF HEATED ABOVE 120F.
         4] THERMAL DECOMPOSITION YIELDS TOXIC CARBON DIOXIDE, CARBON MONOXIDE.
         5] REMOVE ALL SOURCES OF IGNITION IF WITHOUT RISK.
         6] MOVE CONTAINERS FROM AREA IF WITHOUT RISK, COOL EXPOSED CONTAINERS.
         7] DIKE AREA FOR CONTROL AND CONTAINMENT TO PREVENT ENTRY INTO SEWERS, DRAIN, AND WATER WAYS.

LARGE SPILL/NO FIRE/RESCUE: WEAR RUBBER OR NEOPRENE BOOTS, GLOVES.

SPILL CONTROL AND CONTAINMENT:
  HOUSEHOLD SPILL: VENTILATE AREA. WIPE UP LIQUID WITH ABSORBENT MATERIAL. PUT INTO A DISPOSAL CONTAINER. RINSE
  WITH WATER, PICK UP WITH ABSORBENT MATERIAL, PUT INTO DISPOSAL CONTAINER.
  LARGE SPILL: COVER SPILL WITH ABSORBENT, SCOOP UP AND PUT INTO A DISPOSAL CONTAINER. RINSE AREA WITH WATER, PICK
  UP WITH ABSORBENT, PUT INTO DISPOSAL CONTAINER. SEAL AND REMOVE FROM WORK PLACE.

HEALTH HAZARD INFORMATION:
  WHEN USING THIS PRODUCT WEAR: NO SPECIAL REQUIREMENTS.

  A physician should be contacted if anyone develops any signs or symptoms and suspects that they are caused by
  exposure to this product.

FIRST AID:
  Eye Exposure:
  Flush with water for 15 minutes while lifting the upper and lower eye lids. Contact lenses should not be worn
  when working with this product. Get medical attention.

  Skin Exposure:
  Not expected to be a problem.

  Breathing:
  Not expected to be a problem.

  Swallowing:
  Contact the local poison control center for advice. Keep victim warm and at rest. Get medical attention.

PRODUCT: GLADE - RAIN SHOWER FRESH WITH CHLOROPHYLL

INCOMPATIBILITY: NONE KNOWN

INGREDIENTS:

PROPANE	5-10%	TLV: 1000 ppm	CAS#:	74-98-6
ISOBUTANE	5-10%	TLV:	CAS#:	75-28-5
N-BUTANE	5-10%	TLV: 800 ppm	CAS#:	106-97-8
WATER	65-75%	TLV:	CAS#:	7732-18-5

IGNI TEMP: NA FP: 20F LEL: NA UEL: NA VP: NA VD: NA SG: 0.8 PS: LIQUID APPEAR: SPRAY MIST ODOR: FRAGRANCE
PH FACTOR: NA

HAZARD RATINGS: H: F: R: 0 4 0

FIRE FIGHTING:
  HIGH EXPANSION FOAM, LOW EXPANSION FOAM, ALCOHOL FOAM; DRY CHEMICAL, CARBON DIOXIDE, WATER FOG.
  PROTECTIVE CLOTHING, RUBBER GLOVES, AND BREATHING APPARATUS.
WARNING: 1] STRUCTURAL PROTECTIVE CLOTHING IS PERMEABLE, REMAIN CLEAR OF SMOKE, WATER FALL OUT AND WATER RUN OFF.
         2] KEEP OUT OF THE REACH OF CHILDREN.
         3] CONTAINERS MAY EXPLODE IN A FIRE OR IF HEATED ABOVE 120F.
         4] THERMAL DECOMPOSITION YIELDS TOXIC CARBON DIOXIDE, CARBON MONOXIDE.
         5] REMOVE ALL SOURCES OF IGNITION IF WITHOUT RISK.
         6] MOVE CONTAINERS FROM AREA IF WITHOUT RISK, COOL EXPOSED CONTAINERS.
         7] DIKE AREA FOR CONTROL AND CONTAINMENT TO PREVENT ENTRY INTO SEWERS, DRAIN, AND WATER WAYS.

LARGE SPILL/NO FIRE/RESCUE: WEAR RUBBER OR NEOPRENE BOOTS, GLOVES.

SPILL CONTROL AND CONTAINMENT:
  HOUSEHOLD SPILL: VENTILATE AREA. WIPE UP LIQUID WITH ABSORBENT MATERIAL. PUT INTO A DISPOSAL CONTAINER. RINSE
  WITH WATER, PICK UP WITH ABSORBENT MATERIAL, PUT INTO DISPOSAL CONTAINER.
  LARGE SPILL: COVER SPILL WITH ABSORBENT, SCOOP UP AND PUT INTO A DISPOSAL CONTAINER. RINSE AREA WITH WATER, PICK
  UP WITH ABSORBENT, PUT INTO DISPOSAL CONTAINER. SEAL AND REMOVE FROM WORK PLACE.

HEALTH HAZARD INFORMATION:
  WHEN USING THIS PRODUCT WEAR: NO SPECIAL REQUIREMENTS.

  A physician should be contacted if anyone develops any signs or symptoms and suspects that they are caused by
  exposure to this product.

FIRST AID:
  Eye Exposure:
  Flush with water for 15 minutes while lifting the upper and lower eye lids. Contact lenses should not be worn
  when working with this product. Get medical attention.

  Skin Exposure:
  Not expected to be a problem.

  Breathing:
  Not expected to be a problem.

  Swallowing:
  Contact the local poison control center for advice. Keep victim warm and at rest. Get medical attention.

PRODUCT: GLADE - SUMMER BREEZE WITH CHLOROPHYLL

INCOMPATIBILITY: NONE KNOWN

INGREDIENTS:	PROPANE	5-10%	TLV: 1000 ppm	CAS#:	74-98-6
	ISOBUTANE	5-10%	TLV:	CAS#:	75-28-5
	N-BUTANE	5-10%	TLV: 800 ppm	CAS#:	106-97-8
	WATER	65-75%	TLV:	CAS#:	7732-18-5

IGNI TEMP: NA FP: 20F LEL: NA UEL: NA VP: NA VD: NA SG: 0.8 PS: LIQUID APPEAR: SPRAY MIST ODOR: FRAGRANCE
  PH FACTOR: NA

HAZARD RATINGS: H: F: R: 0 4 0

FIRE FIGHTING:
  HIGH EXPANSION FOAM, LOW EXPANSION FOAM, ALCOHOL FOAM; DRY CHEMICAL, CARBON DIOXIDE, WATER FOG.
  PROTECTIVE CLOTHING, RUBBER GLOVES, AND BREATHING APPARATUS.
WARNING: 1] STRUCTURAL PROTECTIVE CLOTHING IS PERMEABLE, REMAIN CLEAR OF SMOKE, WATER FALL OUT AND WATER RUN OFF.
         2] KEEP OUT OF THE REACH OF CHILDREN.
         3] CONTAINERS MAY EXPLODE IN A FIRE OR IF HEATED ABOVE 120F.
         4] THERMAL DECOMPOSITION YIELDS TOXIC CARBON DIOXIDE, CARBON MONOXIDE.
         5] REMOVE ALL SOURCES OF IGNITION IF WITHOUT RISK.
         6] MOVE CONTAINERS FROM AREA IF WITHOUT RISK, COOL EXPOSED CONTAINERS.
         7] DIKE AREA FOR CONTROL AND CONTAINMENT TO PREVENT ENTRY INTO SEWERS, DRAIN, AND WATER WAYS.

LARGE SPILL/NO FIRE/RESCUE: WEAR RUBBER OR NEOPRENE BOOTS, GLOVES.

SPILL CONTROL AND CONTAINMENT:
  HOUSEHOLD SPILL: VENTILATE AREA. WIPE UP LIQUID WITH ABSORBENT MATERIAL. PUT INTO A DISPOSAL CONTAINER. RINSE
  WITH WATER, PICK UP WITH ABSORBENT MATERIAL, PUT INTO DISPOSAL CONTAINER.
  LARGE SPILL: COVER SPILL WITH ABSORBENT, SCOOP UP AND PUT INTO A DISPOSAL CONTAINER. RINSE AREA WITH WATER, PICK
  UP WITH ABSORBENT, PUT INTO DISPOSAL CONTAINER. SEAL AND REMOVE FROM WORK PLACE.

HEALTH HAZARD INFORMATION:
  WHEN USING THIS PRODUCT WEAR: NO SPECIAL REQUIREMENTS.

  A physician should be contacted if anyone develops any signs or symptoms and suspects that they are caused by
  exposure to this product.

FIRST AID:
  Eye Exposure:
  Flush with water for 15 minutes while lifting the upper and lower eye lids. Contact lenses should not be worn
  when working with this product. Get medical attention.

  Skin Exposure:
  Not expected to be a problem.

  Breathing:
  Not expected to be a problem.

  Swallowing:
  Contact the local poison control center for advice. Keep victim warm and at rest. Get medical attention.

PRODUCT: GLADE - SUPER FRESH WITH CHLOROPHYLL

INCOMPATIBILITY: NONE KNOWN

INGREDIENTS:				
PROPANE	5-10%	TLV: 1000 ppm	CAS#:	74-98-6
ISOBUTANE	5-10%	TLV:	CAS#:	75-28-5
N-BUTANE	5-10%	TLV: 800 ppm	CAS#:	106-97-8
WATER	65-75%	TLV:	CAS#:	7732-18-5

IGNI TEMP: NA FP: 20F LEL: NA UEL: NA VP: NA VD: NA SG: 0.8 PS: LIQUID APPEAR: SPRAY MIST ODOR: FRAGRANCE
   PH FACTOR: NA

HAZARD RATINGS: H: F: R: 0 4 0

FIRE FIGHTING:
   HIGH EXPANSION FOAM, LOW EXPANSION FOAM, ALCOHOL FOAM; DRY CHEMICAL, CARBON DIOXIDE, WATER FOG.
   PROTECTIVE CLOTHING, RUBBER GLOVES, AND BREATHING APPARATUS.
WARNING: 1] STRUCTURAL PROTECTIVE CLOTHING IS PERMEABLE, REMAIN CLEAR OF SMOKE, WATER FALL OUT AND WATER RUN OFF.
        2] KEEP OUT OF THE REACH OF CHILDREN.
        3] CONTAINERS MAY EXPLODE IN A FIRE OR IF HEATED ABOVE 120F.
        4] THERMAL DECOMPOSITION YIELDS TOXIC CARBON DIOXIDE, CARBON MONOXIDE.
        5] REMOVE ALL SOURCES OF IGNITION IF WITHOUT RISK.
        6] MOVE CONTAINERS FROM AREA IF WITHOUT RISK, COOL EXPOSED CONTAINERS.
        7] DIKE AREA FOR CONTROL AND CONTAINMENT TO PREVENT ENTRY INTO SEWERS, DRAIN, AND WATER WAYS.

LARGE SPILL/NO FIRE/RESCUE: WEAR RUBBER OR NEOPRENE BOOTS, GLOVES.

SPILL CONTROL AND CONTAINMENT:
   HOUSEHOLD SPILL: VENTILATE AREA. WIPE UP LIQUID WITH ABSORBENT MATERIAL. PUT INTO A DISPOSAL CONTAINER. RINSE
   WITH WATER, PICK UP WITH ABSORBENT MATERIAL, PUT INTO DISPOSAL CONTAINER.
   LARGE SPILL: COVER SPILL WITH ABSORBENT, SCOOP UP AND PUT INTO A DISPOSAL CONTAINER. RINSE AREA WITH WATER, PICK
   UP WITH ABSORBENT, PUT INTO DISPOSAL CONTAINER. SEAL AND REMOVE FROM WORK PLACE.

HEALTH HAZARD INFORMATION:
   WHEN USING THIS PRODUCT WEAR: NO SPECIAL REQUIREMENTS.

   A physician should be contacted if anyone develops any signs or symptoms and suspects that they are caused by
   exposure to this product.

FIRST AID:
   Eye Exposure:
   Flush with water for 15 minutes while lifting the upper and lower eye lids. Contact lenses should not be worn
   when working with this product. Get medical attention.

   Skin Exposure:
   Not expected to be a problem.

   Breathing:
   Not expected to be a problem.

   Swallowing:
   Contact the local poison control center for advice. Keep victim warm and at rest. Get medical attention.

PRODUCT: GLADE II - COUNTRY FRESH

INCOMPATIBILITY: NONE KNOWN

INGREDIENTS: FRAGRANCE   95-99%   TLV: CAS#:
             COLORANT AND SILICA   1-5 %   TLV: CAS#:

IGNI TEMP: NA FP: 140F LEL: NA UEL: NA VP: NA VD: NA SG: .93 PS: LIQUID APPEAR: ODOR: FRAGRANCE
    PH FACTOR: NA

HAZARD RATINGS: H: F: R: 0 2 0

FIRE FIGHTING:
    HIGH EXPANSION FOAM, LOW EXPANSION FOAM, ALCOHOL FOAM; DRY CHEMICAL, CARBON DIOXIDE, WATER FOG.
    PROTECTIVE CLOTHING, RUBBER GLOVES, AND BREATHING APPARATUS.
WARNING: 1] STRUCTURAL PROTECTIVE CLOTHING IS PERMEABLE, REMAIN CLEAR OF SMOKE, WATER FALL OUT AND WATER RUN OFF.
         2] KEEP OUT OF THE REACH OF CHILDREN.
         3] THERMAL DECOMPOSITION YIELDS CARBON DIOXIDE, CARBON MONOXIDE.
         4] CONTAINERS MAY RUPTURE IN A FIRE.
         5] MOVE CONTAINERS FROM AREA IF WITHOUT RISK, COOL EXPOSED CONTAINERS.
         6] DIKE AREA FOR CONTROL AND CONTAINMENT TO PREVENT ENTRY INTO SEWERS, DRAIN, AND WATER WAYS.

LARGE SPILL/NO FIRE/RESCUE: WEAR RUBBER OR NEOPRENE BOOTS, GLOVES.

SPILL CONTROL AND CONTAINMENT:
    HOUSEHOLD SPILL: WIPE UP LIQUID WITH ABSORBENT MATERIAL, PUT INTO A DISPOSAL CONTAINER. WASH AREA WITH SOAP AND
    WATER, PICK UP WITH ABSORBENT MATERIAL, PUT INTO DISPOSAL CONTAINER.
    LARGE SPILL: COVER SPILL AREA WITH ABSORBENT, PICK UP AND PUT INTO A DISPOSAL CONTAINER. RINSE SPILL AREA WITH
    WATER, PICK UP WITH ABSORBENT, PUT INTO DISPOSAL CONTAINER.

HEALTH HAZARD INFORMATION:
    WHEN USING THIS PRODUCT WEAR: NO SPECIAL REQUIREMENTS. Liquid contact may cause eye irritation, tearing, burning
    sensation, itching, swelling, and redness.

    A physician should be contacted if anyone develops any signs or symptoms and suspects that they are caused by
    exposure to this product.

FIRST AID:
Eye Exposure:
Flush with water for at least 15 minutes while lifting the upper and lower eye lids. Contact lenses should not
be worn when working with this product. Get medical attention.

Skin Exposure:
Not expected to be a problem. Wash with soap and water.

Breathing:
Not expected to be a problem.

Swallowing:
Immediately contact the local poison control center for advice. Keep victim warm and at rest. Get medical
attention.

PRODUCT: GLADE II - POWDER FRESH

INCOMPATIBILITY: NONE KNOWN

INGREDIENTS: FRAGRANCE   95-99%   TLV: CAS#:
             COLORANT AND SILICA 1-5 %   TLV: CAS#:

IGNI TEMP: NA FP: 140F LEL: NA UEL: NA VP: NA VD: NA SG: .93 PS: LIQUID APPEAR: ODOR: FRAGRANCE
   PH FACTOR: NA

HAZARD RATINGS: H: F: R: 0 2 0

FIRE FIGHTING:
  HIGH EXPANSION FOAM, LOW EXPANSION FOAM, ALCOHOL FOAM; DRY CHEMICAL, CARBON DIOXIDE, WATER FOG.
  PROTECTIVE CLOTHING, RUBBER GLOVES, AND BREATHING APPARATUS.
WARNING: 1] STRUCTURAL PROTECTIVE CLOTHING IS PERMEABLE, REMAIN CLEAR OF SMOKE, WATER FALL OUT AND WATER RUN OFF.
         2] KEEP OUT OF THE REACH OF CHILDREN.
         3] THERMAL DECOMPOSITION YIELDS CARBON DIOXIDE, CARBON MONOXIDE.
         4] CONTAINERS MAY RUPTURE IN A FIRE.
         5] MOVE CONTAINERS FROM AREA IF WITHOUT RISK, COOL EXPOSED CONTAINERS.
         6] DIKE AREA FOR CONTROL AND CONTAINMENT TO PREVENT ENTRY INTO SEWERS, DRAIN, AND WATER WAYS.

LARGE SPILL/NO FIRE/RESCUE: WEAR RUBBER OR NEOPRENE BOOTS, GLOVES.

SPILL CONTROL AND CONTAINMENT:
  HOUSEHOLD SPILL: WIPE UP LIQUID WITH ABSORBENT MATERIAL, PUT INTO A DISPOSAL CONTAINER. WASH AREA WITH SOAP AND
  WATER, PICK UP WITH ABSORBENT MATERIAL, PUT INTO DISPOSAL CONTAINER.
  LARGE SPILL: COVER SPILL AREA WITH ABSORBENT, PICK UP AND PUT INTO A DISPOSAL CONTAINER. RINSE SPILL AREA WITH
  WATER, PICK UP WITH ABSORBENT, PUT INTO DISPOSAL CONTAINER.

HEALTH HAZARD INFORMATION:
  WHEN USING THIS PRODUCT WEAR: NO SPECIAL REQUIREMENTS. Liquid contact may cause eye irritation, tearing, burning
  sensation, itching, swelling, and redness.

  A physician should be contacted if anyone develops any signs or symptoms and suspects that they are caused by
  exposure to this product.

FIRST AID:
  Eye Exposure:
  Flush with water for at least 15 minutes while lifting the upper and lower eye lids. Contact lenses should not
  be worn when working with this product. Get medical attention.

  Skin Exposure:
  Not expected to be a problem. Wash with soap and water.

  Breathing:
  Not expected to be a problem.

  Swallowing:
  Immediately contact the local poison control center for advice. Keep victim warm and at rest. Get medical
  attention.

PRODUCT: GLADE II - SUPER FRESH

INCOMPATIBILITY: NONE KNOWN

INGREDIENTS: FRAGRANCE   95-99%   TLV: CAS#:
             COLORANT AND SILICA 1-5 %   TLV: CAS#:

IGNI TEMP: NA FP: 140F LEL: NA UEL: NA VP: NA VD: NA SG: .93 PS: LIQUID APPEAR: ODOR: FRAGRANCE
    PH FACTOR: NA

HAZARD RATINGS: H: F: R: 0 2 0

FIRE FIGHTING:
   HIGH EXPANSION FOAM, LOW EXPANSION FOAM, ALCOHOL FOAM; DRY CHEMICAL, CARBON DIOXIDE, WATER FOG.
   PROTECTIVE CLOTHING, RUBBER GLOVES, AND BREATHING APPARATUS.
WARNING: 1] STRUCTURAL PROTECTIVE CLOTHING IS PERMEABLE, REMAIN CLEAR OF SMOKE, WATER FALL OUT AND WATWATER RUN OFF.
         2] KEEP OUT OF THE REACH OF CHILDREN.
         3] THERMAL DECOMPOSITION YIELDS CARBON DIOXIDE, CARBON MONOXIDE.
         4] CONTAINERS MAY RUPTURE IN A FIRE.
         5] MOVE CONTAINERS FROM AREA IF WITHOUT RISK, COOL EXPOSED CONTAINERS.
         6] DIKE AREA FOR CONTROL AND CONTAINMENT TO PREVENT ENTRY INTO SEWERS, DRAIN, AND WATER WAYS.

LARGE SPILL/NO FIRE/RESCUE: WEAR RUBBER OR NEOPRENE BOOTS, GLOVES.

SPILL CONTROL AND CONTAINMENT:
   HOUSEHOLD SPILL: WIPE UP LIQUID WITH ABSORBENT MATERIAL, PUT INTO A DISPOSAL CONTAINER. WASH AREA WITH SOAP AND
   WATER, PICK UP WITH ABSORBENT MATERIAL, PUT INTO DISPOSAL CONTAINER.
   LARGE SPILL: COVER SPILL AREA WITH ABSORBENT, PICK UP AND PUT INTO A DISPOSAL CONTAINER. RINSE SPILL AREA WITH
   WATER, PICK UP WITH ABSORBENT, PUT INTO DISPOSAL CONTAINER.

HEALTH HAZARD INFORMATION:
   WHEN USING THIS PRODUCT WEAR: NO SPECIAL REQUIREMENTS. Liquid contact may cause eye irritation, tearing, burning
   sensation, itching, swelling, and redness.

   A physician should be contacted if anyone develops any signs or symptoms and suspects that they are caused by
   exposure to this product.

FIRST AID:
   Eye Exposure:
   Flush with water for at least 15 minutes while lifting the upper and lower eye lids. Contact lenses should not
   be worn when working with this product. Get medical attention.

   Skin Exposure:
   Not expected to be a problem. Wash with soap and water.

   Breathing:
   Not expected to be a problem.

   Swallowing:
   Immediately contact the local poison control center for advice. Keep victim warm and at rest. Get medical
   attention.

PRODUCT: GLADE COUNTRY - COUNTRY BERRY

INCOMPATIBILITY: NONE KNOWN

INGREDIENTS: NON-HAZARDOUS   100%

IGNI TEMP: NA FP: NA LEL: NA UEL: NA VP: NA VD: NA SG: NA PS: MATTER APPEAR: DRIED BOTANICAL
   ODOR: FRAGRANT PH FACTOR: NA

HAZARD RATINGS: H: F: R: 0 0 0

FIRE FIGHTING:
   HIGH EXPANSION FOAM, LOW EXPANSION FOAM, ALCOHOL FOAM; DRY CHEMICAL, CARBON DIOXIDE, WATER FOG.
   PROTECTIVE CLOTHING, RUBBER GLOVES, AND BREATHING APPARATUS.
WARNING: 1] STRUCTURAL PROTECTIVE CLOTHING IS PERMEABLE, REMAIN CLEAR OF SMOKE, WATER FALL OUT AND WATER RUN OFF.
         2] KEEP OUT OF THE REACH OF CHILDREN.
         3] THERMAL DECOMPOSITION YIELDS TOXIC CARBON DIOXIDE, CARBON MONOXIDE.
         4] MOVE CONTAINERS FROM AREA IF WITHOUT RISK, COOL EXPOSED CONTAINERS.
         5] DIKE AREA FOR CONTROL AND CONTAINMENT TO PREVENT ENTRY INTO SEWERS, DRAIN, AND WATER WAYS.

LARGE SPILL/NO FIRE/RESCUE: WEAR RUBBER OR NEOPRENE BOOTS, GLOVES.

SPILL CONTROL AND CONTAINMENT:
   HOUSEHOLD SPILL: SWEEP UP MATERIAL AND PLACE IN A DISPOSAL CONTAINER.

   LARGE SPILL: SWEEP UP MATERIAL AND PLACE IN A CONTAINER FOR RECYCLE OR DISPOSAL.

HEALTH HAZARD INFORMATION:
   WHEN USING THIS PRODUCT WEAR: NO SPECIAL REQUIREMENTS.

   A physician should be contacted if anyone develops any signs or symptoms and suspects that they are caused by
   exposure to this product.

FIRST AID:
   Eye Exposure:
   Flush material from the eyes. If irritation persists, get medical attention. Contact lenses should not be worn
   when working with this product.

   Skin Exposure:
   Not expected to be a problem.

   Breathing:
   Not expected to be a problem.

   Swallowing:
   Not expected to be a problem. If any signs or symptoms appear, immediately contact the local poison control
   center for advice. Keep victim warm and at rest. Get medical attention.

PRODUCT: GLADE COUNTRY - COUNTRY PEACH

INCOMPATIBILITY: NONE KNOWN

INGREDIENTS: NON-HAZARDOUS   100%

IGNI TEMP: NA FP: NA LEL: NA UEL: NA VP: NA VD: NA SG: NA PS: MATTER APPEAR: DRIED BOTANICAL
     ODOR: FRAGRANT PH FACTOR: NA

HAZARD RATINGS: H: F: R: 0 0 0

FIRE FIGHTING:
   HIGH EXPANSION FOAM, LOW EXPANSION FOAM, ALCOHOL FOAM; DRY CHEMICAL, CARBON DIOXIDE, WATER FOG.
   PROTECTIVE CLOTHING, RUBBER GLOVES, AND BREATHING APPARATUS.
WARNING: 1] STRUCTURAL PROTECTIVE CLOTHING IS PERMEABLE, REMAIN CLEAR OF SMOKE, WATER FALL OUT AND WATER RUN OFF.
         2] KEEP OUT OF THE REACH OF CHILDREN.
         3] THERMAL DECOMPOSITION YIELDS TOXIC CARBON DIOXIDE, CARBON MONOXIDE.
         4] MOVE CONTAINERS FROM AREA IF WITHOUT RISK, COOL EXPOSED CONTAINERS.
         5] DIKE AREA FOR CONTROL AND CONTAINMENT TO PREVENT ENTRY INTO SEWERS, DRAIN, AND WATER WAYS.

LARGE SPILL/NO FIRE/RESCUE: WEAR RUBBER OR NEOPRENE BOOTS, GLOVES.

SPILL CONTROL AND CONTAINMENT:
   HOUSEHOLD SPILL: SWEEP UP MATERIAL AND PLACE IN A DISPOSAL CONTAINER.

   LARGE SPILL: SWEEP UP MATERIAL AND PLACE IN A CONTAINER FOR RECYCLE OR DISPOSAL.

HEALTH HAZARD INFORMATION:
   WHEN USING THIS PRODUCT WEAR: NO SPECIAL REQUIREMENTS.

   A physician should be contacted if anyone develops any signs or symptoms and suspects that they are caused by
   exposure to this product.

FIRST AID:
   Eye Exposure:
   Flush material from the eyes. If irritation persists, get medical attention. Contact lenses should not be worn
   when working with this product.

   Skin Exposure:
   Not expected to be a problem.

   Breathing:
   Not expected to be a problem.

   Swallowing:
   Not expected to be a problem. If any signs or symptoms appear, immediately contact the local poison control
   center for advice. Keep victim warm and at rest. Get medical attention.

PRODUCT: GLADE COUNTRY - COUNTRY SPICE

INCOMPATIBILITY: NONE KNOWN

INGREDIENTS: NON-HAZARDOUS   100%

IGNI TEMP: NA FP: NA LEL: NA UEL: NA VP: NA VD: NA SG: NA PS: MATTER APPEAR: DRIED BOTANICAL
    ODOR: FRAGRANT PH FACTOR: NA

HAZARD RATINGS: H: F: R: 0 0 0

FIRE FIGHTING:
  HIGH EXPANSION FOAM, LOW EXPANSION FOAM, ALCOHOL FOAM; DRY CHEMICAL, CARBON DIOXIDE, WATER FOG.
  PROTECTIVE CLOTHING, RUBBER GLOVES, AND BREATHING APPARATUS.
WARNING: 1] STRUCTURAL PROTECTIVE CLOTHING IS PERMEABLE, REMAIN CLEAR OF SMOKE, WATER FALL OUT AND WATER RUN OFF.
         2] KEEP OUT OF THE REACH OF CHILDREN.
         3] THERMAL DECOMPOSITION YIELDS TOXIC CARBON DIOXIDE, CARBON MONOXIDE.
         4] MOVE CONTAINERS FROM AREA IF WITHOUT RISK, COOL EXPOSED CONTAINERS.
         5] DIKE AREA FOR CONTROL AND CONTAINMENT TO PREVENT ENTRY INTO SEWERS, DRAIN, AND WATER WAYS.

LARGE SPILL/NO FIRE/RESCUE: WEAR RUBBER OR NEOPRENE BOOTS, GLOVES.

SPILL CONTROL AND CONTAINMENT:
  HOUSEHOLD SPILL: SWEEP UP MATERIAL AND PLACE IN A DISPOSAL CONTAINER.

  LARGE SPILL: SWEEP UP MATERIAL AND PLACE IN A CONTAINER FOR RECYCLE OR DISPOSAL.

HEALTH HAZARD INFORMATION:
  WHEN USING THIS PRODUCT WEAR: NO SPECIAL REQUIREMENTS.

  A physician should be contacted if anyone develops any signs or symptoms and suspects that they are caused by exposure to this product.

FIRST AID:
  Eye Exposure:
  Flush material from the eyes. If irritation persists, get medical attention. Contact lenses should not be worn when working with this product.

  Skin Exposure:
  Not expected to be a problem.

  Breathing:
  Not expected to be a problem.

  Swallowing:
  Not expected to be a problem. If any signs or symptoms appear, immediately contact the local poison control center for advice. Keep victim warm and at rest. Get medical attention.

PRODUCT: GLADE COUNTRY - FRENCH VANILLA

INCOMPATIBILITY: NONE KNOWN

INGREDIENTS: NON-HAZARDOUS   100%

IGNI TEMP: NA FP: NA LEL: NA UEL: NA VP: NA VD: NA SG: NA PS: MATTER APPEAR: DRIED BOTANICAL
    ODOR: FRAGRANT PH FACTOR: NA

HAZARD RATINGS: H: F: R: 0 0 0

FIRE FIGHTING:
  HIGH EXPANSION FOAM, LOW EXPANSION FOAM, ALCOHOL FOAM; DRY CHEMICAL, CARBON DIOXIDE, WATER FOG.
  PROTECTIVE CLOTHING, RUBBER GLOVES, AND BREATHING APPARATUS.
WARNING: 1] STRUCTURAL PROTECTIVE CLOTHING IS PERMEABLE, REMAIN CLEAR OF SMOKE, WATER FALL OUT AND WATER RUN OFF.
         2] KEEP OUT OF THE REACH OF CHILDREN.
         3] THERMAL DECOMPOSITION YIELDS TOXIC CARBON DIOXIDE, CARBON MONOXIDE.
         4] MOVE CONTAINERS FROM AREA IF WITHOUT RISK, COOL EXPOSED CONTAINERS.
         5] DIKE AREA FOR CONTROL AND CONTAINMENT TO PREVENT ENTRY INTO SEWERS, DRAIN, AND WATER WAYS.

LARGE SPILL/NO FIRE/RESCUE: WEAR RUBBER OR NEOPRENE BOOTS, GLOVES.

SPILL CONTROL AND CONTAINMENT:
  HOUSEHOLD SPILL: SWEEP UP MATERIAL AND PLACE IN A DISPOSAL CONTAINER.

  LARGE SPILL: SWEEP UP MATERIAL AND PLACE IN A CONTAINER FOR RECYCLE OR DISPOSAL.

HEALTH HAZARD INFORMATION:
  WHEN USING THIS PRODUCT WEAR: NO SPECIAL REQUIREMENTS.

  A physician should be contacted if anyone develops any signs or symptoms and suspects that they are caused by
  exposure to this product.

FIRST AID:
  Eye Exposure:
  Flush material from the eyes. If irritation persists get, medical attention. Contact lenses should not be worn
  when working with this product.

  Skin Exposure:
  Not expected to be a problem.

  Breathing:
  Not expected to be a problem.

  Swallowing:
  Not expected to be a problem. If any signs or symptoms appear, immediately contact the local poison control
  center for advice. Keep victim warm and at rest. Get medical attention.

PRODUCT: GLADE COUNTRY - SPICED APPLE

INCOMPATIBILITY: NONE KNOWN

INGREDIENTS: NON-HAZARDOUS   100%

IGNI TEMP: NA FP: NA LEL: NA UEL: NA VP: NA VD: NA SG: NA PS: MATTER APPEAR: DRIED BOTANICAL
    ODOR: FRAGRANT PH FACTOR: NA

HAZARD RATINGS: H: F: R: 0 0 0

FIRE FIGHTING:
   HIGH EXPANSION FOAM, LOW EXPANSION FOAM, ALCOHOL FOAM; DRY CHEMICAL, CARBON DIOXIDE, WATER FOG.
   PROTECTIVE CLOTHING, RUBBER GLOVES, AND BREATHING APPARATUS.
WARNING: 1] STRUCTURAL PROTECTIVE CLOTHING IS PERMEABLE, REMAIN CLEAR OF SMOKE, WATER FALL OUT AND WATER RUN OFF.
         2] KEEP OUT OF THE REACH OF CHILDREN.
         3] THERMAL DECOMPOSITION YIELDS TOXIC CARBON DIOXIDE, CARBON MONOXIDE.
         4] MOVE CONTAINERS FROM AREA IF WITHOUT RISK, COOL EXPOSED CONTAINERS.
         5] DIKE AREA FOR CONTROL AND CONTAINMENT TO PREVENT ENTRY INTO SEWERS, DRAIN, AND WATER WAYS.

LARGE SPILL/NO FIRE/RESCUE: WEAR RUBBER OR NEOPRENE BOOTS, GLOVES.

SPILL CONTROL AND CONTAINMENT:
   HOUSEHOLD SPILL: SWEEP UP MATERIAL AND PLACE IN A DISPOSAL CONTAINER.

   LARGE SPILL: SWEEP UP MATERIAL AND PLACE IN A CONTAINER FOR RECYCLE OR DISPOSAL.

HEALTH HAZARD INFORMATION:
   WHEN USING THIS PRODUCT WEAR: NO SPECIAL REQUIREMENTS.

   A physician should be contacted if anyone develops any signs or symptoms and suspects that they are caused by
   exposure to this product.

FIRST AID:
   Eye Exposure:
   Flush material from the eyes. If irritation persists, get medical attention. Contact lenses should not be worn
   when working with this product.

   Skin Exposure:
   Not expected to be a problem.

   Breathing:
   Not expected to be a problem.

   Swallowing:
   Not expected to be a problem. If any signs or symptoms appear, immediately contact the local poison control
   center for advice. Keep victim warm and at rest. Get medical attention.

PRODUCT: GLADE LIGHT (ALL SCENTS)

INCOMPATIBILITY: NONE KNOWN

INGREDIENTS: PROPANE_____   TLV: 1000 ppm   CAS#:   74-98-6
             ISOBUTANE_____ 25-35% TLV: CAS#:   75-28-5
             N-BUTANE _____/   TLV: 800 ppm CAS#:   106-97-8

IGNI TEMP: NA FP: 20F LEL: NA UEL: NA VP: NA VD: NA SG: .81 PS: LIQUID APPEAR: SPRAY MIST ODOR: FRAGRANCE
    PH FACTOR: NA

HAZARD RATINGS: H: F: R: 0 4 0

FIRE FIGHTING:
    HIGH EXPANSION FOAM, LOW EXPANSION FOAM, ALCOHOL FOAM; DRY CHEMICAL, CARBON DIOXIDE, WATER FOG.
    PROTECTIVE CLOTHING, RUBBER GLOVES, AND BREATHING APPARATUS.
WARNING: 1] STRUCTURAL PROTECTIVE CLOTHING IS PERMEABLE, REMAIN CLEAR OF SMOKE, WATER FALL OUT AND WATER RUN OFF.
         2] KEEP OUT OF THE REACH OF CHILDREN.
         3] CONTAINERS MAY EXPLODE IN A FIRE OR OF HEATED ABOVE 120F.
         4] THERMAL DECOMPOSITION YIELDS TOXIC CARBON DIOXIDE, CARBON MONOXIDE.
         5] REMOVE ALL SOURCES OF IGNITION IF WITHOUT RISK.
         6] MOVE CONTAINERS FROM AREA IF WITHOUT RISK, COOL EXPOSED CONTAINERS.
         7] DIKE AREA FOR CONTROL AND CONTAINMENT TO PREVENT ENTRY INTO SEWERS, DRAIN, AND WATER WAYS.

LARGE SPILL/NO FIRE/RESCUE: WEAR RUBBER OR NEOPRENE BOOTS, GLOVES.

SPILL CONTROL AND CONTAINMENT:
    HOUSEHOLD SPILL: WIPE UP LIQUID WITH ABSORBENT MATERIAL, PUT INTO DISPOSAL CONTAINER. WASH SPILL AREA WITH SOAP
    AND WATER. PICK UP, PUT INTO DISPOSAL CONTAINER.
    LARGE SPILL: COVER SPILL AREA WITH ABSORBENT, SCOOP UP AND PUT INTO DISPOSAL CONTAINER. RINSE AREA WITH WATER,
    PICK UP WITH ABSORBENT, PUT INTO DISPOSAL CONTAINER.

HEALTH HAZARD INFORMATION:
    WHEN USING THIS PRODUCT WEAR: NO SPECIAL REQUIREMENTS. Contact may cause slight eye irritation.

    A physician should be contacted if anyone develops any signs or symptoms and suspects that they are caused by
    exposure to this product.

FIRST AID:
    Eye Exposure:
    Flush eyes with water for 15 minutes while lifting the upper and lower eye lids. Contact lenses should not be
    worn when working with this product. Get medical attention.

    Skin Exposure:
    Not expected to be a problem.

    Breathing:
    Not expected to be a problem.

    Swallowing:
    If any signs or symptoms appear, immediately contact the local poison control center for advice. Keep victim
    warm and at rest. Get medical attention.

PRODUCT: GLADE LIGHT - FLORAL WHISPER

INCOMPATIBILITY: NONE KNOWN

```
INGREDIENTS: PROPANE_____          TLV: 1000 ppm    CAS#:   74-98-6
             ISOBUTANE_____ 25-35%    TLV:              CAS#:   75-28-5
             N-BUTANE_____/      TLV:  800 ppm    CAS#:   106-97-8
             SURFACTANTS  LESS THAN 1%
             FRAGRANCE    LESS THAN 1%
             WATER        65-75%    TLV:              CAS#: 7732-18-5
```

IGNI TEMP: NA FP: 20F LEL: NA UEL: NA VP: NA VD: NA SG: .81 PS: LIQUID APPEAR: SPRAY MIST ODOR: FRAGRANCE
   PH FACTOR: NA

HAZARD RATINGS: H: F: R: 0 4 0

FIRE FIGHTING:
   HIGH EXPANSION FOAM, LOW EXPANSION FOAM, ALCOHOL FOAM; DRY CHEMICAL, CARBON DIOXIDE, WATER FOG.
   PROTECTIVE CLOTHING, RUBBER GLOVES, AND BREATHING APPARATUS.
WARNING: 1] STRUCTURAL PROTECTIVE CLOTHING IS PERMEABLE, REMAIN CLEAR OF SMOKE, WATER FALL OUT AND WATER RUN OFF.
         2] KEEP OUT OF THE REACH OF CHILDREN.
         3] CONTAINERS MAY EXPLODE IN A FIRE OR IF HEATED ABOVE 120F.
         4] THERMAL DECOMPOSITION YIELDS TOXIC CARBON DIOXIDE, CARBON MONOXIDE.
         5] REMOVE ALL SOURCES OF IGNITION IF WITHOUT RISK.
         6] MOVE CONTAINERS FROM AREA IF WITHOUT RISK, COOL EXPOSED CONTAINERS.
         7] DIKE AREA FOR CONTROL AND CONTAINMENT TO PREVENT ENTRY INTO SEWERS, DRAIN, AND WATER WAYS.

LARGE SPILL/NO FIRE/RESCUE: WEAR RUBBER OR NEOPRENE BOOTS, GLOVES.

SPILL CONTROL AND CONTAINMENT:
   HOUSEHOLD SPILL: WIPE UP LIQUID WITH ABSORBENT MATERIAL. WASH AREA WITH SOAP AND WATER. PICK UP AND PUT INTO
   DISPOSAL CONTAINER.
   LARGE SPILL: COVER SPILL AREA WITH ABSORBENT. SCOOP UP AND PUT INTO A DISPOSAL CONTAINER. RINSE SPILL AREA WITH
   WATER, PICK UP WITH ABSORBENT, PUT INTO A DISPOSAL CONTAINER.

HEALTH HAZARD INFORMATION:
   WHEN USING THIS PRODUCT WEAR: NO SPECIAL REQUIREMENTS. Contact with the eyes may cause slight irritation.

   A physician should be contacted if anyone develops any signs or symptoms and suspects that they are caused by
   exposure to this product.

FIRST AID:
   Eye Exposure:
   Flush eyes with water for 15 minutes while lifting the upper and lower eye lids. Contact lenses should not be
   worn when working with this product. Get medical attention.

   Skin Exposure:
   Not expected to be a problem.

   Breathing:
   Not expected to be a problem.

   Swallowing:
   If any signs or symptoms appear, immediately contact the local poison control center for advice. Keep victim
   warm and at rest. Get medical attention.

PRODUCT: GLADE LIGHT - HINT OF POWDER WITH CHLOROPHYLL

INCOMPATIBILITY: NONE KNOWN

INGREDIENTS: PROPANE   5-10%   TLV: 1000 ppm   CAS#:   74-98-6
             ISOBUTANE   5-10%   TLV: CAS#:   75-28-5
             N-BUTANE   5-10%   TLV: 800 ppm   CAS#:   106-97-8
             WATER   65-75%   TLV: CAS#:   7732-18-5

IGNI TEMP: NA FP: 20F LEL: NA UEL: NA VP: NA VD: NA SG: 0.8 PS: LIQUID APPEAR: SPRAY MIST ODOR: FRAGRANCE
    PH FACTOR: NA

HAZARD RATINGS: H: F: R: 0 4 0

FIRE FIGHTING:
  HIGH EXPANSION FOAM, LOW EXPANSION FOAM, ALCOHOL FOAM; DRY CHEMICAL, CARBON DIOXIDE, WATER FOG.
  PROTECTIVE CLOTHING, RUBBER GLOVES, AND BREATHING APPARATUS.
WARNING: 1] STRUCTURAL PROTECTIVE CLOTHING IS PERMEABLE, REMAIN CLEAR OF SMOKE, WATER FALL OUT AND WATER RUN OFF.
         2] KEEP OUT OF THE REACH OF CHILDREN.
         3] CONTAINERS MAY EXPLODE IN A FIRE OR IF HEATED ABOVE 120F.
         4] THERMAL DECOMPOSITION YIELDS TOXIC CARBON DIOXIDE, CARBON MONOXIDE.
         5] REMOVE ALL SOURCES OF IGNITION IF WITHOUT RISK.
         6] MOVE CONTAINERS FROM AREA IF WITHOUT RISK, COOL EXPOSED CONTAINERS.
         7] DIKE AREA FOR CONTROL AND CONTAINMENT TO PREVENT ENTRY INTO SEWERS, DRAIN, AND WATER WAYS.

LARGE SPILL/NO FIRE/RESCUE: WEAR RUBBER OR NEOPRENE BOOTS, GLOVES.

SPILL CONTROL AND CONTAINMENT:
  HOUSEHOLD SPILL: VENTILATE AREA. WIPE UP LIQUID WITH ABSORBENT MATERIAL. PUT INTO A DISPOSAL CONTAINER. RINSE
  WITH WATER, PICK UP WITH ABSORBENT MATERIAL, PUT INTO DISPOSAL CONTAINER.
  LARGE SPILL: COVER SPILL WITH ABSORBENT, SCOOP UP AND PUT INTO A DISPOSAL CONTAINER. RINSE AREA WITH WATER, PICK
  UP WITH ABSORBENT, PUT INTO DISPOSAL CONTAINER. SEAL AND REMOVE FROM WORK PLACE.

HEALTH HAZARD INFORMATION:
  WHEN USING THIS PRODUCT WEAR: NO SPECIAL REQUIREMENTS.

  A physician should be contacted if anyone develops any signs or symptoms and suspects that they are caused by
  exposure to this product.

FIRST AID:
  Eye Exposure:
  Flush with water for 15 minutes while lifting the upper and lower eye lids. Contact lenses should not be worn
  when working with this product. Get medical attention.

  Skin Exposure:
  Not expected to be a problem.

  Breathing:
  Not expected to be a problem.

  Swallowing:
  Contact the local poison control center for advice. Keep victim warm and at rest. Get medical attention.

PRODUCT: GLADE SPIN FRESH (ALL FRAGRANCE)

INCOMPATIBILITY: NONE KNOWN

INGREDIENTS: ETHYLENE VINYL ACETATE BEADS 60-70% TLV: CAS#:
             TALC   1% TLV: CAS#:
             FRAGRANCE   30-40% TLV: CAS#:

IGNI TEMP: NA FP: 148F LEL: NA UEL: NA VP: NA VD: NA SG: NA PS: BEADS APPEAR: STRAW COLORED PLASTIC
    ODOR: FRAGRANT   PH FACTOR: NA

HAZARD RATINGS: H: F: R:

FIRE FIGHTING:
    HIGH EXPANSION FOAM, LOW EXPANSION FOAM, ALCOHOL FOAM; DRY CHEMICAL, CARBON DIOXIDE, WATER FOG.
    PROTECTIVE CLOTHING, RUBBER GLOVES, AND BREATHING APPARATUS.
WARNING: 1] STRUCTURAL PROTECTIVE CLOTHING IS PERMEABLE, REMAIN CLEAR OF SMOKE, WATER FALL OUT AND WATER RUN OFF.
         2] KEEP OUT OF THE REACH OF CHILDREN.
         3] THERMAL DECOMPOSITION YIELDS TOXIC CARBON DIOXIDE, CARBON MONOXIDE.
         4] MOVE CONTAINERS FROM AREA IF WITHOUT RISK, COOL EXPOSED CONTAINERS.
         5] DIKE AREA FOR CONTROL AND CONTAINMENT TO PREVENT ENTRY INTO SEWERS, DRAIN, AND WATER WAYS.

LARGE SPILL/NO FIRE/RESCUE: WEAR RUBBER OR NEOPRENE BOOTS, GLOVES.

SPILL CONTROL AND CONTAINMENT:
    HOUSEHOLD SPILL: SWEEP UP BEADS AND PUT INTO A DISPOSAL CONTAINER.
    LARGE SPILL: SWEEP UP BEADS AND PUT INTO A DISPOSAL CONTAINER.

HEALTH HAZARD INFORMATION:
    WHEN USING THIS PRODUCT WEAR: NO SPECIAL REQUIREMENT. Ingestion may create a gastrointestinal problem.

    A physician should be contacted if anyone develops any signs or symptoms and suspects that they are caused by
    exposure to this product.

FIRST AID:
  Eye Exposure:
  Not expected to be a problem.

  Skin Exposure:
  Not expected to be a problem.

  Breathing:
  Not expected to be a problem.

  Swallowing:
  Immediately contact the local poison control center for advice. Keep the victim warm and at rest. Get medical
  attention.

PRODUCT: GLO-COAT

INCOMPATIBILITY: NONE KNOWN

INGREDIENTS: STYRENE/ACRYLIC POLYMERS 5-15% TLV: CAS#:
            WAX   1-3% TLV: CAS#:
            DIETHYLENE GLYCOL   Recommended
            MONOETHYL ETHER   3-6% TLV: 30 ppm   CAS#: 111-90-0
            WAX   1-3% TLV: 10 ppm   CAS#: 9010-77-9
            WATER   80-90% TLV: CAS#: 7732-18-5

IGNI TEMP: NA FP: NA LEL: NA UEL: NA VP: NA VD: NA SG: 1.02 PS: LIQUID APPEAR: MILKY WHITE
    ODOR: MILD MONOMERIC   PH FACTOR: 7.3

HAZARD RATINGS: H: F: R: 1   0   0

FIRE FIGHTING:
  HIGH EXPANSION FOAM, LOW EXPANSION FOAM, ALCOHOL FOAM; DRY CHEMICAL, CARBON DIOXIDE, WATER FOG.
  PROTECTIVE CLOTHING, RUBBER GLOVES, AND BREATHING APPARATUS.
WARNING: 1] STRUCTURAL PROTECTIVE CLOTHING IS PERMEABLE, REMAIN CLEAR OF SMOKE, WATER FALL OUT AND WATER RUN OFF.
         2] KEEP OUT OF THE REACH OF CHILDREN.
         3] CONTAINERS MAY BURST IN A FIRE.
         4] THERMAL DECOMPOSITION YIELDS TOXIC CARBON DIOXIDE, CARBON MONOXIDE.
         5] MOVE CONTAINERS FROM AREA IF WITHOUT RISK, COOL EXPOSED CONTAINERS.
         6] DIKE AREA FOR CONTROL AND CONTAINMENT TO PREVENT ENTRY INTO SEWERS, DRAIN, AND WATER WAYS.

LARGE SPILL/NO FIRE/RESCUE: WEAR RUBBER OR NEOPRENE BOOTS, GLOVES.

SPILL CONTROL AND CONTAINMENT:
  HOUSEHOLD SPILL: WIPE UP LIQUID WITH ABSORBENT CLOTH, RINSE OUT IN THE SINK. WIPE UP RESIDUE WITH A DAMP CLOTH,
  RINSE OUT IN THE SINK.
  LARGE SPILL: COVER SPILL WITH ABSORBENT, SWEEP UP AND PUT INTO A DISPOSAL CONTAINER. RINSE WITH WATER, PICK UP
  WITH ABSORBENT, PUT IN DISPOSAL CONTAINER.

HEALTH HAZARD INFORMATION:
  WHEN USING THIS PRODUCT WEAR: NO SPECIAL REQUIREMENT. Direct contact with the eyes may cause stinging, tearing,
  itching, swelling, and redness. Prolonged or repeated contact with the skin may cause skin rash. Wear latex-type
  gloves.

  A physician should be contacted if anyone develops any signs or symptoms and suspects that they are caused by
  exposure to this product.

FIRST AID:
Eye Exposure:
Flush with water for 15 minutes while lifting the upper and lower eye lids. Contact lenses should not be worn
when working with this product. If irritation persists, get medical attention.

  Skin Exposure:
  Wash with soap and water.

  Breathing:
  Not expected to be a problem.

  Swallowing:
  Immediately contact the local poison control center for advice. Keep the victim warm and at rest. Get medical
  attention.

PRODUCT: GLORY RUG CLEANER (AEROSOL)

INCOMPATIBILITY: NONE KNOWN

INGREDIENTS: SODIUM a-OLEFIN SULFONATE UNDER 5% TLV: CAS#:
           MODIFIED ACRYLIC POLYMER   ·2-5% TLV: CAS#:
           ISOBUTANE   5-15% TLV: CAS#:   75-28-5
           WATER   70-85% TLV: CAS#: 7732-18-5

IGNI TEMP: NA FP: 20F LEL: NA UEL: NA VP: NA VD: NA SG: .96 PS: LIQUID APPEAR: WHITE FOAM ODOR: MILD
   PH FACTOR: NA

HAZARD RATINGS: H: F: R: 2 4 0

FIRE FIGHTING:
  HIGH EXPANSION FOAM, LOW EXPANSION FOAM, ALCOHOL FOAM; DRY CHEMICAL, CARBON DIOXIDE, WATER FOG.
  PROTECTIVE CLOTHING, RUBBER GLOVES, AND BREATHING APPARATUS.
WARNING: 1] STRUCTURAL PROTECTIVE CLOTHING IS PERMEABLE, REMAIN CLEAR OF SMOKE, WATER FALL OUT AND WATER RUN OFF.
       2] KEEP OUT OF THE REACH OF CHILDREN.
       3] CONTAINERS WILL EXPLODE IN A FIRE OR IF HEATED ABOVE 140F.
       4] THERMAL DECOMPOSITION YIELDS TOXIC CARBON DIOXIDE, CARBON MONOXIDE.
       5] REMOVE ALL SOURCES OF IGNITION IF WITHOUT RISK.
       6] MOVE CONTAINERS FROM AREA IF WITHOUT RISK, COOL EXPOSED CONTAINERS.
       7] DIKE AREA FOR CONTROL AND CONTAINMENT TO PREVENT ENTRY INTO SEWERS, DRAIN, AND WATER WAYS.

LARGE SPILL/NO FIRE/RESCUE: WEAR RUBBER OR NEOPRENE BOOTS, GLOVES.

SPILL CONTROL AND CONTAINMENT:
  HOUSEHOLD SPILL: SCOOP OR WIPE UP FOAM WITH A DAMP CLOTH. RINSE OUT IN THE SINK. WIPE UP RESIDUE WITH A DAMP
  CLOTH, RINSE OUT IN THE SINK.
  LARGE SPILL: COVER SPILL WITH ABSORBENT. SCOOP UP AND PUT INTO A DISPOSAL CONTAINER. RINSE AREA WITH WATER, PICK
  UP WITH ABSORBENT. PUT IN A DISPOSAL CONTAINER. SEAL AND REMOVE FROM THE WORK AREA.

HEALTH HAZARD INFORMATION:
  WHEN USING THIS PRODUCT WEAR: NO SPECIAL REQUIREMENT. Eye contact may cause stinging, tearing, itching,
  swelling, and redness. Prolonged or repeated skin contact may cause a rash. For prolonged exposure, wear latex
  gloves.

  A physician should be contacted if anyone develops any signs or symptoms and suspects that they are caused by
  exposure to this product.

FIRST AID:
  Eye Exposure:
  Flush with water for 15 minutes while lifting the upper and lower eye lids. Contact lenses should not be worn
  when working with this product. If irritation persists, get medical attention.

  Skin Exposure:
  Wash with soap and water.

  Breathing:
  Not expected to be a problem.

  Swallowing:
  Immediately contact the local poison control center for advice. Keep the victim warm and at rest. Get medical
  attention.

PRODUCT: GLORY RUG SHAMPOO (LIQUID)

INCOMPATIBILITY: NONE KNOWN

INGREDIENTS: SODIUM LAURYL SULFATE 6-10% TLV: CAS#: 151-21-3
             AMMONIUM HYDROXIDE   .1-.5% TLV: 25 ppm(ammonia) CAS#: 1336-21-6
             FORMALDEHYDE   .01-.1% TLV: 1 ppm   CAS#:   50-00-0
             MODIFIED ACRYLIC
             POLYMER   5-10%   TLV: CAS#:
             WATER   79-89% TLV: CAS#: 7732-18-5

IGNI TEMP: NA FP: NA LEL: NA UEL: NA VP: NA VD: NA SG: 1.02 PS: LIQUID APPEAR: STRAW COLORED
     ODOR: MILD DETERGENT PH FACTOR: 9.4

HAZARD RATINGS: H: F: R: 1 0 0

FIRE FIGHTING:
  HIGH EXPANSION FOAM, LOW EXPANSION FOAM, ALCOHOL FOAM; DRY CHEMICAL, CARBON DIOXIDE, WATER FOG.
  PROTECTIVE CLOTHING, RUBBER GLOVES, AND BREATHING APPARATUS.
WARNING: 1] STRUCTURAL PROTECTIVE CLOTHING IS PERMEABLE, REMAIN CLEAR OF SMOKE, WATER FALL OUT AND WATER RUN OFF.
         2] KEEP OUT OF THE REACH OF CHILDREN.
         3] THERMAL DECOMPOSITION YIELDS TOXIC CARBON DIOXIDE, CARBON MONOXIDE.
         4] MOVE CONTAINERS FROM AREA IF WITHOUT RISK, COOL EXPOSED CONTAINERS.
         5] DIKE AREA FOR CONTROL AND CONTAINMENT TO PREVENT ENTRY INTO SEWERS, DRAIN, AND WATER WAYS.

LARGE SPILL/NO FIRE/RESCUE: WEAR RUBBER OR NEOPRENE BOOTS, GLOVES.

SPILL CONTROL AND CONTAINMENT:
  HOUSEHOLD SPILL: WIPE UP LIQUID WITH ABSORBENT CLOTH, RINSE OUT IN SINK. WIPE UP RESIDUE WITH DAMP CLOTH, RINSE
  OUT IN THE SINK.
  LARGE SPILL: COVER WITH ABSORBENT. PICK UP AND PUT INTO A DISPOSAL CONTAINER. RINSE AREA WITH WATER, PICK UP
  WITH ABSORBENT, PUT IN A DISPOSAL CONTAINER. SEAL AND REMOVE FROM THE WORK AREA.

HEALTH HAZARD INFORMATION:
  WHEN USING THIS PRODUCT WEAR: NO SPECIAL REQUIREMENT. Eye contact can cause stinging, tearing, itching,
  swelling, and redness. Prolonged or repeated skin contact may cause itching and redness.

  A physician should be contacted if anyone develops any signs or symptoms and suspects that they are caused by
  exposure to this product.

FIRST AID:
  Eye Exposure:
  Flush with water for 15 minutes while lifting the upper and lower eye lids. Contact lenses should not be worn
  when working with this product. If irritation persists, get medical attention.

  Skin Exposure:
  Wash with soap and water.

  Breathing:
  Not expected to be a problem.

  Swallowing:
  Immediately contact the local poison control center for advice. Keep the victim warm and at rest. Get medical
  attention.

PRODUCT: GODDARD'S ALMOND FURNITURE OIL

INCOMPATIBILITY: NONE KNOWN

INGREDIENTS: ISOPARAFFINIC NAPHTHA   55-65% TLV: CAS#:
             HYDROCARBON OIL   35-45% TLV: CAS#:
             HYDROCARBON POLYMER   2-4% TLV: CAS#:
             FRAGRANCE   1-2% TLV: CAS#:

IGNI TEMP: NA FP: 134F LEL: NA UEL: NA VP: NA VD: NA SG: .8 PS: LIQUID APPEAR: LIGHT BROWN ODOR: ALMOND
   PH FACTOR: NA

HAZARD RATINGS: H: F: R:

FIRE FIGHTING:
   HIGH EXPANSION FOAM, LOW EXPANSION FOAM, ALCOHOL FOAM; DRY CHEMICAL, CARBON DIOXIDE, WATER FOG.
   PROTECTIVE CLOTHING, RUBBER GLOVES, AND BREATHING APPARATUS.
WARNING: 1] STRUCTURAL PROTECTIVE CLOTHING IS PERMEABLE, REMAIN CLEAR OF SMOKE, WATER FALL OUT AND WATER RUN OFF.
         2] KEEP OUT OF THE REACH OF CHILDREN.
         3] REMOVE ALL SOURCES OF IGNITION IF WITHOUT RISK.
         4] THERMAL DECOMPOSITION YIELDS TOXIC CARBON DIOXIDE, CARBON MONOXIDE.
         5] CONTAINERS MAY BURST IN A FIRE.
         6] MOVE CONTAINERS FROM AREA IF WITHOUT RISK, COOL EXPOSED CONTAINERS.
         7] DIKE AREA FOR CONTROL AND CONTAINMENT TO PREVENT ENTRY INTO SEWERS, DRAIN, AND WATER WAYS.

LARGE SPILL/NO FIRE/RESCUE: WEAR RUBBER OR NEOPRENE BOOTS, GLOVES.

SPILL CONTROL AND CONTAINMENT:
   HOUSEHOLD SPILL: WIPE UP LIQUID WITH AN ABSORBENT CLOTH. PUT IN DISPOSAL CONTAINER. WASH SPILL AREA WITH SOAP
   AND WATER, WIPE UP WITH ABSORBENT CLOTH. PUT IN A DISPOSAL CONTAINER.
   LARGE SPILL: COVER SPILL WITH ABSORBENT. PICK UP AND PUT IN A DISPOSAL CONTAINER. SCRUB AREA WITH DETERGENT AND
   WATER, PICK UP WITH ABSORBENT, PUT IN A DISPOSAL CONTAINER. SEAL AND REMOVE FROM THE WORK AREA.

HEALTH HAZARD INFORMATION:
   WHEN USING THIS PRODUCT WEAR: NO SPECIAL REQUIREMENT. Eye contact will cause stinging, tearing, itching,
   swelling, and redness. Repeated or prolonged skin contact may cause irritation. Ingestion may cause abdominal
   pain, nausea, vomiting, and diarrhea. Aspiration into the lungs will cause chemical pneumonia. May be fatal.

   A physician should be contacted if anyone develops any signs or symptoms and suspects that they are caused by
   exposure to this product.

FIRST AID:
   Eye Exposure:
   Flush with water for 15 minutes while lifting the upper and lower eye lids. Contact lenses should not be worn
   when working with this product. If irritation persists, get medical attention.

   Skin Exposure:
   Wash skin with soap and water.

   Breathing:
   Not expected to be a problem.

   Swallowing:
   Do not induce vomiting. Immediately contact the local poison control center for advice. Keep the victim warm and
   at rest. Get medical attention.

PRODUCT: GODDARD'S BRASS AND COPPER CLEANER

INCOMPATIBILITY: STRONG ALKALIS; BASES; CAUSTICS; (e.g. Sodium Hydroxide)

INGREDIENTS: ALUMINUM OXIDE   38-45% TLV: 10 mg/m3(dust) CAS#: 1344-28-1
             SILICA   2-4% TLV: 2.6 mg/m3(dust)CAS#: 63231-67-4
             SURFACTANTS, CLEANERS 40-50% TLV: CAS#:
             FRAGRANCE, COLORANT under .5% TLV: CAS#:
             WATER   15-25% TLV: CAS#: 7732-18-5

IGNI TEMP: NA FP: NA LEL: NA UEL: NA VP: NA VD: NA SG: 1.35 PS: PASTE APPEAR: REDDISH BROWN
     ODOR: CITRUS   PH FACTOR: 2.0

HAZARD RATINGS: H: F: R: 1 0 0

FIRE FIGHTING:
   HIGH EXPANSION FOAM, LOW EXPANSION FOAM, ALCOHOL FOAM; DRY CHEMICAL, CARBON DIOXIDE, WATER FOG.
   PROTECTIVE CLOTHING, RUBBER GLOVES, AND BREATHING APPARATUS.
WARNING: 1] STRUCTURAL PROTECTIVE CLOTHING IS PERMEABLE, REMAIN CLEAR OF SMOKE, WATER FALL OUT AND WATER RUN OFF.
         2] KEEP OUT OF THE REACH OF CHILDREN.
         3] THERMAL DECOMPOSITION YIELDS TOXIC CARBON DIOXIDE, CARBON MONOXIDE.
         4] MOVE CONTAINERS FROM AREA IF WITHOUT RISK, COOL EXPOSED CONTAINERS.
         5] DIKE AREA FOR CONTROL AND CONTAINMENT TO PREVENT ENTRY INTO SEWERS, DRAIN, AND WATER WAYS.

LARGE SPILL/NO FIRE/RESCUE: WEAR RUBBER OR NEOPRENE BOOTS, GLOVES.

SPILL CONTROL AND CONTAINMENT:
   HOUSEHOLD SPILL: WIPE UP LIQUID WITH AN ABSORBENT CLOTH, RINSE OUT IN THE SINK. WIPE UP RESIDUE WITH A DAMP
   CLOTH, RINSE OUT IN THE SINK.
   LARGE SPILL: COVER WITH ABSORBENT. PICK UP AND PUT IN A DISPOSAL CONTAINER. SCRUB AREA WITH DETERGENT AND WATER,
   PICK UP WITH ABSORBENT, PUT IN A DISPOSAL CONTAINER. SEAL AND REMOVE FROM THE WORK AREA.

HEALTH HAZARD INFORMATION:
   WHEN USING THIS PRODUCT WEAR: Eye contact may cause stinging, tearing, itching, swelling, and redness. Prolonged
   skin contact may cause irritation.

   A physician should be contacted if anyone develops any signs or symptoms and suspects that they are caused by
   exposure to this product.

FIRST AID:
Eye Exposure:
Flush with water for 15 minutes while lifting the upper and lower eye lids. Contact lenses should not be worn
when working with this product. If irritation persists, get medical attention.

Skin Exposure:
Wash skin with soap and water.

Breathing:
Not expected to be a problem.

Swallowing:
Immediately contact the local poison control center for advice. Keep the victim warm and at rest. Get medical
attention.

PRODUCT: GODDARD'S CLEANING AND DUSTING SPRAY

INCOMPATIBILITY: NONE KNOWN

INGREDIENTS: ISOPARAFFINIC   Recommended
             HYDROCARBON SOLVENT   10-12% TLV: 400 ppm   CAS#: 64742-48-9
             SURFACTANTS, OILS   3-6% TLV: CAS#:
             FRAGRANCE _   under   .5% TLV: CAS#:
             N-BUTANE   _____ 10-20% TLV: 800 ppm   CAS#:   106-97-8
             PROPANE   _/   TLV: CAS#:   74-98-6
             WATER   TLV: CAS#: 7732-18-5

IGNI TEMP: NA FP: 20F LEL: NA UEL: NA VP: NA VD: NA SG: .72 PS: LIQUID APPEAR: SPRAY MIST ODOR: LEMON
   PH FACTOR: NA

HAZARD RATINGS: H: F: R: 0 4 0

FIRE FIGHTING:
   HIGH EXPANSION FOAM, LOW EXPANSION FOAM, ALCOHOL FOAM; DRY CHEMICAL, CARBON DIOXIDE, WATER FOG.
   PROTECTIVE CLOTHING, RUBBER GLOVES, AND BREATHING APPARATUS.
WARNING: 1] STRUCTURAL PROTECTIVE CLOTHING IS PERMEABLE, REMAIN CLEAR OF SMOKE, WATER FALL OUT AND WATER RUN OFF.
         2] KEEP OUT OF THE REACH OF CHILDREN.
         3] CONTAINERS MAY EXPLODE IN A FIRE OR IF HEATED ABOVE 120F.
         4] REMOVE ALL SOURCES OF IGNITION IF WITHOUT RISK.
         5] THERMAL DECOMPOSITION YIELDS TOXIC, CARBON DIOXIDE, CARBON MONOXIDE.
         6] MOVE CONTAINERS FROM AREA IF WITHOUT RISK, COOL EXPOSED CONTAINERS.
         7] DIKE AREA FOR CONTROL AND CONTAINMENT TO PREVENT ENTRY INTO SEWERS, DRAIN, AND WATER WAYS.

LARGE SPILL/NO FIRE/RESCUE: WEAR RUBBER OR NEOPRENE BOOTS, GLOVES.

SPILL CONTROL AND CONTAINMENT:
   HOUSEHOLD SPILL: WIPE UP LIQUID WITH AN ABSORBENT CLOTH, RINSE OUT IN THE SINK. WASH SPILL AREA WITH SOAP AND
   WATER, WIPE UP WITH DAMP CLOTH, RINSE OUT IN THE SINK.
   LARGE SPILL: COVER SPILL WITH ABSORBENT. SWEEP UP, PUT IN A DISPOSAL CONTAINER. SCRUB AREA WITH A DETERGENT AND
   WATER. PICK UP WITH ABSORBENT, PUT IN A DISPOSAL CONTAINER. SEAL AND REMOVE FROM THE WORK AREA.

HEALTH HAZARD INFORMATION:
   WHEN USING THIS PRODUCT WEAR: NO SPECIAL REQUIREMENT. Eye contact may cause stinging, tearing, itching,
   swelling, and redness. Ingestion may cause stomach pain, nausea, vomiting, and diarrhea.

   A physician should be contacted if anyone develops any signs or symptoms and suspects that they are caused by
   exposure to this product.

FIRST AID:
   Eye Exposure:
   Flush with water for 15 minutes while lifting the upper and lower eye lids. Contact lenses should not be worn
   when working with this product. If irritation persists, get medical attention.

   Skin Exposure:
   Wash skin with soap and water.

   Breathing:
   Not expected to be a problem.

   Swallowing:
   Immediately contact the local poison control center for advice. Keep the victim warm and at rest. Get medical
   attention.

PRODUCT: GODDARD'S DRY CLEAN

INCOMPATIBILITY: STRONG ALKALIS; BASES; CAUSTICS; STRONG OXIDIZERS.

INGREDIENTS: METHYLENE CHLORIDE   20-30% TLV: 100 ppm   CAS#: 75-09-2
           1,1,1-TRICHLOROETHANE 10-20% TLV: 350 ppm   CAS#: 71-55-6
           ISOPROPYL ALCOHOL   5-10% TLV: 400 ppm   CAS#: 67-63-0
           ISOPARAFFINIC   recommended
           HYDROCARBON SOLVENT   3-6% TLV: 400 ppm   CAS#:
           ISOBUTANE AND PROPANE 35-45% TLV: 1000 ppm   CAS#:
    OR N-BUTANE AND PROPANE 35-45% TLV: 800 ppm   CAS#:

IGNI TEMP: NA FP: 20F LEL: NA UEL: NA VP: NA VD: NA SG: 1.2 PS: LIQUID APPEAR: SPRAY
   ODOR: CHLORINATED SOLVENT   PH FACTOR: NA

HAZARD RATINGS: H: F: R: 2 4 0

FIRE FIGHTING:
  HIGH EXPANSION FOAM, LOW EXPANSION FOAM, ALCOHOL FOAM; DRY CHEMICAL, CARBON DIOXIDE, WATER FOG.
  PROTECTIVE CLOTHING, RUBBER GLOVES, AND BREATHING APPARATUS.
WARNING: 1] STRUCTURAL PROTECTIVE CLOTHING IS PERMEABLE, REMAIN CLEAR OF SMOKE, WATER FALL OUT AND WATER RUN OFF.
      2] KEEP OUT OF THE REACH OF CHILDREN.
      3] CONTAINERS WILL EXPLODE IN A FIRE OR IF HEATED ABOVE 120F.
      4] THERMAL DECOMPOSITION YIELDS TOXIC FUMES OF HYDROCHLORIC ACID.
      5] MOVE CONTAINERS FROM AREA IF WITHOUT RISK, COOL EXPOSED CONTAINERS.
      6] DIKE AREA FOR CONTROL AND CONTAINMENT TO PREVENT ENTRY INTO SEWERS, DRAIN, AND WATER WAYS.

LARGE SPILL/NO FIRE/RESCUE: WEAR RUBBER OR NEOPRENE NON-SEALED CHEMICAL PROTECTIVE BOOTS, GLOVES AND BREATHING
                     APPARATUS.

SPILL CONTROL AND CONTAINMENT:
  HOUSEHOLD SPILL: WIPE UP LIQUID WITH AN ABSORBENT CLOTH, RINSE OUT IN THE SINK. WIPE UP ANY RESIDUE WITH A DAMP
  CLOTH, RINSE OUT IN THE SINK.
  LARGE SPILL: COVER WITH ABSORBENT. SWEEP UP AND PUT IN A DISPOSAL CONTAINER. SCRUB AREA WITH A DETERGENT AND
  WATER, PICK UP WITH ABSORBENT. PUT IN A DISPOSAL CONTAINER. SEAL AND REMOVE FROM THE WORK AREA.

HEALTH HAZARD INFORMATION:
  WHEN USING THIS PRODUCT: VENTILATE IMMEDIATE AREA. Inhalation may cause irritation of the nose, throat, and
  lungs. Contact with the eyes can cause irritation. Contact with the skin may cause irritation.

  A physician should be contacted if anyone develops any signs or symptoms and suspects that they are caused by
  exposure to this product.

FIRST AID:
  Eye Exposure:
  Flush with water for 15 minutes while lifting the upper and lower eye lids. Contact lenses should not be worn
  when working with this product. If irritation persists, get medical attention.

  Skin Exposure:
  Wash with soap and water. If irritation appears, get medical attention.

  Breathing:
  If any signs appear move the victim to fresh air.

  Swallowing:
  Immediately contact the local poison control center for advice. Keep the victim warm and at rest. Get medical
  attention.

PRODUCT: GODDARD'S FINE FURNITURE CLEANER

INCOMPATIBILITY: NONE KNOWN

INGREDIENTS: ISOPARAFFINIC   recommended
             HYDROCARBON SOLVENT 10-20%   TLV: 400 ppm   CAS#:
             WAXES, OILS   3-6%   TLV: CAS#:
             SURFACTANTS   under   1%   TLV: CAS#:
             N-BUTANE_____   TLV: 800 ppm   CAS#:   106-97-8
             PROPANE_____ 10-20%   TLV: 1000 ppm   CAS#:   74-98-6
             ISOBUTANE____/   TLV: CAS#:
             WATER   TLV: CAS#: 7732-18-5

IGNI TEMP: NA FP: 20F LEL: NA UEL: NA VP: NA VD: NA SG: .73 PS: LIQUID APPEAR: SPRAY MIST ODOR: CITRUS
    PH FACTOR:

HAZARD RATINGS: H: F: R: 0 4 0

FIRE FIGHTING:
   HIGH EXPANSION FOAM, LOW EXPANSION FOAM, ALCOHOL FOAM; DRY CHEMICAL, CARBON DIOXIDE, WATER FOG.
   PROTECTIVE CLOTHING, RUBBER GLOVES, AND BREATHING APPARATUS.
WARNING: 1] STRUCTURAL PROTECTIVE CLOTHING IS PERMEABLE, REMAIN CLEAR OF SMOKE, WATER FALL OUT AND WATER RUN OFF.
         2] KEEP OUT OF THE REACH OF CHILDREN.
         3] CONTAINERS WILL EXPLODE IN A FIRE OR IF HEATED ABOVE 120F.
         4] THERMAL DECOMPOSITION YIELDS TOXIC CARBON DIOXIDE, CARBON MONOXIDE.
         5] MOVE CONTAINERS FROM AREA IF WITHOUT RISK, COOL EXPOSED CONTAINERS.
         6] DIKE AREA FOR CONTROL AND CONTAINMENT TO PREVENT ENTRY INTO SEWERS, DRAIN, AND WATER WAYS.

LARGE SPILL/NO FIRE/RESCUE: WEAR RUBBER OR NEOPRENE BOOTS, GLOVES.

SPILL CONTROL AND CONTAINMENT:
   HOUSEHOLD SPILL: WIPE UP LIQUID WITH AN ABSORBENT CLOTH. RINSE OUT IN THE SINK. WASH AREA WITH SOAP AND WATER
   WIPE UP DRY WITH AN ABSORBENT CLOTH, RINSE OUT IN THE SINK.
   LARGE SPILL: COVER SPILL WITH ABSORBENT. SWEEP UP AND PUT IN A DISPOSAL CONTAINER. SCRUB AREA WITH A DETERGENT
   AND WATER, PICK UP WITH ABSORBENT. PUT IN A DISPOSAL CONTAINER. SEAL AND REMOVE FROM THE WORK AREA.

HEALTH HAZARD INFORMATION:
   WHEN USING THIS PRODUCT WEAR: NO SPECIAL REQUIREMENT. Eye contact will cause stinging, tearing, itching,
   swelling, and redness. May cause skin irritation if allergic to the product. Ingestion will cause abdominal
   pain, nausea, vomiting, and diarrhea.

   A physician should be contacted if anyone develops any signs or symptoms and suspects that they are caused by
   exposure to this product.

FIRST AID:
   Eye Exposure:
   Flush with water for 15 minutes while lifting the upper and lower eye lids. Contact lenses should not be worn
   when working with this product. If irritation persists, get medical attention.

   Skin Exposure:
   Wash skin with soap and water. If irritation persists, get medical attention.

   Breathing:
   Not expected to be a problem.

   Swallowing:
   Do not induce vomiting. Immediately contact the local poison control center for advice. Keep the victim warm and
   at rest. Get medical attention.

PRODUCT: GODDARD'S   FINE FURNITURE CREME WITH ALMOND OIL

INCOMPATIBILITY: NONE KNOWN

INGREDIENTS: ISOPARAFFINIC NAPHTHA   5-10% TLV: CAS#:
             SILICONE FLUIDS   2-4% TLV: CAS#:
             SURFACTANTS   1-3% TLV: CAS#:
             WAXES   under   1%   TLV: CAS#:
             FRAGRANCE   under   1%   TLV: CAS#:
             WATER   TLV: CAS#: 7732-18-5

IGNI TEMP: NA FP: NA LEL: NA UEL: NA VP: NA VD: NA SG: 1.0 PS: LIQUID APPEAR: WHITE OPAQUE
    ODOR: ALMOND   PH FACTOR: 6.0 to 7.0

HAZARD RATINGS: H: F: R:

FIRE FIGHTING:
    HIGH EXPANSION FOAM, LOW EXPANSION FOAM, ALCOHOL FOAM; DRY CHEMICAL, CARBON DIOXIDE, WATER FOG.
    PROTECTIVE CLOTHING, RUBBER GLOVES, AND BREATHING APPARATUS.
WARNING: 1] STRUCTURAL PROTECTIVE CLOTHING IS PERMEABLE, REMAIN CLEAR OF SMOKE, WATER FALL OUT AND WATER RUN OFF.
         2] KEEP OUT OF THE REACH OF CHILDREN.
         3] THERMAL DECOMPOSITION YIELDS TOXIC CARBON DIOXIDE, CARBON MONOXIDE.
         4] MOVE CONTAINERS FROM AREA IF WITHOUT RISK, COOL EXPOSED CONTAINERS.
         5] DIKE AREA FOR CONTROL AND CONTAINMENT TO PREVENT ENTRY INTO SEWERS, DRAIN, AND WATER WAYS.

LARGE SPILL/NO FIRE/RESCUE: WEAR RUBBER OR NEOPRENE BOOTS, GLOVES.

SPILL CONTROL AND CONTAINMENT:
    HOUSEHOLD SPILL: WIPE UP LIQUID WITH AN ABSORBENT CLOTH, WASH OUT IN THE SINK. WASH SPILL AREA WITH SOAP AND
    WATER, WIPE UP WITH ABSORBENT CLOTH, RINSE OUT IN THE SINK.
    LARGE SPILL: COVER SPILL WITH ABSORBENT. SWEEP UP AND PUT IN A DISPOSAL CONTAINER. SCRUB AREA WITH A DETERGENT
    AND WATER, PICK UP WITH ABSORBENT. PUT IN A DISPOSAL CONTAINER. SEAL AND REMOVE FROM THE WORK AREA.

HEALTH HAZARD INFORMATION:
    WHEN USING THIS PRODUCT WEAR: NO SPECIAL REQUIREMENT. Contact with the eyes can cause stinging, tearing,
    itching, swelling, and redness. Prolonged or repeated contact with the skin may cause irritation. Ingestion will
    cause abdominal pain, nausea, vomiting, and diarrhea.

    A physician should be contacted if anyone develops any signs or symptoms and suspects that they are caused by
    exposure to this product.

FIRST AID:
Eye Exposure:
Flush with water for 15 minutes while lifting the upper and lower eye lids. Contact lenses should not be worn
when working with this product. If irritation persists, get medical attention.

Skin Exposure:
Wash with soap and water. If irritation appears, get medical attention.

Breathing:
Not expected to be a problem.

Swallowing:
Immediately contact the local poison control center for advice. Keep the victim warm and at rest. Get medical
attention.

PRODUCT: GODDARD'S   FINE FURNITURE CREME WITH LEMON BEES WAX

INCOMPATIBILITY: NONE KNOWN

INGREDIENTS: ISOPARAFFINIC NAPHTHA 5-10% TLV: CAS#:
             SILICONE FLUIDS   2-4% TLV: CAS#:
             SURFACTANTS   1-3% TLV: CAS#:
             WAXES   under   1% TLV: CAS#:
             FRAGRANCE   under   1% TLV: CAS#:
             WATER   TLV: CAS#: 7732-18-5

IGNI TEMP: NA FP: NA LEL: NA UEL: NA VP: NA VD: NA SG: 1.0 PS: LIQUID APPEAR: WHITE OPAQUE
   ODOR: ALMOND   PH FACTOR: 6.0 to 7.0

HAZARD RATINGS: H: F: R:

FIRE FIGHTING:
   HIGH EXPANSION FOAM, LOW EXPANSION FOAM, ALCOHOL FOAM; DRY CHEMICAL, CARBON DIOXIDE, WATER FOG.
   PROTECTIVE CLOTHING, RUBBER GLOVES, AND BREATHING APPARATUS.
WARNING: 1] STRUCTURAL PROTECTIVE CLOTHING IS PERMEABLE, REMAIN CLEAR OF SMOKE, WATER FALL OUT AND WATER RUN OFF.
         2] KEEP OUT OF THE REACH OF CHILDREN.
         3] THERMAL DECOMPOSITION YIELDS TOXIC CARBON DIOXIDE, CARBON MONOXIDE.
         4] MOVE CONTAINERS FROM AREA IF WITHOUT RISK, COOL EXPOSED CONTAINERS.
         5] DIKE AREA FOR CONTROL AND CONTAINMENT TO PREVENT ENTRY INTO SEWERS, DRAIN, AND WATER WAYS.

LARGE SPILL/NO FIRE/RESCUE: WEAR RUBBER OR NEOPRENE BOOTS, GLOVES.

SPILL CONTROL AND CONTAINMENT:
   HOUSEHOLD SPILL: WIPE UP LIQUID WITH AN ABSORBENT CLOTH, WASH OUT IN THE SINK. WASH SPILL AREA WITH SOAP AND
   WATER, WIPE UP WITH ABSORBENT CLOTH, RINSE OUT IN THE SINK.
   LARGE SPILL: COVER SPILL WITH ABSORBENT. SWEEP UP AND PUT IN A DISPOSAL CONTAINER. SCRUB AREA WITH A DETERGENT
   AND WATER, PICK UP WITH ABSORBENT. PUT IN A DISPOSAL CONTAINER. SEAL AND REMOVE FROM THE WORK AREA.

HEALTH HAZARD INFORMATION:
   WHEN USING THIS PRODUCT WEAR: NO SPECIAL REQUIREMENT. Contact with the eyes can cause stinging, tearing,
   itching, swelling, and redness. Prolonged or repeated contact with the skin may cause irritation. Ingestion will
   cause abdominal pain, nausea, vomiting, and diarrhea.

   A physician should be contacted if anyone develops any signs or symptoms and suspects that they are caused by
   exposure to this product.

FIRST AID:
Eye Exposure:
Flush with water for 15 minutes while lifting the upper and lower eye lids. Contact lenses should not be worn
when working with this product. If irritation persists, get medical attention.

Skin Exposure:
Wash with soap and water. If irritation appears, get medical attention.

Breathing:
Not expected to be a problem.

Swallowing:
Immediately contact the local poison control center for advice. Keep the victim warm and at rest. Get medical
attention.

PRODUCT: GODDARD'S FINE FURNITURE POLISH WITH ALMOND OIL

INCOMPATIBILITY: NONE KNOWN

INGREDIENTS: ISOPARAFFINIC NAPHTHA 10-20% TLV: CAS#:
             SILICONE FLUIDS   2-4% TLV: CAS#:
             SURFACTANTS   under   1% TLV: CAS#:
             WAXES   under   1% TLV: CAS#:
             ISOBUTANE_____   TLV: CAS#: 75-28-5
             PROPANE_____ 10-20% TLV: 1000 ppm CAS#: 74-98-6
             N-BUTANE_____/   TLV: 800 ppm CAS#: 106-97-8
             WATER   TLV: CAS#: 7732-18-5

IGNI TEMP: NA FP: 20F LEL: NA UEL: NA VP: NA VD: NA SG: .77 PS: LIQUID APPEAR: SPRAY MIST ODOR: ALMOND
    PH FACTOR: NA

HAZARD RATINGS: H: F: R: 0 4 0

FIRE FIGHTING:
    HIGH EXPANSION FOAM, LOW EXPANSION FOAM, ALCOHOL FOAM; DRY CHEMICAL, CARBON DIOXIDE, WATER FOG.
    PROTECTIVE CLOTHING, RUBBER GLOVES, AND BREATHING APPARATUS.
WARNING: 1] STRUCTURAL PROTECTIVE CLOTHING IS PERMEABLE, REMAIN CLEAR OF SMOKE, WATER FALL OUT AND WATER RUN OFF.
         2] KEEP OUT OF THE REACH OF CHILDREN.
         3] CONTAINERS WILL EXPLODE IN A FIRE OR IF HEATED ABOVE 120F.
         4] THERMAL DECOMPOSITION YIELDS TOXIC CARBON DIOXIDE, CARBON MONOXIDE.
         5] MOVE CONTAINERS FROM AREA IF WITHOUT RISK, COOL EXPOSED CONTAINERS.
         6] DIKE AREA FOR CONTROL AND CONTAINMENT TO PREVENT ENTRY INTO SEWERS, DRAIN, AND WATER WAYS.

LARGE SPILL/NO FIRE/RESCUE: WEAR RUBBER OR NEOPRENE BOOTS, GLOVES.

SPILL CONTROL AND CONTAINMENT:
    HOUSEHOLD SPILL: WIPE UP WITH AN ABSORBENT CLOTH. RINSE OUT IN THE SINK. WASH SPILL AREA WITH SOAP AND WATER,
    WIPE UP WITH ABSORBENT CLOTH, RINSE OUT IN THE SINK.
    LARGE SPILL: COVER SPILL WITH ABSORBENT. SWEEP UP AND PUT IN A DISPOSAL CONTAINER. SCRUB AREA WITH DETERGENT AND
    WATER, PICK UP WITH ABSORBENT, PUT IN A DISPOSAL CONTAINER. SEAL AND REMOVE FROM THE WORK AREA.

HEALTH HAZARD INFORMATION:
    WHEN USING THIS PRODUCT WEAR: NO SPECIAL REQUIREMENT. Contact with the eyes can cause stinging, tearing,
    itching, swelling, and redness. Prolonged or repeated contact with the skin may cause irritation. Ingestion will
    cause abdominal pain, nausea, vomiting, and diarrhea.

    A physician should be contacted if anyone develops any signs or symptoms and suspects that they are caused by
    exposure to this product.

FIRST AID:
Eye Exposure:
Flush with water for 15 minutes while lifting the upper and lower eye lids. Contact lenses should not be worn
when working with this product. If irritation persists, get medical attention.

Skin Exposure:
Wash with soap and water. If irritation appears, get medical attention.

Breathing:
Not expected to be a problem.

Swallowing:
Immediately contact the local poison control center for advice. Keep the victim warm and at rest. Get medical
attention.

PRODUCT: GODDARD'S FINE FURNITURE POLISH WITH LEMON BEES WAX

INCOMPATIBILITY: NONE KNOWN

```
INGREDIENTS: ISOPARAFFINIC NAPHTHA  10-20%  TLV:           CAS#:
             SILICONE FLUIDS         2-4%   TLV:           CAS#:
             SURFACTANTS       under   1%   TLV:           CAS#:
             WAXES             under   1%   TLV:           CAS#:
             ISOBUTANE_____              TLV:           CAS#:  75-28-5
             PROPANE_____  10-20%  TLV: 1000 ppm  CAS#:  74-98-6
             N-BUTANE_____/            TLV:  800 ppm  CAS#:  106-97-8
             WATER                      TLV:           CAS#: 7732-18-5
```

IGNI TEMP: NA FP: 20F LEL: NA UEL: NA VP: NA VD: NA SG: .77 PS: LIQUID APPEAR: SPRAY MIST ODOR: ALMOND
  PH FACTOR: NA

HAZARD RATINGS: H: F: R: 0 4 0

FIRE FIGHTING:
  HIGH EXPANSION FOAM, LOW EXPANSION FOAM, ALCOHOL FOAM; DRY CHEMICAL, CARBON DIOXIDE, WATER FOG.
  PROTECTIVE CLOTHING, RUBBER GLOVES, AND BREATHING APPARATUS.
WARNING: 1] STRUCTURAL PROTECTIVE CLOTHING IS PERMEABLE, REMAIN CLEAR OF SMOKE, WATER FALL OUT AND WATER RUN OFF.
         2] KEEP OUT OF THE REACH OF CHILDREN.
         3] CONTAINERS WILL EXPLODE IN A FIRE OR IF HEATED ABOVE 120F.
         4] THERMAL DECOMPOSITION YIELDS TOXIC CARBON DIOXIDE, CARBON MONOXIDE.
         5] MOVE CONTAINERS FROM AREA IF WITHOUT RISK, COOL EXPOSED CONTAINERS.
         6] DIKE AREA FOR CONTROL AND CONTAINMENT TO PREVENT ENTRY INTO SEWERS, DRAIN, AND WATER WAYS.

LARGE SPILL/NO FIRE/RESCUE: WEAR RUBBER OR NEOPRENE BOOTS, GLOVES.

SPILL CONTROL AND CONTAINMENT:
  HOUSEHOLD SPILL: WIPE UP WITH AN ABSORBENT CLOTH. RINSE OUT IN THE SINK. WASH SPILL AREA WITH SOAP AND WATER,
  WIPE UP WITH ABSORBENT CLOTH, RINSE OUT IN THE SINK.
  LARGE SPILL: COVER SPILL WITH ABSORBENT. SWEEP UP AND PUT IN A DISPOSAL CONTAINER. SCRUB AREA WITH DETERGENT AND
  WATER, PICK UP WITH ABSORBENT, PUT IN A DISPOSAL CONTAINER. SEAL AND REMOVE FROM THE WORK AREA.

HEALTH HAZARD INFORMATION:
  WHEN USING THIS PRODUCT WEAR: NO SPECIAL REQUIREMENT. Contact with the eyes can cause stinging, tearing,
  itching, swelling, and redness. Prolonged or repeated contact with the skin may cause irritation. Ingestion will
  cause abdominal pain, nausea, vomiting, and diarrhea.

  A physician should be contacted if anyone develops any signs or symptoms and suspects that they are caused by
  exposure to this product.

FIRST AID:
  Eye Exposure:
  Flush with water for 15 minutes while lifting the upper and lower eye lids. Contact lenses should not be worn
  when working with this product. If irritation persists, get medical attention.

  Skin Exposure:
  Wash with soap and water. If irritation appears, get medical attention.

  Breathing:
  Not expected to be a problem.

  Swallowing:
  Immediately contact the local poison control center for advice. Keep the victim warm and at rest. Get medical
  attention.

PRODUCT: GODDARD'S GLOW BRASS AND COPPER POLISH

INCOMPATIBILITY: NONE KNOWN

INGREDIENTS: MINERAL SEAL OIL   5-10% TLV: 5 mg/m3(mist) CAS#:
             SILICA   35-45% TLV: 10 mg/m3(dust) CAS#: 63231-67-4
             STODDARD SOLVENT   40-50% TLV: 100 ppm   CAS#: 8052-42-3
             FRAGRANCE, COLORANT under 1% TLV: CAS#:
             WATER   1-3%   TLV: CAS#: 7732-18-5

IGNI TEMP: NA FP: 100F LEL: NA UEL: NA VP: NA VD: NA SG: 1.12 PS: PASTE APPEAR: LIGHT YELLOW
    ODOR: LIGHT CHEMICAL   PH FACTOR: 9.3

HAZARD RATINGS: H: F: R: 0 2 0

FIRE FIGHTING:
  HIGH EXPANSION FOAM, LOW EXPANSION FOAM, ALCOHOL FOAM; DRY CHEMICAL, CARBON DIOXIDE, WATER FOG.
  PROTECTIVE CLOTHING, RUBBER GLOVES, AND BREATHING APPARATUS.
WARNING: 1] STRUCTURAL PROTECTIVE CLOTHING IS PERMEABLE, REMAIN CLEAR OF SMOKE, WATER FALL OUT AND WATER RUN OFF.
         2] KEEP OUT OF THE REACH OF CHILDREN.
         3] THERMAL DECOMPOSITION MAY YIELD TOXIC CARBON DIOXIDE, CARBON MONOXIDE.
         4] MOVE CONTAINERS FROM AREA IF WITHOUT RISK, COOL EXPOSED CONTAINERS.
         5] DIKE AREA FOR CONTROL AND CONTAINMENT TO PREVENT ENTRY INTO SEWERS, DRAIN, AND WATER WAYS.

LARGE SPILL/NO FIRE/RESCUE: WEAR RUBBER OR NEOPRENE BOOTS, GLOVES.

SPILL CONTROL AND CONTAINMENT:
  HOUSEHOLD SPILL: WIPE UP PASTE WITH ABSORBENT CLOTH, PUT BACK INTO CONTAINER. WASH CLOTH OUT IN THE SINK, WIPE
  UP RESIDUE WITH DAMP CLOTH, RINSE OUT IN THE SINK.
  LARGE SPILL: SCOOP UP PASTE, PUT IN A DISPOSAL CONTAINER. SCRUB AREA WITH DETERGENT AND WATER, PICK UP WITH
  ABSORBENT, PUT IN A DISPOSAL CONTAINER. SEAL AND REMOVE FROM THE WORK AREA.

HEALTH HAZARD INFORMATION:
  WHEN USING THIS PRODUCT WEAR: LATEX RUBBER GLOVES. Product may cause skin and eye irritation. Ingestion will
  cause abdominal pain, nausea, vomiting, and diarrhea.

  A physician should be contacted if anyone develops any signs or symptoms and suspects that they are caused by
  exposure to this product.

FIRST AID:
  Eye Exposure:
  Flush with water for 15 minutes while lifting the upper and lower eye lids. Contact lenses should not be worn
  when working with this product. If irritation persists, get medical attention.

  Skin Exposure:
  Wash with soap and water. If irritation persists, get medical attention.

  Breathing:
  Not expected to be a problem.

  Swallowing:
  Immediately contact the local poison control center for advice. Keep the victim warm and at rest. Get medical
  attention.

PRODUCT: GODDARD'S LEMON FURNITURE OIL

INCOMPATIBILITY: NONE KNOWN

INGREDIENTS: ISOPARAFFINIC NAPHTHA 55-65% TLV: CAS#:
             HYDROCARBON OIL   40-50% TLV: CAS#:
             HYDROCARBON POLYMER   2-4% TLV: CAS#:
             FRAGRANCE   under   1% TLV: CAS#:

IGNI TEMP: NA FP: 131F LEL: NA UEL: NA VP: NA VD: NA SG: .8 PS: LIQUID APPEAR: CLEAR, LIGHT YELLOW
   ODOR: LEMON   PH FACTOR: NA

HAZARD RATINGS: H: F: R:

FIRE FIGHTING:
  HIGH EXPANSION FOAM, LOW EXPANSION FOAM, ALCOHOL FOAM; DRY CHEMICAL, CARBON DIOXIDE, WATER FOG.
  PROTECTIVE CLOTHING, RUBBER GLOVES, AND BREATHING APPARATUS.
WARNING: 1] STRUCTURAL PROTECTIVE CLOTHING IS PERMEABLE, REMAIN CLEAR OF SMOKE, WATER FALL OUT AND WATER RUN OFF.
         2] KEEP OUT OF THE REACH OF CHILDREN.
         3] ELIMINATE ALL IGNITION SOURCES.
         4] CONTAINER MAY BURST IN A FIRE.
         5] LIQUID RUN-OFF INTO SEWERS CREATES A FIRE/EXPLOSION HAZARD.
         6] THERMAL DECOMPOSITION YIELDS TOXIC CARBON DIOXIDE, CARBON MONOXIDE.
         7] MOVE CONTAINERS FROM AREA IF WITHOUT RISK, COOL EXPOSED CONTAINERS.
         8] DIKE AREA FOR CONTROL AND CONTAINMENT TO PREVENT ENTRY INTO SEWERS, DRAIN, AND WATER WAYS.

LARGE SPILL/NO FIRE/RESCUE: WEAR RUBBER OR NEOPRENE BOOTS, GLOVES.

SPILL CONTROL AND CONTAINMENT:
  HOUSEHOLD SPILL: WIPE UP LIQUID WITH AN ABSORBENT CLOTH, WASH OUT IN THE SINK. WIPE UP RESIDUE WITH A DRY CLOTH,
  WASH OUT IN THE SINK.
  LARGE SPILL: COVER SPILL WITH ABSORBENT. PICK UP AND PUT IN A DISPOSAL CONTAINER. SCRUB AREA WITH DETERGENT AND
  WATER, PICK UP WITH ABSORBENT. PUT IN A DISPOSAL CONTAINER. SEAL AND REMOVE FROM THE WORK AREA.

HEALTH HAZARD INFORMATION:
  WHEN USING THIS PRODUCT WEAR: NO SPECIAL REQUIREMENT. Contact with the eyes can cause stinging, tearing,
  itching, swelling, and redness. Prolonged or repeated skin contact may cause irritation. If ingested, will cause
  abdominal pain, nausea, and vomiting. If it gets into the lungs, it can cause chemical pneumonia.

  A physician should be contacted if anyone develops any signs or symptoms and suspects that they are caused by
  exposure to this product.

FIRST AID:
  Eye Exposure:
  Flush with water for 15 minutes while lifting the upper and lower eye lids. Contact lenses should not be worn
  when working with this product. If irritation persists, get medical attention.

  Skin Exposure:
  Wash with soap and water. If irritation appears, get medical attention.

  Breathing:
  Not expected to be a problem.

  Swallowing:
  Do not induce vomiting. Immediately contact the local poison control center for advice. Keep the victim warm and
  at rest. Get medical attention.

PRODUCT: GODDARD'S LONG SHINE BRASS AND COPPER POLISH

INCOMPATIBILITY: NONE KNOWN

```
INGREDIENTS: AMMONIA           under  1%    TLV: 25 ppm        CAS#: 7664-41-7
             ETHYLENE GLYCOL
             MONOBUTYL ETHER      2-6%     TLV: 25 ppm        CAS#:  111-76-2
             ALUMINUM SILICATE  15-20%     TLV: 10 mg/m3(dust)CAS#:
             ALUMINUM OXIDE       2-4%     TLV: 10 mg/m3(dust)CAS#: 1344-28-1
             SURFACTANT____                TLV:               CAS#:
             FRAGRANCE_____   3-6%    TLV:               CAS#:
             PRESERVATIVE__/              TLV:               CAS#:
             WATER              60-70%     TLV:               CAS#: 7732-18-5
```

IGNI TEMP: NA FP: NA LEL: NA UEL: NA VP: NA VD: NA SG: 1.1 PS: LIQUID APPEAR: LIGHT RED-BROWN
    ODOR: CITRUS   PH FACTOR: 9.7

HAZARD RATINGS: H: F: R: 1 0 0

FIRE FIGHTING:
    HIGH EXPANSION FOAM, LOW EXPANSION FOAM, ALCOHOL FOAM; DRY CHEMICAL, CARBON DIOXIDE, WATER FOG.
    PROTECTIVE CLOTHING, RUBBER GLOVES, AND BREATHING APPARATUS.
WARNING: 1] STRUCTURAL PROTECTIVE CLOTHING IS PERMEABLE, REMAIN CLEAR OF SMOKE, WATER FALL OUT AND WATER RUN OFF.
         2] KEEP OUT OF THE REACH OF CHILDREN.
         3] THIS PRODUCT CONTAINS ABRASIVE MATERIALS, WILL SCRATCH THE EYES.
         4] THERMAL DECOMPOSITION YIELDS TOXIC CARBON DIOXIDE, CARBON MONOXIDE.
         5] MOVE CONTAINERS FROM AREA IF WITHOUT RISK, COOL EXPOSED CONTAINERS.
         6] DIKE AREA FOR CONTROL AND CONTAINMENT TO PREVENT ENTRY INTO SEWERS, DRAIN, AND WATER WAYS.

LARGE SPILL/NO FIRE/RESCUE: WEAR RUBBER OR NEOPRENE BOOTS, GLOVES.

SPILL CONTROL AND CONTAINMENT:
    HOUSEHOLD SPILL: WIPE UP LIQUID WITH AN ABSORBENT CLOTH. RINSE OUT IN THE SINK. WIPE UP RESIDUE WITH DAMP CLOTH,
    RINSE OUT IN THE SINK.
    LARGE SPILL: COVER SPILL WITH ABSORBENT. SWEEP UP AND PUT IN A DISPOSAL CONTAINER. RINSE AREA WITH WATER, PICK
    UP WITH ABSORBENT. PUT IN A DISPOSAL CONTAINER. SEAL AND REMOVE FROM THE WORK AREA.

HEALTH HAZARD INFORMATION:
    WHEN USING THIS PRODUCT WEAR: NO SPECIAL REQUIREMENT. Ingestion may cause burns to the mouth, throat, and
    stomach. Direct contact with the eyes will cause stinging, tearing, itching, swelling, redness and may scratch
    the cornea. Prolonged or repeated skin contact may cause irritation.

    A physician should be contacted if anyone develops any signs or symptoms and suspects that they are caused by
    exposure to this product.

FIRST AID:
Eye Exposure:
Flush with water for 15 minutes while lifting the upper and lower eye lids. Contact lenses should not be worn
when working with this product. If irritation persists, get medical attention.

Skin Exposure:
Wash with soap and water. If irritation appears, get medical attention.

Breathing:
Not expected to be a problem.

Swallowing:
Immediately contact the local poison control center for advice. Keep the victim warm and at rest. Get medical
attention.

PRODUCT: GODDARD'S LONG SHINE SILVER FOAM

INCOMPATIBILITY: NONE KNOWN

INGREDIENTS: ISOPROPYL ALCOHOL   1-3% TLV: 400 ppm   CAS#:   67-63-0
             SILICA   25-35% TLV: 2.6 mg/m3   CAS#:
             SURFACTANTS_____   TLV: CAS#:
             COLORANT_____ 3-5% TLV: CAS#:
             FRAGRANCE_____/   TLV: CAS#:
             WATER   45-69% TLV: CAS#: 7732-18-5

IGNI TEMP: NA FP: NA LEL: NA UEL: NA VP: NA VD: NA SG: 6.4 PS: PASTE APPEAR: BROWN ODOR: FRAGRANT
    PH FACTOR: 7.4

HAZARD RATINGS: H: F: R: 1 0 0

FIRE FIGHTING:
   HIGH EXPANSION FOAM, LOW EXPANSION FOAM, ALCOHOL FOAM; DRY CHEMICAL, CARBON DIOXIDE, WATER FOG.
   PROTECTIVE CLOTHING, RUBBER GLOVES, AND BREATHING APPARATUS.
WARNING: 1] STRUCTURAL PROTECTIVE CLOTHING IS PERMEABLE, REMAIN CLEAR OF SMOKE, WATER FALL OUT AND WATER RUN OFF.
         2] KEEP OUT OF THE REACH OF CHILDREN.
         3] THERMAL DECOMPOSITION YIELDS TOXIC CARBON DIOXIDE, CARBON MONOXIDE.
         4] MOVE CONTAINERS FROM AREA IF WITHOUT RISK, COOL EXPOSED CONTAINERS.
         5] DIKE AREA FOR CONTROL AND CONTAINMENT TO PREVENT ENTRY INTO SEWERS, DRAIN, AND WATER WAYS.

LARGE SPILL/NO FIRE/RESCUE: WEAR RUBBER OR NEOPRENE BOOTS, GLOVES.

SPILL CONTROL AND CONTAINMENT:
   HOUSEHOLD SPILL: SCOOP UP PASTE AND PUT BACK IN CONTAINER. WASH AREA WITH SOAP AND WATER. PICK UP WITH ABSORBENT
   CLOTH, RINSE OUT IN THE SINK.
   LARGE SPILL: SCOOP UP PASTE AND PUT IN A DISPOSAL CONTAINER. SCRUB AREA WITH DETERGENT AND WATER. PICK UP WITH
   ABSORBENT. PUT INTO A DISPOSAL CONTAINER. SEAL AND REMOVE FROM THE WORK AREA.

HEALTH HAZARD INFORMATION:
   WHEN USING THIS PRODUCT WEAR: NO SPECIAL REQUIREMENT. Eye contact causes stinging, tearing, itching, swelling,
   and redness. Ingestion will cause abdominal pain, nausea, and vomiting.

   A physician should be contacted if anyone develops any signs or symptoms and suspects that they are caused by
   exposure to this product.

FIRST AID:
   Eye Exposure:
   Flush with water for 15 minutes while lifting the upper and lower eye lids. Contact lenses should not be worn
   when working with this product. If irritation persists, get medical attention.

   Skin Exposure:
   Not expected to be a problem.

   Breathing:
   Not expected to be a problem.

   Swallowing:
   Do not induce vomiting. Immediately contact the local poison control center for advice. Keep the victim warm and
   at rest. Get medical attention.

PRODUCT: GODDARD'S LONG SHINE SILVER POLISH

INCOMPATIBILITY: NONE KNOWN

INGREDIENTS: ISOPROPYL ALCOHOL   6-12% TLV: 400 ppm   CAS#:   67-63-0
             TITANIUM DIOXIDE about   1% TLV: 5 mg/m3   CAS#: 13463-67-7
             SILICA   10-20% TLV: 2.6 mg/m3   CAS#: 63231-67-4
             SURFACTANTS_____   TLV: CAS#:
             COLORANT_____ 1-3% TLV: CAS#:
             FRAGRANCE_____/   TLV: CAS#:
             WATER   65-75% TLV: CAS#: 7732-18-5

IGNI TEMP: NA FP: 94F LEL: NA UEL: NA VP: NA VD: NA SG: 1.05 PS: LIQUID APPEAR: DARK BROWN ODOR: CHEMICAL
    PH FACTOR: 7.6

HAZARD RATINGS: H: F: R: 0 3 0

FIRE FIGHTING:
    HIGH EXPANSION FOAM, LOW EXPANSION FOAM, ALCOHOL FOAM; DRY CHEMICAL, CARBON DIOXIDE, WATER FOG.
    PROTECTIVE CLOTHING, RUBBER GLOVES, AND BREATHING APPARATUS.
WARNING: 1] STRUCTURAL PROTECTIVE CLOTHING IS PERMEABLE, REMAIN CLEAR OF SMOKE, WATER FALL OUT AND WATER RUN OFF.
         2] KEEP OUT OF THE REACH OF CHILDREN.
         3] THERMAL DECOMPOSITION YIELDS TOXIC CARBON DIOXIDE, CARBON MONOXIDE.
         4] ELIMINATE ALL SOURCES OF IGNITION.
         5] MOVE CONTAINERS FROM AREA IF WITHOUT RISK, COOL EXPOSED CONTAINERS.
         6] DIKE AREA FOR CONTROL AND CONTAINMENT TO PREVENT ENTRY INTO SEWERS, DRAIN, AND WATER WAYS.

LARGE SPILL/NO FIRE/RESCUE: WEAR RUBBER OR NEOPRENE BOOTS, GLOVES.

SPILL CONTROL AND CONTAINMENT:
    HOUSEHOLD SPILL: WIPE UP LIQUID WITH ABSORBENT CLOTH. RINSE OUT IN THE SINK. WASH AREA WITH SOAP AND WATER. PICK
    UP WITH ABSORBENT CLOTH, RINSE OUT IN THE SINK.
    LARGE SPILL: COVER SPILL WITH ABSORBENT. PICK UP AND PUT INTO A DISPOSAL CONTAINER. SCRUB WITH DETERGENT AND
    WATER. PICK UP WITH ABSORBENT. PUT INTO A DISPOSAL CONTAINER. SEAL AND REMOVE FROM THE WORK AREA.

HEALTH HAZARD INFORMATION:
    WHEN USING THIS PRODUCT WEAR: NO SPECIAL REQUIREMENT. Eye contact causes stinging, tearing, itching, swelling,
    and redness. Ingestion will cause abdominal pain, nausea, and vomiting.

    A physician should be contacted if anyone develops any signs or symptoms and suspects that they are caused by
    exposure to this product.

FIRST AID:
    Eye Exposure:
    Flush with water for 15 minutes while lifting the upper and lower eye lids. Contact lenses should not be worn
    when working with this product. If irritation persists, get medical attention.

    Skin Exposure:
    Not expected to be a problem.

    Breathing:
    Not expected to be a problem.

    Swallowing:
    If victim is conscious, give 2-3 glasses of milk to drink. Immediately contact the local poison control center
    for advice. Keep the victim warm and at rest. Get medical attention.

PRODUCT: GODDARD'S PEWTER CLEANER

INCOMPATIBILITY: NONE KNOWN

INGREDIENTS: ISOPROPYL ALCOHOL   1-3% TLV: 400 ppm   CAS#:   67-63-0
             TITANIUM DIOXIDE   20-30% TLV: 10 mg/m3   CAS#: 13463-67-7
             SILICA   10-20% TLV: 2.6 mg/m3(DUST)CAS#: 63231-67-4
             SURFACTANTS   1-3% TLV: CAS#:
             PRESERVATIVE_____under 0.5% TLV: CAS#:
             FRAGRANCE_____/   TLV: CAS#:
             WATER   50-60% TLV: CAS#: 7732-18-5

IGNI TEMP: NA FP: NA LEL: NA UEL: NA VP: NA VD: NA SG: 1.24 PS: PASTE APPEAR: OPAQUE WHITE ODOR: FRAGRANT
   PH FACTOR: 7.6

HAZARD RATINGS: H: F: R: 1 0 0

FIRE FIGHTING:
   HIGH EXPANSION FOAM, LOW EXPANSION FOAM, ALCOHOL FOAM; DRY CHEMICAL, CARBON DIOXIDE, WATER FOG.
   PROTECTIVE CLOTHING, RUBBER GLOVES, AND BREATHING APPARATUS.
WARNING: 1] STRUCTURAL PROTECTIVE CLOTHING IS PERMEABLE, REMAIN CLEAR OF SMOKE, WATER FALL OUT AND WATER RUN OFF.
         2] KEEP OUT OF THE REACH OF CHILDREN.
         3] THERMAL DECOMPOSITION YIELDS TOXIC CARBON DIOXIDE, CARBON MONOXIDE.
         4] MOVE CONTAINERS FROM AREA IF WITHOUT RISK, COOL EXPOSED CONTAINERS.
         5] DIKE AREA FOR CONTROL AND CONTAINMENT TO PREVENT ENTRY INTO SEWERS, DRAIN, AND WATER WAYS.

LARGE SPILL/NO FIRE/RESCUE: WEAR RUBBER OR NEOPRENE BOOTS, GLOVES.

SPILL CONTROL AND CONTAINMENT:
   HOUSEHOLD SPILL: SCOOP UP PASTE AND PUT BACK IN CONTAINER. WASH AREA WITH SOAP AND WATER. PICK UP WITH ABSORBENT
   CLOTH, RINSE OUT IN THE SINK.
   LARGE SPILL: SCOOP UP PASTE AND PUT IN A DISPOSAL CONTAINER. SCRUB AREA WITH DETERGENT AND WATER. PICK UP WITH
   ABSORBENT. PUT INTO A DISPOSAL CONTAINER. SEAL AND REMOVE FROM THE WORK AREA.

HEALTH HAZARD INFORMATION:
   WHEN USING THIS PRODUCT WEAR: NO SPECIAL REQUIREMENT. Eye contact causes stinging, tearing, itching, swelling,
   and redness. Ingestion will cause abdominal pain, nausea, and vomiting.

   A physician should be contacted if anyone develops any signs or symptoms and suspects that they are caused by
   exposure to this product.

FIRST AID:
   Eye Exposure:
   Flush with water for 15 minutes while lifting the upper and lower eye lids. Contact lenses should not be worn
   when working with this product. If irritation persists, get medical attention.

   Skin Exposure:
   Not expected to be a problem.

   Breathing:
   Not expected to be a problem.

   Swallowing:
   If victim is conscious, give 2-3 glasses of milk to drink. Immediately contact the local poison control center
   for advice. Keep the victim warm and at rest. Get medical attention.

PRODUCT: GODDARD'S STAINLESS STEEL CLEANER

INCOMPATIBILITY: NONE KNOWN

INGREDIENTS: ISOPROPYL ALCOHOL   1-3% TLV: 400 ppm   CAS#:   67-63-0
             ALUMINUM OXIDE   30-40% TLV: 10 mg/m3(dust)CAS#: 1344-28-1
             SILICA   1-3% TLV: 2.6 mg/m3   CAS#: 63231-67-4
             SURFACTANTS   1-3% TLV: CAS#:
             PRESERVATIVE__
             COLORANT_____under 0.5% TLV: CAS#:
             FRAGRANCE_____/   TLV: CAS#:
             WATER   50-60% TLV: CAS#: 7732-18-5

IGNI TEMP: NA FP: NA LEL: NA UEL: NA VP: NA VD: NA SG: 1.24 PS: PASTE APPEAR: AQUA BLUE ODOR: FRAGRANT
   PH FACTOR: 8.4

HAZARD RATINGS: H: F: R:

FIRE FIGHTING:
   HIGH EXPANSION FOAM, LOW EXPANSION FOAM, ALCOHOL FOAM; DRY CHEMICAL, CARBON DIOXIDE, WATER FOG.
   PROTECTIVE CLOTHING, RUBBER GLOVES, AND BREATHING APPARATUS.
WARNING: 1] STRUCTURAL PROTECTIVE CLOTHING IS PERMEABLE, REMAIN CLEAR OF SMOKE, WATER FALL OUT AND WATER RUN OFF.
         2] KEEP OUT OF THE REACH OF CHILDREN.
         3] THERMAL DECOMPOSITION YIELDS TOXIC CARBON DIOXIDE, CARBON MONOXIDE.
         4] MOVE CONTAINERS FROM AREA IF WITHOUT RISK, COOL EXPOSED CONTAINERS.
         5] DIKE AREA FOR CONTROL AND CONTAINMENT TO PREVENT ENTRY INTO SEWERS, DRAIN, AND WATER WAYS.

LARGE SPILL/NO FIRE/RESCUE: WEAR RUBBER OR NEOPRENE BOOTS, GLOVES.

SPILL CONTROL AND CONTAINMENT:
   HOUSEHOLD SPILL: SCOOP UP PASTE AND PUT BACK IN CONTAINER. WASH AREA WITH SOAP AND WATER. PICK UP WITH ABSORBENT
   CLOTH, RINSE OUT IN THE SINK.
   LARGE SPILL: SCOOP UP PASTE AND PUT IN A DISPOSAL CONTAINER. SCRUB AREA WITH DETERGENT AND WATER. PICK UP WITH
   ABSORBENT. PUT INTO A DISPOSAL CONTAINER. SEAL AND REMOVE FROM THE WORK AREA.

HEALTH HAZARD INFORMATION:
   WHEN USING THIS PRODUCT WEAR: NO SPECIAL REQUIREMENT. Eye contact causes stinging, tearing, itching, swelling,
   and redness. Ingestion will cause abdominal pain, nausea, and vomiting.

   A physician should be contacted if anyone develops any signs or symptoms and suspects that they are caused by
   exposure to this product.

FIRST AID:
   Eye Exposure:
   Flush with water for 15 minutes while lifting the upper and lower eye lids. Contact lenses should not be worn
   when working with this product. If irritation persists, get medical attention.

   Skin Exposure:
   Not expected to be a problem.

   Breathing:
   Not expected to be a problem.

   Swallowing:
   If victim is conscious, give 2-3 glasses of milk to drink. Immediately contact the local poison control center
   for advice. Keep the victim warm and at rest. Get medical attention.

PRODUCT: HALSA CONDITIONER - CHAMOMILE

INCOMPATIBILITY: NONE KNOWN

INGREDIENTS: ALL ARE NON-HAZARDOUS BY OSHA 1910.1200 STANDARD

IGNI TEMP: NA FP: NA LEL: NA UEL: NA VP: NA VD: NA SG: 1.0 PS: LIQUID APPEAR: OPAQUE   ODOR: FRAGRANT
   PH FACTOR: 5.0

HAZARD RATINGS: H: F: R:

FIRE FIGHTING:
   HIGH EXPANSION FOAM, LOW EXPANSION FOAM, ALCOHOL FOAM; DRY CHEMICAL, CARBON DIOXIDE, WATER FOG.
   PROTECTIVE CLOTHING, RUBBER GLOVES, AND BREATHING APPARATUS.
WARNING: 1] STRUCTURAL PROTECTIVE CLOTHING IS PERMEABLE, REMAIN CLEAR OF SMOKE, WATER FALL OUT AND WATER RUN OFF.
         2] KEEP OUT OF THE REACH OF CHILDREN.
         3] THERMAL DECOMPOSITION MAY YIELD CARBON DIOXIDE, CARBON MONOXIDE.
         4] CONTAINERS WILL MELT AND LEAK IN A FIRE.
         5] MOVE CONTAINERS FROM AREA IF WITHOUT RISK, COOL EXPOSED CONTAINERS.
         6] DIKE AREA FOR CONTROL AND CONTAINMENT TO PREVENT ENTRY INTO SEWERS, DRAIN, AND WATER WAYS.

LARGE SPILL/NO FIRE/RESCUE: WEAR RUBBER OR NEOPRENE BOOTS, GLOVES.

SPILL CONTROL AND CONTAINMENT:
   HOUSEHOLD SPILL: WIPE UP LIQUID WITH AN ABSORBENT CLOTH, RINSE OUT IN THE SINK. WASH AREA WITH WATER, WIPE UP
   WITH ABSORBENT CLOTH, RINSE OUT IN THE SINK.
   LARGE SPILL: SCOOP UP LIQUID AND PUT IN A DISPOSAL CONTAINER. COVER AREA WITH ABSORBENT. SWEEP UP AND PUT IN A
   DISPOSAL CONTAINER. RINSE AREA WITH WATER, PICK UP WITH ABSORBENT AND PUT IN A DISPOSAL CONTAINER.

HEALTH HAZARD INFORMATION:
   WHEN USING THIS PRODUCT WEAR: NO SPECIAL REQUIREMENT. Contact with the eyes may cause slight irritation. May
   cause skin rash if allergic to product. Ingestion will cause abdominal pain, nausea, vomiting, and diarrhea.

   A physician should be contacted if anyone develops any signs or symptoms and suspects that they are caused by
   exposure to this product.

FIRST AID:
Eye Exposure:
Flush with water for 15 minutes while lifting the upper and lower eye lids. Contact lenses should not be worn
when working with this product. If irritation persists, get medical attention.

   Skin Exposure:
Not expected to be a problem.

   Breathing:
Not expected to be a problem.

   Swallowing:
If the victim is conscious, give 2-3 glasses of milk or water to drink. Immediately contact the local poison
control center for advice. Keep the victim warm and at rest. Get medical attention.

PRODUCT: HALSA CONDITIONER - CORNFLOWER

INCOMPATIBILITY: NONE KNOWN

INGREDIENTS: CETYL ALCOHOL   TLV: CAS#: 36653-82-4
             PROPYLENE GLYCOL   TLV: CAS#: 4254-15-3
             WATER   TLV: CAS#: 7732-18-5
             MINERAL OIL   METHYLCHLOROISOTHIAZOLINONE
             HYDROLYZED COLLAGEN   METHYLISOTHIAZOLINONE
             BOTANICAL EXTRACTS   HYDROXYETHYL CELLULOSE
             STEARYLICONIUM   FRAGRANCE
             COLORANTS

IGNI TEMP: NA FP: NA LEL: NA UEL: NA VP: NA VD: NA SG: 1.0 PS: LIQUID APPEAR: OPAQUE   ODOR: FRAGRANT
    PH FACTOR: 5.0

HAZARD RATINGS: H: F: R: 1 0 0

FIRE FIGHTING:
   HIGH EXPANSION FOAM, LOW EXPANSION FOAM, ALCOHOL FOAM; DRY CHEMICAL, CARBON DIOXIDE, WATER FOG.
   PROTECTIVE CLOTHING, RUBBER GLOVES, AND BREATHING APPARATUS.
WARNING: 1] STRUCTURAL PROTECTIVE CLOTHING IS PERMEABLE, REMAIN CLEAR OF SMOKE, WATER FALL OUT AND WATER RUN OFF.
         2] KEEP OUT OF THE REACH OF CHILDREN.
         3] CONTAINERS MAY MELT AND LEAK IN A FIRE.
         4] THERMAL DECOMPOSITION MAY YIELD CARBON DIOXIDE, CARBON MONOXIDE.
         5] MOVE CONTAINERS FROM AREA IF WITHOUT RISK, COOL EXPOSED CONTAINERS.
         6] DIKE AREA FOR CONTROL AND CONTAINMENT TO PREVENT ENTRY INTO SEWERS, DRAIN, AND WATER WAYS.

LARGE SPILL/NO FIRE/RESCUE: WEAR RUBBER OR NEOPRENE BOOTS, GLOVES.

SPILL CONTROL AND CONTAINMENT:
   HOUSEHOLD SPILL: WIPE UP LIQUID WITH AN ABSORBENT CLOTH, RINSE OUT IN THE SINK. WASH AREA WITH WATER, WIPE UP
   WITH AN ABSORBENT CLOTH, RINSE OUT IN THE SINK.
   LARGE SPILL: SCOOP UP LIQUID AND PUT IN A DISPOSAL CONTAINER. COVER AREA WITH ABSORBENT. PICK UP AND PUT IN A
   DISPOSAL CONTAINER. RINSE AREA WITH WATER. PICK UP WITH ABSORBENT AND PUT IN A DISPOSAL CONTAINER.

HEALTH HAZARD INFORMATION:
   WHEN USING THIS PRODUCT WEAR: NO SPECIAL REQUIREMENT. Contact with the eyes may cause irritation. May cause skin
   rash if allergic to the product. Ingestion may cause abdominal pain, nausea, vomiting, and diarrhea.

   A physician should be contacted if anyone develops any signs or symptoms and suspects that they are caused by
   exposure to this product.

FIRST AID:
Eye Exposure:
Flush with water for 15 minutes while lifting the upper and lower eye lids. Contact lenses should not be worn
when working with this product. If irritation persists, get medical attention.

Skin Exposure:
Wash skin with soap and water. If irritation appears, get medical attention.

Breathing:
Not expected to be a problem.

Swallowing:
If victim is conscious, give 2-3 glasses of milk or water to drink. Immediately contact the local poison control
center for advice. Keep the victim warm and at rest. Get medical attention.

PRODUCT: HALSA CONDITIONER - GINGER ROOT

INCOMPATIBILITY: NONE KNOWN

INGREDIENTS: ALL ARE NON-HAZARDOUS BY OSHA 1910.1200 STANDARD

IGNI TEMP: NA FP: NA LEL: NA UEL: NA VP: NA VD: NA SG: 1.0 PS: LIQUID APPEAR: OPAQUE   ODOR: FRAGRANT
   PH FACTOR: 5.0

HAZARD RATINGS: H: F: R:

FIRE FIGHTING:
  HIGH EXPANSION FOAM, LOW EXPANSION FOAM, ALCOHOL FOAM; DRY CHEMICAL, CARBON DIOXIDE, WATER FOG.
  PROTECTIVE CLOTHING, RUBBER GLOVES, AND BREATHING APPARATUS.
WARNING: 1] STRUCTURAL PROTECTIVE CLOTHING IS PERMEABLE, REMAIN CLEAR OF SMOKE, WATER FALL OUT AND WATER RUN OFF.
         2] KEEP OUT OF THE REACH OF CHILDREN.
         3] THERMAL DECOMPOSITION MAY YIELD CARBON DIOXIDE, CARBON MONOXIDE.
         4] CONTAINERS WILL MELT AND LEAK IN A FIRE.
         5] MOVE CONTAINERS FROM AREA IF WITHOUT RISK, COOL EXPOSED CONTAINERS.
         6] DIKE AREA FOR CONTROL AND CONTAINMENT TO PREVENT ENTRY INTO SEWERS, DRAIN, AND WATER WAYS.

LARGE SPILL/NO FIRE/RESCUE: WEAR RUBBER OR NEOPRENE BOOTS, GLOVES.

SPILL CONTROL AND CONTAINMENT:
  HOUSEHOLD SPILL: WIPE UP LIQUID WITH AN ABSORBENT CLOTH, RINSE OUT IN THE SINK. WASH AREA WITH WATER, WIPE UP
  WITH ABSORBENT CLOTH, RINSE OUT IN THE SINK.
  LARGE SPILL: SCOOP UP LIQUID AND PUT IN A DISPOSAL CONTAINER. COVER AREA WITH ABSORBENT. SWEEP UP AND PUT IN A
  DISPOSAL CONTAINER. RINSE AREA WITH WATER, PICK UP WITH ABSORBENT AND PUT IN A DISPOSAL CONTAINER.

HEALTH HAZARD INFORMATION:
  WHEN USING THIS PRODUCT WEAR: NO SPECIAL REQUIREMENT. Contact with the eyes may cause slight irritation. May
  cause skin rash if allergic to product. Ingestion will cause abdominal pain, nausea, vomiting, and diarrhea.

  A physician should be contacted if anyone develops any signs or symptoms and suspects that they are caused by
  exposure to this product.

FIRST AID:
  Eye Exposure:
  Flush with water for 15 minutes while lifting the upper and lower eye lids. Contact lenses should not be worn
  when working with this product. If irritation persists, get medical attention.

  Skin Exposure:
  Not expected to be a problem.

  Breathing:
  Not expected to be a problem.

  Swallowing:
  If the victim is conscious, give 2-3 glasses of milk or water to drink. Immediately contact the local poison
  control center for advice. Keep the victim warm and at rest. Get medical attention.

PRODUCT: HALSA CONDITIONER - MARIGOLD

INCOMPATIBILITY: NONE KNOWN

INGREDIENTS: ALL ARE NON-HAZARDOUS BY OSHA 1910.1200 STANDARD

IGNI TEMP: NA FP: NA LEL: NA UEL: NA VP: NA VD: NA SG: 1.0 PS: LIQUID APPEAR: OPAQUE   ODOR: FRAGRANT
   PH FACTOR: 5.0

HAZARD RATINGS: H: F: R:

FIRE FIGHTING:
   HIGH EXPANSION FOAM, LOW EXPANSION FOAM, ALCOHOL FOAM; DRY CHEMICAL, CARBON DIOXIDE, WATER FOG.
   PROTECTIVE CLOTHING, RUBBER GLOVES, AND BREATHING APPARATUS.
WARNING: 1] STRUCTURAL PROTECTIVE CLOTHING IS PERMEABLE, REMAIN CLEAR OF SMOKE, WATER FALL OUT AND WATER RUN OFF.
         2] KEEP OUT OF THE REACH OF CHILDREN.
         3] THERMAL DECOMPOSITION MAY YIELD CARBON DIOXIDE, CARBON MONOXIDE.
         4] CONTAINERS WILL MELT AND LEAK IN A FIRE.
         5] MOVE CONTAINERS FROM AREA IF WITHOUT RISK, COOL EXPOSED CONTAINERS.
         6] DIKE AREA FOR CONTROL AND CONTAINMENT TO PREVENT ENTRY INTO SEWERS, DRAIN, AND WATER WAYS.

LARGE SPILL/NO FIRE/RESCUE: WEAR RUBBER OR NEOPRENE BOOTS, GLOVES.

SPILL CONTROL AND CONTAINMENT:
   HOUSEHOLD SPILL: WIPE UP LIQUID WITH AN ABSORBENT CLOTH, RINSE OUT IN THE SINK. WASH AREA WITH WATER, WIPE UP
   WITH ABSORBENT CLOTH, RINSE OUT IN THE SINK.
   LARGE SPILL: SCOOP UP LIQUID AND PUT IN A DISPOSAL CONTAINER. COVER AREA WITH ABSORBENT. SWEEP UP AND PUT IN A
   DISPOSAL CONTAINER. RINSE AREA WITH WATER, PICK UP WITH ABSORBENT AND PUT IN A DISPOSAL CONTAINER.

HEALTH HAZARD INFORMATION:
   WHEN USING THIS PRODUCT WEAR: NO SPECIAL REQUIREMENT. Contact with the eyes may cause slight irritation. May
   cause skin rash if allergic to product. Ingestion will cause abdominal pain, nausea, vomiting, and diarrhea.

   A physician should be contacted if anyone develops any signs or symptoms and suspects that they are caused by
   exposure to this product.

FIRST AID:
   Eye Exposure:
   Flush with water for 15 minutes while lifting the upper and lower eye lids. Contact lenses should not be worn
   when working with this product. If irritation persists, get medical attention.

   Skin Exposure:
   Not expected to be a problem.

   Breathing:
   Not expected to be a problem.

   Swallowing:
   If the victim is conscious, give 2-3 glasses of milk or water to drink. Immediately contact the local poison
   control center for advice. Keep the victim warm and at rest. Get medical attention.

PRODUCT: HALSA DANDRUFF SHAMPOO FOR PREMED HAIR

INCOMPATIBILITY: NONE KNOWN

```
INGREDIENTS: SALICYLIC ACID            TLV:            CAS#:      69-72-7
             AMMONIUM LAURYL SULFATE   TLV:            CAS#:
             AMMONIUM LAURETH SULFATE  TLV:            CAS#:
             AMMONIUM CHLORIDE         TLV:            CAS#: 12125-02-9
             COCAMIDOPROPYL BETRINE    TLV:            CAS#:
```

IGNI TEMP: NA FP: NA LEL: NA UEL: NA VP: NA VD: NA SG: 1.02 PS: LIQUID APPEAR: CLEAR GREEN
    ODOR: PERFUME   PH FACTOR: 5.2

HAZARD RATINGS: H: F: R: 1 0 0

FIRE FIGHTING:
  HIGH EXPANSION FOAM, LOW EXPANSION FOAM, ALCOHOL FOAM; DRY CHEMICAL, CARBON DIOXIDE, WATER FOG.
  PROTECTIVE CLOTHING, RUBBER GLOVES, AND BREATHING APPARATUS.
WARNING: 1] STRUCTURAL PROTECTIVE CLOTHING IS PERMEABLE, REMAIN CLEAR OF SMOKE, WATER FALL OUT AND WATER RUN OFF.
         2] KEEP OUT OF THE REACH OF CHILDREN.
         3] THERMAL DECOMPOSITION MAY YIELD CARBON DIOXIDE, CARBON MONOXIDE.
         4] MOVE CONTAINERS FROM AREA IF WITHOUT RISK, COOL EXPOSED CONTAINERS.
         5] DIKE AREA FOR CONTROL AND CONTAINMENT TO PREVENT ENTRY INTO SEWERS, DRAIN, AND WATER WAYS.

LARGE SPILL/NO FIRE/RESCUE: WEAR RUBBER OR NEOPRENE BOOTS, GLOVES.

SPILL CONTROL AND CONTAINMENT:
  HOUSEHOLD SPILL: WIPE UP LIQUID WITH AN ABSORBENT CLOTH, RINSE OUT IN THE SINK. WIPE UP THE RESIDUE WITH A DAMP
  CLOTH. RINSE OUT IN THE SINK.
  LARGE SPILL: COVER SPILL WITH ABSORBENT. SWEEP UP AND PUT IN A DISPOSAL CONTAINER. RINSE AREA WITH WATER, PICK
  UP WITH ABSORBENT, PUT IN A DISPOSAL CONTAINER. SEAL AND REMOVE FROM THE WORK AREA.

HEALTH HAZARD INFORMATION:
  WHEN USING THIS PRODUCT WEAR: NO SPECIAL REQUIREMENT. Contact with the eyes may cause stinging, tearing,
  itching, swelling, and redness. Ingestion may cause abdominal pain, nausea, vomiting, and diarrhea.

  A physician should be contacted if anyone develops any signs or symptoms and suspects that they are caused by
  exposure to this product.

FIRST AID:
Eye Exposure:
Flush with water for 15 minutes while lifting the upper and lower eye lids. Contact lenses should not be worn
when working with this product. If irritation persists, get medical attention.

Skin Exposure:
Wash with soap and water. If irritation appears, get medical attention.

Breathing:
Not expected to be a problem.

Swallowing:
Immediately contact the local poison control center for advice. Keep the victim warm and at rest. Get medical
attention.

PRODUCT: HALSA DANDRUFF SHAMPOO CONCENTRATE - REGULAR

INCOMPATIBILITY: NONE KNOWN

INGREDIENTS: SALICYLIC ACID   TLV: CAS#:   69-72-7
             AMMONIUM LAURYL SULFATE   TLV: CAS#:
             AMMONIUM LAURETH SULFATE   TLV: CAS#:
             AMMONIUM CHLORIDE   TLV: CAS#: 12125-02-9
             COCAMIDOPROPYL BETRINE   TLV: CAS#:

IGNI TEMP: NA FP: NA LEL: NA UEL: NA VP: NA VD: NA SG: 1.03 PS: LIQUID APPEAR: CLEAR GREEN
    ODOR: PERFUME   PH FACTOR: 5.2

HAZARD RATINGS: H: F: R: 1 0 0

FIRE FIGHTING:
  HIGH EXPANSION FOAM, LOW EXPANSION FOAM, ALCOHOL FOAM; DRY CHEMICAL, CARBON DIOXIDE, WATER FOG.
  PROTECTIVE CLOTHING, RUBBER GLOVES, AND BREATHING APPARATUS.
WARNING: 1] STRUCTURAL PROTECTIVE CLOTHING IS PERMEABLE, REMAIN CLEAR OF SMOKE, WATER FALL OUT AND WATER RUN OFF.
         2] KEEP OUT OF THE REACH OF CHILDREN.
         3] THERMAL DECOMPOSITION MAY YIELD CARBON DIOXIDE, CARBON MONOXIDE.
         4] MOVE CONTAINERS FROM AREA IF WITHOUT RISK, COOL EXPOSED CONTAINERS.
         5] DIKE AREA FOR CONTROL AND CONTAINMENT TO PREVENT ENTRY INTO SEWERS, DRAIN, AND WATER WAYS.

LARGE SPILL/NO FIRE/RESCUE: WEAR RUBBER OR NEOPRENE BOOTS, GLOVES.

SPILL CONTROL AND CONTAINMENT:
  HOUSEHOLD SPILL: WIPE UP LIQUID WITH AN ABSORBENT CLOTH, RINSE OUT IN THE SINK. WIPE UP THE RESIDUE WITH A DAMP
  CLOTH. RINSE OUT IN THE SINK.
  LARGE SPILL: COVER SPILL WITH ABSORBENT. SWEEP UP AND PUT IN A DISPOSAL CONTAINER. RINSE AREA WITH WATER, PICK
  UP WITH ABSORBENT, PUT IN A DISPOSAL CONTAINER. SEAL AND REMOVE FROM THE WORK AREA.

HEALTH HAZARD INFORMATION:
  WHEN USING THIS PRODUCT WEAR: NO SPECIAL REQUIREMENT. Contact with the eyes may cause stinging, tearing,
  itching, swelling, and redness. Ingestion may cause abdominal pain, nausea, vomiting, and diarrhea.

  A physician should be contacted if anyone develops any signs or symptoms and suspects that they are caused by
  exposure to this product.

FIRST AID:
  Eye Exposure:
  Flush with water for 15 minutes while lifting the upper and lower eye lids. Contact lenses should not be worn
  when working with this product. If irritation persists, get medical attention.

  Skin Exposure:
  Wash with soap and water. If irritation appears, get medical attention.

  Breathing:
  Not expected to be a problem.

  Swallowing:
  Immediately contact the local poison control center for advice. Keep the victim warm and at rest. Get medical
  attention.

PRODUCT: HALSA DANDRUFF SHAMPOO - REGULAR

INCOMPATIBILITY: NONE KNOWN

INGREDIENTS: SALICYLIC ACID   TLV: CAS#:   69-72-7
             AMMONIUM LAURYL SULFATE   TLV: CAS#:
             AMMONIUM LAURETH SULFATE TLV: CAS#:
             AMMONIUM CHLORIDE   TLV: CAS#: 12125-02-9
             COCAMIDOPROPYL BETRINE   TLV: CAS#:

IGNI TEMP: NA FP: NA LEL: NA UEL: NA VP: NA VD: NA SG: 1.02 PS: LIQUID APPEAR: CLEAR GREEN
   ODOR: PERFUME   PH FACTOR: 5.2

HAZARD RATINGS: H: F: R: 1 0 0

FIRE FIGHTING:
   HIGH EXPANSION FOAM, LOW EXPANSION FOAM, ALCOHOL FOAM; DRY CHEMICAL, CARBON DIOXIDE, WATER FOG.
   PROTECTIVE CLOTHING, RUBBER GLOVES, AND BREATHING APPARATUS.
WARNING: 1] STRUCTURAL PROTECTIVE CLOTHING IS PERMEABLE, REMAIN CLEAR OF SMOKE, WATER FALL OUT AND WATER RUN OFF.
         2] KEEP OUT OF THE REACH OF CHILDREN.
         3] THERMAL DECOMPOSITION MAY YIELD CARBON DIOXIDE, CARBON MONOXIDE.
         4] MOVE CONTAINERS FROM AREA IF WITHOUT RISK, COOL EXPOSED CONTAINERS.
         5] DIKE AREA FOR CONTROL AND CONTAINMENT TO PREVENT ENTRY INTO SEWERS, DRAIN, AND WATER WAYS.

LARGE SPILL/NO FIRE/RESCUE: WEAR RUBBER OR NEOPRENE BOOTS, GLOVES.

SPILL CONTROL AND CONTAINMENT:
   HOUSEHOLD SPILL: WIPE UP LIQUID WITH AN ABSORBENT CLOTH, RINSE OUT IN THE SINK. WIPE UP THE RESIDUE WITH A DAMP
   CLOTH. RINSE OUT IN THE SINK.
   LARGE SPILL: COVER SPILL WITH ABSORBENT. SWEEP UP AND PUT IN A DISPOSAL CONTAINER. RINSE AREA WITH WATER, PICK
   UP WITH ABSORBENT, PUT IN A DISPOSAL CONTAINER. SEAL AND REMOVE FROM THE WORK AREA.

HEALTH HAZARD INFORMATION:
   WHEN USING THIS PRODUCT WEAR: NO SPECIAL REQUIREMENT. Contact with the eyes may cause stinging, tearing,
   itching, swelling, and redness. Ingestion may cause abdominal pain, nausea, vomiting, and diarrhea.

   A physician should be contacted if anyone develops any signs or symptoms and suspects that they are caused by
   exposure to this product.

FIRST AID:
   Eye Exposure:
   Flush with water for 15 minutes while lifting the upper and lower eye lids. Contact lenses should not be worn
   when working with this product. If irritation persists, get medical attention.

   Skin Exposure:
   Wash with soap and water. If irritation appears, get medical attention.

   Breathing:
   Not expected to be a problem.

   Swallowing:
   Immediately contact the local poison control center for advice. Keep the victim warm and at rest. Get medical
   attention.

PRODUCT: HALSA HAIR SPRAY - CHAMOMILE

INCOMPATIBILITY: NONE KNOWN

INGREDIENTS: ETHYL ALCOHOL   60-70%   TLV: 1000 ppm   CAS#: 64-17-5
             PROPANE_____   TLV: 1000 ppm   CAS#: 74-98-6
             ISOBUTANE_____   25-35% TLV: CAS#: 75-28-5
             N-BUTANE_____/   TLV: 800 ppm   CAS#: 106-97-8
             POLYMER,FRAGRANCE,____ 3-6%   TLV: CAS#:
             SURFACTANTS_____/

IGNI TEMP: NA FP: 20F LEL: NA UEL: NA VP: NA VD: NA SG: .72 PS: LIQUID APPEAR: SPRAY MIST ODOR: FRAGRANT
     PH FACTOR: NA

HAZARD RATINGS: H: F: R: 2 4 0

FIRE FIGHTING:
    HIGH EXPANSION FOAM, LOW EXPANSION FOAM, ALCOHOL FOAM; DRY CHEMICAL, CARBON DIOXIDE, WATER FOG.
    PROTECTIVE CLOTHING, RUBBER GLOVES, AND BREATHING APPARATUS.
WARNING: 1] STRUCTURAL PROTECTIVE CLOTHING IS PERMEABLE, REMAIN CLEAR OF SMOKE, WATER FALL OUT AND WATER RUN OFF.
         2] KEEP OUT OF THE REACH OF CHILDREN.
         3] REMOVE ALL SOURCES OF IGNITION IF WITHOUT RISK.
         4] DO NOT USE OR GET NEAR A FLAME OR BURNING MATERIAL WHILE HAIR IS WET.
         5] CONTAINERS WILL EXPLODE IN A FIRE OR IF HEATED ABOVE 120F.
         6] THERMAL DECOMPOSITION YIELDS TOXIC CARBON DIOXIDE, CARBON MONOXIDE.
         7] MOVE CONTAINERS FROM AREA IF WITHOUT RISK, COOL EXPOSED CONTAINERS.
         8] DIKE AREA FOR CONTROL AND CONTAINMENT TO PREVENT ENTRY INTO SEWERS, DRAIN, AND WATER WAYS.

LARGE SPILL/NO FIRE/RESCUE: WEAR NON-SEALED CHEMICAL PROTECTIVE CLOTHING, BOOTS, GLOVES AND BREATHING APPARATUS.

SPILL CONTROL AND CONTAINMENT:
    HOUSEHOLD SPILL: VENTILATE IMMEDIATE SPILL AREA.
    LARGE SPILL: USE NON-SPARKING TOOLS. VENTILATE IMMEDIATE AREA. STOP LEAK IF POSSIBLE. LIQUID: COVER WITH
    NON-COMBUSTIBLE ABSORBENT. PICK UP AND PUT IN A DISPOSAL CONTAINER. SEAL AND REMOVE FROM THE WORK AREA.

HEALTH HAZARD INFORMATION:
    WHEN USING THIS PRODUCT WEAR: NO SPECIAL REQUIREMENT. Ingested ethyl alcohol is highly toxic, causes abdominal
    pain, nausea, and vomiting. Inhalation may cause headache, and drowsiness.

    A physician should be contacted if anyone develops any signs or symptoms and suspects that they are caused by
    exposure to this product.

FIRST AID:
Eye Exposure:
Flush with water for 15 minutes while lifting the upper and lower eye lids. Contact lenses should not be worn
when working with this product. If irritation persists, get medical attention.

    Skin Exposure:
    Wash with soap and water.

    Breathing:
    Stop use of spray. Move victim to fresh air.

    Swallowing:
    If victim is conscious, give 2-3 glasses of water to drink. Immediately contact the local poison control center
    for advice. Keep the victim warm and at rest. Get medical attention.

PRODUCT: HALSA HAIR SPRAY - GINGER ROOT

INCOMPATIBILITY: NONE KNOWN

INGREDIENTS: ETHYL ALCOHOL   60-70%   TLV: 1000 ppm   CAS#: 64-17-5
             PROPANE_____   TLV: 1000 ppm   CAS#: 74-98-6
             ISOBUTANE_____ 25-35% TLV: CAS#: 75-28-5
             N-BUTANE____/   TLV: 800 ppm   CAS#: 106-97-8
             POLYMER,FRAGRANCE,___ 3-6%   TLV: CAS#:
             SURFACTANTS_____/

IGNI TEMP: NA FP: 20F LEL: NA UEL: NA VP: NA VD: NA SG: .72 PS: LIQUID APPEAR: SPRAY MIST ODOR: FRAGRANT
  PH FACTOR: NA

HAZARD RATINGS: H: F: R: 2 4 0

FIRE FIGHTING:
  HIGH EXPANSION FOAM, LOW EXPANSION FOAM, ALCOHOL FOAM; DRY CHEMICAL, CARBON DIOXIDE, WATER FOG.
  PROTECTIVE CLOTHING, RUBBER GLOVES, AND BREATHING APPARATUS.
WARNING: 1] STRUCTURAL PROTECTIVE CLOTHING IS PERMEABLE, REMAIN CLEAR OF SMOKE, WATER FALL OUT AND WATER RUN OFF.
         2] KEEP OUT OF THE REACH OF CHILDREN.
         3] REMOVE ALL SOURCES OF IGNITION IF WITHOUT RISK.
         4] DO NOT USE OR GET NEAR A FLAME OR BURNING MATERIAL WHILE HAIR IS WET.
         5] CONTAINERS WILL EXPLODE IN A FIRE OR IF HEATED ABOVE 120F.
         6] THERMAL DECOMPOSITION YIELDS TOXIC CARBON DIOXIDE, CARBON MONOXIDE.
         7] MOVE CONTAINERS FROM AREA IF WITHOUT RISK, COOL EXPOSED CONTAINERS.
         8] DIKE AREA FOR CONTROL AND CONTAINMENT TO PREVENT ENTRY INTO SEWERS, DRAIN, AND WATER WAYS.

LARGE SPILL/NO FIRE/RESCUE: WEAR NON-SEALED CHEMICAL PROTECTIVE CLOTHING, BOOTS, GLOVES AND BREATHING APPARATUS.

SPILL CONTROL AND CONTAINMENT:
  HOUSEHOLD SPILL: VENTILATE IMMEDIATE SPILL AREA.
  LARGE SPILL: USE NON-SPARKING TOOLS. VENTILATE IMMEDIATE AREA. STOP LEAK IF POSSIBLE. LIQUID: COVER WITH
  NON-COMBUSTIBLE ABSORBENT. PICK UP AND PUT IN A DISPOSAL CONTAINER. SEAL AND REMOVE FROM THE WORK AREA.

HEALTH HAZARD INFORMATION:
  WHEN USING THIS PRODUCT WEAR: NO SPECIAL REQUIREMENT. Ingested ethyl alcohol is highly toxic, causes abdominal
  pain, nausea, and vomiting. Inhalation may cause headache, and drowsiness.

  A physician should be contacted if anyone develops any signs or symptoms and suspects that they are caused by
  exposure to this product.

FIRST AID:
  Eye Exposure:
  Flush with water for 15 minutes while lifting the upper and lower eye lids. Contact lenses should not be worn
  when working with this product. If irritation persists, get medical attention.

  Skin Exposure:
  Wash with soap and water.

  Breathing:
  Stop use of spray. Move victim to fresh air.

  Swallowing:
  If victim is conscious, give 2-3 glasses of water to drink. Immediately contact the local poison control center
  for advice. Keep the victim warm and at rest. Get medical attention.

PRODUCT: HALSA HAIR SPRAY - MARIGOLD

INCOMPATIBILITY: NONE KNOWN

INGREDIENTS: ETHYL ALCOHOL 60-70% TLV: 1000 ppm CAS#: 64-17-5
PROPANE_____ TLV: 1000 ppm CAS#: 74-98-6
ISOBUTANE_____ 25-35% TLV: CAS#: 75-28-5
N-BUTANE____/ TLV: 800 TLV: CAS#: 106-97-8
POLYMER,FRAGRANCE,____ 3-6% TLV: CAS#:
SURFACTANTS_____/

IGNI TEMP: NA FP: 20F LEL: NA UEL: NA VP: NA VD: NA SG: .72 PS: LIQUID APPEAR: SPRAY MIST ODOR: FRAGRANT
PH FACTOR: NA

HAZARD RATINGS: H: F: R: 2 4 0

FIRE FIGHTING:
HIGH EXPANSION FOAM, LOW EXPANSION FOAM, ALCOHOL FOAM; DRY CHEMICAL, CARBON DIOXIDE, WATER FOG.
PROTECTIVE CLOTHING, RUBBER GLOVES, AND BREATHING APPARATUS.
WARNING: 1] STRUCTURAL PROTECTIVE CLOTHING IS PERMEABLE, REMAIN CLEAR OF SMOKE, WATER FALL OUT AND WATER RUN OFF.
2] KEEP OUT OF THE REACH OF CHILDREN.
3] REMOVE ALL SOURCES OF IGNITION IF WITHOUT RISK.
4] DO NOT USE OR GET NEAR A FLAME OR BURNING MATERIAL WHILE HAIR IS WET.
5] CONTAINERS WILL EXPLODE IN A FIRE OR IF HEATED ABOVE 120F.
6] THERMAL DECOMPOSITION YIELDS TOXIC CARBON DIOXIDE, CARBON MONOXIDE.
7] MOVE CONTAINERS FROM AREA IF WITHOUT RISK, COOL EXPOSED CONTAINERS.
8] DIKE AREA FOR CONTROL AND CONTAINMENT TO PREVENT ENTRY INTO SEWERS, DRAIN, AND WATER WAYS.

LARGE SPILL/NO FIRE/RESCUE: WEAR NON-SEALED CHEMICAL PROTECTIVE CLOTHING, BOOTS, GLOVES AND BREATHING APPARATUS.

SPILL CONTROL AND CONTAINMENT:
HOUSEHOLD SPILL: VENTILATE IMMEDIATE SPILL AREA.
LARGE SPILL: USE NON-SPARKING TOOLS. VENTILATE IMMEDIATE AREA. STOP LEAK IF POSSIBLE. LIQUID: COVER WITH
NON-COMBUSTIBLE ABSORBENT. PICK UP AND PUT IN A DISPOSAL CONTAINER. SEAL AND REMOVE FROM THE WORK AREA.

HEALTH HAZARD INFORMATION:
WHEN USING THIS PRODUCT WEAR: NO SPECIAL REQUIREMENT. Ingested ethyl alcohol is highly toxic, causes abdominal
pain, nausea, and vomiting. Inhalation may cause headache, and drowsiness.

A physician should be contacted if anyone develops any signs or symptoms and suspects that they are caused by
exposure to this product.

FIRST AID:
Eye Exposure:
Flush with water for 15 minutes while lifting the upper and lower eye lids. Contact lenses should not be worn
when working with this product. If irritation persists, get medical attention.

Skin Exposure:
Wash with soap and water.

Breathing:
Stop use of spray. Move victim to fresh air.

Swallowing:
If victim is conscious, give 2-3 glasses of water to drink. Immediately contact the local poison control center
for advice. Keep the victim warm and at rest. Get medical attention.

PRODUCT: HALSA HAIR SPRAY - WALNUT LEAVES

INCOMPATIBILITY: NONE KNOWN

INGREDIENTS: ETHYL ALCOHOL   60-70%   TLV: 1000 ppm   CAS#: 64-17-5
             PROPANE_____   TLV: 1000 ppm   CAS#: 74-98-6
             ISOBUTANE_____ 25-35% TLV: CAS#: 75-28-5
             N-BUTANE_____/   TLV: 800 ppm   CAS#: 106-97-8
             POLYMER,FRAGRANCE,____ 3-6%   TLV: CAS#:
             SURFACTANTS_____/

IGNI TEMP: NA FP: 20F LEL: NA UEL: NA VP: NA VD: NA SG: .72 PS: LIQUID APPEAR: SPRAY MIST ODOR: FRAGRANT
   PH FACTOR: NA

HAZARD RATINGS: H: F: R: 2 4 0

FIRE FIGHTING:
   HIGH EXPANSION FOAM, LOW EXPANSION FOAM, ALCOHOL FOAM; DRY CHEMICAL, CARBON DIOXIDE, WATER FOG.
   PROTECTIVE CLOTHING, RUBBER GLOVES, AND BREATHING APPARATUS.
WARNING: 1] STRUCTURAL PROTECTIVE CLOTHING IS PERMEABLE, REMAIN CLEAR OF SMOKE, WATER FALL OUT AND WATER RUN OFF.
         2] KEEP OUT OF THE REACH OF CHILDREN.
         3] REMOVE ALL SOURCES OF IGNITION IF WITHOUT RISK.
         4] DO NOT USE OR GET NEAR A FLAME OR BURNING MATERIAL WHILE HAIR IS WET.
         5] CONTAINERS WILL EXPLODE IN A FIRE OR IF HEATED ABOVE 120F.
         6] THERMAL DECOMPOSITION YIELDS TOXIC CARBON DIOXIDE, CARBON MONOXIDE.
         7] MOVE CONTAINERS FROM AREA IF WITHOUT RISK, COOL EXPOSED CONTAINERS.
         8] DIKE AREA FOR CONTROL AND CONTAINMENT TO PREVENT ENTRY INTO SEWERS, DRAIN, AND WATER WAYS.

LARGE SPILL/NO FIRE/RESCUE: WEAR NON-SEALED CHEMICAL PROTECTIVE CLOTHING, BOOTS, GLOVES AND BREATHING APPARATUS.

SPILL CONTROL AND CONTAINMENT:
   HOUSEHOLD SPILL: VENTILATE IMMEDIATE SPILL AREA.
   LARGE SPILL: USE NON-SPARKING TOOLS. VENTILATE IMMEDIATE AREA. STOP LEAK IF POSSIBLE. LIQUID: COVER WITH
   NON-COMBUSTIBLE ABSORBENT. PICK UP AND PUT IN A DISPOSAL CONTAINER. SEAL AND REMOVE FROM THE WORK AREA.

HEALTH HAZARD INFORMATION:
   WHEN USING THIS PRODUCT WEAR: NO SPECIAL REQUIREMENT. Ingested ethyl alcohol is highly toxic, causes abdominal
   pain, nausea, and vomiting. Inhalation may cause headache, and drowsiness.

   A physician should be contacted if anyone develops any signs or symptoms and suspects that they are caused by
   exposure to this product.

FIRST AID:
   Eye Exposure:
   Flush with water for 15 minutes while lifting the upper and lower eye lids. Contact lenses should not be worn
   when working with this product. If irritation persists, get medical attention.

   Skin Exposure:
   Wash with soap and water.

   Breathing:
   Stop use of spray. Move victim to fresh air.

   Swallowing:
   If victim is conscious, give 2-3 glasses of water to drink. Immediately contact the local poison control center
   for advice. Keep the victim warm and at rest. Get medical attention.

PRODUCT: HALSA MOUSSE - CHAMOMILE

INCOMPATIBILITY: NONE KNOWN

INGREDIENTS: CETYL DIMETHYL HYDROXY ETHYL

AMMONIUM PHOSPHATE	1-5%	TLV:	CAS#:
POLYMER, FRAGRANCE			
SURFACTANT	5-10%	TLV:	CAS#:
PROPANE_____		TLV: 1000 ppm	CAS#: 74-98-7
ISOBUTANE_____	10-20%	TLV:	CAS#: 75-28-5
N-BUTANE_____/		TLV: 800 ppm	CAS#: 106-97-8
WATER	70-80%	TLV:	CAS#: 7732-18-5

IGNI TEMP: NA FP: 20F LEL: NA UEL: NA VP: NA VD: NA SG: .89 PS: LIQUID APPEAR: BEIGE FOAM ODOR: FRAGRANT
   PH FACTOR: 5.6

HAZARD RATINGS: H: F: R: 0 4 -

FIRE FIGHTING:
   HIGH EXPANSION FOAM, LOW EXPANSION FOAM, ALCOHOL FOAM; DRY CHEMICAL, CARBON DIOXIDE, WATER FOG.
   PROTECTIVE CLOTHING, RUBBER GLOVES, AND BREATHING APPARATUS.
WARNING: 1] STRUCTURAL PROTECTIVE CLOTHING IS PERMEABLE, REMAIN CLEAR OF SMOKE, WATER FALL OUT AND WATER RUN OFF.
         2] KEEP OUT OF THE REACH OF CHILDREN.
         3] REMOVE ALL SOURCES OF IGNITION IF WITHOUT RISK.
         4] CONTAINERS WILL EXPLODE IN A FIRE OR IF HEATED ABOVE 120F.
         5] DO NOT USE NEAR A FLAME OR BURNING MATERIAL.
         6] THERMAL DECOMPOSITION YIELDS TOXIC CARBON DIOXIDE, CARBON MONOXIDE.
         7] MOVE CONTAINERS FROM AREA IF WITHOUT RISK, COOL EXPOSED CONTAINERS.
         8] DIKE AREA FOR CONTROL AND CONTAINMENT TO PREVENT ENTRY INTO SEWERS, DRAIN, AND WATER WAYS.

LARGE SPILL/NO FIRE/RESCUE: WEAR RUBBER OR NEOPRENE BOOTS, GLOVES.

SPILL CONTROL AND CONTAINMENT:
   HOUSEHOLD SPILL: WIPE UP FOAM WITH AN ABSORBENT CLOTH, RINSE OUT IN THE SINK. WIPE UP RESIDUE WITH A DAMP CLOTH.
   LARGE SPILL: SCOOP UP FOAM. PUT IN A DISPOSAL CONTAINER. COVER RESIDUE WITH ABSORBENT. SWEEP UP AND PUT IN A
   DISPOSAL CONTAINER, RINSE AREA WITH WATER, PICK UP WITH ABSORBENT. PUT IN A DISPOSAL CONTAINER.

HEALTH HAZARD INFORMATION:
   WHEN USING THIS PRODUCT WEAR: NO SPECIAL REQUIREMENT. Eye contact may cause stinging, tearing, itching,
   swelling, and redness. Ingestion may cause abdominal pain, nausea, and vomiting.

   A physician should be contacted if anyone develops any signs or symptoms and suspects that they are caused by
   exposure to this product.

FIRST AID:
Eye Exposure:
Flush with water for 15 minutes while lifting the upper and lower eye lids. Contact lenses should not be worn
when working with this product. If irritation persists, get medical attention.

   Skin Exposure:
   Not expected to be a problem.

   Breathing:
   Not expected to be a problem.

   Swallowing:
   Immediately contact the local poison control center for advice. Keep the victim warm and at rest. Get medical
   attention.

PRODUCT: HALSA MOUSSE - GINGER ROOT

INCOMPATIBILITY: NONE KNOWN

INGREDIENTS: CETYL DIMETHYL HYDROXY ETHYL

AMMONIUM PHOSPHATE	1-5%	TLV:	CAS#:
POLYMER, FRAGRANCE			
SURFACTANT	5-10%	TLV:	CAS#:
PROPANE_____		TLV: 1000 ppm	CAS#: 74-98-7
ISOBUTANE_____	10-20%	TLV:	CAS#: 75-28-5
N-BUTANE_____/		TLV: 800 ppm	CAS#: 106-97-8
WATER	70-80%	TLV:	CAS#: 7732-18-5

IGNI TEMP: NA FP: 20F LEL: NA UEL: NA VP: NA VD: NA SG: .89 PS: LIQUID APPEAR: BEIGE FOAM ODOR: FRAGRANT
   PH FACTOR: 5.6

HAZARD RATINGS: H: F: R: 0 4 -

FIRE FIGHTING:
   HIGH EXPANSION FOAM, LOW EXPANSION FOAM, ALCOHOL FOAM; DRY CHEMICAL, CARBON DIOXIDE, WATER FOG.
   PROTECTIVE CLOTHING, RUBBER GLOVES, AND BREATHING APPARATUS.
WARNING: 1] STRUCTURAL PROTECTIVE CLOTHING IS PERMEABLE, REMAIN CLEAR OF SMOKE, WATER FALL OUT AND WATER RUN OFF.
         2] KEEP OUT OF THE REACH OF CHILDREN.
         3] REMOVE ALL SOURCES OF IGNITION IF WITHOUT RISK.
         4] CONTAINERS WILL EXPLODE IN A FIRE OR IF HEATED ABOVE 120F.
         5] DO NOT USE NEAR A FLAME OR BURNING MATERIAL.
         6] THERMAL DECOMPOSITION YIELDS TOXIC CARBON DIOXIDE, CARBON MONOXIDE.
         7] MOVE CONTAINERS FROM AREA IF WITHOUT RISK, COOL EXPOSED CONTAINERS.
         8] DIKE AREA FOR CONTROL AND CONTAINMENT TO PREVENT ENTRY INTO SEWERS, DRAIN, AND WATER WAYS.

LARGE SPILL/NO FIRE/RESCUE: WEAR RUBBER OR NEOPRENE BOOTS, GLOVES.

SPILL CONTROL AND CONTAINMENT:
   HOUSEHOLD SPILL: WIPE UP FOAM WITH AN ABSORBENT CLOTH, RINSE OUT IN THE SINK. WIPE UP RESIDUE WITH A DAMP CLOTH.
   LARGE SPILL: SCOOP UP FOAM. PUT IN A DISPOSAL CONTAINER. COVER RESIDUE WITH ABSORBENT. SWEEP UP AND PUT IN A
   DISPOSAL CONTAINER, RINSE AREA WITH WATER, PICK UP WITH ABSORBENT. PUT IN A DISPOSAL CONTAINER.

HEALTH HAZARD INFORMATION:
   WHEN USING THIS PRODUCT WEAR: NO SPECIAL REQUIREMENT. Eye contact may cause stinging, tearing, itching,
   swelling, and redness. Ingestion may cause abdominal pain, nausea, and vomiting.

   A physician should be contacted if anyone develops any signs or symptoms and suspects that they are caused by
   exposure to this product.

FIRST AID:
   Eye Exposure:
   Flush with water for 15 minutes while lifting the upper and lower eye lids. Contact lenses should not be worn
   when working with this product. If irritation persists, get medical attention.

   Skin Exposure:
   Not expected to be a problem.

   Breathing:
   Not expected to be a problem.

   Swallowing:
   Immediately contact the local poison control center for advice. Keep the victim warm and at rest. Get medical
   attention.

PRODUCT: HALSA MOUSSE - MARIGOLD

INCOMPATIBILITY: NONE KNOWN

INGREDIENTS: CETYL DIMETHYL HYDROXY ETHYL

AMMONIUM PHOSPHATE	1-5%	TLV:	CAS#:
POLYMER, FRAGRANCE			
SURFACTANT	5-10%	TLV:	CAS#:
PROPANE_____		TLV: 1000 ppm	CAS#: 74-98-7
ISOBUTANE_____	10-20%	TLV:	CAS#: 75-28-5
N-BUTANE_____/		TLV: 800 ppm	CAS#: 106-97-8
WATER	70-80%	TLV:	CAS#: 7732-18-5

IGNI TEMP: NA FP: 20F LEL: NA UEL: NA VP: NA VD: NA SG: .89 PS: LIQUID APPEAR: BEIGE FOAM ODOR: FRAGRANT
PH FACTOR: 5.6

HAZARD RATINGS: H: F: R: 0 4 -

FIRE FIGHTING:
HIGH EXPANSION FOAM, LOW EXPANSION FOAM, ALCOHOL FOAM; DRY CHEMICAL, CARBON DIOXIDE, WATER FOG.
PROTECTIVE CLOTHING, RUBBER GLOVES, AND BREATHING APPARATUS.
WARNING: 1] STRUCTURAL PROTECTIVE CLOTHING IS PERMEABLE, REMAIN CLEAR OF SMOKE, WATER FALL OUT AND WATER RUN OFF.
         2] KEEP OUT OF THE REACH OF CHILDREN.
         3] REMOVE ALL SOURCES OF IGNITION IF WITHOUT RISK.
         4] CONTAINERS WILL EXPLODE IN A FIRE OR IF HEATED ABOVE 120F.
         5] DO NOT USE NEAR A FLAME OR BURNING MATERIAL.
         6] THERMAL DECOMPOSITION YIELDS TOXIC CARBON DIOXIDE, CARBON MONOXIDE.
         7] MOVE CONTAINERS FROM AREA IF WITHOUT RISK, COOL EXPOSED CONTAINERS.
         8] DIKE AREA FOR CONTROL AND CONTAINMENT TO PREVENT ENTRY INTO SEWERS, DRAIN, AND WATER WAYS.

LARGE SPILL/NO FIRE/RESCUE: WEAR RUBBER OR NEOPRENE BOOTS, GLOVES.

SPILL CONTROL AND CONTAINMENT:
HOUSEHOLD SPILL: WIPE UP FOAM WITH AN ABSORBENT CLOTH, RINSE OUT IN THE SINK. WIPE UP RESIDUE WITH A DAMP CLOTH.
LARGE SPILL: SCOOP UP FOAM. PUT IN A DISPOSAL CONTAINER. COVER RESIDUE WITH ABSORBENT. SWEEP UP AND PUT IN A
DISPOSAL CONTAINER, RINSE AREA WITH WATER, PICK UP WITH ABSORBENT. PUT IN A DISPOSAL CONTAINER.

HEALTH HAZARD INFORMATION:
WHEN USING THIS PRODUCT WEAR: NO SPECIAL REQUIREMENT. Eye contact may cause stinging, tearing, itching,
swelling, and redness. Ingestion may cause abdominal pain, nausea, and vomiting.

A physician should be contacted if anyone develops any signs or symptoms and suspects that they are caused by
exposure to this product.

FIRST AID:
Eye Exposure:
Flush with water for 15 minutes while lifting the upper and lower eye lids. Contact lenses should not be worn
when working with this product. If irritation persists, get medical attention.

Skin Exposure:
Not expected to be a problem.

Breathing:
Not expected to be a problem.

Swallowing:
Immediately contact the local poison control center for advice. Keep the victim warm and at rest. Get medical
attention.

PRODUCT: HALSA MOUSSE - WALNUT LEAVES

INCOMPATIBILITY: NONE KNOWN

INGREDIENTS: CETYL DIMETHYL HYDROXY ETHYL

AMMONIUM PHOSPHATE	1-5%	TLV:	CAS#:
POLYMER, FRAGRANCE			
SURFACTANT	5-10%	TLV:	CAS#:
PROPANE_____		TLV: 1000 ppm	CAS#: 74-98-7
ISOBUTANE_____	10-20%	TLV:	CAS#: 75-28-5
N-BUTANE____/		TLV: 800 ppm	CAS#: 106-97-8
WATER	70-80%	TLV:	CAS#: 7732-18-5

IGNI TEMP: NA FP: 20F LEL: NA UEL: NA VP: NA VD: NA SG: .89 PS: LIQUID APPEAR: BEIGE FOAM ODOR: FRAGRANT
   PH FACTOR: 5.6

HAZARD RATINGS: H: F: R: 0 4 -

FIRE FIGHTING:
   HIGH EXPANSION FOAM, LOW EXPANSION FOAM, ALCOHOL FOAM; DRY CHEMICAL, CARBON DIOXIDE, WATER FOG.
   PROTECTIVE CLOTHING, RUBBER GLOVES, AND BREATHING APPARATUS.
WARNING: 1] STRUCTURAL PROTECTIVE CLOTHING IS PERMEABLE, REMAIN CLEAR OF SMOKE, WATER FALL OUT AND WATER RUN OFF.
         2] KEEP OUT OF THE REACH OF CHILDREN.
         3] REMOVE ALL SOURCES OF IGNITION IF WITHOUT RISK.
         4] CONTAINERS WILL EXPLODE IN A FIRE OR IF HEATED ABOVE 120F.
         5] DO NOT USE NEAR A FLAME OR BURNING MATERIAL.
         6] THERMAL DECOMPOSITION YIELDS TOXIC CARBON DIOXIDE, CARBON MONOXIDE.
         7] MOVE CONTAINERS FROM AREA IF WITHOUT RISK, COOL EXPOSED CONTAINERS.
         8] DIKE AREA FOR CONTROL AND CONTAINMENT TO PREVENT ENTRY INTO SEWERS, DRAIN, AND WATER WAYS.

LARGE SPILL/NO FIRE/RESCUE: WEAR RUBBER OR NEOPRENE BOOTS, GLOVES.

SPILL CONTROL AND CONTAINMENT:
   HOUSEHOLD SPILL: WIPE UP FOAM WITH AN ABSORBENT CLOTH, RINSE OUT IN THE SINK. WIPE UP RESIDUE WITH A DAMP CLOTH.
   LARGE SPILL: SCOOP UP FOAM. PUT IN A DISPOSAL CONTAINER. COVER RESIDUE WITH ABSORBENT. SWEEP UP AND PUT IN A
   DISPOSAL CONTAINER, RINSE AREA WITH WATER, PICK UP WITH ABSORBENT. PUT IN A DISPOSAL CONTAINER.

HEALTH HAZARD INFORMATION:
   WHEN USING THIS PRODUCT WEAR: NO SPECIAL REQUIREMENT. Eye contact may cause stinging, tearing, itching,
   swelling, and redness. Ingestion may cause abdominal pain, nausea, and vomiting.

   A physician should be contacted if anyone develops any signs or symptoms and suspects that they are caused by
   exposure to this product.

FIRST AID:
   Eye Exposure:
   Flush with water for 15 minutes while lifting the upper and lower eye lids. Contact lenses should not be worn
   when working with this product. If irritation persists, get medical attention.

   Skin Exposure:
   Not expected to be a problem.

   Breathing:
   Not expected to be a problem.

   Swallowing:
   Immediately contact the local poison control center for advice. Keep the victim warm and at rest. Get medical
   attention.

PRODUCT: HALSA SHAMPOO - CHAMOMILE

INCOMPATIBILITY: NONE KNOWN

INGREDIENTS: AMMONIUM CHLORIDE   TLV: CAS#: 12125-02-9
             CITRIC ACID   TLV: CAS#:   77-92-9
             WATER   TLV: CAS#: 7732-18-5
             AMMONIUM LAURYL SULFATE   AMMONIUM LAURETH SULFATE
             BOTANICAL EXTRACTS   FRAGRANCE
             METHYLCHLOROISOTHIAZOLINONE   METHYLISOTHIAZOLINONE
             COLORANTS

IGNI TEMP: NA FP: NA LEL: NA UEL: NA VP: NA VD: NA SG: 1.0 PS: LIQUID APPEAR: VISCOUS, CLEAR
    ODOR: FRAGRANT   PH FACTOR: 3.8

HAZARD RATINGS: H: F: R: 2 0 0

FIRE FIGHTING:
   HIGH EXPANSION FOAM, LOW EXPANSION FOAM, ALCOHOL FOAM; DRY CHEMICAL, CARBON DIOXIDE, WATER FOG.
   PROTECTIVE CLOTHING, RUBBER GLOVES, AND BREATHING APPARATUS.
WARNING: 1] STRUCTURAL PROTECTIVE CLOTHING IS PERMEABLE, REMAIN CLEAR OF SMOKE, WATER FALL OUT AND WATER RUN OFF.
         2] KEEP OUT OF THE REACH OF CHILDREN.
         3] THERMAL DECOMPOSITION YIELDS TOXIC CARBON DIOXIDE, CARBON MONOXIDE.
         4] MOVE CONTAINERS FROM AREA IF WITHOUT RISK, COOL EXPOSED CONTAINERS.
         5] DIKE AREA FOR CONTROL AND CONTAINMENT TO PREVENT ENTRY INTO SEWERS, DRAIN, AND WATER WAYS.

LARGE SPILL/NO FIRE/RESCUE: WEAR RUBBER OR NEOPRENE BOOTS, GLOVES.

SPILL CONTROL AND CONTAINMENT:
   HOUSEHOLD SPILL: WIPE UP LIQUID WITH AN ABSORBENT CLOTH. RINSE OUT IN THE SINK. WIPE UP RESIDUE WITH A DAMP
   CLOTH.
   LARGE SPILL: SCOOP UP LIQUID AND PUT IN A DISPOSAL CONTAINER. COVER RESIDUE WITH ABSORBENT. SWEEP UP AND PUT IN
   A DISPOSAL CONTAINER. RINSE AREA WITH WATER, PICK UP WITH ABSORBENT. PUT IN A DISPOSAL CONTAINER. SEAL AND
   REMOVE FROM THE WORK AREA.

HEALTH HAZARD INFORMATION:
   WHEN USING THIS PRODUCT WEAR: NO SPECIAL REQUIREMENT. Contact with the eyes may cause irritation. Ingestion may
   cause abdominal pain, nausea, vomiting, and diarrhea.

   A physician should be contacted if anyone develops any signs or symptoms and suspects that they are caused by
   exposure to this product.

FIRST AID:
   Eye Exposure:
   Flush with water for 15 minutes while lifting the upper and lower eye lids. Contact lenses should not be worn
   when working with this product. If irritation persists, get medical attention.

   Skin Exposure:
   Not expected to be a problem.

   Breathing:
   Not expected to be a problem.

   Swallowing:
   Immediately contact the local poison control center for advice. Keep the victim warm and at rest. Get medical
   attention.

PRODUCT: HALSA SHAMPOO - CORNFLOWER

INCOMPATIBILITY: NONE KNOWN

INGREDIENTS: AMMONIUM CHLORIDE   TLV: CAS#: 12125-02-9
             CITRIC ACID   TLV: CAS#:   77-92-9
             WATER   TLV: CAS#: 7732-18-5
             AMMONIUM LAURYL SULFATE   AMMONIUM LAURETH SULFATE
             BOTANICAL EXTRACTS   FRAGRANCE
             METHYLCHLOROISOTHIAZOLINONE   METHYLISOTHIAZOLINONE
             COLORANTS   COCAMIDE DEA

IGNI TEMP: NA FP: NA LEL: NA UEL: NA VP: NA VD: NA SG: 1.0 PS: LIQUID APPEAR: VISCOUS, CLEAR
     ODOR: FRAGRANT   PH FACTOR: 3.8

HAZARD RATINGS: H: F: R: 2 0 0

FIRE FIGHTING:
   HIGH EXPANSION FOAM, LOW EXPANSION FOAM, ALCOHOL FOAM; DRY CHEMICAL, CARBON DIOXIDE, WATER FOG.
   PROTECTIVE CLOTHING, RUBBER GLOVES, AND BREATHING APPARATUS.
WARNING: 1] STRUCTURAL PROTECTIVE CLOTHING IS PERMEABLE, REMAIN CLEAR OF SMOKE, WATER FALL OUT AND WATER RUN OFF.
         2] KEEP OUT OF THE REACH OF CHILDREN.
         3] THERMAL DECOMPOSITION YIELDS TOXIC CARBON DIOXIDE, CARBON MONOXIDE.
         4] MOVE CONTAINERS FROM AREA IF WITHOUT RISK, COOL EXPOSED CONTAINERS.
         5] DIKE AREA FOR CONTROL AND CONTAINMENT TO PREVENT ENTRY INTO SEWERS, DRAIN, AND WATER WAYS.

LARGE SPILL/NO FIRE/RESCUE: WEAR RUBBER OR NEOPRENE BOOTS, GLOVES.

SPILL CONTROL AND CONTAINMENT:
   HOUSEHOLD SPILL: WIPE UP LIQUID WITH AN ABSORBENT CLOTH. RINSE OUT IN THE SINK. WIPE UP RESIDUE WITH A DAMP
   CLOTH.
   LARGE SPILL: SCOOP UP LIQUID AND PUT IN A DISPOSAL CONTAINER. COVER RESIDUE WITH ABSORBENT. SWEEP UP AND PUT IN
   A DISPOSAL CONTAINER. RINSE AREA WITH WATER, PICK UP WITH ABSORBENT. PUT IN A DISPOSAL CONTAINER. SEAL AND
   REMOVE FROM THE WORK AREA.

HEALTH HAZARD INFORMATION:
   WHEN USING THIS PRODUCT WEAR: NO SPECIAL REQUIREMENT. Contact with the eyes may cause irritation. Ingestion may
   cause abdominal pain, nausea, vomiting, and diarrhea.

   A physician should be contacted if anyone develops any signs or symptoms and suspects that they are caused by
   exposure to this product.

FIRST AID:
   Eye Exposure:
   Flush with water for 15 minutes while lifting the upper and lower eye lids. Contact lenses should not be worn
   when working with this product. If irritation persists, get medical attention.

   Skin Exposure:
   Not expected to be a problem.

   Breathing:
   Not expected to be a problem.

   Swallowing:
   Immediately contact the local poison control center for advice. Keep the victim warm and at rest. Get medical
   attention.

PRODUCT: HALSA SHAMPOO - GINGER ROOT

INCOMPATIBILITY: NONE KNOWN

INGREDIENTS: AMMONIUM CHLORIDE   TLV: CAS#: 12125-02-9
             CITRIC ACID   TLV: CAS#:   77-92-9
             WATER   TLV: CAS#: 7732-18-5
             AMMONIUM LAURYL SULFATE   AMMONIUM LAURETH SULFATE
             BOTANICAL EXTRACTS   FRAGRANCE
             METHYLCHLOROISOTHIAZOLINONE   METHYLISOTHIAZOLINONE
             COLORANTS

IGNI TEMP: NA FP: NA LEL: NA UEL: NA VP: NA VD: NA SG: 1.0 PS: LIQUID APPEAR: VISCOUS, CLEAR
    ODOR: FRAGRANT   PH FACTOR: 3.8

HAZARD RATINGS: H: F: R: 2 0 0

FIRE FIGHTING:
  HIGH EXPANSION FOAM, LOW EXPANSION FOAM, ALCOHOL FOAM; DRY CHEMICAL, CARBON DIOXIDE, WATER FOG.
  PROTECTIVE CLOTHING, RUBBER GLOVES, AND BREATHING APPARATUS.
WARNING: 1] STRUCTURAL PROTECTIVE CLOTHING IS PERMEABLE, REMAIN CLEAR OF SMOKE, WATER FALL OUT AND WATER RUN OFF.
         2] KEEP OUT OF THE REACH OF CHILDREN.
         3] THERMAL DECOMPOSITION YIELDS TOXIC CARBON DIOXIDE, CARBON MONOXIDE.
         4] MOVE CONTAINERS FROM AREA IF WITHOUT RISK, COOL EXPOSED CONTAINERS.
         5] DIKE AREA FOR CONTROL AND CONTAINMENT TO PREVENT ENTRY INTO SEWERS, DRAIN, AND WATER WAYS.

LARGE SPILL/NO FIRE/RESCUE: WEAR RUBBER OR NEOPRENE BOOTS, GLOVES.

SPILL CONTROL AND CONTAINMENT:
  HOUSEHOLD SPILL: WIPE UP LIQUID WITH AN ABSORBENT CLOTH. RINSE OUT IN THE SINK. WIPE UP RESIDUE WITH A DAMP
  CLOTH.
  LARGE SPILL: SCOOP UP LIQUID AND PUT IN A DISPOSAL CONTAINER. COVER RESIDUE WITH ABSORBENT. SWEEP UP AND PUT IN
  A DISPOSAL CONTAINER. RINSE AREA WITH WATER, PICK UP WITH ABSORBENT. PUT IN A DISPOSAL CONTAINER. SEAL AND
  REMOVE FROM THE WORK AREA.

HEALTH HAZARD INFORMATION:
  WHEN USING THIS PRODUCT WEAR: NO SPECIAL REQUIREMENT. Contact with the eyes may cause irritation. Ingestion may
  cause abdominal pain, nausea, vomiting, and diarrhea.

  A physician should be contacted if anyone develops any signs or symptoms and suspects that they are caused by
  exposure to this product.

FIRST AID:
  Eye Exposure:
  Flush with water for 15 minutes while lifting the upper and lower eye lids. Contact lenses should not be worn
  when working with this product. If irritation persists, get medical attention.

  Skin Exposure:
  Not expected to be a problem.

  Breathing:
  Not expected to be a problem.

  Swallowing:
  Immediately contact the local poison control center for advice. Keep the victim warm and at rest. Get medical
  attention.

PRODUCT: HALSA SHAMPOO - MARIGOLD

INCOMPATIBILITY: NONE KNOWN

INGREDIENTS: AMMONIUM CHLORIDE   TLV: CAS#: 12125-02-9
             CITRIC ACID   TLV: CAS#:   77-92-9
             WATER   TLV: CAS#: 7732-18-5
             AMMONIUM LAURYL SULFATE   AMMONIUM LAURETH SULFATE
             BOTANICAL EXTRACTS   FRAGRANCE
             METHYLCHLOROISOTHIAZOLINONE   METHYLISOTHIAZOLINONE
             COLORANTS

IGNI TEMP: NA FP: NA LEL: NA UEL: NA VP: NA VD: NA SG: 1.0 PS: LIQUID APPEAR: VISCOUS, CLEAR
   ODOR: FRAGRANT   PH FACTOR: 3.8

HAZARD RATINGS: H: F: R: 2 0 0

FIRE FIGHTING:
  HIGH EXPANSION FOAM, LOW EXPANSION FOAM, ALCOHOL FOAM; DRY CHEMICAL, CARBON DIOXIDE, WATER FOG.
  PROTECTIVE CLOTHING, RUBBER GLOVES, AND BREATHING APPARATUS.
WARNING: 1] STRUCTURAL PROTECTIVE CLOTHING IS PERMEABLE, REMAIN CLEAR OF SMOKE, WATER FALL OUT AND WATER RUN OFF.
         2] KEEP OUT OF THE REACH OF CHILDREN.
         3] THERMAL DECOMPOSITION YIELDS TOXIC CARBON DIOXIDE, CARBON MONOXIDE.
         4] MOVE CONTAINERS FROM AREA IF WITHOUT RISK, COOL EXPOSED CONTAINERS.
         5] DIKE AREA FOR CONTROL AND CONTAINMENT TO PREVENT ENTRY INTO SEWERS, DRAIN, AND WATER WAYS.

LARGE SPILL/NO FIRE/RESCUE: WEAR RUBBER OR NEOPRENE BOOTS, GLOVES.

SPILL CONTROL AND CONTAINMENT:
  HOUSEHOLD SPILL: WIPE UP LIQUID WITH AN ABSORBENT CLOTH. RINSE OUT IN THE SINK. WIPE UP RESIDUE WITH A DAMP
  CLOTH.
  LARGE SPILL: SCOOP UP LIQUID AND PUT IN A DISPOSAL CONTAINER. COVER RESIDUE WITH ABSORBENT. SWEEP UP AND PUT IN
  A DISPOSAL CONTAINER. RINSE AREA WITH WATER, PICK UP WITH ABSORBENT. PUT IN A DISPOSAL CONTAINER. SEAL AND
  REMOVE FROM THE WORK AREA.

HEALTH HAZARD INFORMATION:
  WHEN USING THIS PRODUCT WEAR: NO SPECIAL REQUIREMENT. Contact with the eyes may cause irritation. Ingestion may
  cause abdominal pain, nausea, vomiting, and diarrhea.

  A physician should be contacted if anyone develops any signs or symptoms and suspects that they are caused by
  exposure to this product.

FIRST AID:
  Eye Exposure:
  Flush with water for 15 minutes while lifting the upper and lower eye lids. Contact lenses should not be worn
  when working with this product. If irritation persists, get medical attention.

  Skin Exposure:
  Not expected to be a problem.

  Breathing:
  Not expected to be a problem.

  Swallowing:
  Immediately contact the local poison control center for advice. Keep the victim warm and at rest. Get medical
  attention.

PRODUCT: HALSA SHAMPOO - WALNUT LEAVES

INCOMPATIBILITY: NONE KNOWN

INGREDIENTS: AMMONIUM CHLORIDE   TLV: CAS#: 12125-02-9
             CITRIC ACID   TLV: CAS#:   77-92-9
             WATER   TLV: CAS#: 7732-18-5
             AMMONIUM LAURYL SULFATE   AMMONIUM LAURETH SULFATE
             BOTANICAL EXTRACTS   FRAGRANCE
             METHYLCHLOROISOTHIAZOLINONE   METHYLISOTHIAZOLINONE
             COLORANTS

IGNI TEMP: NA FP: NA LEL: NA UEL: NA VP: NA VD: NA SG: 1.0 PS: LIQUID APPEAR: VISCOUS, CLEAR
    ODOR: FRAGRANT   PH FACTOR: 3.8

HAZARD RATINGS: H: F: R: 2 0 0

FIRE FIGHTING:
   HIGH EXPANSION FOAM, LOW EXPANSION FOAM, ALCOHOL FOAM; DRY CHEMICAL, CARBON DIOXIDE, WATER FOG.
   PROTECTIVE CLOTHING, RUBBER GLOVES, AND BREATHING APPARATUS.
WARNING: 1] STRUCTURAL PROTECTIVE CLOTHING IS PERMEABLE, REMAIN CLEAR OF SMOKE, WATER FALL OUT AND WATER RUN OFF.
         2] KEEP OUT OF THE REACH OF CHILDREN.
         3] THERMAL DECOMPOSITION YIELDS TOXIC CARBON DIOXIDE, CARBON MONOXIDE.
         4] MOVE CONTAINERS FROM AREA IF WITHOUT RISK, COOL EXPOSED CONTAINERS.
         5] DIKE AREA FOR CONTROL AND CONTAINMENT TO PREVENT ENTRY INTO SEWERS, DRAIN, AND WATER WAYS.

LARGE SPILL/NO FIRE/RESCUE: WEAR RUBBER OR NEOPRENE BOOTS, GLOVES.

SPILL CONTROL AND CONTAINMENT:
   HOUSEHOLD SPILL: WIPE UP LIQUID WITH AN ABSORBENT CLOTH. RINSE OUT IN THE SINK. WIPE UP RESIDUE WITH A DAMP
   CLOTH.
   LARGE SPILL: SCOOP UP LIQUID AND PUT IN A DISPOSAL CONTAINER. COVER RESIDUE WITH ABSORBENT. SWEEP UP AND PUT IN
   A DISPOSAL CONTAINER. RINSE AREA WITH WATER, PICK UP WITH ABSORBENT. PUT IN A DISPOSAL CONTAINER. SEAL AND
   REMOVE FROM THE WORK AREA.

HEALTH HAZARD INFORMATION:
   WHEN USING THIS PRODUCT WEAR: NO SPECIAL REQUIREMENT. Contact with the eyes may cause irritation. Ingestion may
   cause abdominal pain, nausea, vomiting, and diarrhea.

   A physician should be contacted if anyone develops any signs or symptoms and suspects that they are caused by
   exposure to this product.

FIRST AID:
   Eye Exposure:
   Flush with water for 15 minutes while lifting the upper and lower eye lids. Contact lenses should not be worn
   when working with this product. If irritation persists, get medical attention.

   Skin Exposure:
   Not expected to be a problem.

   Breathing:
   Not expected to be a problem.

   Swallowing:
   Immediately contact the local poison control center for advice. Keep the victim warm and at rest. Get medical
   attention.

PRODUCT: HOT SHOTS

INCOMPATIBILITY: NONE KNOWN

INGREDIENTS: IRON POWDER   TLV: CAS#: 7439-89-6
             VERMICULITE   TLV: CAS#:
             ACTIVATED CARBON   TLV: CAS#:
             SODIUM CHLORIDE   TLV: CAS#: 7647-14-5
             WATER   TLV: CAS#: 7732-18-5

IGNI TEMP: NA FP: NA LEL: NA UEL: NA VP: NA VD: NA SG: .7 PS: POWDER APPEAR: IN A WHITE POUCH
    ODOR: PH FACTOR: NA

HAZARD RATINGS: H: F: R:

FIRE FIGHTING:
   HIGH EXPANSION FOAM, LOW EXPANSION FOAM, ALCOHOL FOAM; DRY CHEMICAL, CARBON DIOXIDE, WATER FOG.
   PROTECTIVE CLOTHING, RUBBER GLOVES, AND BREATHING APPARATUS.
WARNING: 1] STRUCTURAL PROTECTIVE CLOTHING IS PERMEABLE, REMAIN CLEAR OF SMOKE, WATER FALL OUT AND WATER RUN OFF.
         2] KEEP OUT OF THE REACH OF CHILDREN.
         3] WHEN WET PRODUCT PRODUCES HEAT AS IN A HOT PACK.
         4] MOVE CONTAINERS FROM AREA IF WITHOUT RISK, COOL EXPOSED CONTAINERS.
         5] DIKE AREA FOR CONTROL AND CONTAINMENT TO PREVENT ENTRY INTO SEWERS, DRAIN, AND WATER WAYS.

LARGE SPILL/NO FIRE/RESCUE: WEAR RUBBER OR NEOPRENE BOOTS, GLOVES.

SPILL CONTROL AND CONTAINMENT:
   HOUSEHOLD SPILL: PICK UP POUCH, PUT IN A DISPOSAL CONTAINER. SWEEP UP RESIDUE, PUT IN A DISPOSAL CONTAINER. WIPE
   UP RESIDUE WITH A DAMP CLOTH. RINSE OUT IN THE SINK.
   LARGE SPILL: SCOOP UP POWDER, PUT IN A DISPOSAL CONTAINER. RINSE AREA WITH WATER, PICK UP WITH ABSORBENT. PUT IN
   A SEPARATE DISPOSAL CONTAINER. DO NOT PUT WET ABSORBENT INTO CONTAINER WITH THE POWDER. SEAL AND REMOVE FROM THE
   WORK AREA.

HEALTH HAZARD INFORMATION:
   WHEN USING THIS PRODUCT: Place a cloth between skin and pouch to minimize chances of minor burns.

   A physician should be contacted if anyone develops any signs or symptoms and suspects that they are caused by
   exposure to this product.

FIRST AID:
   Eye Exposure:
   Not expected to be a problem.

   Skin Exposure:
   Not expected to be a problem.

   Breathing:
   Not expected to be a problem.

   Swallowing:
   Immediately contact the local poison control center for advice. Keep the victim warm and at rest. Get medical
   attention.

PRODUCT: IVORY, LIQUID - TOILET SOAP

INCOMPATIBILITY: NONE KNOWN

INGREDIENTS:	ANIMAL FAT	TLV:	CAS#:	
	VEGETABLE FATS	TLV:	CAS#:	
	SALT	TLV:	CAS#:	
	GLYCERINE	TLV:	CAS#:	56-81-5
	FRAGRANCE	TLV:	CAS#:	

IGNI TEMP: NA FP: 200F LEL: NA UEL: NA VP: NA VD: NA SG: 1.05 PS: LIQUID APPEAR: WHITE OPAQUE
     ODOR: GRASSY   PH FACTOR:

HAZARD RATINGS: H: F: R:

FIRE FIGHTING:
  HIGH EXPANSION FOAM, LOW EXPANSION FOAM, ALCOHOL FOAM; DRY CHEMICAL, CARBON DIOXIDE, WATER FOG.
  PROTECTIVE CLOTHING, RUBBER GLOVES, AND BREATHING APPARATUS.
WARNING: 1] STRUCTURAL PROTECTIVE CLOTHING IS PERMEABLE, REMAIN CLEAR OF SMOKE, WATER FALL OUT AND WATER RUN OFF.
         2] KEEP OUT OF THE REACH OF CHILDREN.
         3] HAZARDOUS DECOMPOSITION PRODUCT ARE UNKNOWN.
         4] MOVE CONTAINERS FROM AREA IF WITHOUT RISK, COOL EXPOSED CONTAINERS.
         5] DIKE AREA FOR CONTROL AND CONTAINMENT TO PREVENT ENTRY INTO SEWERS, DRAIN, AND WATER WAYS.

LARGE SPILL/NO FIRE/RESCUE: WEAR RUBBER OR NEOPRENE BOOTS, GLOVES.

SPILL CONTROL AND CONTAINMENT:
  HOUSEHOLD SPILL: WIPE UP WITH AN ABSORBENT CLOTH, RINSE OUT IN THE SINK. WIPE UP RESIDUE WITH A DAMP CLOTH,
  RINSE OUT IN THE SINK.
  LARGE SPILL: PICK UP WITH ABSORBENT, PUT IN A DISPOSAL CONTAINER. RINSE AREA WITH WATER, PICK UP WITH ABSORBENT,
  PUT IN A DISPOSAL CONTAINER. SEAL AND REMOVE FROM THE WORK AREA.

HEALTH HAZARD INFORMATION:
  WHEN USING THIS PRODUCT WEAR: NO SPECIAL REQUIREMENT. Eye contact will cause stinging, tearing, itching,
  swelling, and redness. Prolonged contact may cause skin irritation. Ingestion may result in abdominal pain,
  nausea, vomiting, and diarrhea.

  A physician should be contacted if anyone develops any signs or symptoms and suspects that they are caused by
  exposure to this product.

FIRST AID:
Eye Exposure:
Flush with water for 15 minutes while lifting the upper and lower eye lids. Contact lenses should not be worn
when working with this product. Get medical attention.

Skin Exposure:
Wash with water. If irritation develops, get medical attention.

Breathing:
Not expected to be a problem.

Swallowing:
If the victim is conscious, give 1-2 glasses of water to drink. Immediately contact the local poison control
center for advice. Keep the victim warm and at rest. Get medical attention.

PRODUCT: IVORY BAR SOAP - TOILET SOAP

INCOMPATIBILITY: NONE KNOWN

INGREDIENTS: ANIMAL FAT   TLV: CAS#:
             VEGETABLE FAT   TLV: CAS#:
             FRAGRANCE   TLV: CAS#:

IGNI TEMP: NA FP: NA LEL: NA UEL: NA VP: NA VD: NA SG: .9 PS: SOLID APPEAR: WHITE BAR
   ODOR: GRASSY COLOGNE   PH FACTOR: NA

HAZARD RATINGS: H: F: R:

FIRE FIGHTING:
   HIGH EXPANSION FOAM, LOW EXPANSION FOAM, ALCOHOL FOAM; DRY CHEMICAL, CARBON DIOXIDE, WATER FOG.
   PROTECTIVE CLOTHING, RUBBER GLOVES, AND BREATHING APPARATUS.
WARNING: 1] STRUCTURAL PROTECTIVE CLOTHING IS PERMEABLE, REMAIN CLEAR OF SMOKE, WATER FALL OUT AND WATER RUN OFF.
         2] KEEP OUT OF THE REACH OF CHILDREN.
         3] PRODUCT WILL CAUSE A SLIPPING HAZARD.
         4] MOVE CONTAINERS FROM AREA IF WITHOUT RISK, COOL EXPOSED CONTAINERS.
         5] DIKE AREA FOR CONTROL AND CONTAINMENT TO PREVENT ENTRY INTO SEWERS, DRAIN, AND WATER WAYS.

LARGE SPILL/NO FIRE/RESCUE: WEAR RUBBER OR NEOPRENE BOOTS, GLOVES.

SPILL CONTROL AND CONTAINMENT:
   HOUSEHOLD SPILL: PICK UP THE BAR, WIPE UP ANY RESIDUE WITH A CLOTH.
   LARGE SPILL: SCOOP UP BAR AND PUT IN A CONTAINER. RINSE AREA, COVER WITH ABSORBENT. PICK UP AND PUT IN A
   DISPOSAL CONTAINER. SEAL AND REMOVE FROM THE WORK AREA.

HEALTH HAZARD INFORMATION:
   WHEN USING THIS PRODUCT WEAR: NO SPECIAL REQUIREMENT. If soap gets into the eyes, it may cause irritation.

A physician should be contacted if anyone develops any signs or symptoms and suspects that they are caused by
exposure to this product.

FIRST AID:
   Eye Exposure:
   Flush with water for 15 minutes while lifting the upper and lower eye lids. Contact lenses should not be worn
   when working with this product. If irritation persists, get medical attention.

   Skin Exposure:
   Not expected to be a problem.

   Breathing:
   Not expected to be a problem.

   Swallowing:
   If the victim is conscious, give 2-3 glasses of water to drink. Immediately contact the local poison control
   center for advice. Keep the victim warm and at rest. Get medical attention.

PRODUCT: IVORY LIQUID   (DISH WASHING LIQUID)

INCOMPATIBILITY: NONE KNOWN

INGREDIENTS:	ANIONIC SURFACTANTS	TLV:	CAS#:
	NONIONIC SURFACTANTS	TLV:	CAS#:
	AMPHOTERIC SURFACTANTS	TLV:	CAS#:
	ETHYL ALCOHOL	TLV: 1900 mg/m3	CAS#: 64-17-5
	WATER	TLV:	CAS#: 7732-18-5

IGNI TEMP: NA FP: 108F LEL: NA UEL: NA VP: NA VD: NA SG: NA PS: LIQUID APPEAR: WHITE OPAQUE ODOR: PERFUME
   PH FACTOR: NA

HAZARD RATINGS: H: F: R:

FIRE FIGHTING:
   HIGH EXPANSION FOAM, LOW EXPANSION FOAM, ALCOHOL FOAM; DRY CHEMICAL, CARBON DIOXIDE, WATER FOG.
   PROTECTIVE CLOTHING, RUBBER GLOVES, AND BREATHING APPARATUS.
WARNING: 1] STRUCTURAL PROTECTIVE CLOTHING IS PERMEABLE, REMAIN CLEAR OF SMOKE, WATER FALL OUT AND WATER RUN OFF.
         2] KEEP OUT OF THE REACH OF CHILDREN.
         3] PRODUCT DOES NOT SUSTAIN COMBUSTION.
         4] MOVE CONTAINERS FROM AREA IF WITHOUT RISK, COOL EXPOSED CONTAINERS.
         5] DIKE AREA FOR CONTROL AND CONTAINMENT TO PREVENT ENTRY INTO SEWERS, DRAIN, AND WATER WAYS.

LARGE SPILL/NO FIRE/RESCUE: WEAR RUBBER OR NEOPRENE BOOTS, GLOVES.

SPILL CONTROL AND CONTAINMENT:
   HOUSEHOLD SPILL: WIPE UP LIQUID WITH AN ABSORBENT CLOTH. RINSE OUT IN THE SINK. WIPE UP RESIDUE WITH A DAMP
   CLOTH.
   LARGE SPILL: COVER SPILL WITH ABSORBENT. PICK UP AND PUT IN A DISPOSAL CONTAINER. RINSE AREA WITH WATER, PICK UP
   WITH ABSORBENT, PUT IN A DISPOSAL CONTAINER. SEAL AND REMOVE FROM THE WORK AREA.

HEALTH HAZARD INFORMATION:
   WHEN USING THIS PRODUCT WEAR: NO SPECIAL REQUIREMENT. Contact with the eyes may cause stinging, tearing,
   itching, swelling, and redness. Ingestion may result in nausea, vomiting, and diarrhea. Prolonged skin contact
   may cause drying of the skiskin.

   A physician should be contacted if anyone develops any signs or symptoms and suspects that they are caused by
   exposure to this product.

FIRST AID:
Eye Exposure:
Flush with water for 15 minutes while lifting the upper and lower eye lids. Contact lenses should not be worn
when working with this product. If irritation persists, get medical attention.

   Skin Exposure:
Wash the skin with water.

   Breathing:
Not expected to be a problem.

   Swallowing:
If the victim is conscious, give 2-3 glasses of water to drink. Immediately contact the local poison control
center for advice. Keep the victim warm and at rest. Get medical attention.

PRODUCT: IVORY SNOW   (SOAP GRANULES)

INCOMPATIBILITY: NONE KNOWN

INGREDIENTS: PURE IVORY SOAP   TLV: CAS#:
             FABRIC WHITENERS   TLV: CAS#:
             PERFUME   TLV: CAS#:

IGNI TEMP: NA FP: NA LEL: NA UEL: NA VP: NA VD: NA SG: NA PS: POWDER APPEAR: WHITE GRANULES
   ODOR: PERFUMED   PH FACTOR: NA

HAZARD RATINGS: H: F: R:

FIRE FIGHTING:
   HIGH EXPANSION FOAM, LOW EXPANSION FOAM, ALCOHOL FOAM; DRY CHEMICAL, CARBON DIOXIDE, WATER FOG.
   PROTECTIVE CLOTHING, RUBBER GLOVES, AND BREATHING APPARATUS.
WARNING: 1] STRUCTURAL PROTECTIVE CLOTHING IS PERMEABLE, REMAIN CLEAR OF SMOKE, WATER FALL OUT AND WATER RUN OFF.
         2] KEEP OUT OF THE REACH OF CHILDREN.
         3] MOVE CONTAINERS FROM AREA IF WITHOUT RISK, COOL EXPOSED CONTAINERS.
         4] DIKE AREA FOR CONTROL AND CONTAINMENT TO PREVENT ENTRY INTO SEWERS, DRAIN, AND WATER WAYS.

LARGE SPILL/NO FIRE/RESCUE: WEAR RUBBER OR NEOPRENE BOOTS, GLOVES.

SPILL CONTROL AND CONTAINMENT:
   HOUSEHOLD SPILL: SWEEP UP AND PUT IN A DISPOSAL CONTAINER. RINSE AREA, WIPE UP WITH A CLOTH.
   LARGE SPILL: AVOID CREATING A DUST. GENTLY SWEEP UP THE POWDER. PUT INTO SALVAGE OR DISPOSAL CONTAINER. SEAL AND
   REMOVE FROM THE AREA. RINSE SPILL AREA WITH WATER, PICK UP WITH ABSORBENT. PUT IN A DISPOSAL CONTAINER. SEAL AND
   REMOVE FROM THE WORK AREA.

HEALTH HAZARD INFORMATION:
   WHEN USING THIS PRODUCT WEAR: NO SPECIAL REQUIREMENT. Prolonged inhalation of dust may cause respiratory
   irritation, coughing, sore throat, wheezing, and shortness of breath. Eye contact may cause stinging, tearing,
   itching, swelling, and redness. Ingestion may cause abdominal pain, nausea, vomiting, and diarrhea.

   A physician should ld be contacted if anyone develops any signs or symptoms and suspects that they are caused by
   exposure to this product.

FIRST AID:
   Eye Exposure:
   Flush with water for 15 minutes while lifting the upper and lower eye lids. Contact lenses should not be worn
   when working with this product. If irritation persists, get medical attention.

   Skin Exposure:
   Flush powder from the skin with water.

   Breathing:
   If problems develop, move the victim to fresh air.

   Swallowing:
   If the victim is conscious, give 2-3 glasses of water to drink. Immediately contact the local poison control
   center for advice. Keep the victim warm and at rest. Get medical attention.

PRODUCT: IVORY SNOW, LIQUID (LAUNDRY DETERGENT)

INCOMPATIBILITY: ACIDS AND ACID FUMES.

INGREDIENTS: ANIONIC SURFACTANTS   TLV: CAS#:
             NONIONIC SURFACTANTS   TLV: CAS#:
             ETHYL ALCOHOL   TLV: 1900 mg/m3   CAS#:   64-17-5

IGNI TEMP: NA FP: 200F LEL: NA UEL: NA VP: NA VD: NA SG: 1.06 PS: LIQUID APPEAR: DARK AMBER
      ODOR: PERFUME   PH FACTOR: NA

HAZARD RATINGS: H: F: R:

FIRE FIGHTING:
   HIGH EXPANSION FOAM, LOW EXPANSION FOAM, ALCOHOL FOAM; DRY CHEMICAL, CARBON DIOXIDE, WATER FOG.
   PROTECTIVE CLOTHING, RUBBER GLOVES, AND BREATHING APPARATUS.
WARNING: 1] STRUCTURAL PROTECTIVE CLOTHING IS PERMEABLE, REMAIN CLEAR OF SMOKE, WATER FALL OUT AND WATER RUN OFF.
         2] KEEP OUT OF THE REACH OF CHILDREN.
         3] PRODUCT IA A AQUEOUS SOLUTION CONTAINING ETHYL ALCOHOL AND DOES NOT SUSTAIN COMBUSTION.
         4] THERMAL DECOMPOSITION MAY YIELD CARBON DIOXIDE, CARBON MONOXIDE.
         5] MOVE CONTAINERS FROM AREA IF WITHOUT RISK, COOL EXPOSED CONTAINERS.
         6] DIKE AREA FOR CONTROL AND CONTAINMENT TO PREVENT ENTRY INTO SEWERS, DRAIN, AND WATER WAYS.

LARGE SPILL/NO FIRE/RESCUE: WEAR RUBBER OR NEOPRENE BOOTS, GLOVES.

SPILL CONTROL AND CONTAINMENT:
   HOUSEHOLD SPILL: WIPE UP WITH AN ABSORBENT CLOTH, RINSE OUT IN THE SINK. WIPE UP RESIDUE WITH A DAMP CLOTH,
   RINSE OUT IN THE SINK.
   LARGE SPILL: PICK UP WITH ABSORBENT, PUT IN A DISPOSAL CONTAINER. RINSE AREA WITH WATER, PICK UP WITH ABSORBENT.
   PUT IN A DISPOSAL CONTAINER. SEAL AND REMOVE FROM THE WORK AREA.

HEALTH HAZARD INFORMATION:
   WHEN USING THIS PRODUCT WEAR: NO SPECIAL REQUIREMENT. Eye contact may cause stinging, tearing, itching,
   swelling, and redness. Prolonged contact with liquid concentrate may cause drying of the skin and irritation.
   Ingestion may result in abdominal pain, nausea, vomiting, and diarrhea.

   A physician should be contacted if anyone develops any signs or symptoms and suspects that they are caused by
   exposure to this product.

FIRST AID:
Eye Exposure:
Flush with water for 15 minutes while lifting the upper and lower eye lids. Contact lenses should not be worn
when working with this product. Get medical attention.

Skin Exposure:
Wash with water. If irritation develops, get medical attention.

Breathing:
Not expected to be a problem.

Swallowing:
If the victim is conscious, give 1-2 glasses of water to drink. Immediately contact the local poison control
center for advice. Keep the victim warm and at rest. Get medical attention.

PRODUCT: J/WAX AUTO CARPET CLEANER

INCOMPATIBILITY: STRONG ACIDS

INGREDIENTS: SURFACTANTS_____   2-6%   TLV: CAS#:
             CLEANERS_____/   TLV: CAS#:
             ETHYLENE GLYCOL MONO
             BUTYL ETHER   1-3%   TLV: CAS#:   111-76-2
             DIETHYLENE GLYCOL
             MONO BUTYL ETHER   2-4%   TLV: CAS#:   112-34-5
             ISOBUTANE   5-10%   TLV: CAS#:   75-28-5
             WATER   80-9-% TLV: CAS#: 7732-18-5

IGNI TEMP: NA FP: 20F LEL: NA UEL: NA VP: NA VD: NA SG: 1.0 PS: LIQUID APPEAR: FOAM   ODOR: MILD
   PH FACTOR: 12.3

HAZARD RATINGS: H: F: R:

FIRE FIGHTING:
   HIGH EXPANSION FOAM, LOW EXPANSION FOAM, ALCOHOL FOAM; DRY CHEMICAL, CARBON DIOXIDE, WATER FOG.
   PROTECTIVE CLOTHING, RUBBER GLOVES, AND BREATHING APPARATUS.
WARNING: 1] STRUCTURAL PROTECTIVE CLOTHING IS PERMEABLE, REMAIN CLEAR OF SMOKE, WATER FALL OUT AND WATER RUN OFF.
         2] KEEP OUT OF THE REACH OF CHILDREN.
         3] CONTAINERS WILL EXPLODE IN A FIRE OR IF HEATED ABOVE 120F.
         4] THERMAL DECOMPOSITION YIELDS TOXIC CARBON DIOXIDE, CARBON MONOXIDE.
         5] MOVE CONTAINERS FROM AREA IF WITHOUT RISK, COOL EXPOSED CONTAINERS.
         6] DIKE AREA FOR CONTROL AND CONTAINMENT TO PREVENT ENTRY INTO SEWERS, DRAIN, AND WATER WAYS.

LARGE SPILL/NO FIRE/RESCUE: WEAR RUBBER OR NEOPRENE BOOTS, GLOVES.

SPILL CONTROL AND CONTAINMENT:
   HOUSEHOLD SPILL: WIPE UP FOAM WITH ABSORBENT CLOTH. RINSE OUT IN THE SINK. RINSE SPILL AREA WITH WATER, PICK UP
   WITH ABSORBENT CLOTH.
   LARGE SPILL: COVER SPILL AREA WITH ABSORBENT. PICK UP AND PUT IN A DISPOSAL CONTAINER. RINSE AREA WITH WATER.
   PICK UP WITH ABSORBENT. PUT IN A DISPOSAL CONTAINER. SEAL AND REMOVE FROM THE WORK AREA.

HEALTH HAZARD INFORMATION:
   WHEN USING THIS PRODUCT WEAR: NO SPECIAL REQUIREMENT. Direct contact with the eyes can cause stinging, tearing,
   itching, swelling, and redness. Contact with the skin may cause itching and redness.

   A physician should be contacted if anyone develops any signs or symptoms and suspects that they are caused by
   exposure to this product.

FIRST AID:
   Eye Exposure:
   Flush with water for 15 minutes while lifting the upper and lower eye lids. Contact lenses should not be worn
   when working with this product. If irritation persists, get medical attention.

   Skin Exposure:
   Wash skin with soap and water. If irritation appears, get medical attention.

   Breathing:
   Not expected to be a problem.

   Swallowing:
   Immediately contact the local poison control center for advice. Keep the victim warm and at rest. Get medical
   attention.

PRODUCT: J/WAX AUTO FRESHENER

INCOMPATIBILITY: NONE KNOWN

INGREDIENTS: FRAGRANCE   35-45%   TLV: CAS#:
             ALUMINA BEADS   55-65%   TLV: CAS#:

IGNI TEMP: NA FP: 142F LEL: NA UEL: NA VP: NA VD: NA SG: NA PS: BEADS APPEAR: OFF-WHITE   ODOR: FRAGRANT
   PH FACTOR: NA

HAZARD RATINGS: H: F: R:

FIRE FIGHTING:
   HIGH EXPANSION FOAM, LOW EXPANSION FOAM, ALCOHOL FOAM; DRY CHEMICAL, CARBON DIOXIDE, WATER FOG.
   PROTECTIVE CLOTHING, RUBBER GLOVES, AND BREATHING APPARATUS.
WARNING: 1] STRUCTURAL PROTECTIVE CLOTHING IS PERMEABLE, REMAIN CLEAR OF SMOKE, WATER FALL OUT AND WATER RUN OFF.
         2] KEEP OUT OF THE REACH OF CHILDREN.
         3] THERMAL DECOMPOSITION MAY YIELD CARBON DIOXIDE, CARBON MONOXIDE.
         4] MOVE CONTAINERS FROM AREA IF WITHOUT RISK, COOL EXPOSED CONTAINERS.
         5] DIKE AREA FOR CONTROL AND CONTAINMENT TO PREVENT ENTRY INTO SEWERS, DRAIN, AND WATER WAYS.

LARGE SPILL/NO FIRE/RESCUE: WEAR RUBBER OR NEOPRENE BOOTS, GLOVES.

SPILL CONTROL AND CONTAINMENT:
   HOUSEHOLD SPILL: SWEEP UP BEADS AND PUT IN A DISPOSAL CONTAINER.
   LARGE SPILL: SWEEP UP BEADS AND PUT IN A DISPOSAL CONTAINER. SEAL AND REMOVE FROM THE WORK AREA.

HEALTH HAZARD INFORMATION:
   WHEN USING THIS PRODUCT WEAR: NO SPECIAL REQUIREMENT. Direct contact with the eyes may cause stinging, tearing,
   itching, swelling, and redness. Prolonged or repeated skin contact may cause irritation. Ingestion may cause
   gastrointestinal irritation with nausea and vomiting.

   A physician should be contacted if anyone develops any signs or symptoms and suspects that they are caused by
   exposure to this product.

FIRST AID:
Eye Exposure:
Flush with water for 15 minutes while lifting the upper and lower eye lids. Contact lenses should not be worn
when working with this product. If irritation persists, get medical attention.

   Skin Exposure:
   Wash the skin with soap and water.

   Breathing:
   Not expected to be a problem.

   Swallowing:
   Immediately contact the local poison control center for advice. Keep the victim warm and at rest. Get medical
   attention.

PRODUCT: J/WAX BASIC BLACK

INCOMPATIBILITY: NONE KNOWN

INGREDIENTS: SILICONE   10-20%   TLV: CAS#:
             AMMONIA   under .1%   TLV: 25 ppm   CAS#: 7664-41-7
             WATER   80-90%   TLV: CAS#: 7732-18-5

IGNI TEMP: NA FP: NA LEL: NA UEL: NA VP: NA VD: NA SG: 1.0 PS: LIQUID APPEAR: WHITE OPAQUE
   ODOR: MILD   PH FACTOR: 9.6

HAZARD RATINGS: H: F: R:

FIRE FIGHTING:
   HIGH EXPANSION FOAM, LOW EXPANSION FOAM, ALCOHOL FOAM; DRY CHEMICAL, CARBON DIOXIDE, WATER FOG.
   PROTECTIVE CLOTHING, RUBBER GLOVES, AND BREATHING APPARATUS.
WARNING: 1] STRUCTURAL PROTECTIVE CLOTHING IS PERMEABLE, REMAIN CLEAR OF SMOKE, WATER FALL OUT AND WATER RUN OFF.
         2] KEEP OUT OF THE REACH OF CHILDREN.
         3] CONTAINERS MAY MELT AND LEAK IN A FIRE.
         4] THERMAL DECOMPOSITION MAY YIELD CARBON DIOXIDE, CARBON MONOXIDE.
         5] MOVE CONTAINERS FROM AREA IF WITHOUT RISK, COOL EXPOSED CONTAINERS.
         6] DIKE AREA FOR CONTROL AND CONTAINMENT TO PREVENT ENTRY INTO SEWERS, DRAIN, AND WATER WAYS.

LARGE SPILL/NO FIRE/RESCUE: WEAR RUBBER OR NEOPRENE BOOTS, GLOVES.

SPILL CONTROL AND CONTAINMENT:
   HOUSEHOLD SPILL: WIPE UP LIQUID WITH AN ABSORBENT CLOTH, RINSE OUT IN THE SINK. RINSE SPILL AREA WITH WATER,
   WIPE UP WITH AN ABSORBENT CLOTH, RINSE OUT IN THE SINK.
   LARGE SPILL: COVER SPILL WITH ABSORBENT. SWEEP UP AND PUT IN A DISPOSAL CONTAINER. RINSE AREA WITH WATER, PICK
   UP WITH ABSORBENT, PUT IN A DISPOSAL CONTAINER. SEAL AND REMOVE FROM THE WORK AREA.

HEALTH HAZARD INFORMATION:
   WHEN USING THIS PRODUCT WEAR: NO SPECIAL REQUIREMENT. Direct contact with the eyes or skin may cause irritation.
   Ingestion may cause gastrointestinal irritation.

   A physician should be contacted if anyone develops any signs or symptoms and suspects that they are caused by
   exposure to this product.

FIRST AID:
   Eye Exposure:
   Flush with water for 15 minutes while lifting the upper and lower eye lids. Contact lenses should not be worn
   when working with this product. If irritation persists, get medical attention.

   Skin Exposure:
   Wash the skin with soap and water.

   Breathing:
   Not expected to be a problem.

   Swallowing:
   Immediately contact the local poison control center for advice. Keep the victim warm and at rest. Get medical
   attention.

PRODUCT: J/WAX CARNU

INCOMPATIBILITY: NONE KNOWN

```
INGREDIENTS: ALUMINUM SILICATE      20-30%    TLV:             CAS#:
             SILICONE               5-10%     TLV:             CAS#:
             ALIPHATIC NAPHTHA      25-32%    TLV: 100 ppm     CAS#:
             AMMONIA               under .1%  TLV: 25 ppm      CAS#: 7664-41-7
             WATER                  30-45%    TLV:             CAS#: 7732-18-5
```

IGNI TEMP: NA FP: NA LEL: NA UEL: NA VP: NA VD: NA SG: 1.0 PS: LIQUID APPEAR: TAN OPAQUE ODOR: MILD
  PH FACTOR: 9.8

HAZARD RATINGS: H: F: R:

FIRE FIGHTING:
  HIGH EXPANSION FOAM, LOW EXPANSION FOAM, ALCOHOL FOAM; DRY CHEMICAL, CARBON DIOXIDE, WATER FOG.
  PROTECTIVE CLOTHING, RUBBER GLOVES, AND BREATHING APPARATUS.
WARNING: 1] STRUCTURAL PROTECTIVE CLOTHING IS PERMEABLE, REMAIN CLEAR OF SMOKE, WATER FALL OUT AND WATER RUN OFF.
         2] KEEP OUT OF THE REACH OF CHILDREN.
         3] CONTAINERS MAY MELT\AND LEAK IN A FIRE.
         4] THERMAL DECOMPOSITION MAY YIELD CARBON DIOXIDE, CARBON MONOXIDE.
         5] MOVE CONTAINERS FROM AREA IF WITHOUT RISK, COOL EXPOSED CONTAINERS.
         6] DIKE AREA FOR CONTROL AND CONTAINMENT TO PREVENT ENTRY INTO SEWERS, DRAIN, AND WATER WAYS.

LARGE SPILL/NO FIRE/RESCUE: WEAR RUBBER OR NEOPRENE BOOTS, GLOVES.

SPILL CONTROL AND CONTAINMENT:
  HOUSEHOLD SPILL: WIPE UP WITH AN ABSORBENT CLOTH. RINSE OUT IN THE SINK. WIPE UP RESIDUE WITH A DAMP CLOTH.
  LARGE SPILL: COVER SPILL WITH ABSORBENT. SWEEP UP AND PUT IN A DISPOSAL CONTAINER. RINSE WITH WATER, PICK UP
  WITH ABSORBENT. SEAL AND REMOVE FROM THE WORK AREA.

HEALTH HAZARD INFORMATION:
  WHEN USING THIS PRODUCT WEAR: NO SPECIAL REQUIREMENT. Direct contact with the eyes may cause stinging, tearing,
  itching, swelling, and redness. Prolonged or repeated skin contact may cause irritation. Ingestion may cause
  gastrointestinal irritation with nausea and vomiting.

  A physician should be contacted if anyone develops any signs or symptoms and suspects that they are caused by
  exposure to this product.

FIRST AID:
Eye Exposure:
Flush with water for 15 minutes while lifting the upper and lower eye lids. Contact lenses should not be worn
when working with this product. If irritation persists, get medical attention.

Skin Exposure:
Wash the skin with soap and water.

Breathing:
Not expected to be a problem.

Swallowing:
Immediately contact the local poison control center for advice. Keep the victim warm and at rest. Get medical
attention.

PRODUCT: J/WAX CAR WASH

INCOMPATIBILITY: NONE KNOWN

INGREDIENTS: SURFACTANTS   10-20% TLV: CAS#:
            FORMALDEHYDE   under .15% TLV: 1 ppm   CAS#:   50-00-0
            WATER   80-90% TLV: CAS#: 7732-18-5

IGNI TEMP: NA FP: NA LEL: NA UEL: NA VP: NA VD: NA SG: 1.0 PS: LIQUID APPEAR: CLEAR AQUA ODOR: MILD
    PH FACTOR: 6.5

HAZARD RATINGS: H: F: R:

FIRE FIGHTING:
   HIGH EXPANSION FOAM, LOW EXPANSION FOAM, ALCOHOL FOAM; DRY CHEMICAL, CARBON DIOXIDE, WATER FOG.
   PROTECTIVE CLOTHING, RUBBER GLOVES, AND BREATHING APPARATUS.
WARNING: 1] STRUCTURAL PROTECTIVE CLOTHING IS PERMEABLE, REMAIN CLEAR OF SMOKE, WATER FALL OUT AND WATER RUN OFF.
         2] KEEP OUT OF THE REACH OF CHILDREN.
         3] CONTAINERS MAY MELT AND LEAK IN A FIRE.
         4] THERMAL DECOMPOSITION MAY YIELD CARBON DIOXIDE, CARBON MONOXIDE.
         5] MOVE CONTAINERS FROM AREA IF WITHOUT RISK, COOL EXPOSED CONTAINERS.
         6] DIKE AREA FOR CONTROL AND CONTAINMENT TO PREVENT ENTRY INTO SEWERS, DRAIN, AND WATER WAYS.

LARGE SPILL/NO FIRE/RESCUE: WEAR RUBBER OR NEOPRENE BOOTS, GLOVES.

SPILL CONTROL AND CONTAINMENT:
   HOUSEHOLD SPILL: WIPE UP LIQUID WITH AN ABSORBENT CLOTH, RINSE OUT IN THE SINK. WIPE UP RESIDUE WITH A DAMP
   CLOTH, RINSE OUT IN THE SINK.
   LARGE SPILL: COVER SPILL WITH ABSORBENT. SWEEP UP AND PUT IN A DISPOSAL CONTAINER. RINSE AREA WITH WATER, PICK
   UP WITH ABSORBENT. PUT IN A DISPOSAL CONTAINER. SEAL AND REMOVE FROM THE WORK AREA.

HEALTH HAZARD INFORMATION:
   WHEN USING THIS PRODUCT WEAR: NO SPECIAL REQUIREMENT. Direct contact with the eyes can cause stinging, tearing,
   itching, swelling, and redness. Prolonged or repeated skin contact may cause irritation. Ingestion may cause
   gastrointestinal irritation with nausea and vomiting.

   A physician should be contacted if anyone develops any signs or symptoms and suspects that they are caused by
   exposure to this product.

FIRST AID:
   Eye Exposure:
   Flush with water for 15 minutes while lifting the upper and lower eye lids. Contact lenses should not be worn
   when working with this product. If irritation persists, get medical attention.

   Skin Exposure:
   Wash the skin with soap and water.

   Breathing:
   Not expected to be a problem.

   Swallowing:
   Immediately contact the local poison control center for advice. Keep the victim warm and at rest. Get medical
   attention.

```
PRODUCT: J/WAX  CHROME CLEANER

INCOMPATIBILITY: NONE KNOWN

INGREDIENTS: ABRASIVE              20-30%  TLV:              CAS#:
             ALIPHATIC NAPHTHA     20-30%  TLV: 100 ppm     CAS#:
             AMMONIA            under .2%  TLV:  25 ppm     CAS#: 7664-41-7
             WATER                 45-60%  TLV:              CAS#: 7732-18-5

IGNI TEMP: NA  FP: NA  LEL: NA  UEL: NA  VP: NA  VD: NA  SG: 1.0  PS: LIQUID  APPEAR: WHITE OPAQUE
     ODOR: AMMONIACAL       PH FACTOR: 9.7

HAZARD RATINGS:   H:  F:  R:
```

FIRE FIGHTING:
HIGH EXPANSION FOAM, LOW EXPANSION FOAM, ALCOHOL FOAM; DRY CHEMICAL, CARBON DIOXIDE, WATER FOG.
PROTECTIVE CLOTHING, RUBBER GLOVES, AND BREATHING APPARATUS.
WARNING: 1] STRUCTURAL PROTECTIVE CLOTHING IS PERMEABLE, REMAIN CLEAR OF SMOKE, WATER FALL OUT AND WATER RUN OFF.
         2] KEEP OUT OF THE REACH OF CHILDREN.
         3] CONTAINERS MAY MELT AND LEAK IN A FIRE.
         4] THERMAL DECOMPOSITION MAY YIELD CARBON DIOXIDE, CARBON MONOXIDE.
         5] MOVE CONTAINERS FROM AREA IF WITHOUT RISK, COOL EXPOSED CONTAINERS.
         6] DIKE AREA FOR CONTROL AND CONTAINMENT TO PREVENT ENTRY INTO SEWERS, DRAIN, AND WATER WAYS.

LARGE SPILL/NO FIRE/RESCUE: WEAR RUBBER OR NEOPRENE BOOTS, GLOVES.

SPILL CONTROL AND CONTAINMENT:
HOUSEHOLD SPILL: WIPE UP MATERIAL WITH AN ABSORBENT CLOTH. RINSE OUT IN THE SINK. WIPE UP RESIDUE WITH A DAMP
CLOTH, RINSE OUT IN THE SINK.
LARGE SPILL: COVER SPILL WITH ABSORBENT. SWEEP UP AND PUT IN A DISPOSAL CONTAINER. RINSE AREA WITH WATER, PICK
UP WITH ABSORBENT. PUT IN A DISPOSAL CONTAINER. SEAL AND REMOVE FROM THE WORK AREA.

HEALTH HAZARD INFORMATION:
WHEN USING THIS PRODUCT WEAR: NO SPECIAL REQUIREMENT. Direct contact with the eyes can cause stinging, tearing,
itching, swelling, and redness. Direct contact with the skin can cause irritation.

A physician should be contacted if anyone develops any signs or symptoms and suspects that they are caused by
exposure to this product.

FIRST AID:
Eye Exposure:
Flush with water for 15 minutes while lifting the upper and lower eye lids. Contact lenses should not be worn
when working with this product. If irritation persists, get medical attention.

Skin Exposure:
Wash the skin with soap and water.

Breathing:
Not expected to be a problem.

Swallowing:
Immediately contact the local poison control center for advice. Keep the victim warm and at rest. Get medical
attention.

PRODUCT: J/WAX FINE FABRIC CLEANER

INCOMPATIBILITY: STRONG ACIDS.

INGREDIENTS: SURFACTANTS_____ 2.6% TLV: CAS#:
            CLEANERS_____/   TLV: CAS#:
            ETHYLENE GLYCOL
            MONO BUTYL ETHER   1-3% TLV: CAS#: 111-76-2
            DIETHYLENE GLYCOL
            MONO BUTYL ETHER   2-4% TLV: 25 ppm CAS#: 112-34-5
            ISOBUTANE   5-10% TLV: CAS#: 75-28-5
            WATER   80-90% TLV: CAS#: 7732-18-5

IGNI TEMP: NA FP: 20F LEL: NA UEL: NA VP: NA VD: NA SG: 1.0 PS: LIQUID APPEAR: FORM   ODOR: MILD
    PH FACTOR: 12.3

HAZARD RATINGS: H: F: R:

FIRE FIGHTING:
    HIGH EXPANSION FOAM, LOW EXPANSION FOAM, ALCOHOL FOAM; DRY CHEMICAL, CARBON DIOXIDE, WATER FOG.
    PROTECTIVE CLOTHING, RUBBER GLOVES, AND BREATHING APPARATUS.
WARNING: 1] STRUCTURAL PROTECTIVE CLOTHING IS PERMEABLE, REMAIN CLEAR OF SMOKE, WATER FALL OUT AND WATER RUN OFF.
         2] KEEP OUT OF THE REACH OF CHILDREN.
         3] CONTAINERS WILL EXPLODE IN A FIRE OR IF HEATED ABOVE 120F.
         4] THERMAL DECOMPOSITION MAY YIELD CARBON DIOXIDE, CARBON MONOXIDE.
         5] MOVE CONTAINERS FROM AREA IF WITHOUT RISK, COOL EXPOSED CONTAINERS.
         6] DIKE AREA FOR CONTROL AND CONTAINMENT TO PREVENT ENTRY INTO SEWERS, DRAIN, AND WATER WAYS.

LARGE SPILL/NO FIRE/RESCUE: WEAR RUBBER OR NEOPRENE BOOTS, GLOVES.

SPILL CONTROL AND CONTAINMENT:
    HOUSEHOLD SPILL: WIPE UP FOAM WITH AN ABSORBENT CLOTH, RINSE OUT IN THE SINK. WIPE UP RESIDUE WITH A DAMP CLOTH,
    RINSE OUT IN THE SINK.
    LARGE SPILL: COVER SPILL WITH ABSORBENT. PICK UP AND PUT IN A DISPOSAL CONTAINER. RINSE AREA WITH WATER. PICK UP
    WITH ABSORBENT, PUT IN A DISPOSAL CONTAINER. SEAL AND REMOVE FROM THE WORK AREA.

HEALTH HAZARD INFORMATION:
    WHEN USING THIS PRODUCT WEAR: NO SPECIAL REQUIREMENT. Direct contact with eyes can cause stinging, tearing,
    itching, swelling, and redness. Direct contact with the skin can cause irritation. Ingestion may cause
    gastrointestinal irritation with nausea and vomiting.

    A physician should be contacted if anyone develops any signs or symptoms and suspects that they are caused by
    exposure to this product.

FIRST AID:
Eye Exposure:
Flush with water for 15 minutes while lifting the upper and lower eye lids. Contact lenses should not be worn
when working with this product. If irritation persists, get medical attention.

    Skin Exposure:
Wash the skin with soap and water.

    Breathing:
Not expected to be a problem.

    Swallowing:
Immediately contact the local poison control center for advice. Keep the victim warm and at rest. Get medical
attention.

PRODUCT: J/WAX KIT PASTE WAX

INCOMPATIBILITY: NONE KNOWN

INGREDIENTS: ABRASIVE   6-12%   TLV: CAS#:
             FORMALDEHYDE   under 0.15%   TLV: 1 ppm   CAS#:   50-00-0
             AMMONIA   under 0.2%   TLV: 25 ppm   CAS#: 7664-41-7
             STODDARD SOLVENT   25-40%   TLV: 100 ppm   CAS#: 8052-41-3
             WATER   40-50%   TLV: CAS#: 7732-18-5
             SILICONE, WAXES, SURFACTANTS   10-15%

IGNI TEMP: NA FP: 186F LEL: NA UEL: NA VP: NA VD: NA SG: 1.0 PS: PASTE APPEAR: TAN, OPAQUE ODOR: NAPHTHA
   PH FACTOR: 8.8

HAZARD RATINGS: H: F: R:

FIRE FIGHTING:
   HIGH EXPANSION FOAM, LOW EXPANSION FOAM, ALCOHOL FOAM; DRY CHEMICAL, CARBON DIOXIDE, WATER FOG.
   PROTECTIVE CLOTHING, RUBBER GLOVES, AND BREATHING APPARATUS.
WARNING: 1] STRUCTURAL PROTECTIVE CLOTHING IS PERMEABLE, REMAIN CLEAR OF SMOKE, WATER FALL OUT AND WATER RUN OFF.
         2] KEEP OUT OF THE REACH OF CHILDREN.
         3] CONTAINERS MAY MELT AND LEAK IN A FIRE.
         4] THERMAL DECOMPOSITION MAY YIELD CARBON DIOXIDE, CARBON MONOXIDE.
         5] MOVE CONTAINERS FROM AREA IF WITHOUT RISK, COOL EXPOSED CONTAINERS.
         6] DIKE AREA FOR CONTROL AND CONTAINMENT TO PREVENT ENTRY INTO SEWERS, DRAIN, AND WATER WAYS.

LARGE SPILL/NO FIRE/RESCUE: WEAR RUBBER OR NEOPRENE BOOTS, GLOVES.

SPILL CONTROL AND CONTAINMENT:
   HOUSEHOLD SPILL: SWEEP OR SCRAPE UP PASTE AND PUT IN THE CONTAINER. WASH SPILL AREA THOROUGHLY WITH SOAP AND
   WATER, PICK UP WITH AN ABSORBENT CLOTH.
   LARGE SPILL: SWEEP OR SCRAPE UP MATERIAL, PUT IN A DISPOSAL CONTAINER. SCRUB AREA WITH DETERGENT AND WATER, PICK
   UP WITH ABSORBENT. PUT IN A DISPOSAL CONTAINER. SEAL AND REMOVE FROM THE WORK AREA.

HEALTH HAZARD INFORMATION:
   WHEN USING THIS PRODUCT WEAR: NO SPECIAL REQUIREMENT. Direct contact with eyes can cause severe irritation.
   Product has abrasive material in it. Direct skin contact can cause irritation. It is unlikely that the TLV for
   formaldehyde will be exceeded while using this product.

   A physician should be contacted if anyone develops any signs or symptoms and suspects that they are caused by
   exposure to this product.

FIRST AID:
   Eye Exposure:
   Flush with water for 15 minutes while lifting the upper and lower eye lids. Contact lenses should not be worn
   when working with this product. Get medical attention.

   Skin Exposure:
   Wash the skin with soap and water.

   Breathing:
   Not expected to be a problem.

   Swallowing:
   Immediately contact the local poison control center for advice. Keep the victim warm and at rest. Get medical
   attention.

PRODUCT: J/WAX LIQUID KIT WAX

INCOMPATIBILITY: NONE KNOWN

INGREDIENTS: ABRASIVE   8-14% TLV: CAS#:
             ALIPHATIC NAPHTHA 20-30% TLV: 100 ppm CAS#:
             WATER   50-60% TLV: CAS#: 7732-18-5
             SILICONE, WAXES, SURFACTANTS 10-15%

IGNI TEMP: NA FP: 145F LEL: NA UEL: NA VP: NA VD: NA SG: 1.0 PS: LIQUID APPEAR: TAN, OPAQUE, CREAMY
    ODOR: NAPHTHA   PH FACTOR: 9.1

HAZARD RATINGS: H: F: R:

FIRE FIGHTING:
    HIGH EXPANSION FOAM, LOW EXPANSION FOAM, ALCOHOL FOAM; DRY CHEMICAL, CARBON DIOXIDE, WATER FOG.
    PROTECTIVE CLOTHING, RUBBER GLOVES, AND BREATHING APPARATUS.
WARNING: 1] STRUCTURAL PROTECTIVE CLOTHING IS PERMEABLE, REMAIN CLEAR OF SMOKE, WATER FALL OUT AND WATER RUN OFF.
         2] KEEP OUT OF THE REACH OF CHILDREN.
         3] CONTAINERS MAY MELT AND LEAK IN A FIRE.
         4] LIQUID RUN-OFF INTO SEWERS MAY CREATE A FIRE OR EXPLOSION HAZARD.
         5] THERMAL DECOMPOSITION MAY YIELD CARBON DIOXIDE, CARBON MONOXIDE.
         6] MOVE CONTAINERS FROM AREA IF WITHOUT RISK, COOL EXPOSED CONTAINERS.
         7] DIKE AREA FOR CONTROL AND CONTAINMENT TO PREVENT ENTRY INTO SEWERS, DRAIN, AND WATER WAYS.

LARGE SPILL/NO FIRE/RESCUE: WEAR RUBBER OR NEOPRENE BOOTS, GLOVES.

SPILL CONTROL AND CONTAINMENT:
    HOUSEHOLD SPILL: WIPE UP LIQUID WITH AN ABSORBENT CLOTH, RINSE OUT IN THE SINK. WIPE UP RESIDUE WITH A DAMP
    CLOTH, RINSE OUT IN THE SINK.
    LARGE SPILL: COVER SPILL WITH ABSORBENT. PICK UP AND PUT IN A DISPOSAL CONTAINER. RINSE AREA WITH WATER, PICK UP
    WITH ABSORBENT, PUT IN A DISPOSAL CONTAINER. SEAL AND REMOVE FROM THE WORK AREA.

HEALTH HAZARD INFORMATION:
    WHEN USING THIS PRODUCT WEAR: NO SPECIAL REQUIREMENT. Direct eye contact can cause stinging, tearing, itching,
    swelling, and redness. Skin contact may cause irritation.

    A physician should be contacted if anyone develops any signs or symptoms and suspects that they are caused by
    exposure to this product.

FIRST AID:
    Eye Exposure:
    Flush with water for 15 minutes while lifting the upper and lower eye lids. Contact lenses should not be worn
    when working with this product. If irritation persists, get medical attention.

    Skin Exposure:
    Wash the skin with soap and water.

    Breathing:
    Not expected to be a problem.

    Swallowing:
    Immediately contact the local poison control center for advice. Keep the victim warm and at rest. Get medical
    attention.

PRODUCT: J/WAX SPRINT WAX

INCOMPATIBILITY: NONE KNOWN

INGREDIENTS: FORMALDEHYDE   under 0.15% TLV: 1 ppm   CAS#:   50-00-0
             AMMONIA   under 0.1%   TLV: 25 ppm   CAS#: 7664-41-7
             ISOPARAFFINIC NAPHTHA 40-50% TLV: 300 ppm   CAS#:
             WATER   40-50% TLV: CAS#: 7732-18-5
             SILICONE, WAXES, SURFACTANT   5-10%

IGNI TEMP: NA FP: 165F LEL: NA UEL: NA VP: NA VD: NA SG: .9 PS: LIQUID APPEAR: THICK, LIGHT YELLOW
     ODOR: NAPHTHA   PH FACTOR: 6.5

HAZARD RATINGS: H: F: R:

FIRE FIGHTING:
   HIGH EXPANSION FOAM, LOW EXPANSION FOAM, ALCOHOL FOAM; DRY CHEMICAL, CARBON DIOXIDE, WATER FOG.
   PROTECTIVE CLOTHING, RUBBER GLOVES, AND BREATHING APPARATUS.
WARNING: 1] STRUCTURAL PROTECTIVE CLOTHING IS PERMEABLE, REMAIN CLEAR OF SMOKE, WATER FALL OUT AND WATER RUN OFF.
         2] KEEP OUT OF THE REACH OF CHILDREN.
         3] CONTAINERS MAY MELT AND LEAK LIQUID IN A FIRE.
         4] LIQUID RUN-OFF INTO SEWERS MAY CREATE A FIRE OR EXPLOSION HAZARD.
         5] THERMAL DECOMPOSITION MAY YIELD CARBON DIOXIDE, CARBON MONOXIDE.
         6] MOVE CONTAINERS FROM AREA IF WITHOUT RISK, COOL EXPOSED CONTAINERS.
         7] DIKE AREA FOR CONTROL AND CONTAINMENT TO PREVENT ENTRY INTO SEWERS, DRAIN, AND WATER WAYS.

LARGE SPILL/NO FIRE/RESCUE: WEAR RUBBER OR NEOPRENE BOOTS, GLOVES.

SPILL CONTROL AND CONTAINMENT:
   HOUSEHOLD SPILL: WIPE UP LIQUID WITH AN ABSORBENT CLOTH, RINSE OUT IN THE SINK. WIPE UP RESIDUE WITH A DAMP
   CLOTH, RINSE OUT IN THE SINK.
   LARGE SPILL: COVER SPILL WITH ABSORBENT. PICK UP AND PUT IN A DISPOSAL CONTAINER. RINSE AREA WITH WATER, PICK UP
   WITH ABSORBENT. PUT IN A DISPOSAL CONTAINER. SEAL AND REMOVE FROM THE WORK AREA.

HEALTH HAZARD INFORMATION:
   WHEN USING THIS PRODUCT WEAR: NO SPECIAL REQUIREMENT. Direct eye contact can cause irritation. Skin contact may
   cause irritation. It is unlikely the TLV for formaldehyde will be exceeded while using this product.

   A physician should be contacted if anyone develops any signs or symptoms and suspects that they are caused by
   exposure to this product.

FIRST AID:
   Eye Exposure:
   Flush with water for 15 minutes while lifting the upper and lower eye lids. Contact lenses should not be worn
   when working with this product. If irritation persists, get medical attention.

   Skin Exposure:
   Wash the skin with soap and water.

   Breathing:
   Not expected to be a problem.

   Swallowing:
   Immediately contact the local poison control center for advice. Keep the victim warm and at rest. Get medical
   attention.

PRODUCT: J/WAX VINYL TOP AND INTERIOR CLEANER

INCOMPATIBILITY: STRONG ACIDS.

INGREDIENTS: ETHYLENE GLYCOL
MONO BUTYL ETHER	2-3%	TLV: 25 ppm	CAS#: 111-76-2
WATER	85-95%	TLV:	CAS#: 7732-18-5
ISOBUTANE	5-15%	TLV:	CAS#: 75-28-5
SURFACTANTS, CLEANERS	3-6%		

IGNI TEMP: NA FP: 20F LEL: NA UEL: NA VP: NA VD: NA SG: 1.0 PS: LIQUID APPEAR: FOAM   ODOR: MILD
  PH FACTOR: 12.3

HAZARD RATINGS: H: F: R:

FIRE FIGHTING:
  HIGH EXPANSION FOAM, LOW EXPANSION FOAM, ALCOHOL FOAM; DRY CHEMICAL, CARBON DIOXIDE, WATER FOG.
  PROTECTIVE CLOTHING, RUBBER GLOVES, AND BREATHING APPARATUS.
WARNING: 1] STRUCTURAL PROTECTIVE CLOTHING IS PERMEABLE, REMAIN CLEAR OF SMOKE, WATER FALL OUT AND WATER RUN OFF.
         2] KEEP OUT OF THE REACH OF CHILDREN.
         3] REMOVE ALL SOURCES OF IGNITION IF WITHOUT RISK.
         4] CONTAINERS WILL EXPLODE IN A FIRE OR IF HEATED ABOVE 120F.
         5] THERMAL DECOMPOSITION MAY YIELD CARBON DIOXIDE, CARBON MONOXIDE.
         6] MOVE CONTAINERS FROM AREA IF WITHOUT RISK, COOL EXPOSED CONTAINERS.
         7] DIKE AREA FOR CONTROL AND CONTAINMENT TO PREVENT ENTRY INTO SEWERS, DRAIN, AND WATER WAYS.

LARGE SPILL/NO FIRE/RESCUE: WEAR RUBBER OR NEOPRENE BOOTS, GLOVES.

SPILL CONTROL AND CONTAINMENT:
  HOUSEHOLD SPILL: WIPE UP FOAM WITH AN ABSORBENT CLOTH. RINSE OUT IN THE SINK. RINSE SPOT WITH WATER, WIPE UP
  WITH AN ABSORBENT CLOTH, RINSE OUT IN THE SINK.
  LARGE SPILL: COVER SPILL WITH ABSORBENT. SWEEP UP AND PUT IN A DISPOSAL CONTAINER. RINSE AREA WITH WATER, PICK
  UP WITH ABSORBENT. PUT IN A DISPOSAL CONTAINER. SEAL AND REMOVE FROM THE WORK AREA.

HEALTH HAZARD INFORMATION:
  WHEN USING THIS PRODUCT WEAR: NO SPECIAL REQUIREMENT. Direct contact with eyes can cause irritation. Contact
  with the skin can cause irritation.

  A physician should be contacted if anyone develops any signs or symptoms and suspects that they are caused by
  exposure to this product.

FIRST AID:
  Eye Exposure:
  Flush with water for 15 minutes while lifting the upper and lower eye lids. Contact lenses should not be worn
  when working with this product. If irritation persists, get medical attention.

  Skin Exposure:
  Wash the skin with soap and water.

  Breathing:
  Not expected to be a problem.

  Swallowing:
  Immediately contact the local poison control center for advice. Keep the victim warm and at rest. Get medical
  attention.

PRODUCT: J/WAX WHITE SIDE WALL CLEANER

INCOMPATIBILITY: STRONG ACIDS (e.g. MURIATIC ACID)

INGREDIENTS: AMMONIA   under 0.5% TLV: 25 ppm   CAS#: 7664-41-7
             PROPANE_____ 5-15% TLV: 1000 ppm   CAS#:   75-98-6
             N-BUTANE_____/   TLV: 800 ppm   CAS#: 106-97-8
             WATER   75-85% TLV: CAS#: 7732-18-5
             SURFACTANT   2-5%   TLV: CAS#:
             ALKALI METASILICATE 2-5%   TLV: CAS#:
             GLYCOL ETHER   1-3%   TLV: CAS#:

IGNI TEMP: NA FP: 20F LEL: NA UEL: NA VP: NA VD: NA SG: .9 PS: LIQUID APPEAR: FOAM   ODOR: AMMONIACAL
    PH FACTOR: 12.2

HAZARD RATINGS: H: F: R:

FIRE FIGHTING:
   HIGH EXPANSION FOAM, LOW EXPANSION FOAM, ALCOHOL FOAM; DRY CHEMICAL, CARBON DIOXIDE, WATER FOG.
   PROTECTIVE CLOTHING, RUBBER GLOVES, AND BREATHING APPARATUS.
WARNING: 1] STRUCTURAL PROTECTIVE CLOTHING IS PERMEABLE, REMAIN CLEAR OF SMOKE, WATER FALL OUT AND WATER RUN OFF.
         2] KEEP OUT OF THE REACH OF CHILDREN.
         3] REMOVE ALL IGNITION SOURCES IF WITHOUT RISK.
         4] CONTAINERS WILL EXPLODE IN A FIRE OR IF HEATED ABOVE 120F.
         5] THERMAL DECOMPOSITION MAY YIELD CARBON DIOXIDE, CARBON MONOXIDE.
         6] MOVE CONTAINERS FROM AREA IF WITHOUT RISK, COOL EXPOSED CONTAINERS.
         7] DIKE AREA FOR CONTROL AND CONTAINMENT TO PREVENT ENTRY INTO SEWERS, DRAIN, AND WATER WAYS.

LARGE SPILL/NO FIRE/RESCUE: WEAR RUBBER OR NEOPRENE BOOTS, GLOVES.

SPILL CONTROL AND CONTAINMENT:
   HOUSEHOLD SPILL: OPEN WINDOWS. WIPE UP LIQUID WITH AN ABSORBENT CLOTH.
   LARGE SPILL: VENTILATE ENCLOSED AREAS. PICK UP LIQUID WITH ABSORBENT, PUT IN A DISPOSAL CONTAINER. SEAL AND
   REMOVE FROM THE WORK AREA.

HEALTH HAZARD INFORMATION:
   WHEN USING THIS PRODUCT WEAR: NO SPECIAL REQUIREMENT. Eye contact will cause stinging, tearing, itching,
   swelling, and redness.

   A physician should be contacted if anyone develops any signs or symptoms and suspects that they are caused by
   exposure to this product.

FIRST AID:
Eye Exposure:
Flush with water for 15 minutes while lifting the upper and lower eye lids. Contact lenses should not be worn
when working with this product. Get medical attention.

   Skin Exposure:
   Not expected to be a problem.

   Breathing:
   If any symptoms develop, move victim to fresh air.

   Swallowing:
   Immediately contact the local poison control center for advice. Keep the victim warm and at rest. Get medical
   attention.

PRODUCT: JOY

INCOMPATIBILITY: CHLORINE BLEACH.

INGREDIENTS:	ANIONIC SURFACTANT	TLV:	CAS#:	
	NONIONIC SURFACTANT	TLV:	CAS#:	
	ETHYL ALCOHOL	TLV: 1900 mg/m3	CAS#:	64-17-5
	WATER	TLV:	CAS#:	7732-18-5

IGNI TEMP: NA FP: 132F LEL: NA UEL: NA VP: NA VD: NA SG: 1.02 PS: LIQUID APPEAR: CLEAR YELLOW
   ODOR: PERFUMED   PH FACTOR: NA

HAZARD RATINGS: H: F: R:

FIRE FIGHTING:
   HIGH EXPANSION FOAM, LOW EXPANSION FOAM, ALCOHOL FOAM; DRY CHEMICAL, CARBON DIOXIDE, WATER FOG.
   PROTECTIVE CLOTHING, RUBBER GLOVES, AND BREATHING APPARATUS.
WARNING: 1] STRUCTURAL PROTECTIVE CLOTHING IS PERMEABLE, REMAIN CLEAR OF SMOKE, WATER FALL OUT AND WATER RUN OFF.
         2] KEEP OUT OF THE REACH OF CHILDREN.
         3] DO NOT MIX WITH CHLORINE BLEACH, YIELDS HAZARDOUS FUMES.
         4] MOVE CONTAINERS FROM AREA IF WITHOUT RISK, COOL EXPOSED CONTAINERS.
         5] DIKE AREA FOR CONTROL AND CONTAINMENT TO PREVENT ENTRY INTO SEWERS, DRAIN, AND WATER WAYS.

LARGE SPILL/NO FIRE/RESCUE: WEAR RUBBER OR NEOPRENE BOOTS, GLOVES.

SPILL CONTROL AND CONTAINMENT:
   HOUSEHOLD SPILL: WIPE UP LIQUID WITH AN ABSORBENT CLOTH, RINSE OUT IN THE SINK. WIPE UP RESIDUE WITH A DAMP
   CLOTH. RINSE OUT IN THE SINK.
   LARGE SPILL: PICK UP LIQUID WITH ABSORBENT, PUT IN A DISPOSAL CONTAINER. RINSE AREA WITH WATER, PICK UP WITH
   ABSORBENT. PUT IN A DISPOSAL CONTAINER. SEAL AND REMOVE FROM THE WORK AREA.

HEALTH HAZARD INFORMATION:
   WHEN USING THIS PRODUCT WEAR: NO SPECIAL REQUIREMENT. Eye contact will cause stinging, tearing, itching,
   swelling, and redness. Prolonged skin contact may cause drying of the skin or irritation. Ingestion may result
   in nausea, vomiting, and diarrhea.

   A physician should be contacted if anyone develops any signs or symptoms and suspects that they are caused by
   exposure to this product.

FIRST AID:
   Eye Exposure:
   Flush with water for 15 minutes while lifting the upper and lower eye lids. Contact lenses should not be worn
   when working with this product. Get medical attention.

   Skin Exposure:
   Rinse off concentrate with water.

   Breathing:
   Not expected to be a problem.

   Swallowing:
   If victim is conscious, give 2-3 glasses of water to drink. Immediately contact the local poison control center
   for advice. Keep the victim warm and at rest. Get medical attention.

```
PRODUCT: JUBILEE LIQUID

INCOMPATIBILITY: NONE KNOWN

INGREDIENTS: SURFACTANT_____        3-7%     TLV:              CAS#:
             WAX_____/                     TLV:              CAS#:
             ALIPHATIC NAPHTHA     30-40%     TLV: 100 ppm      CAS#:
             FRAGRANCE      under    .1%      TLV:              CAS#:
             ORTHO BENZYL PARA
             CHLOROPHENOL            .1%      TLV:              CAS#:
             WATER                             TLV:              CAS#: 7732-18-5
```

IGNI TEMP: NA FP: 200F LEL: NA UEL: NA VP: NA VD: NA SG: .9 PS: LIQUID APPEAR: WHITE ODOR: PLEASANT
   PH FACTOR: NA

HAZARD RATINGS: H: F: R:

FIRE FIGHTING:
   HIGH EXPANSION FOAM, LOW EXPANSION FOAM, ALCOHOL FOAM; DRY CHEMICAL, CARBON DIOXIDE, WATER FOG.
   PROTECTIVE CLOTHING, RUBBER GLOVES, AND BREATHING APPARATUS.
WARNING: 1] STRUCTURAL PROTECTIVE CLOTHING IS PERMEABLE, REMAIN CLEAR OF SMOKE, WATER FALL OUT AND WATER RUN OFF.
         2] KEEP OUT OF THE REACH OF CHILDREN.
         3] CONTAINERS MAY BURST IN A FIRE.
         4] THERMAL DECOMPOSITION MAY YIELD CARBON DIOXIDE, CARBON MONOXIDE.
         5] MOVE CONTAINERS FROM AREA IF WITHOUT RISK, COOL EXPOSED CONTAINERS.
         6] DIKE AREA FOR CONTROL AND CONTAINMENT TO PREVENT ENTRY INTO SEWERS, DRAIN, AND WATER WAYS.

LARGE SPILL/NO FIRE/RESCUE: WEAR RUBBER OR NEOPRENE BOOTS, GLOVES.

SPILL CONTROL AND CONTAINMENT:
   HOUSEHOLD SPILL: WIPE UP LIQUID WITH AN ABSORBENT CLOTH, RINSE OUT IN THE SINK. WIPE UP THE RESIDUE WITH A DAMP
   CLOTH, RINSE OUT IN THE SINK.
   LARGE SPILL: PICK UP LIQUID WITH ABSORBENT, PUT IN A DISPOSAL CONTAINER. RINSE AREA WITH WATER, PICK UP WITH
   ABSORBENT. PUT IN A DISPOSAL CONTAINER SEAL AND REMOVE FROM THE WORK AREA.

HEALTH HAZARD INFORMATION:
   WHEN USING THIS PRODUCT WEAR: NO SPECIAL REQUIREMENT. Eye contact may cause stinging, tearing, itching,
   swelling, and redness. Ingestion may cause abdominal pain, nausea, and vomiting.

   A physician should be contacted if anyone develops any signs or symptoms and suspects that they are caused by
   exposure to this product.

FIRST AID:
   Eye Exposure:
   Flush with water for 15 minutes while lifting the upper and lower eye lids. Contact lenses should not be worn
   when working with this product. Get medical attention.

   Skin Exposure:
   Wash skin with soap and water.

   Breathing:
   Not expected to be a problem.

   Swallowing:
   Immediately contact the local poison control center for advice. Keep the victim warm and at rest. Get medical
   attention.

PRODUCT: JUBILEE SPRAY

INCOMPATIBILITY: NONE KNOWN

INGREDIENTS: ISOPARAFFINIC   Recommended

HYDROCARBON SOLVENT	3-6%	TLV: 400 ppm	CAS#:	64742-48-9
WATER	65-75%	TLV:	CAS#:	7732-18-5
PROPANE_____		TLV: 1000 ppm	CAS#:	74-98-6
ISOBUTANE_____	10-20%	TLV:	CAS#:	75-28-5
N-BUTANE____/		TLV: 800 ppm	CAS#:	106-97-8

IGNI TEMP: NA FP: 20F LEL: NA UEL: NA VP: NA VD: NA SG: .9 PS: LIQUID APPEAR: SPRAY MIST ODOR: PLEASANT
   PH FACTOR: NA

HAZARD RATINGS: H: F: R: 0 4 0

FIRE FIGHTING:
   HIGH EXPANSION FOAM, LOW EXPANSION FOAM, ALCOHOL FOAM; DRY CHEMICAL, CARBON DIOXIDE, WATER FOG.
   PROTECTIVE CLOTHING, RUBBER GLOVES, AND BREATHING APPARATUS.
WARNING: 1] STRUCTURAL PROTECTIVE CLOTHING IS PERMEABLE, REMAIN CLEAR OF SMOKE, WATER FALL OUT AND WATER RUN OFF.
         2] KEEP OUT OF THE REACH OF CHILDREN.
         3] REMOVE ALL SOURCES OF IGNITION IF WITHOUT RISK.
         4] CONTAINERS WILL EXPLODE IN A FIRE OR IF HEATED ABOVE 120F.
         5] THERMAL DECOMPOSITION YIELDS TOXIC CARBON DIOXIDE, CARBON MONOXIDE.
         6] MOVE CONTAINERS FROM AREA IF WITHOUT RISK, COOL EXPOSED CONTAINERS.
         7] DIKE AREA FOR CONTROL AND CONTAINMENT TO PREVENT ENTRY INTO SEWERS, DRAIN, AND WATER WAYS.

LARGE SPILL/NO FIRE/RESCUE: WEAR RUBBER OR NEOPRENE BOOTS, GLOVES.

SPILL CONTROL AND CONTAINMENT:
   HOUSEHOLD SPILL: OPEN WINDOWS. WIPE UP LIQUID WITH AN ABSORBENT CLOTH, RINSE OUT IN THE SINK. WIPE UP RESIDUE
   WITH A DAMP CLOTH, RINSE OUT IN THE SINK.
   LARGE SPILL: VENTILATE ENCLOSED SPACES. PICK UP LIQUID WITH ABSORBENT, PUT IN A DISPOSAL CONTAINER. RINSE AREA
   WITH WATER, PICK UP WITH ABSORBENT. PUT IN A DISPOSAL CONTAINER. SEAL AND REMOVE FROM THE WORK AREA.

HEALTH HAZARD INFORMATION:
   WHEN USING THIS PRODUCT WEAR: NO SPECIAL REQUIREMENT. Eye contact will cause stinging, tearing, itching,
   swelling, and redness. Ingestion may cause gastrointestinal irritation.

A physician should be contacted if anyone develops any signs or symptoms and suspects that they are caused by
exposure to this product.

FIRST AID:
Eye Exposure:
Flush with water for 15 minutes while lifting the upper and lower eye lids. Contact lenses should not be worn
when working with this product. Get medical attention.

Skin Exposure:
Rinse off liquid with water. wash with soap and water.

Breathing:
If anything develops, move the victim to fresh air.

Swallowing:
Immediately contact the local poison control center for advice. Keep the victim warm and at rest. Get medical
attention.

```
PRODUCT: KINGSFORD BRIQUETS

INCOMPATIBILITY:

INGREDIENTS: CHAR DUST                TLV: 2 mg/m3(dust) CAS#: 16291-96-6
             SAW DUST        .5-5%    TLV: 1 mg/m3(dust) CAS#:

IGNI TEMP: NA  FP: NA  LEL: NA  UEL: NA  VP: NA  VD: NA  SG: NA  PS: SOLID  APPEAR: BLACK  ODOR: NONE
   PH FACTOR: NA

HAZARD RATINGS:   H:  F:  R:   2  1  0
```

FIRE FIGHTING:
   HIGH EXPANSION FOAM, LOW EXPANSION FOAM, ALCOHOL FOAM; DRY CHEMICAL, CARBON DIOXIDE, WATER FOG.
   PROTECTIVE CLOTHING, RUBBER GLOVES, AND BREATHING APPARATUS.
WARNING: 1] STRUCTURAL PROTECTIVE CLOTHING IS PERMEABLE, REMAIN CLEAR OF SMOKE, WATER FALL OUT AND WATER RUN OFF.
         2] KEEP OUT OF THE REACH OF CHILDREN.
         3] DO NOT ALLOW STORED CHARCOAL TO BECOME WET. IF WET MOVE AWAY FROM ALL COMBUSTIBLE MATERIAL PREFERABLY
            TO A SAFE OUTSIDE AREA AND ALLOW TO TOTALLY DRY.
         4] DO NOT USE CHARCOAL FOR COOKING OR HEATING IN AN ENCLOSED AREA WITHOUT GOOD OUTSIDE VENTING OF FUMES.
         5] MOVE CONTAINERS FROM AREA IF WITHOUT RISK, COOL EXPOSED CONTAINERS.
         6] DIKE AREA FOR CONTROL AND CONTAINMENT TO PREVENT ENTRY INTO SEWERS, DRAIN, AND WATER WAYS.

LARGE SPILL/NO FIRE/RESCUE: WEAR RUBBER OR NEOPRENE BOOTS, GLOVES, DUST MASK OR BREATHING APPARATUS.

SPILL CONTROL AND CONTAINMENT:
   HOUSEHOLD SPILL: PICK UP BRIQUETS AND PLACE BACK IN THE BAG. GENTLY SWEEP UP RESIDUE AND PUT IN BBQ. DAMP MOP OR
   WIPE ANY REMAINING RESIDUE.
   LARGE SPILL: AVOID CREATING A DUST. SCOOP UP BRIQUETS AND PUT IN A CONTAINER. GENTLY SWEEP OR VACUUM UP RESIDUE.
   RINSE AREA WITH WATER, PICK UP WITH ABSORBENT AND PUT IN A DISPOSAL CONTAINER.

HEALTH HAZARD INFORMATION:
   WHEN USING THIS PRODUCT WEAR: NO SPECIAL REQUIREMENT. Dust may cause irritation of the eye, nose, and throat.
   Possible sensitizer.

   A physician should be contacted if anyone develops any signs or symptoms and suspects that they are caused by
   exposure to this product.

FIRST AID:
   Eye Exposure:
   Flush with water for 15 minutes while lifting the upper and lower eye lids. Contact lenses should not be worn
   when working with this product. If irritation persists, get medical attention.

   Skin Exposure:
   Rinse skin with water, wash with soap and water.

   Breathing:
   If any problem develops, move the victim to fresh air.

   Swallowing:
   Not expected to be a problem.

PRODUCT: KINGSFORD WITH MESQUITE, CHARCOAL BRIQUETS

INCOMPATIBILITY:

INGREDIENTS: CHAR DUST   TLV: 2 mg/m3(dust) CAS#: 16291-96-6
             MESQUITE SAW DUST   5-25%   TLV: 1 mg/m3(dust) CAS#:

IGNI TEMP: NA FP: NA LEL: NA UEL: NA VP: NA VD: NA SG: NA PS: SOLID APPEAR: BLACK ODOR: NONE
   PH FACTOR: NA

HAZARD RATINGS: H: F: R: 2 1 0

FIRE FIGHTING:
  HIGH EXPANSION FOAM, LOW EXPANSION FOAM, ALCOHOL FOAM; DRY CHEMICAL, CARBON DIOXIDE, WATER FOG.
  PROTECTIVE CLOTHING, RUBBER GLOVES, AND BREATHING APPARATUS.
WARNING: 1] STRUCTURAL PROTECTIVE CLOTHING IS PERMEABLE, REMAIN CLEAR OF SMOKE, WATER FALL OUT AND WATER RUN OFF.
         2] KEEP OUT OF THE REACH OF CHILDREN.
         3] DO NOT ALLOW STORED CHARCOAL TO BECOME WET. IF WET MOVE AWAY FROM ALL COMBUSTIBLE MATERIAL PREFERABLY
            TO A SAFE OUTSIDE AREA AND ALLOW TO TOTALLY DRY.
         4] DO NOT USE CHARCOAL FOR COOKING OR HEATING IN AN ENCLOSED AREA WITHOUT GOOD OUTSIDE VENTING OF FUMES.
         5] MOVE CONTAINERS FROM AREA IF WITHOUT RISK, COOL EXPOSED CONTAINERS.
         6] DIKE AREA FOR CONTROL AND CONTAINMENT TO PREVENT ENTRY INTO SEWERS, DRAIN, AND WATER WAYS.

LARGE SPILL/NO FIRE/RESCUE: WEAR RUBBER OR NEOPRENE BOOTS, GLOVES, DUST MASK OR BREATHING APPARATUS.

SPILL CONTROL AND CONTAINMENT:
  HOUSEHOLD SPILL: PICK UP BRIQUETS AND PLACE BACK IN THE BAG. GENTLY SWEEP UP RESIDUE AND PUT IN BBQ. DAMP MOP OR
  WIPE ANY REMAINING RESIDUE.
  LARGE SPILL: AVOID CREATING A DUST. SCOOP UP BRIQUETS AND PUT IN A CONTAINER. GENTLY SWEEP OR VACUUM UP RESIDUE.
  RINSE AREA WITH WATER, PICK UP WITH ABSORBENT, AND PUT IN A DISPOSAL CONTAINER.

HEALTH HAZARD INFORMATION:
  WHEN USING THIS PRODUCT WEAR: NO SPECIAL REQUIREMENT. Dust may cause irritation of the eye, nose, and throat.
  Possible sensitizer.

  A physician should be contacted if anyone develops any signs or symptoms and suspects that they are caused by
  exposure to this product.

FIRST AID:
  Eye Exposure:
  Flush with water for 15 minutes while lifting the upper and lower eye lids. Contact lens should not be worn when
  working with this product. If irritation persists, get medical attention.

  Skin Exposure:
  Rinse skin with water, wash with soap and water.

  Breathing:
  If any problem develops, move the victim to fresh air.

  Swallowing:
  Not expected to be a problem.

PRODUCT: KINGSFORD   BBQ BAG

INCOMPATIBILITY:

INGREDIENTS: PETROLEUM
              HYDROCARBON BLEND   5-25%   TLV: 50 ppm   CAS#: 64742-47-8
              CHAR DUST   TLV: 2 mg/m3(dust) CAS#: 16291-96-6
              SAW DUST   .5-5%   TLV: 1 mg/m3(dust) CAS#:

IGNI TEMP: 400F FP: 140F LEL: NA UEL: NA VP: NA VD: NA SG: NA PS: SOLID APPEAR: BLACK ODOR: NONE
     PH FACTOR: NA

HAZARD RATINGS: H: F: R: 2 2 0

FIRE FIGHTING:
    HIGH EXPANSION FOAM, LOW EXPANSION FOAM, ALCOHOL FOAM; DRY CHEMICAL, CARBON DIOXIDE, WATER FOG.
    PROTECTIVE CLOTHING, RUBBER GLOVES, AND BREATHING APPARATUS.
WARNING: 1] STRUCTURAL PROTECTIVE CLOTHING IS PERMEABLE, REMAIN CLEAR OF SMOKE, WATER FALL OUT AND WATER RUN OFF.
         2] KEEP OUT OF THE REACH OF CHILDREN.
         3] DO NOT ALLOW STORED CHARCOAL TO BECOME WET. IF WET MOVE AWAY FROM ALL COMBUSTIBLE MATERIAL PREFERABLY
            TO A SAFE OUTSIDE AREA AND ALLOW TO TOTALLY DRY.
         4] DO NOT USE CHARCOAL FOR COOKING OR HEATING IN AN ENCLOSED AREA WITHOUT GOOD OUTSIDE VENTING OF FUMES.
         5] DO NOT USE STARTER FLUID OR ELECTRIC STARTER.
         6] MOVE CONTAINERS FROM AREA IF WITHOUT RISK, COOL EXPOSED CONTAINERS.
         7] DIKE AREA FOR CONTROL AND CONTAINMENT TO PREVENT ENTRY INTO SEWERS, DRAIN, AND WATER WAYS.

LARGE SPILL/NO FIRE/RESCUE: WEAR RUBBER OR NEOPRENE BOOTS, GLOVES, DUST MASK OR BREATHING APPARATUS.

SPILL CONTROL AND CONTAINMENT:
    HOUSEHOLD SPILL: PICK UP BRIQUETS AND PLACE BACK IN THE BAG. GENTLY SWEEP UP RESIDUE AND PUT IN BBQ. DAMP MOP OR
    WIPE ANY REMAINING RESIDUE.
    LARGE SPILL: AVOID CREATING A DUST. SCOOP UP BRIQUETS AND PUT IN A CONTAINER. GENTLY SWEEP OR VACUUM UP RESIDUE.
    RINSE AREA WITH WATER, PICK UP WITH ABSORBENT AND PUT IN A DISPOSAL CONTAINER.

HEALTH HAZARD INFORMATION:
    WHEN USING THIS PRODUCT WEAR: NO SPECIAL REQUIREMENT. Dust may cause irritation of the eye, nose, and throat.
    Possible sensitizer.

    A physician should be contacted if anyone develops any signs or symptoms and suspects that they are caused by
    exposure to this product.

FIRST AID:
    Eye Exposure:
    Flush with water for 15 minutes while lifting the upper and lower eye lids. Contact lenses should not be worn
    when working with this product. If irritation persists, get medical attention.

    Skin Exposure:
    Rinse skin with water, wash with soap and water.

    Breathing:
    If any problem develops, move the victim to fresh air.

    Swallowing:
    Not expected to be a problem.

PRODUCT: KINGSFORD FIRE RINGS - CHARCOAL

INCOMPATIBILITY:

INGREDIENTS: CHAR DUST   TLV: 2 mg/m3(dust) CAS#: 16291-96-6
             SAW DUST   .5-5%   TLV: 1 mg/m3(dust) CAS#:

IGNI TEMP: NA FP: NA LEL: NA UEL: NA VP: NA VD: NA SG: NA PS: SOLID APPEAR: BLACK ODOR: NONE
   PH FACTOR: NA

HAZARD RATINGS: H: F: R: 2 1 0

FIRE FIGHTING:
   HIGH EXPANSION FOAM, LOW EXPANSION FOAM, ALCOHOL FOAM; DRY CHEMICAL, CARBON DIOXIDE, WATER FOG.
   PROTECTIVE CLOTHING, RUBBER GLOVES, AND BREATHING APPARATUS.
WARNING: 1] STRUCTURAL PROTECTIVE CLOTHING IS PERMEABLE, REMAIN CLEAR OF SMOKE, WATER FALL OUT AND WATER RUN OFF.
         2] KEEP OUT OF THE REACH OF CHILDREN.
         3] DO NOT ALLOW STORED CHARCOAL TO BECOME WET. IF WET MOVE AWAY FROM ALL COMBUSTIBLE MATERIAL PREFERABLY
            TO A SAFE OUTSIDE AREA AND ALLOW TO TOTALLY DRY.
         4] DO NOT USE CHARCOAL FOR COOKING OR HEATING IN AN ENCLOSED AREA WITHOUT GOOD OUTSIDE VENTING OF FUMES.
         5] MOVE CONTAINERS FROM AREA IF WITHOUT RISK, COOL EXPOSED CONTAINERS.
         6] DIKE AREA FOR CONTROL AND CONTAINMENT TO PREVENT ENTRY INTO SEWERS, DRAIN, AND WATER WAYS.

LARGE SPILL/NO FIRE/RESCUE: WEAR RUBBER OR NEOPRENE BOOTS, GLOVES, DUST MASK OR BREATHING APPARATUS.

SPILL CONTROL AND CONTAINMENT:
   HOUSEHOLD SPILL: PICK UP BRIQUETS AND PLACE BACK IN THE BAG. GENTLY SWEEP UP RESIDUE AND PUT IN BBQ. DAMP MOP OR
   WIPE ANY REMAINING RESIDUE.
   LARGE SPILL: AVOID CREATING A DUST. SCOOP UP BRIQUETS AND PUT IN A CONTAINER. GENTLY SWEEP OR VACUUM UP RESIDUE.
   RINSE AREA WITH WATER, PICK UP WITH ABSORBENT AND PUT IN A DISPOSAL CONTAINER.

HEALTH HAZARD INFORMATION:
   WHEN USING THIS PRODUCT WEAR: NO SPECIAL REQUIREMENT. Dust may cause irritation of the eye, nose, and throat.
   Possible sensitizer.

   A physician should be contacted if anyone develops any signs or symptoms and suspects that they are caused by
   exposure to this product.

FIRST AID:
   Eye Exposure:
   Flush with water for 15 minutes while lifting the upper and lower eye lids. Contact lenses should not be worn
   when working with this product. If irritation persists, get medical attention.

   Skin Exposure:
   Rinse skin with water, wash with soap and water.

   Breathing:
   If any problem develops, move the victim to fresh air.

   Swallowing:
   Not expected to be a problem.

PRODUCT: KINGSFORD ODORLESS CHARCOAL LIGHTER

INCOMPATIBILITY: STRONG OXIDIZERS; (e.g. LIQUID CHLORINE; SODIUM OR CALCIUM HYPOCHLORITE.)

INGREDIENTS: PARAFFINIC

HYDROCARBON SOLVENT__		TLV: 100 ppm	CAS#:
NAPHTHENIC_____ 100%		TLV:	CAS#:
HYDROCARBONS_____/		TLV: 50 ppm	CAS#: 64742-47-8

IGNI TEMP: NA FP: 145F LEL: 1% UEL: 6% VP: 2 mmHg @ 100F VD: 6.2 SG: .80 PS: LIQUID APPEAR: CLEAR
    ODOR: NONE   PH FACTOR: NA

HAZARD RATINGS: H: F: R: 2 2 1

FIRE FIGHTING:
  HIGH EXPANSION FOAM, LOW EXPANSION FOAM, ALCOHOL FOAM; DRY CHEMICAL, CARBON DIOXIDE, WATER FOG.
  PROTECTIVE CLOTHING, RUBBER GLOVES, AND BREATHING APPARATUS.
WARNING: 1] STRUCTURAL PROTECTIVE CLOTHING IS PERMEABLE, REMAIN CLEAR OF SMOKE, WATER FALL OUT AND WATER RUN OFF.
         2] KEEP OUT OF THE REACH OF CHILDREN.
         3] DO NOT STORE NEAR OPEN FLAME.
         4] DO NOT USE IN CONJUNCTION WITH AN ELECTRIC CHARCOAL STARTER.
         5] REMOVE ALL SOURCES OF IGNITION IF WITHOUT RISK.
         6] MOVE CONTAINERS FROM AREA IF WITHOUT RISK, COOL EXPOSED CONTAINERS.
         7] DIKE AREA FOR CONTROL AND CONTAINMENT TO PREVENT ENTRY INTO SEWERS, DRAIN, AND WATER WAYS.

LARGE SPILL/NO FIRE/RESCUE: WEAR RUBBER OR NEOPRENE BOOTS, GLOVES.

SPILL CONTROL AND CONTAINMENT:
  HOUSEHOLD SPILL: REMOVE ALL SOURCES OF IGNITION AND VENTILATE THE AREA. WIPE UP LIQUID WITH AN ABSORBENT CLOTH.
  TAKE OUTSIDE TO DRY. WASH SPILL AREA WITH SOAP AND WATER.
  LARGE SPILL: USE NON-SPARKING TOOLS. COVER LIQUID WITH A NON-COMBUSTIBLE ABSORBENT. SWEEP UP AND PUT IN A
  DISPOSAL CONTAINER. SCRUB AREA WITH DETERGENT AND WATER, PICK UP WITH ABSORBENT. PUT IN A DISPOSAL CONTAINER.
  SEAL AND REMOVE FROM THE WORK AREA.

HEALTH HAZARD INFORMATION:
  WHEN USING THIS PRODUCT WEAR: NO SPECIAL REQUIREMENTS. Product will cause irritation of the eyes and skin.
  Inhalation may cause dizziness, headache, unconsciousness, and convulsions. If aspirated into the lungs, may
  cause severe pulmonary injury or death.

  A physician should be contacted if anyone develops any signs or symptoms and suspects that they are caused by
  exposure to this product.

FIRST AID:
  Eye Exposure:
  Flush with water for 15 minutes while lifting the upper and lower eye lids. Contact lenses should not be worn
  when working with this product. If irritation persists, get medical attention.

  Skin Exposure:
  Wash with soap and water. If irritation appears, get medical attention.

  Breathing:
  Move the victim to fresh air. If breathing becomes labored give oxygen. Treat other symptoms as they appear.
  Keep the victim warm and at rest. Get medical attention.

  Swallowing:
  Do not induce vomiting. Immediately contact the local poison control center for advice. Keep victim warm and at
  rest. Get medical attention.

PRODUCT: KIRK'S CoCo HARDWATER CASTILE SOAP

INCOMPATIBILITY: NONE KNOWN

INGREDIENTS: COCONUT OIL   TLV: CAS#:
             GLYCERINE   TLV: CAS#:
             FATTY ESTERS   TLV: CAS#:
             COCONUT ACIDS   TLV: CAS#:
             PERFUME   TLV: CAS#:

IGNI TEMP: NA FP: NA LEL: NA UEL: NA VP: NA VD: NA SG: 1.25 PS: SOLID APPEAR: CREAM WHITE
     ODOR: LIGHT PERFUME   PH FACTOR: NA

HAZARD RATINGS: H: F: R:

FIRE FIGHTING:
   HIGH EXPANSION FOAM, LOW EXPANSION FOAM, ALCOHOL FOAM; DRY CHEMICAL, CARBON DIOXIDE, WATER FOG.
   PROTECTIVE CLOTHING, RUBBER GLOVES, AND BREATHING APPARATUS.
WARNING: 1] STRUCTURAL PROTECTIVE CLOTHING IS PERMEABLE, REMAIN CLEAR OF SMOKE, WATER FALL OUT AND WATER RUN OFF.
         2] KEEP OUT OF THE REACH OF CHILDREN.
         3] THERMAL DECOMPOSITION MAY YIELD CARBON DIOXIDE, CARBON MONOXIDE.
         4] MOVE CONTAINERS FROM AREA IF WITHOUT RISK, COOL EXPOSED CONTAINERS.
         5] DIKE AREA FOR CONTROL AND CONTAINMENT TO PREVENT ENTRY INTO SEWERS, DRAIN, AND WATER WAYS.

LARGE SPILL/NO FIRE/RESCUE: WEAR RUBBER OR NEOPRENE BOOTS, GLOVES.

SPILL CONTROL AND CONTAINMENT:
   HOUSEHOLD SPILL: PICK UP MATERIAL. WIPE UP RESIDUE WITH A DAMP CLOTH.
   LARGE SPILL: SCOOP UP MATERIAL AND PUT IN A DISPOSAL CONTAINER. RINSE AREA WITH WATER, PICK UP WITH ABSORBENT.
   PUT IN A DISPOSAL CONTAINER.

HEALTH HAZARD INFORMATION:
   WHEN USING THIS PRODUCT WEAR: NO SPECIAL REQUIREMENT. Product is an eye irritant causing stinging, tearing,
   itching, swelling, and redness. May cause irritation of the skin. Ingestion will cause abdominal pain, nausea,
   vomiting, and diarrhea.

   A physician should be contacted if anyone develops any signs or symptoms and suspects that they are caused by
   exposure to this product.

FIRST AID:
   Eye Exposure:
   Flush with water for 15 minutes while lifting the upper and lower eye lids. Contact lenses should not be worn
   when working with this product. Get medical attention.

   Skin Exposure:
   Rinse skin with water. If irritation appears, get medical attention.

   Breathing:
   Not expected to be a problem.

   Swallowing:
   If the victim is conscious, give 2-3 glasses of water to drink. Immediately contact the local poison control
   center for advice. Keep the victim warm and at rest. Get medical attention.

PRODUCT: KLEAN 'N SHINE

INCOMPATIBILITY: NONE KNOWN

INGREDIENTS: ISOPARAFFINIC   Recommended
             HYDROCARBON SOLVENT   5-10%   TLV: 300 ppm   CAS#: 65751-65-7
             PROPANE_____   TLV: 1000 ppm   CAS#:   74-98-6
             ISOBUTANE_____   5-15%   TLV: CAS#:   75-28-5
             N-BUTANE____/   TLV: 800 ppm   CAS#: 106-97-8
             WATER   70-80%   TLV: CAS#: 7732-18-5

IGNI TEMP: NA FP: 20F LEL: NA UEL: NA VP: NA VD: NA SG: .9 PS: LIQUID APPEAR: FOAM   ODOR: CITRUS
    PH FACTOR: NA

HAZARD RATINGS: H: F: R: 0 4 0

FIRE FIGHTING:
    HIGH EXPANSION FOAM, LOW EXPANSION FOAM, ALCOHOL FOAM; DRY CHEMICAL, CARBON DIOXIDE, WATER FOG.
    PROTECTIVE CLOTHING, RUBBER GLOVES, AND BREATHING APPARATUS.
WARNING: 1] STRUCTURAL PROTECTIVE CLOTHING IS PERMEABLE, REMAIN CLEAR OF SMOKE, WATER FALL OUT AND WATER RUN OFF.
         2] KEEP OUT OF THE REACH OF CHILDREN.
         3] CONTAINERS WILL EXPLODE IN A FIRE OF IF HEATED ABOVE 120F.
         4] THERMAL DECOMPOSITION YIELDS TOXIC CARBON DIOXIDE, CARBON MONOXIDE.
         5] MOVE CONTAINERS FROM AREA IF WITHOUT RISK, COOL EXPOSED CONTAINERS.
         6] DIKE AREA FOR CONTROL AND CONTAINMENT TO PREVENT ENTRY INTO SEWERS, DRAIN, AND WATER WAYS.

LARGE SPILL/NO FIRE/RESCUE: WEAR RUBBER OR NEOPRENE BOOTS, GLOVES.

SPILL CONTROL AND CONTAINMENT:
    HOUSEHOLD SPILL: WIPE UP FOAM WITH AN ABSORBENT CLOTH. WIPE UP RESIDUE WITH A DAMP CLOTH.
    LARGE SPILL: COVER WITH ABSORBENT. SCOOP UP AND PUT IN A DISPOSAL CONTAINER. RINSE AREA WITH WATER. PICK UP WITH
    ABSORBENT. PUT IN A DISPOSAL CONTAINER. SEAL AND REMOVE FROM THE WORK AREA.

HEALTH HAZARD INFORMATION:
    WHEN USING THIS PRODUCT WEAR: NO SPECIAL REQUIREMENT. Contact may cause eye irritation. Ingestion may cause
    abdominal pain, nausea, and vomiting.

    A physician should be contacted if anyone develops any signs or symptoms and suspects that they are caused by
    exposure to this product.

FIRST AID:
    Eye Exposure:
    Flush with water for 15 minutes while lifting the upper and lower eye lids. Contact lenses should not be worn
    when working with this product. Get medical attention.

    Skin Exposure:
    Not expected to be a problem.

    Breathing:
    Not expected to be a problem.

    Swallowing:
    Immediately contact the local poison control center for advice. Keep the victim warm and at rest. Get medical
    attention.

PRODUCT: KLEAR

INCOMPATIBILITY: NONE KNOWN

```
INGREDIENTS: STYRENE POLYMER_____    5-10%   TLV:                CAS#:
             ACRYLIC POLYMER__/              TLV:                CAS#:
             TRI BUTOXY
             ETHYL PHOSPHATE       1-3%    TLV:                CAS#:    78-51-3
             DIETHYLENEGLYCOL
             MONOETHYL ETHER       3-5%    TLV:                CAS#:    111-90-0
             WAX                   1-3%    TLV: 10 mg/m3(dust) CAS#:    9010-77-9
```

IGNI TEMP: NA FP: NA LEL: NA UEL: NA VP: NA VD: NA SG: 1.02 PS: LIQUID APPEAR: MILK WHITE
    ODOR: MONOMERIC   PH FACTOR: 7.3

HAZARD RATINGS: H: F: R:

FIRE FIGHTING:
  HIGH EXPANSION FOAM, LOW EXPANSION FOAM, ALCOHOL FOAM; DRY CHEMICAL, CARBON DIOXIDE, WATER FOG.
  PROTECTIVE CLOTHING, RUBBER GLOVES, AND BREATHING APPARATUS.
WARNING: 1] STRUCTURAL PROTECTIVE CLOTHING IS PERMEABLE, REMAIN CLEAR OF SMOKE, WATER FALL OUT AND WATER RUN OFF.
         2] KEEP OUT OF THE REACH OF CHILDREN.
         3] THERMAL DECOMPOSITION MAY YIELD CARBON DIOXIDE, CARBON MONOXIDE.
         4] MOVE CONTAINERS FROM AREA IF WITHOUT RISK, COOL EXPOSED CONTAINERS.
         5] DIKE AREA FOR CONTROL AND CONTAINMENT TO PREVENT ENTRY INTO SEWERS, DRAIN, AND WATER WAYS.

LARGE SPILL/NO FIRE/RESCUE: WEAR RUBBER OR NEOPRENE BOOTS, GLOVES.

SPILL CONTROL AND CONTAINMENT:
  HOUSEHOLD SPILL: WIPE UP LIQUID WITH AN ABSORBENT CLOTH. RINSE OUT IN THE SINK. WIPE UP RESIDUE WITH A DAMP
  CLOTH. RINSE OUT IN THE SINK.
  LARGE SPILL: COVER WITH ABSORBENT. PICK UP AND PUT IN A DISPOSAL CONTAINER. RINSE AREA WITH WATER. PICK UP WITH
  ABSORBENT. PUT IN A DISPOSAL CONTAINER. SEAL AND REMOVE FROM THE WORK AREA.

HEALTH HAZARD INFORMATION:
  WHEN USING THIS PRODUCT WEAR: NO SPECIAL REQUIREMENT. Eye contact may cause stinging, tearing, itching,
  swelling, and redness. Prolonged skin contact may cause irritation.

  A physician should be contacted if anyone develops any signs or symptoms and suspects that they are caused by
  exposure to this product.

FIRST AID:
  Eye Exposure:
  Flush with water for 15 minutes while lifting the upper and lower eye lids. Contact lenses should not be worn
  when working with this product. Get medical attention.

  Skin Exposure:
  Wash with soap and water. If irritation appears, get medical attention.

  Breathing:
  Not expected to be a problem.

  Swallowing:
  Immediately contact the local poison control center for advice. Keep the victim warm and at rest. Get medical
  attention.

PRODUCT: LAVA TOILET SOAP

INCOMPATIBILITY: NONE KNOWN

INGREDIENTS: GLYCERINE   TLV: CAS#:   56-81-5
             TITANIUM DIOXIDE   TLV: CAS#: 13463-67-7
             COCONUT OIL   PUMICE
             COCONUT ACID   CHROMIUM HYDROXIDE GREEN
             FRAGRANCE

IGNI TEMP: NA FP: NA LEL: NA UEL: NA VP: NA VD: NA SG: 1.18 PS: SOLID APPEAR: GREEN BAR ODOR: PERFUMED
   PH FACTOR: NA

HAZARD RATINGS: H: F: R:

FIRE FIGHTING:
   HIGH EXPANSION FOAM, LOW EXPANSION FOAM, ALCOHOL FOAM; DRY CHEMICAL, CARBON DIOXIDE, WATER FOG.
   PROTECTIVE CLOTHING, RUBBER GLOVES, AND BREATHING APPARATUS.
WARNING: 1] STRUCTURAL PROTECTIVE CLOTHING IS PERMEABLE, REMAIN CLEAR OF SMOKE, WATER FALL OUT AND WATER RUN OFF.
         2] KEEP OUT OF THE REACH OF CHILDREN.
         3] MOVE CONTAINERS FROM AREA IF WITHOUT RISK, COOL EXPOSED CONTAINERS.
         4] DIKE AREA FOR CONTROL AND CONTAINMENT TO PREVENT ENTRY INTO SEWERS, DRAIN, AND WATER WAYS.

LARGE SPILL/NO FIRE/RESCUE: WEAR RUBBER OR NEOPRENE BOOTS, GLOVES.

SPILL CONTROL AND CONTAINMENT:
   HOUSEHOLD SPILL: PICK UP THE BAR. WIPE UP RESIDUE WITH A DAMP CLOTH. RINSE OUT IN THE SINK.
   LARGE SPILL: SCOOP UP SOLID AND PUT IN A DISPOSAL CONTAINER. RINSE AREA WITH WATER, COVER WITH ABSORBENT. PICK
   UP AND PUT IN A DISPOSAL CONTAINER. SEAL AND REMOVE FROM THE WORK AREA.

HEALTH HAZARD INFORMATION:
   WHEN USING THIS PRODUCT WEAR: NO SPECIAL REQUIREMENT. Eye contact may cause stinging, tearing, itching,
   swelling, and redness. Ingestion may cause nausea and vomiting.

   A physician should be contacted if anyone develops any signs or symptoms and suspects that they are caused by
   exposure to this product.

FIRST AID:
   Eye Exposure:
   Flush with water for 15 minutes while lifting the upper and lower eye lids. Contact lenses should not be worn
   when working with this product. If irritation persists, get medical attention.

   Skin Exposure:
   Not expected to be a problem.

   Breathing:
   Not expected to be a problem.

   Swallowing:
   Immediately contact the local poison control center for advice. Keep the victim warm and at rest. Get medical
   attention.

PRODUCT: LAVA, LIQUID SOAP - TOILET SOAP

INCOMPATIBILITY: NONE KNOWN

INGREDIENTS: ANIMAL FAT   TLV: CAS#:
             VEGETABLE FAT   TLV: CAS#:
             PUMICE   TLV: CAS#:
             GLYCERINE   TLV: CAS#:   56-81-5
             CLAY; SALT, FRAGRANCE

IGNI TEMP: NA FP: 200F LEL: NA UEL: NA VP: NA VD: NA SG: 1.07 PS: LIQUID APPEAR: GREEN & ABRASIVE
    ODOR: PERFUMED   PH FACTOR: NA

HAZARD RATINGS: H: F: R:

FIRE FIGHTING:
   HIGH EXPANSION FOAM, LOW EXPANSION FOAM, ALCOHOL FOAM; DRY CHEMICAL, CARBON DIOXIDE, WATER FOG.
   PROTECTIVE CLOTHING, RUBBER GLOVES, AND BREATHING APPARATUS.
WARNING: 1] STRUCTURAL PROTECTIVE CLOTHING IS PERMEABLE, REMAIN CLEAR OF SMOKE, WATER FALL OUT AND WATER RUN OFF.
         2] KEEP OUT OF THE REACH OF CHILDREN.
         3] PRODUCT HAS AN ABRASIVE COMPOUND (PUMICE).
         4] THERMAL DECOMPOSITION PRODUCTS ARE UNKNOWN.
         5] MOVE CONTAINERS FROM AREA IF WITHOUT RISK, COOL EXPOSED CONTAINERS.
         6] DIKE AREA FOR CONTROL AND CONTAINMENT TO PREVENT ENTRY INTO SEWERS, DRAIN, AND WATER WAYS.

LARGE SPILL/NO FIRE/RESCUE: WEAR RUBBER OR NEOPRENE BOOTS, GLOVES.

SPILL CONTROL AND CONTAINMENT:
   HOUSEHOLD SPILL: WIPE UP WITH AN ABSORBENT CLOTH, RINSE OUT IN THE SINK. WIPE UP RESIDUE WITH A DAMP CLOTH,
   RINSE OUT IN THE SINK.
   LARGE SPILL: PICK UP WITH ABSORBENT, PUT IN A DISPOSAL CONTAINER. RINSE AREA WITH WATER, PICK UP WITH ABSORBENT.
   PUT IN A DISPOSAL CONTAINER, SEAL AND REMOVE FROM THE WORK AREA.

HEALTH HAZARD INFORMATION:
   WHEN USING THIS PRODUCT WEAR: NO SPECIAL REQUIREMENT. Eye contact will cause stinging, tearing, itching,
   swelling, redness and may scratch the cornea. Prolonged contact with concentrated liquid may cause drying of the
   skin and irritation. Ingestion will cause abdominal pain, nausea, vomiting, and diarrhea.

   A physician should be contacted if anyone develops any signs or symptoms and suspects that they are caused by
   exposure to this product.

FIRST AID:
   Eye Exposure:
   Flush with water for 15 minutes while lifting the upper and lower eye lids. Contact lenses should not be worn
   when working with this product. Get medical attention.

   Skin Exposure:
   Wash with water. If irritation develops, get medical attention.

   Breathing:
   Not expected to be a problem.

   Swallowing:
   If the victim is conscious, give 2-3 glasses of milk or water to drink. Immediately contact the local poison
   control center for advice. Keep the victim warm and at rest. Get medical attention.

PRODUCT: L'ENVIE CONDITIONER - CAPTURE

INCOMPATIBILITY: NONE KNOWN

INGREDIENTS: CETYL ALCOHOL   TLV: CAS#: 36653-82-4
             PROPYLENE GLYCOL   TLV: CAS#: 4254-15-3
             CITRIC ACID   TLV: CAS#:   77-92-9
             CETRIMONIUM CHLORIDE   TLV: CAS#:   112-02-7
             DIMETHYL STEARAMINE   STEARALKONIUM CHLORIDE
             HYDROXYETHYL CELLULOSE   METHYLISOTHIAZOLINONE
             METHYLCHLOROISOTHIAZOLINONE   DEIONIZED WATER
             FRAGRANCE   COLORANT

IGNI TEMP: NA FP: NA LEL: NA UEL: NA VP: NA VD: NA SG: 1.0 PS: LIQUID APPEAR: AMBER OPAQUE
     ODOR: FRAGRANT   PH FACTOR: 3.2

HAZARD RATINGS: H: F: R: 1   0   0

FIRE FIGHTING:
   HIGH EXPANSION FOAM, LOW EXPANSION FOAM, ALCOHOL FOAM; DRY CHEMICAL, CARBON DIOXIDE, WATER FOG.
   PROTECTIVE CLOTHING, RUBBER GLOVES, AND BREATHING APPARATUS.
WARNING: 1] STRUCTURAL PROTECTIVE CLOTHING IS PERMEABLE, REMAIN CLEAR OF SMOKE, WATER FALL OUT AND WATER RUN OFF.
         2] KEEP OUT OF THE REACH OF CHILDREN.
         3] CONTAINERS MAY MELT AND LEAK IN A FIRE.
         4] THERMAL DECOMPOSITION MAY YIELD CARBON DIOXIDE, CARBON MONOXIDE.
         5] MOVE CONTAINERS FROM AREA IF WITHOUT RISK, COOL EXPOSED CONTAINERS.
         6] DIKE AREA FOR CONTROL AND CONTAINMENT TO PREVENT ENTRY INTO SEWERS, DRAIN, AND WATER WAYS.

LARGE SPILL/NO FIRE/RESCUE: WEAR RUBBER OR NEOPRENE BOOTS, GLOVES.

SPILL CONTROL AND CONTAINMENT:
   HOUSEHOLD SPILL: WIPE UP LIQUID WITH AN ABSORBENT CLOTH. RINSE OUT IN THE SINK. WIPE UP RESIDUE WITH A DAMP
   CLOTH, RINSE OUT IN THE SINK.
   LARGE SPILL: COVER SPILL WITH ABSORBENT. SWEEP UP AND PUT IN A DISPOSAL CONTAINER. RINSE AREA WITH WATER, PICK
   UP WITH ABSORBENT. PUT IN A DISPOSAL CONTAINER. SEAL AND REMOVE FROM THE WORK AREA.

HEALTH HAZARD INFORMATION:
   WHEN USING THIS PRODUCT WEAR: NO SPECIAL REQUIREMENT. Eye contact may cause stinging, tearing, itching,
   swelling, and redness. Ingestion may cause nausea and vomiting.

   A physician should be contacted if anyone develops any signs or symptoms and suspects that they are caused by
   exposure to this product.

FIRST AID:
Eye Exposure:
Flush with water for 15 minutes while lifting the upper and lower eye lids. Contact lenses should not be worn
when working with this product. If irritation persists, get medical attention.

Skin Exposure:
Not expected to be a problem.

Breathing:
Not expected to be a problem.

Swallowing:
Immediately contact the local poison control center for advice. Keep the victim warm and at rest. Get medical
attention.

PRODUCT: L'ENVIE CONDITIONER - CYPRESS

INCOMPATIBILITY: NONE KNOWN

INGREDIENTS:			
CETYL ALCOHOL	TLV:		CAS#: 36653-82-4
PROPYLENE GLYCOL	TLV:		CAS#: 4254-15-3
CITRIC ACID	TLV:		CAS#: 77-92-9
CETRIMONIUM CHLORIDE	TLV:		CAS#: 112-02-7
DIMETHYL STEARAMINE		STEARALKONIUM CHLORIDE	
HYDROXYETHYL CELLULOSE		METHYLISOTHIAZOLINONE	
METHYLCHLOROISOTHIAZOLINONE		DEIONIZED WATER	
FRAGRANCE		COLORANT	

IGNI TEMP: NA FP: NA LEL: NA UEL: NA VP: NA VD: NA SG: 1.0 PS: LIQUID APPEAR: AMBER OPAQUE
   ODOR: FRAGRANT   PH FACTOR: 3.2

HAZARD RATINGS: H: F: R: 1 0 0

FIRE FIGHTING:
   HIGH EXPANSION FOAM, LOW EXPANSION FOAM, ALCOHOL FOAM; DRY CHEMICAL, CARBON DIOXIDE, WATER FOG.
   PROTECTIVE CLOTHING, RUBBER GLOVES, AND BREATHING APPARATUS.
WARNING: 1] STRUCTURAL PROTECTIVE CLOTHING IS PERMEABLE, REMAIN CLEAR OF SMOKE, WATER FALL OUT AND WATER RUN OFF.
        2] KEEP OUT OF THE REACH OF CHILDREN.
        3] CONTAINERS MAY MELT AND LEAK IN A FIRE.
        4] THERMAL DECOMPOSITION MAY YIELD CARBON DIOXIDE, CARBON MONOXIDE.
        5] MOVE CONTAINERS FROM AREA IF WITHOUT RISK, COOL EXPOSED CONTAINERS.
        6] DIKE AREA FOR CONTROL AND CONTAINMENT TO PREVENT ENTRY INTO SEWERS, DRAIN, AND WATER WAYS.

LARGE SPILL/NO FIRE/RESCUE: WEAR RUBBER OR NEOPRENE BOOTS, GLOVES.

SPILL CONTROL AND CONTAINMENT:
   HOUSEHOLD SPILL: WIPE UP LIQUID WITH AN ABSORBENT CLOTH. RINSE OUT IN THE SINK. WIPE UP RESIDUE WITH A DAMP
   CLOTH, RINSE OUT IN THE SINK.
   LARGE SPILL: COVER SPILL WITH ABSORBENT. SWEEP UP AND PUT IN A DISPOSAL CONTAINER. RINSE AREA WITH WATER, PICK
   UP WITH ABSORBENT. PUT IN A DISPOSAL CONTAINER. SEAL AND REMOVE FROM THE WORK AREA.

HEALTH HAZARD INFORMATION:
   WHEN USING THIS PRODUCT WEAR: NO SPECIAL REQUIREMENT. Eye contact may cause stinging, tearing, itching,
   swelling, and redness. Ingestion may cause nausea and vomiting.

   A physician should be contacted if anyone develops any signs or symptoms and suspects that they are caused by
   exposure to this product.

FIRST AID:
   Eye Exposure:
   Flush with water for 15 minutes while lifting the upper and lower eye lids. Contact lenses should not be worn
   when working with this product. If irritation persists, get medical attention

   Skin Exposure:
   Not expected to be a problem.

   Breathing:
   Not expected to be a problem.

   Swallowing:
   Immediately contact the local poison control center for advice. Keep the victim warm and at rest. Get medical
   attention.

PRODUCT: L'ENVIE CONDITIONER - EMBER MUSK

INCOMPATIBILITY: NONE KNOWN

INGREDIENTS: CETYL ALCOHOL   TLV: CAS#: 36653-82-4
             PROPYLENE GLYCOL   TLV: CAS#: 4254-15-3
             CITRIC ACID   TLV: CAS#:   77-92-9
             CETRIMONIUM CHLORIDE   TLV: CAS#:   112-02-7
             DIMETHYL STEARAMINE   STEARALKONIUM CHLORIDE
             HYDROXYETHYL CELLULOSE   METHYLISOTHIAZOLINONE
             METHYLCHLOROISOTHIAZOLINONE   DEIONIZED WATER
             FRAGRANCE   COLORANT

IGNI TEMP: NA FP: NA LEL: NA UEL: NA VP: NA VD: NA SG: 1.0 PS: LIQUID APPEAR: AMBER OPAQUE
     ODOR: FRAGRANT   PH FACTOR: 3.2

HAZARD RATINGS: H: F: R: 1   0   0

FIRE FIGHTING:
    HIGH EXPANSION FOAM, LOW EXPANSION FOAM, ALCOHOL FOAM; DRY CHEMICAL, CARBON DIOXIDE, WATER FOG.
    PROTECTIVE CLOTHING, RUBBER GLOVES, AND BREATHING APPARATUS.
WARNING: 1] STRUCTURAL PROTECTIVE CLOTHING IS PERMEABLE, REMAIN CLEAR OF SMOKE, WATER FALL OUT AND WATER RUN OFF.
         2] KEEP OUT OF THE REACH OF CHILDREN.
         3] CONTAINERS MAY MELT AND LEAK IN A FIRE.
         4] THERMAL DECOMPOSITION MAY YIELD CARBON DIOXIDE, CARBON MONOXIDE.
         5] MOVE CONTAINERS FROM AREA IF WITHOUT RISK, COOL EXPOSED CONTAINERS.
         6] DIKE AREA FOR CONTROL AND CONTAINMENT TO PREVENT ENTRY INTO SEWERS, DRAIN, AND WATER WAYS.

LARGE SPILL/NO FIRE/RESCUE: WEAR RUBBER OR NEOPRENE BOOTS, GLOVES.

SPILL CONTROL AND CONTAINMENT:
    HOUSEHOLD SPILL: WIPE UP LIQUID WITH AN ABSORBENT CLOTH. RINSE OUT IN THE SINK. WIPE UP RESIDUE WITH A DAMP
    CLOTH, RINSE OUT IN THE SINK.
    LARGE SPILL: COVER SPILL WITH ABSORBENT. SWEEP UP AND PUT IN A DISPOSAL CONTAINER. RINSE AREA WITH WATER, PICK
    UP WITH ABSORBENT. PUT IN A DISPOSAL CONTAINER. SEAL AND REMOVE FROM THE WORK AREA.

HEALTH HAZARD INFORMATION:
    WHEN USING THIS PRODUCT WEAR: NO SPECIAL REQUIREMENT. Eye contact may cause stinging, tearing, itching,
    swelling, and redness. Ingestion may cause nausea and vomiting.

    A physician should be contacted if anyone develops any signs or symptoms and suspects that they are caused by
    exposure to this product.

FIRST AID:
Eye Exposure:
Flush with water for 15 minutes while lifting the upper and lower eye lids. Contact lenses should not be worn
when working with this product. If irritation persists, get medical attention

Skin Exposure:
Not expected to be a problem.

Breathing:
Not expected to be a problem.

Swallowing:
Immediately contact the local poison control center for advice. Keep the victim warm and at rest. Get medical
attention.

PRODUCT: L'ENVIE CONDITIONER - LAGACE'

INCOMPATIBILITY: NONE KNOWN

INGREDIENTS: CETYL ALCOHOL   TLV: CAS#: 36653-82-4
             PROPYLENE GLYCOL   TLV: CAS#: 4254-15-3
             CITRIC ACID   TLV: CAS#:   77-92-9
             CETRIMONIUM CHLORIDE   TLV: CAS#:   112-02-7
             DIMETHYL STEARAMINE   STEARALKONIUM CHLORIDE
             HYDROXYETHYL CELLULOSE   METHYLISOTHIAZOLINONE
             METHYLCHLOROISOTHIAZOLINONE DEIONIZED WATER
             FRAGRANCE   COLORANT

IGNI TEMP: NA FP: NA LEL: NA UEL: NA VP: NA VD: NA SG: 1.0 PS: LIQUID APPEAR: AMBER OPAQUE
     ODOR: FRAGRANT   PH FACTOR: 3.2

HAZARD RATINGS: H: F: R: 1 0 0

FIRE FIGHTING:
  HIGH EXPANSION FOAM, LOW EXPANSION FOAM, ALCOHOL FOAM; DRY CHEMICAL, CARBON DIOXIDE, WATER FOG.
  PROTECTIVE CLOTHING, RUBBER GLOVES, AND BREATHING APPARATUS.
WARNING: 1] STRUCTURAL PROTECTIVE CLOTHING IS PERMEABLE, REMAIN CLEAR OF SMOKE, WATER FALL OUT AND WATER RUN OFF.
         2] KEEP OUT OF THE REACH OF CHILDREN.
         3] CONTAINERS MAY MELT AND LEAK IN A FIRE.
         4] THERMAL DECOMPOSITION MAY YIELD CARBON DIOXIDE, CARBON MONOXIDE.
         5] MOVE CONTAINERS FROM AREA IF WITHOUT RISK, COOL EXPOSED CONTAINERS.
         6] DIKE AREA FOR CONTROL AND CONTAINMENT TO PREVENT ENTRY INTO SEWERS, DRAIN, AND WATER WAYS.

LARGE SPILL/NO FIRE/RESCUE: WEAR RUBBER OR NEOPRENE BOOTS, GLOVES.

SPILL CONTROL AND CONTAINMENT:
  HOUSEHOLD SPILL: WIPE UP LIQUID WITH AN ABSORBENT CLOTH. RINSE OUT IN THE SINK. WIPE UP RESIDUE WITH A DAMP
  CLOTH, RINSE OUT IN THE SINK.
  LARGE SPILL: COVER SPILL WITH ABSORBENT. SWEEP UP AND PUT IN A DISPOSAL CONTAINER. RINSE AREA WITH WATER, PICK
  UP WITH ABSORBENT. PUT IN A DISPOSAL CONTAINER. SEAL AND REMOVE FROM THE WORK AREA.

HEALTH HAZARD INFORMATION:
  WHEN USING THIS PRODUCT WEAR: NO SPECIAL REQUIREMENT. Eye contact may cause stinging, tearing, itching,
  swelling, and redness. Ingestion may cause nausea and vomiting.

  A physician should be contacted if anyone develops any signs or symptoms and suspects that they are caused by
  exposure to this product.

FIRST AID:
  Eye Exposure:
  Flush with water for 15 minutes while lifting the upper and lower eye lids. Contact lenses should not be worn
  when working with this product. If irritation persists, get medical attention.

  Skin Exposure:
  Not expected to be a problem.

  Breathing:
  Not expected to be a problem.

  Swallowing:
  Immediately contact the local poison control center for advice. Keep the victim warm and at rest. Get medical
  attention.

PRODUCT: L'ENVIE CONDITIONER - MILANO

INCOMPATIBILITY: NONE KNOWN

INGREDIENTS: CETYL ALCOHOL   TLV: CAS#: 36653-82-4
             PROPYLENE GLYCOL   TLV: CAS#: 4254-15-3
             CITRIC ACID   TLV: CAS#:   77-92-9
             CETRIMONIUM CHLORIDE   TLV: CAS#:   112-02-7
             DIMETHYL STEARAMINE   STEARALKONIUM CHLORIDE
             HYDROXYETHYL CELLULOSE   METHYLISOTHIAZOLINONE
             METHYLCHLOROISOTHIAZOLINONE DEIONIZED WATER
             FRAGRANCE   COLORANT

IGNI TEMP: NA FP: NA LEL: NA UEL: NA VP: NA VD: NA SG: 1.0 PS: LIQUID APPEAR: AMBER OPAQUE
    ODOR: FRAGRANT   PH FACTOR: 3.2

HAZARD RATINGS: H: F: R: 1 0 0

FIRE FIGHTING:
  HIGH EXPANSION FOAM, LOW EXPANSION FOAM, ALCOHOL FOAM; DRY CHEMICAL, CARBON DIOXIDE, WATER FOG.
  PROTECTIVE CLOTHING, RUBBER GLOVES, AND BREATHING APPARATUS.
WARNING: 1] STRUCTURAL PROTECTIVE CLOTHING IS PERMEABLE, REMAIN CLEAR OF SMOKE, WATER FALL OUT AND WATER RUN OFF.
         2] KEEP OUT OF THE REACH OF CHILDREN.
         3] CONTAINERS MAY MELT AND LEAK IN A FIRE.
         4] THERMAL DECOMPOSITION MAY YIELD CARBON DIOXIDE, CARBON MONOXIDE.
         5] MOVE CONTAINERS FROM AREA IF WITHOUT RISK, COOL EXPOSED CONTAINERS.
         6] DIKE AREA FOR CONTROL AND CONTAINMENT TO PREVENT ENTRY INTO SEWERS, DRAIN, AND WATER WAYS.

LARGE SPILL/NO FIRE/RESCUE: WEAR RUBBER OR NEOPRENE BOOTS, GLOVES.

SPILL CONTROL AND CONTAINMENT:
  HOUSEHOLD SPILL: WIPE UP LIQUID WITH AN ABSORBENT CLOTH. RINSE OUT IN THE SINK. WIPE UP RESIDUE WITH A DAMP
  CLOTH, RINSE OUT IN THE SINK.
  LARGE SPILL: COVER SPILL WITH ABSORBENT. SWEEP UP AND PUT IN A DISPOSAL CONTAINER. RINSE AREA WITH WATER, PICK
  UP WITH ABSORBENT. PUT IN A DISPOSAL CONTAINER. SEAL AND REMOVE FROM THE WORK AREA.

HEALTH HAZARD INFORMATION:
  WHEN USING THIS PRODUCT WEAR: NO SPECIAL REQUIREMENT. Eye contact may cause stinging, tearing, itching,
  swelling, and redness. Ingestion may cause nausea and vomiting.

  A physician should be contacted if anyone develops any signs or symptoms and suspects that they are caused by
  exposure to this product.

FIRST AID:
  Eye Exposure:
  Flush with water for 15 minutes while lifting the upper and lower eye lids. Contact lenses should not be worn
  when working with this product. If irritation persists, get medical attention

  Skin Exposure:
  Not expected to be a problem.

  Breathing:
  Not expected to be a problem.

  Swallowing:
  Immediately contact the local poison control center for advice. Keep the victim warm and at rest. Get medical
  attention.

PRODUCT: L'ENVIE CONDITIONER - SIAM

INCOMPATIBILITY: NONE KNOWN

INGREDIENTS: CETYL ALCOHOL   TLV: CAS#: 36653-82-4
             PROPYLENE GLYCOL   TLV: CAS#: 4254-15-3
             CITRIC ACID   TLV: CAS#:   77-92-9
             CETRIMONIUM CHLORIDE   TLV: CAS#: 112-02-7
             DIMETHYL STEARAMINE   STEARALKONIUM CHLORIDE
             HYDROXYETHYL CELLULOSE   METHYLISOTHIAZOLINONE
             METHYLCHLOROISOTHIAZOLINONE DEIONIZED WATER
             FRAGRANCE   COLORANT

IGNI TEMP: NA FP: NA LEL: NA UEL: NA VP: NA VD: NA SG: 1.0 PS: LIQUID APPEAR: AMBER OPAQUE
    ODOR: FRAGRANT   PH FACTOR: 3.2

HAZARD RATINGS: H: F: R: 1 0 0

FIRE FIGHTING:
    HIGH EXPANSION FOAM, LOW EXPANSION FOAM, ALCOHOL FOAM; DRY CHEMICAL, CARBON DIOXIDE, WATER FOG.
    PROTECTIVE CLOTHING, RUBBER GLOVES, AND BREATHING APPARATUS.
WARNING: 1] STRUCTURAL PROTECTIVE CLOTHING IS PERMEABLE, REMAIN CLEAR OF SMOKE, WATER FALL OUT AND WATER RUN OFF.
         2] KEEP OUT OF THE REACH OF CHILDREN.
         3] CONTAINERS MAY MELT AND LEAK IN A FIRE.
         4] THERMAL DECOMPOSITION MAY YIELD CARBON DIOXIDE, CARBON MONOXIDE.
         5] MOVE CONTAINERS FROM AREA IF WITHOUT RISK, COOL EXPOSED CONTAINERS.
         6] DIKE AREA FOR CONTROL AND CONTAINMENT TO PREVENT ENTRY INTO SEWERS, DRAIN, AND WATER WAYS.

LARGE SPILL/NO FIRE/RESCUE: WEAR RUBBER OR NEOPRENE BOOTS, GLOVES.

SPILL CONTROL AND CONTAINMENT:
    HOUSEHOLD SPILL: WIPE UP LIQUID WITH AN ABSORBENT CLOTH. RINSE OUT IN THE SINK. WIPE UP RESIDUE WITH A DAMP
    CLOTH, RINSE OUT IN THE SINK.
    LARGE SPILL: COVER SPILL WITH ABSORBENT. SWEEP UP AND PUT IN A DISPOSAL CONTAINER. RINSE AREA WITH WATER, PICK
    UP WITH ABSORBENT. PUT IN A DISPOSAL CONTAINER. SEAL AND REMOVE FROM THE WORK AREA.

HEALTH HAZARD INFORMATION:
    WHEN USING THIS PRODUCT WEAR: NO SPECIAL REQUIREMENT. Eye contact may cause stinging, tearing, itching,
    swelling, and redness. Ingestion may cause nausea and vomiting.

    A physician should be contacted if anyone develops any signs or symptoms and suspects that they are caused by
    exposure to this product.

FIRST AID:
    Eye Exposure:
    Flush with water for 15 minutes while lifting the upper and lower eye lids. Contact lenses should not be worn
    when working with this product. If irritation persists, get medical attention

    Skin Exposure:
    Not expected to be a problem.

    Breathing:
    Not expected to be a problem.

    Swallowing:
    Immediately contact the local poison control center for advice. Keep the victim warm and at rest. Get medical
    attention.

PRODUCT: L'ENVIE SHAMPOO - CAPTURE

INCOMPATIBILITY: NONE KNOWN

INGREDIENTS: SODIUM LAURETH SULFATE   TLV: CAS#:
           AMMONIUM CHLORIDE   TLV: CAS#: 12125-02-9
           CITRIC ACID   TLV: CAS#:   77-92-7
           TETRASODIUM EDTA   TLV: CAS#: 10378-23-1
           AMMONIUM LAURYL SULFATE   AMMONIUM LAURETH SULFATE
           COCAMIDE DEA   COCAMIDE MEA
           GLYCOL DISTEARATE   METHYLCHLOROISOTHIAZOLINONE
           METHYLISOTHIAZOLINONE   COLORANT
           DEIONIZED WATER   FRAGRANCE

IGNI TEMP: NA FP: NA LEL: NA UEL: NA VP: NA VD: NA SG: 1.02 PS: LIQUID APPEAR: AMBER   ODOR: FRAGRANT
  PH FACTOR: 5.3

HAZARD RATINGS: H: F: R: 1 0 0

FIRE FIGHTING:
  HIGH EXPANSION FOAM, LOW EXPANSION FOAM, ALCOHOL FOAM; DRY CHEMICAL, CARBON DIOXIDE, WATER FOG.
  PROTECTIVE CLOTHING, RUBBER GLOVES, AND BREATHING APPARATUS.
WARNING: 1] STRUCTURAL PROTECTIVE CLOTHING IS PERMEABLE, REMAIN CLEAR OF SMOKE, WATER FALL OUT AND WATER RUN OFF.
       2] KEEP OUT OF THE REACH OF CHILDREN.
       3] THERMAL DECOMPOSITION YIELDS TOXIC CARBON DIOXIDE, CARBON MONOXIDE.
       4] CONTAINERS MAY MELT AND LEAK IN A FIRE.
       5] MOVE CONTAINERS FROM AREA IF WITHOUT RISK, COOL EXPOSED CONTAINERS.
       6] DIKE AREA FOR CONTROL AND CONTAINMENT TO PREVENT ENTRY INTO SEWERS, DRAIN, AND WATER WAYS.

LARGE SPILL/NO FIRE/RESCUE: WEAR RUBBER OR NEOPRENE BOOTS, GLOVES.

SPILL CONTROL AND CONTAINMENT:
  HOUSEHOLD SPILL: WIPE UP LIQUID WITH AN ABSORBENT CLOTH, RINSE OUT IN THE SINK. WIPE UP RESIDUE WITH A DAMP
CLOTH, RINSE OUT IN THE SINK.
  LARGE SPILL: COVER SPILL WITH ABSORBENT, SWEEP UP AND PUT IN A DISPOSAL CONTAINER. RINSE AREA WITH WATER, PICK
UP WITH ABSORBENT, PUT IN A DISPOSAL CONTAINER. SEAL AND REMOVE CONTAINER FROM THE WORK AREA.

HEALTH HAZARD INFORMATION:
  WHEN USING THIS PRODUCT WEAR: NO SPECIAL REQUIREMENT. Eye contact may cause stinging, tearing, itching,
swelling, and redness. Ingestion may cause nausea and vomiting.

  A physician should be contacted if anyone develops any signs or symptoms and suspects that they are caused by
exposure to this product.

FIRST AID:
Eye Exposure:
Flush with water for 15 minutes while lifting the upper and lower eye lids. Contact lenses should not be worn
when working with this product. If irritation persists, get medical attention.

  Skin Exposure:
Not expected to be a problem.

  Breathing:
Not expected to be a problem.

  Swallowing:
Immediately contact the local poison control center for advice. Keep the victim warm and at rest. Get medical
attention.

PRODUCT: L'ENVIE SHAMPOO - CYPRESS

INCOMPATIBILITY: NONE KNOWN

INGREDIENTS: SODIUM LAURETH SULFATE   TLV: CAS#:
            AMMONIUM CHLORIDE   TLV: CAS#: 12125-02-9
            CITRIC ACID   TLV: CAS#:   77-92-7
            TETRASODIUM EDTA   TLV: CAS#: 10378-23-1

AMMONIUM LAURYL SULFATE	AMMONIUM LAURETH SULFATE
COCAMIDE DEA	COCAMIDE MEA
GLYCOL DISTEARATE	METHYLCHLOROISOTHIAZOLINONE
METHYLISOTHIAZOLINONE	COLORANT
DEIONIZED WATER	FRAGRANCE

IGNI TEMP: NA   FP: NA   LEL: NA   UEL: NA   VP: NA   VD: NA   SG: 1.02   PS: LIQUID   APPEAR: AMBER   ODOR: FRAGRANT
   PH FACTOR: 5.3

HAZARD RATINGS: H: F: R: 1   0   0

FIRE FIGHTING:
   HIGH EXPANSION FOAM, LOW EXPANSION FOAM, ALCOHOL FOAM; DRY CHEMICAL, CARBON DIOXIDE, WATER FOG.
   PROTECTIVE CLOTHING, RUBBER GLOVES, AND BREATHING APPARATUS.
WARNING: 1) STRUCTURAL PROTECTIVE CLOTHING IS PERMEABLE, REMAIN CLEAR OF SMOKE, WATER FALL OUT AND WATER RUN OFF.
          2) KEEP OUT OF THE REACH OF CHILDREN.
          3) THERMAL DECOMPOSITION YIELDS TOXIC CARBON DIOXIDE, CARBON MONOXIDE.
          4) CONTAINERS   MAY MELT AND LEAK IN A FIRE.
          5) MOVE CONTAINERS FROM AREA IF WITHOUT RISK, COOL EXPOSED CONTAINERS.
          6) DIKE AREA FOR CONTROL AND CONTAINMENT TO PREVENT ENTRY INTO SEWERS, DRAIN, AND WATER WAYS.

LARGE SPILL/NO FIRE/RESCUE: WEAR RUBBER OR NEOPRENE BOOTS, GLOVES.

SPILL CONTROL AND CONTAINMENT:
   HOUSEHOLD SPILL: WIPE UP LIQUID WITH AN ABSORBENT CLOTH, RINSE OUT IN THE SINK. WIPE UP RESIDUE WITH A DAMP
   CLOTH, RINSE OUT IN THE SINK.
   LARGE SPILL: COVER SPILL WITH ABSORBENT, SWEEP UP AND PUT IN A DISPOSAL CONTAINER. RINSE AREA WITH WATER, PICK
   UP WITH ABSORBENT, PUT IN A DISPOSAL CONTAINER. SEAL AL AND REMOVE CONTAINER FROM THE WORK AREA.

HEALTH HAZARD INFORMATION:
   WHEN USING THIS PRODUCT WEAR: NO SPECIAL REQUIREMENT. Eye contact may cause stinging, tearing, itching,
   swelling, and redness. Ingestion may cause nausea and vomiting.

   A physician should be contacted if anyone develops any signs or symptoms and suspects that they are caused by
   exposure to this product.

FIRST AID:
   Eye Exposure:
   Flush with water for 15 minutes while lifting the upper and lower eye lids. Contact lenses should not be worn
   when working with this product. If irritation persists, get medical attention.

   Skin Exposure:
   Not expected to be a problem.

   Breathing:
   Not expected to be a problem.

   Swallowing:
   Immediately contact the local poison control center for advice. Keep the victim warm and at rest. Get medical
   attention.

PRODUCT: L'ENVIE SHAMPOO - EMBER MUSK

INCOMPATIBILITY: NONE KNOWN

INGREDIENTS: SODIUM LAURETH SULFATE   TLV: CAS#:
             AMMONIUM CHLORIDE   TLV: CAS#: 12125-02-9
             CITRIC ACID   TLV: CAS#: 77-92-7
             TETRASODIUM EDTA   TLV: CAS#: 10378-23-1
             AMMONIUM LAURYL SULFATE   AMMONIUM LAURETH SULFATE
             COCAMIDE DEA   COCAMIDE MEA
             GLYCOL DISTEARATE   METHYLCHLOROISOTHIAZOLINONE
             METHYLISOTHIAZOLINONE   COLORANT
             DEIONIZED WATER   FRAGRANCE

IGNI TEMP: NA FP: NA LEL: NA UEL: NA VP: NA VD: NA SG: 1.02 PS: LIQUID APPEAR: AMBER   ODOR: FRAGRANT
    PH FACTOR: 5.3

HAZARD RATINGS: H: F: R: 1   0   0

FIRE FIGHTING:
    HIGH EXPANSION FOAM, LOW EXPANSION FOAM, ALCOHOL FOAM; DRY CHEMICAL, CARBON DIOXIDE, WATER FOG.
    PROTECTIVE CLOTHING, RUBBER GLOVES, AND BREATHING APPARATUS.
WARNING: 1] STRUCTURAL PROTECTIVE CLOTHING IS PERMEABLE, REMAIN CLEAR OF SMOKE, WATER FALL OUT AND WATER RUN OFF.
         2] KEEP OUT OF THE REACH OF CHILDREN.
         3] THERMAL DECOMPOSITION YIELDS TOXIC CARBON DIOXIDE, CARBON MONOXIDE.
         4] CONTAINERS MAY MELT AND LEAK IN A FIRE.
         5] MOVE CONTAINERS FROM AREA IF WITHOUT RISK, COOL EXPOSED CONTAINERS.
         6] DIKE AREA FOR CONTROL AND CONTAINMENT TO PREVENT ENTRY INTO SEWERS, DRAIN, AND WATER WAYS.

LARGE SPILL/NO FIRE/RESCUE: WEAR RUBBER OR NEOPRENE BOOTS, GLOVES.

SPILL CONTROL AND CONTAINMENT:
    HOUSEHOLD SPILL: WIPE UP LIQUID WITH AN ABSORBENT CLOTH, RINSE OUT IN THE SINK. WIPE UP RESIDUE WITH A DAMP
    CLOTH, RINSE OUT IN THE SINK.
    LARGE SPILL: COVER SPILL WITH ABSORBENT, SWEEP UP AND PUT IN A DISPOSAL CONTAINER. RINSE AREA WITH WATER, PICK
    UP WITH ABSORBENT, PUT IN A DISPOSAL CONTAINER. SEAL AND REMOVE CONTAINER FROM THE WORK AREA.

HEALTH HAZARD INFORMATION:
    WHEN USING THIS PRODUCT WEAR: NO SPECIAL REQUIREMENT. Eye contact may cause stinging, tearing, itching,
    swelling, and redness. Ingestion may cause nausea and vomiting.

    A physician should be contacted if anyone develops any signs or symptoms and suspects that they are caused by
    exposure to this product.

FIRST AID:
Eye Exposure:
Flush with water for 15 minutes while lifting the upper and lower eye lids. Contact lenses should not be worn
when working with this product. If irritation persists, get medical attention.

    Skin Exposure:
    Not expected to be a problem.

    Breathing:
    Not expected to be a problem.

    Swallowing:
    Immediately contact the local poison control center for advice. Keep the victim warm and at rest. Get medical
    attention.

PRODUCT: L'ENVIE SHAMPOO - LEGACE'

INCOMPATIBILITY: NONE KNOWN

INGREDIENTS: SODIUM LAURETH SULFATE   TLV: CAS#:
            AMMONIUM CHLORIDE   TLV: CAS#: 12125-02-9
            CITRIC ACID   TLV: CAS#:   77-92-7
            TETRASODIUM EDTA   TLV: CAS#: 10378-23-1
            AMMONIUM LAURYL SULFATE   AMMONIUM LAURETH SULFATE
            COCAMIDE DEA   COCAMIDE MEA
            GLYCOL DISTEARATE   METHYLCHLOROISOTHIAZOLINONE
            METHYLISOTHIAZOLINONE   COLORANT
            DEIONIZED WATER   FRAGRANCE

IGNI TEMP: NA FP: NA LEL: NA UEL: NA VP: NA VD: NA SG: 1.02 PS: LIQUID APPEAR: AMBER   ODOR: FRAGRANT
    PH FACTOR: 5.3

HAZARD RATINGS: H: F: R: 1 0 0

FIRE FIGHTING:
    HIGH EXPANSION FOAM, LOW EXPANSION FOAM, ALCOHOL FOAM; DRY CHEMICAL, CARBON DIOXIDE, WATER FOG.
    PROTECTIVE CLOTHING, RUBBER GLOVES, AND BREATHING APPARATUS.
WARNING: 1] STRUCTURAL PROTECTIVE CLOTHING IS PERMEABLE, REMAIN CLEAR OF SMOKE, WATER FALL OUT AND WATER RUN OFF.
         2] KEEP OUT OF THE REACH OF CHILDREN.
         3] THERMAL DECOMPOSITION YIELDS TOXIC CARBON DIOXIDE, CARBON MONOXIDE.
         4] CONTAINERS MAY MELT AND LEAK IN A FIRE.
         5] MOVE CONTAINERS FROM AREA IF WITHOUT RISK, COOL EXPOSED CONTAINERS.
         6] DIKE AREA FOR CONTROL AND CONTAINMENT TO PREVENT ENTRY INTO SEWERS, DRAIN, AND WATER WAYS.

LARGE SPILL/NO FIRE/RESCUE: WEAR RUBBER OR NEOPRENE BOOTS, GLOVES.

SPILL CONTROL AND CONTAINMENT:
    HOUSEHOLD SPILL: WIPE UP LIQUID WITH AN ABSORBENT CLOTH, RINSE OUT IN THE SINK. WIPE UP RESIDUE WITH A DAMP
    CLOTH, RINSE OUT IN THE SINK.
    LARGE SPILL: COVER SPILL WITH ABSORBENT, SWEEP UP AND PUT IN A DISPOSAL CONTAINER. RINSE AREA WITH WATER, PICK
    UP WITH ABSORBENT, PUT IN A DISPOSAL CONTAINER. SEAL AND REMOVE CONTAINER FROM THE WORK AREA.

HEALTH HAZARD INFORMATION:
    WHEN USING THIS PRODUCT WEAR: NO SPECIAL REQUIREMENT. Eye contact may cause stinging, tearing, itching,
    swelling, and redness. Ingestion may cause nausea and vomiting.

    A physician should be contacted if anyone develops any signs or symptoms and suspects that they are caused by
    exposure to this product.

FIRST AID:
    Eye Exposure:
    Flush with water for 15 minutes while lifting the upper and lower eye lids. Contact lenses should not be worn
    when working with this product. If irritation persists, get medical attention.

    Skin Exposure:
    Not expected to be a problem.

    Breathing:
    Not expected to be a problem.

    Swallowing:
    Immediately contact the local poison control center for advice. Keep the victim warm and at rest. Get medical
    attention.

PRODUCT: L'ENVIE SHAMPOO - MILANO

INCOMPATIBILITY: NONE KNOWN

INGREDIENTS: SODIUM LAURETH SULFATE   TLV: CAS#:
             AMMONIUM CHLORIDE   TLV: CAS#: 12125-02-9
             CITRIC ACID   TLV: CAS#:   77-92-7
             TETRASODIUM EDTA   TLV: CAS#: 10378-23-1
             AMMONIUM LAURYL SULFATE   AMMONIUM LAURETH SULFATE
             COCAMIDE DEA   COCAMIDE MEA
             GLYCOL DISTEARATE   METHYLCHLOROISOTHIAZOLINONE
             METHYLISOTHIAZOLINONE   COLORANT
             DEIONIZED WATER   FRAGRANCE

IGNI TEMP: NA FP: NA LEL: NA UEL: NA VP: NA VD: NA SG: 1.02 PS: LIQUID APPEAR: AMBER   ODOR: FRAGRANT
    PH FACTOR: 5.3

HAZARD RATINGS: H: F: R: 1 0 0

FIRE FIGHTING:
    HIGH EXPANSION FOAM, LOW EXPANSION FOAM, ALCOHOL FOAM; DRY CHEMICAL, CARBON DIOXIDE, WATER FOG.
    PROTECTIVE CLOTHING, RUBBER GLOVES, AND BREATHING APPARATUS.
WARNING: 1] STRUCTURAL PROTECTIVE CLOTHING IS PERMEABLE, REMAIN CLEAR OF SMOKE, WATER FALL OUT AND WATER RUN OFF.
         2] KEEP OUT OF THE REACH OF CHILDREN.
         3] THERMAL DECOMPOSITION YIELDS TOXIC CARBON DIOXIDE, CARBON MONOXIDE.
         4] CONTAINERS MAY MELT AND LEAK IN A FIRE.
         5] MOVE CONTAINERS FROM AREA IF WITHOUT RISK, COOL EXPOSED CONTAINERS.
         6] DIKE AREA FOR CONTROL AND CONTAINMENT TO PREVENT ENTRY INTO SEWERS, DRAIN, AND WATER WAYS.

LARGE SPILL/NO FIRE/RESCUE: WEAR RUBBER OR NEOPRENE BOOTS, GLOVES.

SPILL CONTROL AND CONTAINMENT:
    HOUSEHOLD SPILL: WIPE UP LIQUID WITH AN ABSORBENT CLOTH, RINSE OUT IN THE SINK. WIPE UP RESIDUE WITH A DAMP
    CLOTH, RINSE OUT IN THE SINK.
    LARGE SPILL: COVER SPILL WITH ABSORBENT, SWEEP UP AND PUT IN A DISPOSAL CONTAINER. RINSE AREA WITH WATER, PICK
    UP WITH ABSORBENT, PUT IN A DISPOSAL CONTAINER. SEAL AND REMOVE CONTAINER FROM THE WORK AREA.

HEALTH HAZARD INFORMATION:
    WHEN USING THIS PRODUCT WEAR: NO SPECIAL REQUIREMENT. Eye contact may cause stinging, tearing, itching,
    swelling, and redness. Ingestion may cause nausea and vomiting.

    A physician should be contacted if anyone develops any signs or symptoms and suspects that they are caused by
    exposure to this product.

FIRST AID:
Eye Exposure:
Flush with water for 15 minutes while lifting the upper and lower eye lids. Contact lenses should not be worn
when working with this product. If irritation persists, get medical attention.

Skin Exposure:
Not expected to be a problem.

Breathing:
Not expected to be a problem.

Swallowing:
Immediately contact the local poison control center for advice. Keep the victim warm and at rest. Get medical
attention.

PRODUCT: L'ENVIE SHAMPOO - SIAM

INCOMPATIBILITY: NONE KNOWN

INGREDIENTS: SODIUM LAURETH SULFATE   TLV: CAS#:
            AMMONIUM CHLORIDE   TLV: CAS#: 12125-02-9
            CITRIC ACID   TLV: CAS#:   77-92-7
            TETRASODIUM EDTA   TLV: CAS#: 10378-23-1
            AMMONIUM LAURYL SULFATE   AMMONIUM LAURETH SULFATE
            COCAMIDE DEA   COCAMIDE MEA
            GLYCOL DISTEARATE   METHYLCHLOROISOTHIAZOLINONE
            METHYLISOTHIAZOLINONE   COLORANT
            DEIONIZED WATER   FRAGRANCE

IGNI TEMP: NA FP: NA LEL: NA UEL: NA VP: NA VD: NA SG: 1.02 PS: LIQUID APPEAR: AMBER   ODOR: FRAGRANT
    PH FACTOR: 5.3

HAZARD RATINGS: H: F: R: 1 0 0

FIRE FIGHTING:
  HIGH EXPANSION FOAM, LOW EXPANSION FOAM, ALCOHOL FOAM; DRY CHEMICAL, CARBON DIOXIDE, WATER FOG.
  PROTECTIVE CLOTHING, RUBBER GLOVES, AND BREATHING APPARATUS.
WARNING: 1] STRUCTURAL PROTECTIVE CLOTHING IS PERMEABLE, REMAIN CLEAR OF SMOKE, WATER FALL OUT AND WATER RUN OFF.
         2] KEEP OUT OF THE REACH OF CHILDREN.
         3] THERMAL DECOMPOSITION YIELDS TOXIC CARBON DIOXIDE, CARBON MONOXIDE.
         4] CONTAINERS MAY MELT AND LEAK IN A FIRE.
         5] MOVE CONTAINERS FROM AREA IF WITHOUT RISK, COOL EXPOSED CONTAINERS.
         6] DIKE AREA FOR CONTROL AND CONTAINMENT TO PREVENT ENTRY INTO SEWERS, DRAIN, AND WATER WAYS.

LARGE SPILL/NO FIRE/RESCUE: WEAR RUBBER OR NEOPRENE BOOTS, GLOVES.

SPILL CONTROL AND CONTAINMENT:
  HOUSEHOLD SPILL: WIPE UP LIQUID WITH AN ABSORBENT CLOTH, RINSE OUT IN THE SINK. WIPE UP RESIDUE WITH A DAMP
  CLOTH, RINSE OUT IN THE SINK.
  LARGE SPILL: COVER SPILL WITH ABSORBENT, SWEEP UP AND PUT IN A DISPOSAL CONTAINER. RINSE AREA WITH WATER, PICK
  UP WITH ABSORBENT, PUT IN A DISPOSAL CONTAINER. SEAL AND REMOVE CONTAINER FROM THE WORK AREA.

HEALTH HAZARD INFORMATION:
  WHEN USING THIS PRODUCT WEAR: NO SPECIAL REQUIREMENT. Eye contact may cause stinging, tearing, itching,
  swelling, and redness. Ingestion may cause nausea and vomiting.

  A physician should be contacted if anyone develops any signs or symptoms and suspects that they are caused by
  exposure to this product.

FIRST AID:
  Eye Exposure:
  Flush with water for 15 minutes while lifting the upper and lower eye lids. Contact lenses should not be worn
  when working with this product. If irritation persists, get medical attention.

  Skin Exposure:
  Not expected to be a problem.

  Breathing:
  Not expected to be a problem.

  Swallowing:
  Immediately contact the local poison control center for advice. Keep the victim warm and at rest. Get medical
  attention.

PRODUCT: LESTOIL DEODORIZING RUG SHAMPOO

INCOMPATIBILITY: NONE KNOWN

INGREDIENTS: AMMONIA   TLV: 25 ppm   CAS#: 7664-41-7
             SODIUM LAURYL SULFATE   TLV: CAS#:
             TRISODIUM
             PHOSPHATE DODECAHYDRATE   TLV: CAS#: 10101-89-0
             ISOBUTANE   TLV: CAS#:   75-28-5
             N-BUTANE   TLV: CAS#:   106-97-8
             WATER   TLV: CAS#: 7732-18-5
             SODIUM LAURYL SARCOSINATE; COCONUT DIETHANOLAMIDE; FRAGRANCE.

IGNI TEMP: NA FP: 20F LEL: NA UEL: NA VP: NA VD: NA SG: .95 PS: FOAM APPEAR: SPRAY ODOR: FRAGRANCE
    PH FACTOR: 9.5

HAZARD RATINGS: H: F: R: 0   4   0

FIRE FIGHTING:
    HIGH EXPANSION FOAM, LOW EXPANSION FOAM, ALCOHOL FOAM; DRY CHEMICAL, CARBON DIOXIDE, WATER FOG.
    PROTECTIVE CLOTHING, RUBBER GLOVES, AND BREATHING APPARATUS.
WARNING: 1] STRUCTURAL PROTECTIVE CLOTHING IS PERMEABLE, REMAIN CLEAR OF SMOKE, WATER FALL OUT AND WATER RUN OFF.
         2] KEEP OUT OF THE REACH OF CHILDREN.
         3] CONTAINERS WILL EXPLODE IN A FIRE OR IF HEATED ABOVE 120F.
         4] FLAMMABLE GAS - WATER BASED FOAM SELF EXTINGUISHING.
         5] MOVE CONTAINERS FROM AREA IF WITHOUT RISK, COOL EXPOSED CONTAINERS.
         6] DIKE AREA FOR CONTROL AND CONTAINMENT TO PREVENT ENTRY INTO SEWERS, DRAIN, AND WATER WAYS.

LARGE SPILL/NO FIRE/RESCUE: WEAR RUBBER OR NEOPRENE BOOTS, GLOVES.

SPILL CONTROL AND CONTAINMENT:
    HOUSEHOLD SPILL: DILUTE FOAM WITH WATER, WIPE UP WITH AN ABSORBENT CLOTH. RINSE OUT IN THE SINK. WIPE UP RESIDUE
    WITH A DAMP CLOTH, RINSE IN THE SINK.
    LARGE SPILL: VENTILATE ENCLOSED SPACES. PICK UP WITH ABSORBENT, PUT IN A DISPOSAL CONTAINER. RINSE AREA WITH
    WATER, PICK UP WITH ABSORBENT. PUT IN A DISPOSAL CONTAINER, SEAL AND REMOVE FROM THE WORK AREA.

HEALTH HAZARD INFORMATION:
    WHEN USING THIS PRODUCT WEAR: NO SPECIAL REQUIREMENT. Eye contact may cause stinging, tearing, itching,
    swelling, and redness. Ingestion may result in nausea and vomiting.

    A physician should be contacted if anyone develops any signs or symptoms and suspects that they are caused by
    exposure to this product.

FIRST AID:
Eye Exposure:
Flush with water for 15 minutes while lifting the upper and lower eye lids. Contact lenses should not be worn
when working with this product. Get medical attention.

Skin Exposure:
Wash with soap and water.

Breathing:
Not expected to be a problem.

Swallowing:
Immediately contact the local poison control center for advice. Keep the victim warm and at rest. Get medical
attention.

PRODUCT: LESTOIL FLOOR CLEANER

INCOMPATIBILITY: NONE KNOWN

INGREDIENTS: POTASSIUM HYDROXIDE   TLV: CAS#: 1310-58-3
             TETRASODIUM EDTA SOLUTION   TLV: CAS#: 10378-23-1
             TRIETHANOLAMINE LAURYL SULFATE   COCONUT DIETHANOLAMIDE
             TALL OIL FATTY ACID   DEIONIZED WATER
             DMDM HYDANTION   PERFUME OIL
             RED SUPERNYLITE T-3B   YELLOW, ACID 114

IGNI TEMP: NA FP: NA LEL: NA UEL: NA VP: NA VD: NA SG: NA PS: LIQUID APPEAR: PALE ORANGE
    ODOR: PERFUME   PH FACTOR: 10.5

HAZARD RATINGS: H: F: R:

FIRE FIGHTING:
  HIGH EXPANSION FOAM, LOW EXPANSION FOAM, ALCOHOL FOAM; DRY CHEMICAL, CARBON DIOXIDE, WATER FOG.
  PROTECTIVE CLOTHING, RUBBER GLOVES, AND BREATHING APPARATUS.
WARNING: 1] STRUCTURAL PROTECTIVE CLOTHING IS PERMEABLE, REMAIN CLEAR OF SMOKE, WATER FALL OUT AND WATER RUN OFF.
         2] KEEP OUT OF THE REACH OF CHILDREN.
         3] THERMAL DECOMPOSITION PRODUCTS ARE UNKNOWN.
         4] MOVE CONTAINERS FROM AREA IF WITHOUT RISK, COOL EXPOSED CONTAINERS.
         5] DIKE AREA FOR CONTROL AND CONTAINMENT TO PREVENT ENTRY INTO SEWERS, DRAIN, AND WATER WAYS.

LARGE SPILL/NO FIRE/RESCUE: WEAR RUBBER OR NEOPRENE BOOTS, GLOVES.

SPILL CONTROL AND CONTAINMENT:
  HOUSEHOLD SPILL: WIPE UP LIQUID WITH AN ABSORBENT CLOTH. RINSE OUT IN THE SINK. WIPE UP RESIDUE WITH A DAMP
  CLOTH, RINSE OUT IN THE SINK.
  LARGE SPILL: PICK UP WITH AN ABSORBENT, PUT IN A DISPOSAL CONTAINER. RINSE AREA WITH WATER, PICK UP WITH
  ABSORBENT. PUT IN A DISPOSAL CONTAINER. SEAL AND REMOVE FROM THE WORK AREA.

HEALTH HAZARD INFORMATION:
  WHEN USING THIS PRODUCT WEAR: NO SPECIAL REQUIREMENT. Eye contact may cause stinging, tearing, itching,
  swelling, and redness. Ingestion may result in nausea and vomiting

  A physician should be contacted if anyone develops any signs or symptoms and suspects that they are caused by
  exposure to this product.

FIRST AID:
  Eye Exposure:
  Flush with water for 15 minutes while lifting the upper and lower eye lids. Contact lenses should not be worn
  when working with this product. Get medical attention.

  Skin Exposure:
  Not expected to be a problem.

  Breathing:
  Not expected to be a problem.

  Swallowing:
  Immediately contact the local poison control center for advice. Keep the victim warm and at rest. Get medical
  attention.

PRODUCT: LESTOIL HEAVY DUTY CLEANER

INCOMPATIBILITY: RUBBER TIRES; ASPHALT TILE; PLASTIC; CLOTHING; GLASS; PAINT.

INGREDIENTS: TETRASODIUM EDTA   TLV: CAS#: 10378-23-1
             SODIUM HYDROXIDE   TLV: CAS#: 1310-73-2
             WATER   TLV: CAS#: 7732-18-5
             PINE OIL   TLV: CAS#: 8002-09-3
             PETROLEUM DISTILLATE   TALL OIL FATTY ACID
             TMPD   YELLOW, ACID 114

IGNI TEMP: NA FP: 103F LEL: NA UEL: NA VP: NA VD: NA SG: .95 PS: LIQUID APPEAR: THICK,VISCOUS
   ODOR PETROLEUM   PH FACTOR: 12

HAZARD RATINGS: H: F: R:

FIRE FIGHTING:
   HIGH EXPANSION FOAM, LOW EXPANSION FOAM, ALCOHOL FOAM; DRY CHEMICAL, CARBON DIOXIDE, WATER FOG.
   PROTECTIVE CLOTHING, RUBBER GLOVES, AND BREATHING APPARATUS.
WARNING: 1] STRUCTURAL PROTECTIVE CLOTHING IS PERMEABLE, REMAIN CLEAR OF SMOKE, WATER FALL OUT AND WATER RUN OFF.
         2] KEEP OUT OF THE REACH OF CHILDREN.
         3] CHECK THE INCOMPATIBLE MATERIALS.
         4] MOVE CONTAINERS FROM AREA IF WITHOUT RISK, COOL EXPOSED CONTAINERS.
         5] DIKE AREA FOR CONTROL AND CONTAINMENT TO PREVENT ENTRY INTO SEWERS, DRAIN, AND WATER WAYS.

LARGE SPILL/NO FIRE/RESCUE: WEAR RUBBER OR NEOPRENE BOOTS, GLOVES.

SPILL CONTROL AND CONTAINMENT:
   HOUSEHOLD SPILL: WIPE UP WITH AN ABSORBENT CLOTH. RINSE AREA WITH WATER, WIPE UP WITH ABSORBENT CLOTH. RINSE OUT
   IN THE SINK. WIPE UP RESIDUE WITH A DAMP CLOTH.
   LARGE SPILL: PICK UP WITH ABSORBENT, PUT IN A DISPOSAL CONTAINER. RINSE AREA WITH WATER PICK UP WITH ABSORBENT.
   PUT IN A DISPOSAL CONTAINER. SEAL AND REMOVE FROM THE WORK AREA.

HEALTH HAZARD INFORMATION:
   WHEN USING THIS PRODUCT WEAR: NEOPRENE OR OTHER PROTECTIVE GLOVES. Eye contact may cause stinging, tearing,
   itching, swelling, and redness. Ingestion will cause abdominal pain, nausea, vomiting, and diarrhea. May cause
   central nervous system depression and respiratory arrest.

   A physician should be contacted if anyone develops any signs or symptoms and suspects that they are caused by
   exposure to this product.

FIRST AID:
Eye Exposure:
Flush with water for 15 minutes while lifting the upper and lower eye lids. Contact lenses should not be worn
when working with this product. Get medical attention.

Skin Exposure:
Wash off with water.

Breathing:
Not expected to be a problem.

Swallowing:
Do not induce vomiting. If victim is conscious, give 2-3 glasses of milk to drink. Immediately contact the local
poison control center for advice. Keep the victim warm and at rest. Get medical attention.

PRODUCT: LIFE BUOY DEODORANT SOAP

INCOMPATIBILITY: NONE KNOWN

INGREDIENTS: NONE IDENTIFIED BY MANUFACTURER

IGNI TEMP: NA FP: NA LEL: NA UEL: NA VP: NA VD: NA SG: 1.04 PS: SOLID   APPEAR: WHITE,CORAL,OR GOLD BAR
    ODOR: PERFUME   PH FACTOR: 10

HAZARD RATINGS: H: F: R:

FIRE FIGHTING:
  HIGH EXPANSION FOAM, LOW EXPANSION FOAM, ALCOHOL FOAM; DRY CHEMICAL, CARBON DIOXIDE, WATER FOG.
  PROTECTIVE CLOTHING, RUBBER GLOVES, AND BREATHING APPARATUS.
WARNING: 1] STRUCTURAL PROTECTIVE CLOTHING IS PERMEABLE, REMAIN CLEAR OF SMOKE, WATER FALL OUT AND WATER RUN OFF.
         2] KEEP OUT OF THE REACH OF CHILDREN.
         3] MOVE CONTAINERS FROM AREA IF WITHOUT RISK, COOL EXPOSED CONTAINERS.
         4] DIKE AREA FOR CONTROL AND CONTAINMENT TO PREVENT ENTRY INTO SEWERS, DRAIN, AND WATER WAYS.

LARGE SPILL/NO FIRE/RESCUE: WEAR RUBBER OR NEOPRENE BOOTS, GLOVES, DUST MASK.

SPILL CONTROL AND CONTAINMENT:
  HOUSEHOLD SPILL: PICK UP THE BAR AND WIPE UP RESIDUE WITH A DAMP CLOTH.
  LARGE SPILL: SCOOP UP MATERIAL AND PUT IN A DISPOSAL CONTAINER. RINSE AREA WITH WATER, PICK UP WITH ABSORBENT.
  PUT IN A DISPOSAL CONTAINER. SEAL AND REMOVE FROM WORK AREA.

HEALTH HAZARD INFORMATION:
  WHEN USING THIS PRODUCT WEAR: NO SPECIAL REQUIREMENT. Eye contact will cause stinging, tearing, itching,
  swelling, andand redness. Prolonged skin contact may cause redness and itching. Ingestion will cause abdominal
  pain, nausea, vomiting, and mild diarrhea.

  A physician should be contacted if anyone develops any signs or symptoms and suspects that they are caused by
  exposure to this product.

FIRST AID:
  Eye Exposure:
  Flush with water for 15 minutes while lifting the upper and lower eye lids. Contact lenses should not be worn
  when working with this product. Get medical attention.

  Skin Exposure:
  Wash off with water.

  Breathing:
  Not expected to be a problem.

  Swallowing:
  If the victim is conscious, give 2-3 glasses of milk or water to drink. Immediately contact the local poison
  control center for advice. Keep the victim warm and at rest. Get medical attention.

PRODUCT: LIQUID PLUMBER and PROFESSIONAL STRENGTH LIQUID PLUMBER

INCOMPATIBILITY: TOILET BOWL CLEANERS; DRAIN CLEANERS; ACIDS; AMMONIA; PRODUCT CONTACTING AMMONIA.

INGREDIENTS: SODIUM HYDROXIDE   .5-2% TLV: 2 mg/m3(ceiling) CAS#: 1310-73-2
             SODIUM HYPOCHLORITE   5-10% TLV: CAS#: 7681-52-9

IGNI TEMP: NA FP: NA LEL: NA UEL: NA VP: NA VD: NA SG: 1.1 PS: LIQUID APPEAR: CLEAR   ODOR: CHLORINE
   PH FACTOR: 13.2

HAZARD RATINGS: H: F: R: 3 0 1   HAZARD CLASS   8

FIRE FIGHTING:
   HIGH EXPANSION FOAM, LOW EXPANSION FOAM, ALCOHOL FOAM; DRY CHEMICAL, CARBON DIOXIDE, WATER FOG.
   PROTECTIVE CLOTHING, RUBBER GLOVES, AND BREATHING APPARATUS.
WARNING: 1] STRUCTURAL PROTECTIVE CLOTHING IS PERMEABLE, REMAIN CLEAR OF SMOKE, WATER FALL OUT AND WATER RUN OFF.
         2] KEEP OUT OF THE REACH OF CHILDREN.
         3] MIXING WITH OTHER TYPE OF DRAIN CLEANERS IN A CLOGGED DRAIN, MAY CAUSE AN EXPLOSIVE BLOWING OF LIQUID
            BACK OUT OF THE DRAIN.
         4] DO NOT MIX WITH ANY INCOMPATIBLE PRODUCTS LISTED ABOVE, YIELDS TOXIC CHLORINE OR AMMONIA GAS.
         5] MOVE CONTAINERS FROM AREA IF WITHOUT RISK, COOL EXPOSED CONTAINERS.
         6] DIKE AREA FOR CONTROL AND CONTAINMENT TO PREVENT ENTRY INTO SEWERS, DRAIN, AND WATER WAYS.

LARGE SPILL/NO FIRE/RESCUE: WEAR RUBBER OR NEOPRENE BOOTS, GLOVES AND BREATHING APPARATUS.

SPILL CONTROL AND CONTAINMENT:
   HOUSEHOLD SPILL: OPEN WINDOWS. WEAR RUBBER GLOVES. WIPE UP WITH AN ABSORBENT CLOTH. RINSE OUT IN THE SINK. USE A
   SOLUTION 2 oz VINEGAR AND 8 oz WATER. APPLY ON TOP OF LIQUID PLUMBER TO NEUTRALIZE IT. WIPE UP WITH AN ABSORBENT
   CLOTH. REPEAT PROCESS. RINSE CLOTH OUT IN THE SINK.
   LARGE SPILL: COVER SPILL WITH A 4 PARTS WATER TO 1 PART VINEGAR SOLUTION. PICK UP WITH ABSORBENT, PUT IN A
   DISPOSAL CONTAINER. REPEAT PROCESS. SEAL AND REMOVE ALL CONTAINERS FROM THE WORK AREA.

HEALTH HAZARD INFORMATION:
   WHEN USING THIS PRODUCT WEAR: RUBBER GLOVES. Eye contact with this corrosive will cause severe burns. Skin
   contact will cause burns. Ingestion causes severe burns to the mouth, throat, and stomach. (During normal
   household use conditions, the likelihood of any adverse health effects are low.)

   A physician should be contacted if anyone develops any signs or symptoms and suspects that they are caused by
   exposure to this product.

FIRST AID:
   Eye Exposure:
   Flush with water for 15 minutes while lifting the upper and lower eye lids. Contact lenses should not be worn
   when working with this product. Get medical attention.

   Skin Exposure:
   Remove contaminated clothing, rinse skin with water. If irritation develops, get medical attention.

   Breathing:
   Any developments, move the victim to fresh air.

   Swallowing:
   Do not induce vomiting. Immediately contact the local poison control center for advice. Keep the victim warm and
   at rest. Get medical attention.

PRODUCT: LITTER GREEN

INCOMPATIBILITY: NONE KNOWN

INGREDIENTS: ALFALFA PELLETS_____   100% TLV: CAS#:
             OAT HULLS_____/   TLV: CAS#:

IGNI TEMP: NA FP: NA LEL: NA UEL: NA VP: NA VD: NA SG: NA PS: PELLETS APPEAR: GREEN/YELLOW   ODOR:
   PH FACTOR: NA

HAZARD RATINGS: H: F: R: 2 0 0

FIRE FIGHTING:
   HIGH EXPANSION FOAM, LOW EXPANSION FOAM, ALCOHOL FOAM; DRY CHEMICAL, CARBON DIOXIDE, WATER FOG.
   PROTECTIVE CLOTHING, RUBBER GLOVES, AND BREATHING APPARATUS.
WARNING: 1] STRUCTURAL PROTECTIVE CLOTHING IS PERMEABLE, REMAIN CLEAR OF SMOKE, WATER FALL OUT AND WATER RUN OFF.
         2] KEEP OUT OF THE REACH OF CHILDREN.
         3] DUST IS CAPABLE OF FORMING AN EXPLOSIVE MIXTURE IN THE AIR.
         4] MOVE CONTAINERS FROM AREA IF WITHOUT RISK, COOL EXPOSED CONTAINERS.
         5] DIKE AREA FOR CONTROL AND CONTAINMENT TO PREVENT ENTRY INTO SEWERS, DRAIN, AND WATER WAYS.

LARGE SPILL/NO FIRE/RESCUE: WEAR RUBBER OR NEOPRENE BOOTS, GLOVES.

SPILL CONTROL AND CONTAINMENT:
   HOUSEHOLD SPILL: AVOID CREATING A DUST. GENTLY SWEEP UP AND PUT BACK IN THE BAG OR PUT IN THE LITTER BOX.
   LARGE SPILL: AVOID CREATING A DUST. PICK UP MATERIAL AND PUT IN A DISPOSAL CONTAINER. RINSE AREA WITH WATER,
   PICK UP WITH AN ABSORBENT. PUT IN A DISPOSAL CONTAINER. SEAL AND REMOVE FROM THE WORK AREA.

HEALTH HAZARD INFORMATION:
   WHEN USING THIS PRODUCT WEAR: NO SPECIAL REQUIREMENT. Dust inhalation may cause sensitization with symptoms from
   hay fever to asthma. Eye contact may cause stinging, tearing, itching, swelling, and redness.

   A physician should be contacted if anyone develops any signs or symptoms and suspects that they are caused by
   exposure to this product.

FIRST AID:
   Eye Exposure:
   Flush with water for 15 minutes while lifting the upper and lower eye lids. Contact lenses should not be worn
   when working with this product. Get medical attention.

   Skin Exposure:
   Not expected to be a problem.

   Breathing:
   If symptoms appear, move the victim to fresh air.

   Swallowing:
   Not expected to be a health hazard.

PRODUCT: LUX BEAUTY SOAP

INCOMPATIBILITY: NONE KNOWN

INGREDIENTS: NOT LISTED BY MANUFACTURER

IGNI TEMP: NA FP: NA LEL: NA UEL: NA VP: NA VD: NA SG: 1.04 PS: SOLID APPEAR: BAR   ODOR: NONE
    PH FACTOR:

HAZARD RATINGS: H: F: R:

FIRE FIGHTING:
  HIGH EXPANSION FOAM, LOW EXPANSION FOAM, ALCOHOL FOAM; DRY CHEMICAL, CARBON DIOXIDE, WATER FOG.
  PROTECTIVE CLOTHING, RUBBER GLOVES, AND BREATHING APPARATUS.
WARNING: 1] STRUCTURAL PROTECTIVE CLOTHING IS PERMEABLE, REMAIN CLEAR OF SMOKE, WATER FALL OUT AND WATER RUN OFF.
         2] KEEP OUT OF THE REACH OF CHILDREN.
         3] M] MOVE CONTAINERS FROM AREA IF WITHOUT RISK, COOL EXPOSED CONTAINERS.
         4] DIKE AREA FOR CONTROL AND CONTAINMENT TO PREVENT ENTRY INTO SEWERS, DRAIN, AND WATER WAYS.

LARGE SPILL/NO FIRE/RESCUE: WEAR RUBBER OR NEOPRENE BOOTS, GLOVES.

SPILL CONTROL AND CONTAINMENT:
  HOUSEHOLD SPILL: PICK UP MATERIAL AND WIPE UP RESIDUE WITH A DAMP CLOTH.
  LARGE SPILL: PICK UP MATERIAL AND PUT IN A DISPOSAL CONTAINER. RINSE AREA WITH WATER. PICK UP WITH ABSORBENT.
  PUT IN A DISPOSAL CONTAINER. SEAL AND REMOVE FROM THE WORK AREA.

HEALTH HAZARD INFORMATION:
  WHEN USING THIS PRODUCT WEAR: NO SPECIAL REQUIREMENT. Eye contact may cause stinging, tearing, itching,
  swelling, and redness. Prolonged contact may cause skin irritation. Ingestion may cause abdominal pain, nausea,
  vomiting, and mild diarrhea.

  A physician should be contacted if anyone develops any signs or symptoms and suspects that they are caused by
  exposure to this product.

FIRST AID:
  Eye Exposure:
  Flush with water for 15 minutes while lifting the upper and lower eye lids. Contact lenses should not be worn
  when working with this product. Get medical attention.

  Skin Exposure:
  Wash off with water. If irritation appears, get medical attention.

  Breathing:
  Not expected to be a problem.

  Swallowing:
  If the victim is conscious, give 1-2 glasses or milk or water to drink. Immediately contact the local poison
  control center for advice. Keep the victim warm and at rest. Get medical attention.

PRODUCT: LUX LIGHT DUTY LIQUID DISH WASHING DETERGENT

INCOMPATIBILITY: ALL PRODUCTS CONTAINING CHLORINE (e.g. BLEACH, WINDEX ETC.)

INGREDIENTS: NOT LISTED BY MANUFACTURER

IGNI TEMP: NA FP: NA LEL: NA UEL: NA VP: NA VD: NA SG: 1.03 PS: LIQUID APPEAR: PINK   ODOR: PERFUMED
   PH FACTOR: 6.4 - 6.8

HAZARD RATINGS: H: F: R: 2 0 0

FIRE FIGHTING:
   HIGH EXPANSION FOAM, LOW EXPANSION FOAM, ALCOHOL FOAM; DRY CHEMICAL, CARBON DIOXIDE, WATER FOG.
   PROTECTIVE CLOTHING, RUBBER GLOVES, AND BREATHING APPARATUS.
WARNING: 1] STRUCTURAL PROTECTIVE CLOTHING IS PERMEABLE, REMAIN CLEAR OF SMOKE, WATER FALL OUT AND WATER RUN OFF.
         2] KEEP OUT OF THE REACH OF CHILDREN.
         3] THERMAL DECOMPOSITION MAY YIELD OXIDE OF NITROGEN AND OXIDES OF SULFUR.
         4] MOVE CONTAINERS FROM AREA IF WITHOUT RISK, COOL EXPOSED CONTAINERS.
         5] DIKE AREA FOR CONTROL AND CONTAINMENT TO PREVENT ENTRY INTO SEWERS, DRAIN, AND WATER WAYS.

LARGE SPILL/NO FIRE/RESCUE: WEAR RUBBER OR NEOPRENE BOOTS, GLOVES.

SPILL CONTROL AND CONTAINMENT:
   HOUSEHOLD SPILL: WIPE UP WITH AN ABSORBENT CLOTH. RINSE OUT IN THE SINK. WIPE UP RESIDUE WITH A DAMP CLOTH.
   RINSE OUT IN THE SINK.
   LARGE SPILL: PICK UP WITH ABSORBENT. PUT IN A DISPOSAL CONTAINER. RINSE AREA WITH WATER. PICK UP WITH ABSORBENT.
   PUT IN A DISPOSAL CONTAINER, SEAL AND REMOVE FROM THE WORK AREA.

HEALTH HAZARD INFORMATION:
   WHEN USING THIS PRODUCT WEAR: NO SPECIAL REQUIREMENT. Eye contact may cause stinging, tearing, itching,
   swelling, and redness. Prolonged contact may cause skin irritation. Ingestion may cause abdominal pain, nausea,
   vomiting, and diarrhea. Inhalation may produce irritation of the respiratory tract.

   A physician should be contacted if anyone develops any signs or symptoms and suspects that they are caused by
   exposure to this product.

FIRST AID:
   Eye Exposure:
   Flush with water for 15 minutes while lifting the upper and lower eye lids. Contact lenses should not be worn
   when working with this product. Get medical attention.

   Skin Exposure:
   Wash with water. If irritation develops, get medical attention.

   Breathing:
   If irritation develops, move the victim to fresh air.

   Swallowing:
   If the victim is conscious, give 1-2 glasses of milk or water to drink. Immediately contact the local poison
   control center for advice. Keep the victim warm and at rest. Get medical attention.

PRODUCT: LYSOL BRAND BATHROOM TOUCH-UPS (DISINFECTANT CLEANING WIPES)

INCOMPATIBILITY: NONE KNOWN

INGREDIENTS: ALKYL DIMETHYL ETHYL
        BENZYL AMMONIUM CHLORIDE .9% TLV: CAS#: 68956-79-6
        ALKYL DIMETHYL BENZYL
        AMMONIUM CHLORIDE   .9% TLV: CAS#: 68391-01-5
        ALKYLOXY POLYETHYLENE
        OXYETHANOL   3.39% TLV: CAS#: 68131-40-8

IGNI TEMP: NA FP: NA LEL: NA UEL: NA VP: NA VD: NA SG: NA PS: LIQUID IN PAPER APPEAR: BLUE TOWEL ODOR:
PH FACTOR: NA

HAZARD RATINGS: H: F: R: 0 1 0

FIRE FIGHTING:
  HIGH EXPANSION FOAM, LOW EXPANSION FOAM, ALCOHOL FOAM; DRY CHEMICAL, CARBON DIOXIDE, WATER FOG.
  PROTECTIVE CLOTHING, RUBBER GLOVES, AND BREATHING APPARATUS.
WARNING: 1] STRUCTURAL PROTECTIVE CLOTHING IS PERMEABLE, REMAIN CLEAR OF SMOKE, WATER FALL OUT AND WATER RUN OFF.
        2] KEEP OUT OF THE REACH OF CHILDREN.
        3] THERMAL DECOMPOSITION MAY YIELD CARBON DIOXIDE, CARBON MONOXIDE AND OTHER UNDETERMINED ORGANIC
          COMPOUNDS.
        4] MOVE CONTAINERS FROM AREA IF WITHOUT RISK, COOL EXPOSED CONTAINERS.
        5] DIKE AREA FOR CONTROL AND CONTAINMENT TO PREVENT ENTRY INTO SEWERS, DRAIN, AND WATER WAYS.

LARGE SPILL/NO FIRE/RESCUE: WEAR RUBBER OR NEOPRENE BOOTS, GLOVES.

SPILL CONTROL AND CONTAINMENT:
  <u>HOUSEHOLD</u> <u>SPILL</u>: PICK UP PACKETS AND PUT BACK IN THE BOX. IF LEAKING PUT IN TRASH CONTAINER.
  <u>LARGE</u> <u>SPILL</u>: SWEEP UP PACKETS AND PUT IN DISPOSAL CONTAINER. COVER LIQUID WITH ABSORBENT. PICK UP AND PUT IN A
  DISPOSAL CONTAINER. SEAL AND REMOVE FROM THE WORK AREA.

HEALTH HAZARD INFORMATION:
  WHEN USING THIS PRODUCT WEAR: NO SPECIAL REQUIREMENT. Eye contact may cause stinging, tearing, itching,
  swelling, and redness.

  A physician should be contacted if anyone develops any signs or symptoms and suspects that they are caused by
  exposure to this product.

FIRST AID:
Eye Exposure:
Flush with water for 15 minutes while lifting the upper and lower eye lids. Contact lenses should not be worn
when working with this product. Get medical attention.

Skin Exposure:
Not expected to be a problem.

Breathing:
Not expected to be a problem.

Swallowing:
Not expected to be a health hazard. Immediately contact the local poison control center for advice. Keep the
victim warm and at rest. Get medical attention.

PRODUCT: LYSOL BRAND CLING THICK LIQUID TOILET BOWEL CLEANER - FRESH & PINE

INCOMPATIBILITY: CHLORINE BLEACH; DRAIN CLEANERS; RUST REMOVERS; AMMONIA AND PRODUCTS CONTAINING AMMONIA.

INGREDIENTS: ALKYL DIMETHYL BENZYL
            AMMONIUM CHLORIDE   .45% TLV: CAS#: 68424-95-0
            OCTYLDECYL, DIOCTYL &
            DIDECYL DIMETHYL
            AMMONIUM CHLORIDE   .67% TLV: CAS#: 8001-54-5

IGNI TEMP: NA FP: 200F LEL: NA UEL: NA VP: NA VD: NA SG: 1.0 PS: LIQUID
   APPEAR: CLEAR BLUE   ODOR: FRESH SCENT   PH FACTOR: 3.4
           CLEAR GREEN ODOR: PINE SCENT

HAZARD RATINGS: H: F: R: 3 0 0

FIRE FIGHTING:
   HIGH EXPANSION FOAM, ALCOHOL FOAM; DRY CHEMICAL, CARBON DIOXIDE, WATER FOG.
   PROTECTIVE CLOTHING, RUBBER GLOVES, AND BREATHING APPARATUS.
WARNING: 1] STRUCTURAL PROTECTIVE CLOTHING IS PERMEABLE, REMAIN CLEAR OF SMOKE, WATER FALL OUT AND WATER RUN OFF.
         2] KEEP OUT OF THE REACH OF CHILDREN.
         3] THERMAL DECOMPOSITION MAY YIELD TOXIC FUMES.
         4] MOVE CONTAINERS FROM AREA IF WITHOUT RISK, COOL EXPOSED CONTAINERS.
         5] DIKE AREA FOR CONTROL AND CONTAINMENT TO PREVENT ENTRY INTO SEWERS, DRAIN, AND WATER WAYS.

LARGE SPILL/NO FIRE/RESCUE: WEAR RUBBER OR NEOPRENE BOOTS, GLOVES.

SPILL CONTROL AND CONTAINMENT:
   HOUSEHOLD SPILL: OPEN WINDOWS. WEAR RUBBER GLOVES. WIPE UP LIQUID WITH AN ABSORBENT CLOTH. RINSE OUT IN THE
   SINK. RINSE SPILL AREA WITH WATER. WIPE UP WITH AN ABSORBENT CLOTH. RINSE OUT IN THE SINK. WIPE UP RESIDUE WITH
   A DAMP CLOTH.
   LARGE SPILL: PICK UP LIQUID WITH ABSORBENT. PUT IN A DISPOSAL CONTAINER. RINSE AREA WITH WATER. PICK UP WITH
   ABSORBENT. PUT IN A DISPOSAL CONTAINER. SEAL AND REMOVE FROM THE WORK AREA.

HEALTH HAZARD INFORMATION:
   WHEN USING THIS PRODUCT WEAR: RUBBER GLOVES. Skin contact may cause moderate to severe skin irritation, redness,
   and swelling. Wash hands after handling. Eye contact may cause severe irritation, burns, and permanent eye
   damage. Ingestion may cause irritation of the mouth, throat, and gastrointestinal tract.

   A physician should be contacted if anyone develops any signs or symptoms and suspects that they are caused by
   exposure to this product.

FIRST AID:
  Eye Exposure:
  Flush with water for 15 minutes while lifting the upper and lower eye lids. Contact lenses should not be worn
  when working with this product. Get medical attention.

  Skin Exposure:
  Rinse off with water. If irritation develops, get medical attention.

  Breathing:
  Not expected to be a problem.

  Swallowing:
  If the victim is conscious, give 2-3 glasses of milk or water to drink. Immediately contact the local poison
  control center for advice. Keep the victim warm and at rest. Get medical attention.

PRODUCT: LYSOL BRAND DISINFECTANT

INCOMPATIBILITY: CHLORINE BLEACH

INGREDIENTS: ETHANOL   1.8-2.0% TLV: 1000 ppm   CAS#:   64-17-5
           ORTHO-BENZYL-p-
           CHLOROPHENOL   2.7% TLV: CAS#:   120-32-1
           ORTHO-PHENYLPHENOL 2.8% TLV: CAS#:   90-43-7

IGNI TEMP: NA FP: 200F LEL: NA UEL: NA VP: NA VD: 1 SG: 1.3 PS: LIQUID APPEAR: CLEAR RED   ODOR:
  PH FACTOR: 10.7

HAZARD RATINGS: H: F: R: 2 0 0

FIRE FIGHTING:
    HIGH EXPANSION FOAM, LOW EXPANSION FOAM, ALCOHOL FOAM; DRY CHEMICAL, CARBON DIOXIDE, WATER FOG.
    PROTECTIVE CLOTHING, RUBBER GLOVES, AND BREATHING APPARATUS.
WARNING: 1] STRUCTURAL PROTECTIVE CLOTHING IS PERMEABLE, REMAIN CLEAR OF SMOKE, WATER FALL OUT AND WATER RUN OFF.
       2] KEEP OUT OF THE REACH OF CHILDREN.
       3] PRODUCT IS PACKAGED IN P.V.C. CONTAINERS.
       4] THERMAL DECOMPOSITION YIELDS TOXIC CARBON DIOXIDE, CARBON MONOXIDE, AND OTHER UNKNOWN ORGANIC
          COMPOUNDS.
       5] MOVE CONTAINERS FROM AREA IF WITHOUT RISK, COOL EXPOSED CONTAINERS.
       6] DIKE AREA FOR CONTROL AND CONTAINMENT TO PREVENT ENTRY INTO SEWERS, DRAIN, AND WATER WAYS.

LARGE SPILL/NO FIRE/RESCUE: WEAR RUBBER OR NEOPRENE BOOTS, GLOVES, BREATHING APPARATUS.

SPILL CONTROL AND CONTAINMENT:
    HOUSEHOLD SPILL: OPEN WINDOWS. WEAR RUBBER GLOVES. WIPE UP WITH AN ABSORBENT CLOTH. RINSE OUT IN THE SINK. WIPE
    UP RESIDUE WITH A DAMP CLOTH, RINSE OUT IN THE SINK.
    LARGE SPILL: PICK UP WITH ABSORBENT, PUT IN A DISPOSAL CONTAINER. RINSE AREA WITH WATER. PICK UP WITH ABSORBENT.
    PUT IN A DISPOSAL CONTAINER, SEAL AND REMOVE FROM THE WORK AREA.

HEALTH HAZARD INFORMATION:
    WHEN USING THIS PRODUCT WEAR: OPEN WINDOWS. RUBBER GLOVES. May cause irritation. Wash hands after use. Eye
    contact may cause severe stinging, tearing, itching, swelling, and redness. Ingestion may cause abdominal pain,
    nausea, vomiting, and diarrhea.

    A physician should be contacted if anyone develops any signs or symptoms and suspects that they are caused by
    exposure to this product.

FIRST AID:
    Eye Exposure:
    Flush with water for 15 minutes while lifting the upper and lower eye lids. Contact lenses should not be worn
    when working with this product. Get medical attention.

    Skin Exposure:
    Wash with soap and water. If irritation develops, get medical attention.

    Breathing:
    Not expected to be a problem. If problem develops move victim to fresh air.

    Swallowing:
    Do not induce vomiting. If victim is conscious, give 2-3 glasses of water to drink. Immediately contact the
    local poison control center for advice. Keep the victim warm and at rest. Get medical attention.

PRODUCT: LYSOL BRAND DISINFECTANT BASIN, TUB AND TILE CLEANER (PUMP SPRAY)

INCOMPATIBILITY: NONE KNOWN

INGREDIENTS: DIETHYLENE GLYCOL
              MONO BUTYL ETHER   6.0%   TLV: CAS#: 112-34-5

IGNI TEMP: NA FP: 200F LEL: NA UEL: NA VP: NA VD: NA SG: 1.02 PS: LIQUID APPEAR: CLEAR STRAW COLOR
     ODOR: PH FACTOR: 12.2

HAZARD RATINGS: H: F: R: 1 0 0

FIRE FIGHTING:
  HIGH EXPANSION FOAM, LOW EXPANSION FOAM, ALCOHOL FOAM; DRY CHEMICAL, CARBON DIOXIDE, WATER FOG.
  PROTECTIVE CLOTHING, RUBBER GLOVES, AND BREATHING APPARATUS.
WARNING: 1] STRUCTURAL PROTECTIVE CLOTHING IS PERMEABLE, REMAIN CLEAR OF SMOKE, WATER FALL OUT AND WATER RUN OFF.
         2] KEEP OUT OF THE REACH OF CHILDREN.
         3] THERMAL DECOMPOSITION MAY YIELD TOXIC FUMES AND VAPORS.
         4] MOVE CONTAINERS FROM AREA IF WITHOUT RISK, COOL EXPOSED CONTAINERS.
         5] DIKE AREA FOR CONTROL AND CONTAINMENT TO PREVENT ENTRY INTO SEWERS, DRAIN, AND WATER WAYS.

LARGE SPILL/NO FIRE/RESCUE: WEAR RUBBER OR NEOPRENE BOOTS, GLOVES.

SPILL CONTROL AND CONTAINMENT:
  HOUSEHOLD SPILL: OPEN WINDOWS. WEAR RUBBER GLOVES. WIPE UP WITH AN ABSORBENT CLOTH. RINSE OUT IN THE SINK. WIPE
  UP RESIDUE WITH A DAMP CLOTH. RINSE OUT IN THE SINK.
  LARGE SPILL: PICK UP WITH ABSORBENT. PUT IN A DISPOSAL CONTAINER. RINSE AREA WITH WATER. PICK UP WITH ABSORBENT.
  PUT IN A DISPOSAL CONTAINER, SEAL AND REMOVE FROM THE WORK AREA.

HEALTH HAZARD INFORMATION:
  WHEN USING THIS PRODUCT WEAR: OPEN WINDOWS. WEAR RUBBER GLOVES. Prolonged skin contact may cause irritation with
  redness or rose colored rash. Wash hands after handling. Eye contact causes stinging, tearing, itching,
  swelling, and redness. Ingestion causes irritation of the gastrointestinal tract with nausea, vomiting, and
  diarrhea.

  A physician should be contacted if anyone develops any signs or symptoms and suspects that they are caused by
  exposure to this product.

FIRST AID:
  Eye Exposure:
  Flush with water for 15 minutes while lifting the upper and lower eye lids. Contact lenses should not be worn
  when working with this product. Get medical attention.

  Skin Exposure:
  Wash with soap and water. If irritation develops, get medical attention.

  Breathing:
  Not expected to be a problem.

  Swallowing:
  If the victim is conscious, give 2-3 glasses of water to drink. Immediately contact the local poison control
  center for advice. Keep the victim warm and at rest. Get medical attention.

PRODUCT: LYSOL BRAND DISINFECTANT SPRAY - COUNTRY

INCOMPATIBILITY: STRONG OXIDIZERS.

INGREDIENTS: CARBON DIOXIDE   4.0% TLV: 5000 ppm   CAS#:   124-38-9
             ETHANOL   79.0% TLV: 1000 ppm   CAS#:   64-17-5
             ORTHO-PHENYLPHENOL   .1% TLV: CAS#:

IGNI TEMP: NA FP: 70F LEL: NA UEL: NA VP: 105 psig VD: 1 SG: .835 PS: LIQUID APPEAR: AEROSOL SPRAY
    ODOR: ORIGINAL   PH FACTOR: 10

HAZARD RATINGS: H: F: R: 1 1 0

FIRE FIGHTING:
  HIGH EXPANSION FOAM, LOW EXPANSION FOAM, ALCOHOL FOAM; DRY CHEMICAL, CARBON DIOXIDE, WATER FOG.
  PROTECTIVE CLOTHING, RUBBER GLOVES, AND BREATHING APPARATUS.
WARNING: 1] STRUCTURAL PROTECTIVE CLOTHING IS PERMEABLE, REMAIN CLEAR OF SMOKE, WATER FALL OUT AND WATER RUN OFF.
         2] KEEP OUT OF THE REACH OF CHILDREN.
         3] THERMAL DECOMPOSITION YIELDS TOXIC CARBON DIOXIDE, CARBON MONOXIDE, AND OTHER UNKNOWN ORGANIC
            COMPOUNDS.
         4] CONTAINERS WILL EXPLODE IN A FIRE OR IF HEATED ABOVE 120F.
         5] MOVE CONTAINERS FROM AREA IF WITHOUT RISK, COOL EXPOSED CONTAINERS.
         6] DIKE AREA FOR CONTROL AND CONTAINMENT TO PREVENT ENTRY INTO SEWERS, DRAIN, AND WATER WAYS.

LARGE SPILL/NO FIRE/RESCUE: WEAR RUBBER OR NEOPRENE BOOTS, GLOVES.

SPILL CONTROL AND CONTAINMENT:
  HOUSEHOLD SPILL: WIPE UP LIQUID WITH AN ABSORBENT CLOTH. RINSE OUT IN THE SINK. WIPE UP RESIDUE WITH A DAMP
  CLOTH. RINSE OUT IN THE SINK.
  LARGE SPILL: PICK UP LIQUID WITH AN ABSORBENT, PUT IN A DISPOSAL CONTAINER. RINSE AREA WITH WATER, PICK UP WITH
  AN ABSORBENT, PUT IN A DISPOSAL CONTAINER. SEAL AND REMOVE FROM THE WORK AREA.

HEALTH HAZARD INFORMATION:
  WHEN USING THIS PRODUCT WEAR: NO SPECIAL REQUIREMENT. Inhalation may cause dizziness, drowsiness, nausea, and
  vomiting. Eye contact may cause stinging, tearing, itching, swelling, and redness. Ingestion is unlikely through
  normal anticipated use. Ingesting small quantities is not expected to cause any significant adverse effects.

  A physician should be contacted if anyone develops any signs or symptoms and suspects that they are caused by
  exposure to this product.

FIRST AID:
Eye Exposure:
Flush with water for 15 minutes while lifting the upper and lower eye lids. Contact lenses should not be worn
when working with this product. Get medical attention.

Skin Exposure:
Wash with soap and water.

Breathing:
If symptoms appear, move the victim to fresh air, treat symptoms. Keep the victim warm and at rest, get medical
attention.

Swallowing:
If the victim is conscious, give 1-2 glasses of water to drink. Immediately contact the local poison control
center for advice. Keep the victim warm and at rest. Get medical attention.

PRODUCT: LYSOL BRAND DISINFECTANT DEODORIZING CLEANER

INCOMPATIBILITY: NONE KNOWN

INGREDIENTS: ALKYL (C12-C16) DIMETHYL
              BENZYL AMMONIUM CHLORIDE 2.7% TLV: CAS#: 68424-85-1

IGNI TEMP: NA FP: 200F LEL: NA UEL: NA VP: NA VD: NA SG: 1.02 PS: LIQUID APPEAR: FLUORESCENT AMBER
    ODOR: PH FACTOR: 8.35

HAZARD RATINGS: H: F: R: 1 0 0

FIRE FIGHTING:
    HIGH EXPANSION FOAM, LOW EXPANSION FOAM, ALCOHOL FOAM; DRY CHEMICAL, CARBON DIOXIDE, WATER FOG.
    PROTECTIVE CLOTHING, RUBBER GLOVES, AND BREATHING APPARATUS.
WARNING: 1] STRUCTURAL PROTECTIVE CLOTHING IS PERMEABLE, REMAIN CLEAR OF SMOKE, WATER FALL OUT AND WATER RUN OFF.
         2] KEEP OUT OF THE REACH OF CHILDREN.
         3] THERMAL DECOMPOSITION YIELDS TOXIC CARBON DIOXIDE, CARBON MONOXIDE, AND OTHER UNKNOWN ORGANIC
            COMPOUNDS.
         4] MOVE CONTAINERS FROM AREA IF WITHOUT RISK, COOL EXPOSED CONTAINERS.
         5] DIKE AREA FOR CONTROL AND CONTAINMENT TO PREVENT ENTRY INTO SEWERS, DRAIN, AND WATER WAYS.

LARGE SPILL/NO FIRE/RESCUE: WEAR RUBBER OR NEOPRENE BOOTS, GLOVES.

SPILL CONTROL AND CONTAINMENT:
    HOUSEHOLD SPILL: OPEN WINDOWS. WIPE UP WITH AN ABSORBENT CLOTH. RINSE OUT IN THE SINK. WIPE UP RESIDUE WITH A
    DAMP CLOTH. WASH HANDS AFTER USE.
    LARGE SPILL: PICK UP WITH ABSORBENT. PUT IN A DISPOSAL CONTAINER. RINSE AREA WITH WATER. PICK UP WITH ABSORBENT.
    PUT IN A DISPOSAL CONTAINER. SEAL AND REMOVE FROM THE WORK AREA.

HEALTH HAZARD INFORMATION:
    WHEN USING THIS PRODUCT WEAR: NO SPECIAL REQUIREMENT. OPEN WINDOWS. Prolonged or repeated skin contact may cause
    irritation. Eye contact will cause stinging, tearing, itching, swelling, and redness. Ingestion will cause
    abdominal pain, nausea, vomiting, and diarrhea.

    A physician should be contacted if anyone develops any signs or symptoms and suspects that they are caused by
    exposure to this product.

FIRST AID:
    Eye Exposure:
    Flush with water for 15 minutes while lifting the upper and lower eye lids. Contact lenses should not be worn
    when working with this product. Get medical attention.

    Skin Exposure:
    Wash hands with soap and water. If irritation develops, get medical attention.

    Breathing:
    Not expected to be a problem.

    Swallowing:
    Do not induce vomiting. If victim is conscious, give 2-3 glasses of water to drink. Immediately contact the
    local poison control center for advice. Keep the victim warm and at rest. Get medical attention.

PRODUCT: LYSOL BRAND DISINFECTANT DIRECT MULTI PURPOSE CLEANER

INCOMPATIBILITY: NONE KNOWN

INGREDIENTS: DIETHYLENE GLYCOL MONO-
                      BUTYL ETHER   5.99% TLV: CAS#: 112-34-5

IGNI TEMP: NA FP: 200F LEL: NA UEL: NA VP: NA VD: NA SG: 1.01 PS: LIQUID
    APPEAR: CLEAR BLUE   ODOR: PH FACTOR: 12.1
                CLEAR VIOLET   LAVENDER SCENT
                CLEAR RED   FLORAL SCENT

HAZARD RATINGS: H: F: R:

FIRE FIGHTING:
   HIGH EXPANSION FOAM, LOW EXPANSION FOAM, ALCOHOL FOAM; DRY CHEMICAL, CARBON DIOXIDE, WATER FOG.
   PROTECTIVE CLOTHING, RUBBER GLOVES, AND BREATHING APPARATUS.
WARNING: 1] STRUCTURAL PROTECTIVE CLOTHING IS PERMEABLE, REMAIN CLEAR OF SMOKE, WATER FALL OUT AND WATER RUN OFF.
              2] KEEP OUT OF THE REACH OF CHILDREN.
              3] THERMAL DECOMPOSITION YIELDS TOXIC CARBON DIOXIDE, CARBON MONOXIDE, AND OTHER UNKNOWN ORGANIC
                  COMPOUNDS.
              4] MOVE CONTAINERS FROM AREA IF WITHOUT RISK, COOL EXPOSED CONTAINERS.
              5] DIKE AREA FOR CONTROL AND CONTAINMENT TO PREVENT ENTRY INTO SEWERS, DRAIN, AND WATER WAYS.

LARGE SPILL/NO FIRE/RESCUE: WEAR RUBBER OR NEOPRENE BOOTS, GLOVES.

SPILL CONTROL AND CONTAINMENT:
   HOUSEHOLD SPILL: WIPE UP WITH AN ABSORBENT CLOTH. RINSE OUT IN THE SINK. WIPE UP RESIDUE WITH A DAMP CLOTH.
   RINSE OUT IN THE SINK.
   LARGE SPILL: PICK UP WITH ABSORBENT. PUT IN A DISPOSAL CONTAINER. RINSE AREA WITH WATER. PICK UP WITH ABSORBENT.
   PUT IN A DISPOSAL CONTAINER. SEAL AND REMOVE FROM THE WORK AREA.

HEALTH HAZARD INFORMATION:
   WHEN USING THIS PRODUCT WEAR: Eye contact may cause stinging, tearing, itching, swelling, and redness. Ingestion
   may cause abdominal pain, nausea, vomiting, and diarrhea.

   A physician should be contacted if anyone develops any signs or symptoms and suspects that they are caused by
   exposure to this product.

FIRST AID:
   Eye Exposure:
   Flush with water for 15 minutes while lifting the upper and lower eye lids. Contact lenses should not be worn
   when working with this product. Get medical attention.

   Skin Exposure:
   Not expected to be a problem Wash hands with soap and water.

   Breathing:
   Not expected to be a problem.

   Swallowing:
   Immediately contact the local poison control center for advice. Keep the victim warm and at rest. Get medical
   attention.

PRODUCT: LYSOL BRAND DISINFECTANT - PINE ACTION

INCOMPATIBILITY: NONE KNOWN

INGREDIENTS: ISOPROPANOL   12.0% TLV: CAS#:   67-63-0

IGNI TEMP: NA FP: 85F LEL: NA UEL: NA VP: NA VD: NA SG: .97 PS: LIQUID APPEAR: CLEAR AMBER
    ODOR: PINE   PH FACTOR: 6.5

HAZARD RATINGS: H: F: R: 2 1 1

FIRE FIGHTING:
  HIGH EXPANSION FOAM, LOW EXPANSION FOAM, ALCOHOL FOAM; DRY CHEMICAL, CARBON DIOXIDE, WATER FOG.
  PROTECTIVE CLOTHING, RUBBER GLOVES, AND BREATHING APPARATUS.
WARNING: 1] STRUCTURAL PROTECTIVE CLOTHING IS PERMEABLE, REMAIN CLEAR OF SMOKE, WATER FALL OUT AND WATER RUN OFF.
         2] KEEP OUT OF THE REACH OF CHILDREN.
         3] REMOVE ALL SOURCES OF IGNITION IF WITHOUT RISK.
         4] CONTAINERS MAY MELT AND LEAK IN A FIRE.
         5] THERMAL DECOMPOSITION YIELDS CARBON DIOXIDE, CARBON MONOXIDE, AND OTHER UNKNOWN ORGANIC COMPOUNDS.
         6] MOVE CONTAINERS FROM AREA IF WITHOUT RISK, COOL EXPOSED CONTAINERS.
         7] DIKE AREA FOR CONTROL AND CONTAINMENT TO PREVENT ENTRY INTO SEWERS, DRAIN, AND WATER WAYS.

LARGE SPILL/NO FIRE/RESCUE: WEAR RUBBER OR NEOPRENE BOOTS, GLOVES.

SPILL CONTROL AND CONTAINMENT:
  HOUSEHOLD SPILL: OPEN WINDOWS. WIPE UP WITH AN ABSORBENT CLOTH. RINSE OUT IN THE SINK. WIPE UP RESIDUE WITH A
  DAMP CLOTH. RINSE OUT IN THE SINK.
  LARGE SPILL: USE NON-SPARKING TOOLS. PICK UP WITH NON-COMBUSTIBLE ABSORBENT. PUT IN A DISPOSAL CONTAINER. RINSE
  AREA WITH WATER, PICK UP WITH ABSORBENT. PUT IN A DISPOSAL CONTAINER. SEAL AND REMOVE FROM THE WORK AREA.

HEALTH HAZARD INFORMATION:
  WHEN USING THIS PRODUCT WEAR: Prolonged contact may cause skin irritation. Eye contact may cause stinging,
  tearing, itching, swelling, and redness. Ingestion may cause abdominal pain, nausea, vomiting, and diarrhea.

    A physician should be contacted if anyone develops any signs or symptoms and suspects that they are caused by
  exposure to this product.

FIRST AID:
  Eye Exposure:
  Flush with water for 15 minutes while lifting the upper and lower eye lids. Contact lenses should not be worn
  when working with this product. Get medical attention.

  Skin Exposure:
  Rinse skin with water. Wash hands after use.

  Breathing:
  If any problems develop, move victim to fresh air.

  Swallowing:
  Do not induce vomiting. If victim is conscious, give 2-3 glasses of milk or water to drink. Immediately contact
  the local poison control center for advice. Keep the victim warm and at rest. Get medical attention.

PRODUCT: LYSOL BRAND DISINFECTANT - PINE SCENT

INCOMPATIBILITY: CHLORINE BLEACH.

INGREDIENTS: HEXYLENE GLYCOL   8.0% TLV: 25 ppm   CAS#: 107-41-5
             ISOPROPANOL   1.5% TLV: 400 ppm   CAS#: 67-63-0
             ORTHO BENZYL-p-
             CHLOROPHENOL   4.5% TLV: CAS#: 120-32-1

IGNI TEMP: NA FP: 200F LEL: NA UEL: NA VP: NA VD: 1 SG: 1.02 PS: LIQUID APPEAR: RED/AMBER
    ODOR: PINE SCENT   PH FACTOR: 10.9

HAZARD RATINGS: H: F: R: 2 0 0

FIRE FIGHTING:
   HIGH EXPANSION FOAM, LOW EXPANSION FOAM, ALCOHOL FOAM; DRY CHEMICAL, CARBON DIOXIDE, WATER FOG.
   PROTECTIVE CLOTHING, RUBBER GLOVES, AND BREATHING APPARATUS.
WARNING: 1] STRUCTURAL PROTECTIVE CLOTHING IS PERMEABLE, REMAIN CLEAR OF SMOKE, WATER FALL OUT AND WATER RUN OFF.
         2] KEEP OUT OF THE REACH OF CHILDREN.
         3] THERMAL DECOMPOSITION YIELDS TOXIC CARBON DIOXIDE, CARBON MONOXIDE, AND OTHER UNKNOWN ORGANIC
            COMPOUNDS.
         4] MOVE CONTAINERS FROM AREA IF WITHOUT RISK, COOL EXPOSED CONTAINERS.
         5] DIKE AREA FOR CONTROL AND CONTAINMENT TO PREVENT ENTRY INTO SEWERS, DRAIN, AND WATER WAYS.

LARGE SPILL/NO FIRE/RESCUE: WEAR RUBBER OR NEOPRENE BOOTS, GLOVES.

SPILL CONTROL AND CONTAINMENT:
   HOUSEHOLD SPILL: WEAR RUBBER GLOVES. WIPE UP WITH AN ABSORBENT CLOTH. RINSE OUT IN THE SINK. WIPE UP RESIDUE
   WITH A DAMP CLOTH, RINSE OUT IN THE SINK.
   LARGE SPILL: PICK UP WITH ABSORBENT. PUT IN A DISPOSAL CONTAINER, RINSE AREA WITH WATER. PICK UP WITH ABSORBENT.
   SEAL AND REMOVE FROM THE WORK AREA.

HEALTH HAZARD INFORMATION:
   WHEN USING THIS PRODUCT WEAR: Skin contact may cause irritation. Wash hands after use. Eye contact may cause
   severe stinging, tearing, itching, swelling, and redness. Ingestion may cause abdominal pain, nausea, vomiting,
   and diarrhea.

   A physician should be contacted if anyone develops any signs or symptoms and suspects that they are caused by
   exposure to this product.

FIRST AID:
   Eye Exposure:
   Flush with water for 15 minutes while lifting the upper and lower eye lids. Contact lenses should not be worn
   when working with this product. Get medical attention.

   Skin Exposure:
   Wash with soap and water. If irritation develops, get medical attention.

   Breathing:
   If irritation develops, move victim to fresh air.

   Swallowing:
   Do not induce vomiting. If victim is conscious give 2-3 glasses of water to drink. Immediately contact the local
   poison control center for advice. Keep the victim warm and at rest. Get medical attention.

PRODUCT: LYSOL BRAND DISINFECTANT SPRAY - FRESH

INCOMPATIBILITY: STRONG OXIDIZERS.

INGREDIENTS: CARBON DIOXIDE   4.0% TLV: 5000 ppm   CAS#:   124-38-9
             ETHANOL   79.0% TLV: 1000 ppm   CAS#:   64-17-5
             ORTHO-PHENYLPHENOL   .1% TLV: CAS#:

IGNI TEMP: NA FP: 70F LEL: NA UEL: NA VP: 105 psig VD: 1 SG: .835 PS: LIQUID   APPEAR: AEROSOL SPRAY
   ODOR: ORIGINAL   PH FACTOR: 10

HAZARD RATINGS: H: F: R: 1 1 0

FIRE FIGHTING:
   HIGH EXPANSION FOAM, LOW EXPANSION FOAM, ALCOHOL FOAM; DRY CHEMICAL, CARBON DIOXIDE, WATER FOG.
   PROTECTIVE CLOTHING, RUBBER GLOVES, AND BREATHING APPARATUS.
WARNING: 1] STRUCTURAL PROTECTIVE CLOTHING IS PERMEABLE, REMAIN CLEAR OF SMOKE, WATER FALL OUT AND WATER RUN OFF.
         2] KEEP OUT OF THE REACH OF CHILDREN.
         3] THERMAL DECOMPOSITION YIELDS TOXIC CARBON DIOXIDE, CARBON MONOXIDE, AND OTHER UNKNOWN ORGANIC
            COMPOUNDS.
         4] CONTAINERS WILL EXPLODE IN A FIRE OR IF HEATED ABOVE 120F.
         5] MOVE CONTAINERS FROM AREA IF WITHOUT RISK, COOL EXPOSED CONTAINERS.
         6] DIKE AREA FOR CONTROL AND CONTAINMENT TO PREVENT ENTRY INTO SEWERS, DRAIN, AND WATER WAYS.

LARGE SPILL/NO FIRE/RESCUE: WEAR RUBBER OR NEOPRENE BOOTS, GLOVES.

SPILL CONTROL AND CONTAINMENT:
   HOUSEHOLD SPILL: WIPE UP LIQUID WITH AN ABSORBENT CLOTH. RINSE OUT IN THE SINK. WIPE UP RESIDUE WITH A DAMP
   CLOTH. RINSE OUT IN THE SINK.
   LARGE SPILL: PICK UP LIQUID WITH AN ABSORBENT, PUT IN A DISPOSAL CONTAINER. RINSE AREA WITH WATER, PICK UP WITH
   AN ABSORBENT, PUT IN A DISPOSAL CONTAINER. SEAL AND REMOVE FROM THE WORK AREA.

HEALTH HAZARD INFORMATION:
   WHEN USING THIS PRODUCT WEAR: NO SPECIAL REQUIREMENT. Inhalation may cause dizziness, drowsiness, nausea, and
   vomiting. Eye contact may cause stinging, tearing, itching, swelling, and redness. Ingestion is unlikely through
   normal anticipated use. Ingesting small quantities is not expected to cause any significant adverse effects.

   A physician should be contacted if anyone develops any signs or symptoms and suspects that they are caused by
   exposure to this product.

FIRST AID:
   Eye Exposure:
   Flush with water for 15 minutes while lifting the upper and lower eye lids. Contact lenses should not be worn
   when working with this product. Get medical attention.

   Skin Exposure:
   Wash with soap and water.

   Breathing:
   If symptoms appear, move the victim to fresh air, treat symptoms. Keep the victim warm and at rest, get medical
   attention.

   Swallowing:
   If the victim is conscious, give 1-2 glasses of water to drink. Immediately contact the local poison control
   center for advice. Keep the victim warm and at rest. Get medical attention.

PRODUCT: LYSOL BRAND DISINFECTANT SPRAY - LIGHT

INCOMPATIBILITY: STRONG OXIDIZERS.

INGREDIENTS: CARBON DIOXIDE   4.0%   TLV: 5000 ppm   CAS#:   124-38-9
ETHANOL   79.0%   TLV: 1000 ppm   CAS#:   64-17-5
ORTHO-PHENYLPHENOL   .1%   TLV: CAS#:

IGNI TEMP: NA   FP: 70F   LEL: NA   UEL: NA   VP: 105 psig   VD: 1   SG: .835   PS: LIQUID   APPEAR: AEROSOL SPRAY
   ODOR: ORIGINAL   PH FACTOR: 10

HAZARD RATINGS: H: F: R: 1   1   0

FIRE FIGHTING:
   HIGH EXPANSION FOAM, LOW EXPANSION FOAM, ALCOHOL FOAM; DRY CHEMICAL, CARBON DIOXIDE, WATER FOG.
   PROTECTIVE CLOTHING, RUBBER GLOVES, AND BREATHING APPARATUS.
WARNING: 1] STRUCTURAL PROTECTIVE CLOTHING IS PERMEABLE, REMAIN CLEAR OF SMOKE, WATER FALL OUT AND WATER RUN OFF.
      2] KEEP OUT OF THE REACH OF CHILDREN.
      3] THERMAL DECOMPOSITION YIELDS TOXIC CARBON DIOXIDE, CARBON MONOXIDE, AND OTHER UNKNOWN ORGANIC
         COMPOUNDS.
      4] CONTAINERS WILL EXPLODE IN A FIRE OR IF HEATED ABOVE 120F.
      5] MOVE CONTAINERS FROM AREA IF WITHOUT RISK, COOL EXPOSED CONTAINERS.
      6] DIKE AREA FOR CONTROL AND CONTAINMENT TO PREVENT ENTRY INTO SEWERS, DRAIN, AND WATER WAYS.

LARGE SPILL/NO FIRE/RESCUE: WEAR RUBBER OR NEOPRENE BOOTS, GLOVES.

SPILL CONTROL AND CONTAINMENT:
   HOUSEHOLD SPILL: WIPE UP LIQUID WITH AN ABSORBENT CLOTH. RINSE OUT IN THE SINK. WIPE UP RESIDUE WITH A DAMP
   CLOTH. RINSE OUT IN THE SINK.
   LARGE SPILL: PICK UP LIQUID WITH AN ABSORBENT, PUT IN A DISPOSAL CONTAINER. RINSE AREA WITH WATER, PICK UP WITH
   AN ABSORBENT, PUT IN A DISPOSAL CONTAINER. SEAL AND REMOVE FROM THE WORK AREA.

HEALTH HAZARD INFORMATION:
   WHEN USING THIS PRODUCT WEAR: NO SPECIAL REQUIREMENT. Inhalation may cause dizziness, drowsiness, nausea, and
   vomiting. Eye contact may cause stinging, tearing, itching, swelling, and redness. Ingestion is unlikely through
   normal anticipated use. Ingesting small quantities is not expected to cause any significant adverse effects.

   A physician should be contacted if anyone develops any signs or symptoms and suspects that they are caused by
   exposure to this product.

FIRST AID:
Eye Exposure:
Flush with water for 15 minutes while lifting the upper and lower eye lids. Contact lenses should not be worn
when working with this product. Get medical attention.

Skin Exposure:
Wash with soap and water.

Breathing:
If symptoms appear, move the victim to fresh air, treat symptoms. Keep the victim warm and at rest, get medical
attention.

Swallowing:
If the victim is conscious, give 1-2 glasses of water to drink. Immediately contact the local poison control
center for advice. Keep the victim warm and at rest. Get medical attention.

PRODUCT: LYSOL BRAND DISINFECTANT SPRAY - ORIGINAL

INCOMPATIBILITY: STRONG OXIDIZERS.

INGREDIENTS: CARBON DIOXIDE   4.0% TLV: 5000 ppm   CAS#:   124-38-9
             ETHANOL   79.0% TLV: 1000 ppm   CAS#:   64-17-5
             ORTHO-PHENYLPHENOL   .1% TLV: CAS#:

IGNI TEMP: NA FP: 70F LEL: NA UEL: NA VP: 105 psig VD: 1 SG: .835 PS: LIQUID APPEAR: AEROSOL SPRAY
   ODOR: ORIGINAL   PH FACTOR: 10

HAZARD RATINGS: H: F: R: 1 1 0

FIRE FIGHTING:
  HIGH EXPANSION FOAM, LOW EXPANSION FOAM, ALCOHOL FOAM; DRY CHEMICAL, CARBON DIOXIDE, WATER FOG.
  PROTECTIVE CLOTHING, RUBBER GLOVES, AND BREATHING APPARATUS.
WARNING: 1] STRUCTURAL PROTECTIVE CLOTHING IS PERMEABLE, REMAIN CLEAR OF SMOKE, WATER FALL OUT AND WATER RUN OFF.
         2] KEEP OUT OF THE REACH OF CHILDREN.
         3] THERMAL DECOMPOSITION YIELDS TOXIC CARBON DIOXIDE, CARBON MONOXIDE, AND OTHER UNKNOWN ORGANIC
            COMPOUNDS.
         4] CONTAINERS WILL EXPLODE IN A FIRE OR IF HEATED ABOVE 120F.
         5] MOVE CONTAINERS FROM AREA IF WITHOUT RISK, COOL EXPOSED CONTAINERS.
         6] DIKE AREA FOR CONTROL AND CONTAINMENT TO PREVENT ENTRY INTO SEWERS, DRAIN, AND WATER WAYS.

LARGE SPILL/NO FIRE/RESCUE: WEAR RUBBER OR NEOPRENE BOOTS, GLOVES.

SPILL CONTROL AND CONTAINMENT:
  HOUSEHOLD SPILL: WIPE UP LIQUID WITH AN ABSORBENT CLOTH. RINSE OUT IN THE SINK. WIPE UP RESIDUE WITH A DAMP
  CLOTH. RINSE OUT IN THE SINK.
  LARGE SPILL: PICK UP LIQUID WITH AN ABSORBENT, PUT IN A DISPOSAL CONTAINER. RINSE AREA WITH WATER, PICK UP WITH
  AN ABSORBENT, PUT IN A DISPOSAL CONTAINER. SEAL AND REMOVE FROM THE WORK AREA.

HEALTH HAZARD INFORMATION:
  WHEN USING THIS PRODUCT WEAR: NO SPECIAL REQUIREMENT. Inhalation may cause dizziness, drowsiness, nausea, and
  vomiting. Eye contact may cause stinging, tearing, itching, swelling, and redness. Ingestion is unlikely through
  normal anticipated use. Ingesting small quantities is not expected to cause any significant adverse effects.

  A physician should be contacted if anyone develops any signs or symptoms and suspects that they are caused by
  exposure to this product.

FIRST AID:
  Eye Exposure:
  Flush with water for 15 minutes while lifting the upper and lower eye lids. Contact lenses should not be worn
  when working with this product. Get medical attention.

  Skin Exposure:
  Wash with soap and water.

  Breathing:
  If symptoms appear, move the victim to fresh air, treat symptoms. Keep the victim warm and at rest, get medical
  attention.

  Swallowing:
  If the victim is conscious, give 1-2 glasses of water to drink. Immediately contact the local poison control
  center for advice. Keep the victim warm and at rest. Get medical attention.

PRODUCT: LYSOL BRAND FOAMING DISINFECTANT BASIN, TUB AND TILE CLEANER II

INCOMPATIBILITY: CHLORINE BLEACH; DRAIN CLEANERS, RUST REMOVERS, AMMONIA AND PRODUCTS CONTAINING AMMONIA.

INGREDIENTS: DIETHYLENE GLYCOL MONO-
                 BUTYL ETHER   5.6% TLV: CAS#: 112-34-5
                 ISOBUTANE   7.0% TLV: CAS#: 75-28-5

IGNI TEMP: NA FP: 20F LEL: NA UEL: NA VP: 34-40 psig @ 70F VD: NA SG: 1.01-1.02   PS: LIQUID   APPEAR: FOAM
      ODOR: PH FACTOR: 12.1

HAZARD RATINGS: H: F: R: 1 4 0

FIRE FIGHTING:
    HIGH EXPANSION FOAM, LOW EXPANSION FOAM, ALCOHOL FOAM; DRY CHEMICAL, CARBON DIOXIDE, WATER FOG.
    PROTECTIVE CLOTHING, RUBBER GLOVES, AND BREATHING APPARATUS.
WARNING: 1] STRUCTURAL PROTECTIVE CLOTHING IS PERMEABLE, REMAIN CLEAR OF SMOKE, WATER FALL OUT AND WATER RUN OFF.
         2] KEEP OUT OF THE REACH OF CHILDREN.
         3] CONTAINER WILL EXPLODE IN A FIRE OR IF HEATED ABOVE 120F.
         4] THERMAL DECOMPOSITION MAY YIELD CARBON DIOXIDE, CARBON MONOXIDE AND OTHER UNKNOWN ORGANIC COMPOUNDS.
         5] MOVE CONTAINERS FROM AREA IF WITHOUT RISK, COOL EXPOSED CONTAINERS.
         6] DIKE AREA FOR CONTROL AND CONTAINMENT TO PREVENT ENTRY INTO SEWERS, DRAIN, AND WATER WAYS.

LARGE SPILL/NO FIRE/RESCUE: WEAR RUBBER OR NEOPRENE BOOTS, GLOVES.

SPILL CONTROL AND CONTAINMENT:
    HOUSEHOLD SPILL: WIPE UP FOAM WITH AN ABSORBENT CLOTH, RINSE OUT IN THE SINK. WIPE UP RESIDUE WITH A DAMP
    CLOTH, RINSE OUT IN THE SINK.
    LARGE SPILL: PICK UP WITH ABSORBENT, PUT IN A DISPOSAL CONTAINER. RINSE AREA WITH WATER, PICK UP WITH ABSORBENT.
    PUT IN A DISPOSAL CONTAINER. SEAL AND REMOVE FROM THE WORK AREA.

HEALTH HAZARD INFORMATION:
    WHEN USING THIS PRODUCT WEAR: NO SPECIAL REQUIREMENT. Prolonged contact may cause irritation of the skin. Eye
    contact may cause stinging, tearing, itching, swelling, and redness. Ingestion may cause nausea, vomiting, and
    diarrhea.

    A physician should be contacted if anyone develops any signs or symptoms and suspects that they are caused by
    exposure to this product.

FIRST AID:
Eye Exposure:
Flush with water for 15 minutes while lifting the upper and lower eye lids. Contact lenses should not be worn
when working with this product. Get medical attention.

    Skin Exposure:
    Rinse off with water.

    Breathing:
    Not expected to be a problem.

    Swallowing:
    If the victim is conscious, give 1-2 glasses of water to drink. Immediately contact the local poison control
    center for advice. Keep the victim warm and at rest. Get medical attention.

PRODUCT: MATCH LIGHT   (CHARCOAL)

INCOMPATIBILITY: NONE KNOWN

INGREDIENTS: MINERAL SPIRITS   5-25%   TLV: 50 ppm   CAS#: 64742-47-8
             CHAR DUST   TLV: 2 mg/m3   CAS#: 16291-96-6
             SAW DUST   .5-5%   TLV: 1 mg/m3   CAS#:

IGNI TEMP: 400F FP: 105F LEL: NA UEL: NA VP: NA VD: NA SG: NA PS: SOLID APPEAR: BLACK CUBES   ODOR:
    PH FACTOR: NA

HAZARD RATINGS: H: F: R: 2 2 0

FIRE FIGHTING:
    HIGH EXPANSION FOAM, LOW EXPANSION FOAM, ALCOHOL FOAM; DRY CHEMICAL, CARBON DIOXIDE, WATER FOG.
    PROTECTIVE CLOTHING, RUBBER GLOVES, AND BREATHING APPARATUS.
WARNING: 1] STRUCTURAL PROTECTIVE CLOTHING IS PERMEABLE, REMAIN CLEAR OF SMOKE, WATER FALL OUT AND WATER RUN OFF.
         2] KEEP OUT OF THE REACH OF CHILDREN.
         3] DO NOT ALLOW STORED CHARCOAL TO BE COME WET. IF WET MOVE AWAY FROM ALL COMBUSTIBLE MATERIAL,
            PREFERABLY TO A SAFE OUTSIDE AREA AND ALLOW TO TOTALLY DRY.
         4] DO NOT USE CHARCOAL FOR COOKING OR HEATING IN A ENCLOSED AREA WITHOUT GOOD OUTSIDE VENTING OF THE
            FUMES.
         5] DO NOT USE STARTER FLUID OR AN ELECTRIC CHARCOAL STARTER.
         6] MOVE CONTAINERS FROM AREA IF WITHOUT RISK, COOL EXPOSED CONTAINERS.
         7] DIKE AREA FOR CONTROL AND CONTAINMENT TO PREVENT ENTRY INTO SEWERS, DRAIN, AND WATER WAYS.

LARGE SPILL/NO FIRE/RESCUE: WEAR RUBBER OR NEOPRENE BOOTS, GLOVES.

SPILL CONTROL AND CONTAINMENT:
    HOUSEHOLD SPILL: PICK UP BRIQUETS, PUT BACK IN THE BAG. GENTLY SWEEP UP RESIDUE AND PUT IN THE BARBECUE. WIPE UP
    REMAINING RESIDUE WITH A DAMP CLOTH.
    LARGE SPILL: AVOID CREATING A DUST. SCOOP UP BRIQUETS AND PUT IN A DISPOSAL CONTAINER. GENTLY SWEEP OR VACUUM UP
    RESIDUE. PUT IN DISPOSAL CONTAINER. RINSE AREA WITH WATER, PICK UP WITH ABSORBENT. PUT IN DISPOSAL CONTAINER.
    SEAL AND REMOVE FROM THE WORK AREA.

HEALTH HAZARD INFORMATION:
    WHEN USING THIS PRODUCT WEAR: NO SPECIAL REQUIREMENT. Dust may cause irritation of the eyes, nose, and throat.
    May possibly be a sensitizer.

    A physician should be contacted if anyone develops any signs or symptoms and suspects that they are caused by
    exposure to this product.

FIRST AID:
    Eye Exposure:
    Flush with water for 15 minutes while lifting the upper and lower eye lids. Contact lenses should not be worn
    when working with this product. Get medical attention.

    Skin Exposure:
    Rinse dust off with water. If irritation develops, get medical attention.

    Breathing:
    If any problem develops, move the victim to fresh air.

    Swallowing:
    Immediately contact the local poison control center for advice. Keep the victim warm and at rest. Get medical
    attention.

PRODUCT: MAXFORCE PHAROAH ANT KILLER

INCOMPATIBILITY: NONE KNOWN.

INGREDIENTS: TETRAHYDRO-5,5-DIMETHYL-2
             (1H)-PYRIMIDINONE   1.0% TLV: 1.4 mg/m3   CAS#: 67485-29-4

IGNI TEMP: NA FP: 200F LEL: NA UEL: NA VP: NA VD: NA SG: 1.44 PS: WAXY SOLID
    APPEAR: PALE YELLOW - YELLOWISH-BROWN   ODOR: NA PH: NA

HAZARD RATINGS: H: F: R:

FIRE FIGHTING:
    HIGH EXPANSION FOAM, LOW EXPANSION FOAM, ALCOHOL FOAM; DRY CHEMICAL, CARBON DIOXIDE, WATER FOG.
    PROTECTIVE CLOTHING, RUBBER GLOVES, AND BREATHING APPARATUS.
WARNING: 1] STRUCTURAL PROTECTIVE CLOTHING IS PERMEABLE, REMAIN CLEAR OF SMOKE, WATER FALL OUT AND WATER RUN OFF.
         2] KEEP OUT OF THE REACH OF CHILDREN.
         3] THERMAL DECOMPOSITION YIELDS TOXIC CARBON DIOXIDE, CARBON MONOXIDE, METHANE, OXIDES OF NITROGEN, AND
            OXIDES OF HYDROGEN FLUORIDE.
         4] MOVE CONTAINERS FROM AREA IF WITHOUT RISK, COOL EXPOSED CONTAINERS.
         5] DIKE AREA FOR CONTROL AND CONTAINMENT TO PREVENT ENTRY INTO SEWERS, DRAIN, AND WATER WAYS.

LARGE SPILL/NO FIRE/RESCUE: WEAR RUBBER OR NEOPRENE BOOTS, GLOVES, BREATHING APPARATUS.

SPILL CONTROL AND CONTAINMENT:
    HOUSEHOLD SPILL: WIPE UP WAXY MATERIAL WITH A DRY CLOTH. PUT IN A DISPOSAL CONTAINER. SCRUB AREA WITH SOAP AND
    WATER. PICK UP WITH AN ABSORBENT CLOTH. RINSE OUT IN THE SINK. WASH HANDS WITH SOAP AND WATER.
    LARGE SPILL: SCOOP UP MATERIAL AND PUT IN A DISPOSAL CONTAINER. PICK UP RESIDUE WITH ABSORBENT. USE A STIFF
    BRUSH SWEEP UP AND PUT IN A DISPOSAL CONTAINER. SEAL AND REMOVE FROM THE WORK AREA.

HEALTH HAZARD INFORMATION:
    WHEN USING THIS PRODUCT WEAR: NO SPECIAL REQUIREMENT. Not expected to be a problem by any exposure route.
    Effects resulting from overexposure are not anticipated to occur. Formulation is packaged in a child-resistant
    container.

    A physician should be contacted if anyone develops any signs or symptoms and suspects that they are caused by
    exposure to this product.

FIRST AID:
Eye Exposure:
Flush with water for 15 minutes while lifting the upper and lower eye lids. Contact lenses should not be worn
when working with this product. Get medical attention.

Skin Exposure:
Wash with plenty of soap and water.

Breathing:
Not expected to be a problem.

Swallowing:
Immediately contact the local poison control center for advice. Keep the victim warm and at rest. Get medical
attention.

PRODUCT: MAXFORCE ROACH CONTROL SYSTEM

INCOMPATIBILITY: NONE KNOWN.

INGREDIENTS: TETRAHYDRO-5,5-DIMETHYL-2
             (1H)-PYRIMIDINONE   2.0% TLV: 1.4 mg/m3   CAS#: 67485-29-4

IGNI TEMP: NA FP: 200F LEL: NA UEL: NA VP: NA VD: NA SG: 1.44 PS: WAXY SOLID
   APPEAR: PALE YELLOW - YELLOWISH-BROWN   ODOR: NA PH: NA

HAZARD RATINGS: H: F: R:

FIRE FIGHTING:
  HIGH EXPANSION FOAM, LOW EXPANSION FOAM, ALCOHOL FOAM; DRY CHEMICAL, CARBON DIOXIDE, WATER FOG.
  PROTECTIVE CLOTHING, RUBBER GLOVES, AND BREATHING APPARATUS.
WARNING: 1] STRUCTURAL PROTECTIVE CLOTHING IS PERMEABLE, REMAIN CLEAR OF SMOKE, WATER FALL OUT AND WATER RUN OFF.
         2] KEEP OUT OF THE REACH OF CHILDREN.
         3] THERMAL DECOMPOSITION YIELDS TOXIC CARBON DIOXIDE, CARBON MONOXIDE, METHANE, OXIDES OF NITROGEN, AND
            OXIDES OF HYDROGEN FLUORIDE.
         4] MOVE CONTAINERS FROM AREA IF WITHOUT RISK, COOL EXPOSED CONTAINERS.
         5] DIKE AREA FOR CONTROL AND CONTAINMENT TO PREVENT ENTRY INTO SEWERS, DRAIN, AND WATER WAYS.

LARGE SPILL/NO FIRE/RESCUE: WEAR RUBBER OR NEOPRENE BOOTS, GLOVES, BREATHING APPARATUS.

SPILL CONTROL AND CONTAINMENT:
  HOUSEHOLD SPILL: WIPE UP WAXY MATERIAL WITH A DRY CLOTH. PUT IN A DISPOSAL CONTAINER. SCRUB AREA WITH SOAP AND
  WATER. PICK UP WITH AN ABSORBENT CLOTH. RINSE OUT IN THE SINK. WASH HANDS WITH SOAP AND WATER.
  LARGE SPILL: SCOOP UP MATERIAL AND PUT IN A DISPOSAL CONTAINER. PICK UP RESIDUE WITH ABSORBENT. USE A STIFF
  BRUSH SWEEP UP AND PUT IN A DISPOSAL CONTAINER. SEAL AND REMOVE FROM THE WORK AREA.

HEALTH HAZARD INFORMATION:
  WHEN USING THIS PRODUCT WEAR: NO SPECIAL REQUIREMENT. Not expected to be a problem by any exposure route.
  Effects resulting from overexposure are not anticipated to occur. Formulation is packaged in a child-resistant
  container.

  A physician should be contacted if anyone develops any signs or symptoms and suspects that they are caused by
  exposure to this product.

FIRST AID:
  Eye Exposure:
  Flush with water for 15 minutes while lifting the upper and lower eye lids. Contact lenses should not be worn
  when working with this product. Get medical attention.

  Skin Exposure:
  Wash with plenty of soap and water.

  Breathing:
  Not expected to be a problem.

  Swallowing:
  Immediately contact the local poison control center for advice. Keep the victim warm and at rest. Get medical
  attention.

PRODUCT: MONCHEL TOILET SOAP

INCOMPATIBILITY: NONE KNOWN

INGREDIENTS: TITANIUM DIOXIDE   TLV: CAS#: 1346-67-7
             GLYCERINE   TLV: CAS#:   56-81-5
             SODIUM CHLORIDE   TLV: CAS#: 7647-14-5
             TETRASODIUM EDTA   TLV: CAS#: 10378-23-1
             WATER   TLV: CAS#: 7732-18-5
             SODIUM TALLOWATE, SODIUM COCOATE, COCONUT ACID, FRAGRANCE.

IGNI TEMP: NA FP: NA LEL: NA UEL: NA VP: NA VD: NA SG: 1.08 PS: SOLID APPEAR: WHITE BAR
     ODOR: SCENTED OR UNSCENTED   PH FACTOR: NA

HAZARD RATINGS: H: F: R:

FIRE FIGHTING:
   HIGH EXPANSION FOAM, LOW EXPANSION FOAM, ALCOHOL FOAM; DRY CHEMICAL, CARBON DIOXIDE, WATER FOG.
   PROTECTIVE CLOTHING, RUBBER GLOVES, AND BREATHING APPARATUS.
WARNING: 1] STRUCTURAL PROTECTIVE CLOTHING IS PERMEABLE, REMAIN CLEAR OF SMOKE, WATER FALL OUT AND WATER RUN OFF.
         2] KEEP OUT OF THE REACH OF CHILDREN.
         3] PRODUCT RESIDUE WILL CAUSE A SLIPPING HAZARD.
         4] MOVE CONTAINERS FROM AREA IF WITHOUT RISK, COOL EXPOSED CONTAINERS.
         5] DIKE AREA FOR CONTROL AND CONTAINMENT TO PREVENT ENTRY INTO SEWERS, DRAIN, AND WATER WAYS.

LARGE SPILL/NO FIRE/RESCUE: WEAR RUBBER OR NEOPRENE BOOTS, GLOVES.

SPILL CONTROL AND CONTAINMENT:
   HOUSEHOLD SPILL: PICK UP BAR. WIPE UP RESIDUE WITH AN ABSORBENT CLOTH. RINSE OUT IN THE SINK.
   LARGE SPILL: SCOOP UP BARS AND PUT IN A DISPOSAL CONTAINER. SWEEP UP RESIDUE AND PUT IN A DISPOSAL CONTAINER.
   SCRUB AREA WITH WATER. PICK UP WITH ABSORBENT. PUT IN A DISPOSAL CONTAINER. SEAL AND REMOVE FROM THE WORK AREA.

HEALTH HAZARD INFORMATION:
   WHEN USING THIS PRODUCT WEAR: NO SPECIAL REQUIREMENT. Eye contact causes stinging, tearing, itching, swelling,
   and redness. Ingestion will cause mild gastrointestinal irritation with nausea, vomiting, and diarrhea. Use on
   irritated or extremely dry skin may aggravate it.

   A physician should be contacted if anyone develops any signs or symptoms and suspects that they are caused by
   exposure to this product.

FIRST AID:
   Eye Exposure:
 Flush with water for 15 minutes while lifting the upper and lower eye lids. Contact lenses should not be worn
 when working with this product. Get medical attention.

   Skin Exposure:
   Not expected to be a problem, If irritation develops, do not use. If persists get medical attention.

   Breathing:
   Not expected to be a problem.

   Swallowing:
   If the victim is conscious, give 2-3 glasses of water to drink. Immediately contact the local poison control
   center for advice. Keep the victim warm and at rest. Get medical attention.

PRODUCT: MR. CLEAN - LIQUID SYNTHETIC DETERGENT (NON-PHOSPHATE)

INCOMPATIBILITY: NONE KNOWN.

INGREDIENTS: ANIONIC SURFACTANT   TLV: CAS#:
             BUTYL DIGLYCOL   TLV: CAS#:
             SODIUM CITRATE   TLV: CAS#: 6132-04-3
             SODIUM CARBONATE   TLV: CAS#: 497-19-8
             COLORANT, PERFUME.

IGNI TEMP: NA FP: 200F LEL: NA UEL: NA VP: NA VD: NA SG: 1.04-1.06 PS: LIQUID   APPEAR: CLEAR YELLOW
     ODOR: LEMON   PH FACTOR: NA

HAZARD RATINGS: H: F: R:

FIRE FIGHTING:
   HIGH EXPANSION FOAM, LOW EXPANSION FOAM, ALCOHOL FOAM; DRY CHEMICAL, CARBON DIOXIDE, WATER FOG.
   PROTECTIVE CLOTHING, RUBBER GLOVES, AND BREATHING APPARATUS.
WARNING: 1] STRUCTURAL PROTECTIVE CLOTHING IS PERMEABLE, REMAIN CLEAR OF SMOKE, WATER FALL OUT AND WATER RUN OFF.
         2] KEEP OUT OF THE REACH OF CHILDREN.
         3] THERMAL DECOMPOSITION MAY YIELD TOXIC FUMES AND VAPORS.
         4] MOVE CONTAINERS FROM AREA IF WITHOUT RISK, COOL EXPOSED CONTAINERS.
         5] DIKE AREA FOR CONTROL AND CONTAINMENT TO PREVENT ENTRY INTO SEWERS, DRAIN, AND WATER WAYS.

LARGE SPILL/NO FIRE/RESCUE: WEAR RUBBER OR NEOPRENE BOOTS, GLOVES.

SPILL CONTROL AND CONTAINMENT:
   HOUSEHOLD SPILL: WIPE UP WITH AN ABSORBENT CLOTH. RINSE OUT IN THE SINK. WIPE UP RESIDUE WITH A DAMP CLOTH,
   RINSE OUT IN THE SINK.
   LARGE SPILL: PICK UP WITH ABSORBENT. PUT IN A DISPOSAL CONTAINER. RINSE AREA WITH WATER. PICK UP WITH ABSORBENT.
   PUT IN A DISPOSAL CONTAINER. SEAL AND REMOVE FROM THE WORK AREA.

HEALTH HAZARD INFORMATION:
   WHEN USING THIS PRODUCT WEAR: NO SPECIAL REQUIREMENT. Eye contact will cause stinging, tearing, itching,
   swelling, and redness. Prolonged contact may cause drying of the skin and irritation. Ingestion will cause
   abdominal pain, nausea, vomiting, and diarrhea.

   A physician should be contacted if anyone develops any signs or symptoms and suspects that they are caused by
   exposure to this product.

FIRST AID:
Eye Exposure:
Flush with water for 15 minutes while lifting the upper and lower eye lids. Contact lenses should not be worn
when working with this product. Get medical attention.

   Skin Exposure:
   Wash off with water. If irritation develops, get medical attention.

   Breathing:
   Not expected to be a problem.

   Swallowing:
   If the victim is conscious, give 2-3 glasses of water to drink. Immediately contact the local poison control
   center for advice. Keep the victim warm and at rest. Get medical attention.

PRODUCT: MR. CLEAN SOFT CLEANER - LIQUID ABRASIVE CLEANSER

INCOMPATIBILITY: NONE KNOWN.

INGREDIENTS: CALCIUM CARBONATE   TLV: CAS#: 471-34-1
             SURFACTANTS   TLV: CAS#:
             SOLVENT   TLV: CAS#:
             PERFUME   TLV: CAS#:
             COLORANTS   TLV: CAS#:

IGNI TEMP: NA FP: 200F LEL: NA UEL: NA VP: NA VD: NA SG: 1.28 PS: LIQUID APPEAR: LIGHT YELLOW
     ODOR: CITRUS   PH FACTOR: NA

HAZARD RATINGS: H: F: R:

FIRE FIGHTING:
  HIGH EXPANSION FOAM, LOW EXPANSION FOAM, ALCOHOL FOAM; DRY CHEMICAL, CARBON DIOXIDE, WATER FOG.
  PROTECTIVE CLOTHING, RUBBER GLOVES, AND BREATHING APPARATUS.
WARNING: 1] STRUCTURAL PROTECTIVE CLOTHING IS PERMEABLE, REMAIN CLEAR OF SMOKE, WATER FALL OUT AND WATER RUN OFF.
         2] KEEP OUT OF THE REACH OF CHILDREN.
         3] THERMAL DECOMPOSITION MAY YIELD TOXIC FUMES AND VAPORS.
         4] MOVE CONTAINERS FROM AREA IF WITHOUT RISK, COOL EXPOSED CONTAINERS.
         5] DIKE AREA FOR CONTROL AND CONTAINMENT TO PREVENT ENTRY INTO SEWERS, DRAIN, AND WATER WAYS.

LARGE SPILL/NO FIRE/RESCUE: WEAR RUBBER OR NEOPRENE BOOTS, GLOVES.

SPILL CONTROL AND CONTAINMENT:
  HOUSEHOLD SPILL: WIPE UP WITH AN ABSORBENT CLOTH. RINSE OUT IN THE SINK. WIPE UP RESIDUE WITH A DAMP CLOTH.
  RINSE OUT IN THE SINK.
  LARGE SPILL: PICK UP WITH ABSORBENT. PUT IN A DISPOSAL CONTAINER. RINSE AREA WITH WATER, PICK UP WITH ABSORBENT.
  PUT IN A DISPOSAL CONTAINER. SEAL AND REMOVE FROM THE WORK AREA.

HEALTH HAZARD INFORMATION:
  WHEN USING THIS PRODUCT WEAR: NO SPECIAL REQUIREMENT. Eye contact will cause stinging, tearing, itching,
  swelling, and redness. Prolonged contact will cause drying of the skin and irritation. Ingestion will cause
  abdominal pain, nausea, vomiting, and diarrhea.

  A physician should be contacted if anyone develops any signs or symptoms and suspects that they are caused by
  exposure to this product.

FIRST AID:
  Eye Exposure:
  Flush with water for 15 minutes while lifting the upper and lower eye lids. Contact lenses should not be worn
  when working with this product. Get medical attention.

  Skin Exposure:
  Wash off with water. If irritation develops, get medical attention.

  Breathing:
  Not expected to be a problem.

  Swallowing:
  If the victim is conscious, give 2-3 glasses of water to drink. Immediately contact the local poison control
  center for advice. Keep the victim warm and at rest. Get medical attention.

PRODUCT: MR. CLEAN SPRAY - ALL PURPOSE CLEANER

INCOMPATIBILITY: NONE KNOWN.

INGREDIENTS: SODIUM CITRATE   TLV: CAS#: 6132-04-3
             SODIUM CARBONATE   TLV: CAS#: 497-19-8
             WATER   TLV: CAS#: 7732-18-5
             ANIONIC SURFACTANT   TLV: CAS#:
             BUTYL DIGLYCOL   TLV: CAS#:
             PERFUME   TLV: CAS#:

IGNI TEMP: NA FP: 200F LEL: NA UEL: NA VP: NA VD: NA SG: 1.03 PS: LIQUID APPEAR: COLORLESS ODOR: LEMON
  PH FACTOR: NA

HAZARD RATINGS: H: F: R:

FIRE FIGHTING:
  HIGH EXPANSION FOAM, LOW EXPANSION FOAM, ALCOHOL FOAM; DRY CHEMICAL, CARBON DIOXIDE, WATER FOG.
  PROTECTIVE CLOTHING, RUBBER GLOVES, AND BREATHING APPARATUS.
WARNING: 1] STRUCTURAL PROTECTIVE CLOTHING IS PERMEABLE, REMAIN CLEAR OF SMOKE, WATER FALL OUT AND WATER RUN OFF.
         2] KEEP OUT OF THE REACH OF CHILDREN.
         3] THERMAL DECOMPOSITION MAY YIELD TOXIC FUMES AND VAPORS.
         4] MOVE CONTAINERS FROM AREA IF WITHOUT RISK, COOL EXPOSED CONTAINERS.
         5] DIKE AREA FOR CONTROL AND CONTAINMENT TO PREVENT ENTRY INTO SEWERS, DRAIN, AND WATER WAYS.

LARGE SPILL/NO FIRE/RESCUE: WEAR RUBBER OR NEOPRENE BOOTS, GLOVES.

SPILL CONTROL AND CONTAINMENT:
  HOUSEHOLD SPILL: WIPE UP WITH AN ABSORBENT CLOTH. RINSE OUT IN THE SINK. WIPE UP RESIDUE WITH A DAMP CLOTH.
  RINSE OUT IN THE SINK.
  LARGE SPILL: PICK UP WITH ABSORBENT. PUT IN A DISPOSAL CONTAINER. RINSE AREA WITH WATER. PUT UP WITH ABSORBENT.
  PUT IN A DISPOSAL CONTAINER. SEAL AND REMOVE FROM THE WORK AREA.

HEALTH HAZARD INFORMATION:
  WHEN USING THIS PRODUCT WEAR: NO SPECIAL REQUIREMENT. Eye contact will cause stinging, tearing, itching,
  swelling, and redness. Ingestion may cause abdominal pain, nausea, vomiting, and diarrhea.

  A physician should be contacted if anyone develops any signs or symptoms and suspects that they are caused by
  exposure to this product.

FIRST AID:
  Eye Exposure:
  Flush with water for 15 minutes while lifting the upper and lower eye lids. Contact lenses should not be worn
  when working with this product. Get medical attention.

  Skin Exposure:
  Rinse off with water.

  Breathing:
  Not expected to be a problem.

  Swallowing:
  If the victim is conscious, give 2-3 glasses of water to drink. Immediately contact the local poison control
  center for advice. Keep the victim warm and at rest. Get medical attention.

PRODUCT: MR. MUSCLE   OVEN CLEANER

INCOMPATIBILITY: PRODUCTS WITH AMMONIA; TOILET BOWL/DRAIN & HOUSEHOLD CLEANERS; AMMONIA; ACIDS; VINEGAR.

INGREDIENTS: SODIUM HYDROXIDE   under 5% TLV: 2 mg/m3 ceil CAS#: 1310-73-2
             MONOETHANOLAMINE   under 5% TLV: 3 ppm   CAS#: 141-43-5
             ISOBUTANE   under 2% TLV: CAS#: 75-28-5
             BUTANE   under 5% TLV: 800 ppm   CAS#: 106-97-8

IGNI TEMP: NA FP: 20F LEL: NA UEL: NA VP: 17 mmHg VD: 1.0 SG: 1.1 PS: LIQUID   APPEAR: MILKY SPRAY
     ODOR: CITRUS   PH FACTOR: 13.0

HAZARD RATINGS: H: F: R:

FIRE FIGHTING:
   HIGH EXPANSION FOAM, LOW EXPANSION FOAM, ALCOHOL FOAM; DRY CHEMICAL, CARBON DIOXIDE, WATER FOG.
   PROTECTIVE CLOTHING, RUBBER GLOVES, AND BREATHING APPARATUS.
WARNING: 1] STRUCTURAL PROTECTIVE CLOTHING IS PERMEABLE, REMAIN CLEAR OF SMOKE, WATER FALL OUT AND WATER RUN OFF.
         2] KEEP OUT OF THE REACH OF CHILDREN.
         3] CONTAINERS WILL EXPLODE IF A FIRE OR IF HEATED ABOVE 120F.
         4] REMOVE ALL IGNITION SOURCES IF WITHOUT RISK.
         5] FOR A LARGE SPILL VENTILATE ENCLOSED SPACES AND EVACUATE.
         6] MOVE CONTAINERS FROM AREA IF WITHOUT RISK, COOL EXPOSED CONTAINERS.
         7] DIKE AREA FOR CONTROL AND CONTAINMENT TO PREVENT ENTRY INTO SEWERS, DRAIN, AND WATER WAYS.

LARGE SPILL/NO FIRE/RESCUE: WEAR RUBBER OR NEOPRENE BOOTS, GLOVES.

SPILL CONTROL AND CONTAINMENT:
   HOUSEHOLD SPILL: WEAR RUBBER GLOVES. WIPE UP DRY WITH AN ABSORBENT CLOTH. RINSE OUT IN THE SINK. WIPE UP RESIDUE
   WITH A DAMP CLOTH. RINSE OUT IN THE SINK. WASH HANDS WITH SOAP AND WATER.
   LARGE SPILL: PICK UP WITH ABSORBENT. RINSE AREA WITH WATER. PICK UP WITH ABSORBENT. PUT ALL ABSORBENT IN A
   DISPOSAL CONTAINER.

HEALTH HAZARD INFORMATION:
   WHEN USING THIS PRODUCT WEAR: NO SPECIAL REQUIREMENT. Eye contact will cause burning sensation, tearing, and
   tissue damage. Inhalation may cause irritation of the nose, throat, and lungs. Ingestion may cause burning on
   the mouth, throat, and stomach and spontaneous vomiting may occur. Skin contact may cause burns.

   A physician should be contacted if anyone develops any signs or symptoms and suspects that they are caused by
   exposure to this product.

FIRST AID:
   Eye Exposure:
   Flush with water for 15 minutes while lifting the upper and lower eye lids. Contact lenses should not be worn
   when working with this product. Get medical attention.

   Skin Exposure:
   Remove contaminated cloth flush skin with water. If irritation develops, get medical attention.

   Breathing:
   If irritation develops, move victim to fresh air.

   Swallowing:
   Do not induce vomiting. If victim is conscious, rinse out the mouth, give 2-3 glasses of milk or water to drink.
   Immediately contact the local poison control center for advice. Keep the victim warm and at rest. Get medical
   attention.

PRODUCT: MURPHY LAUNDRY AID

INCOMPATIBILITY: STRONG OXIDIZERS; e.g. Chlorine bleach, Swimming pool acid etc.

INGREDIENTS: MORPHOLINE SOAP
             OF VEGETABLE OIL   5-15% TLV: CAS#:
             DIETHANOLAMINE SOAP
             OF VEGETABLE OIL   0-10% TLV: CAS#:
             SODIUM EDTA   0-1% TLV: CAS#:
             WATER   84-95% TLV: CAS#: 7732-18-5

IGNI TEMP: NA FP: 250F LEL: NA UEL: NA VP: NA VD: NA SG: NA PS: LIQUID APPEAR: AMBER ODOR: CITRUS
  PH FACTOR: 9

HAZARD RATINGS: H: F: R: 1 1 0

FIRE FIGHTING:
  HIGH EXPANSION FOAM, LOW EXPANSION FOAM, ALCOHOL FOAM; DRY CHEMICAL, CARBON DIOXIDE, WATER FOG.
  PROTECTIVE CLOTHING, RUBBER GLOVES, AND BREATHING APPARATUS.
WARNING: 1] STRUCTURAL PROTECTIVE CLOTHING IS PERMEABLE, REMAIN CLEAR OF SMOKE, WATER FALL OUT AND WATER RUN OFF.
        2] KEEP OUT OF THE REACH OF CHILDREN.
        3] CAUTION IN THE SPILL AREA. LIQUID IS A SLIPPING HAZARD.
        4] THERMAL DECOMPOSITION MAY YIELD CARBON DIOXIDE, CARBON MONOXIDE.
        5] AVOID PROLONGED CONTACT WITH THE MORE REACTIVE ALLOYS (e.g. ALUMINUM, BRASS, BRONZE, etc.)
        6] MOVE CONTAINERS FROM AREA IF WITHOUT RISK, COOL EXPOSED CONTAINERS.
        7] DIKE AREA FOR CONTROL AND CONTAINMENT TO PREVENT ENTRY INTO SEWERS, DRAIN, AND WATER WAYS.

LARGE SPILL/NO FIRE/RESCUE: WEAR RUBBER OR NEOPRENE BOOTS, GLOVES.

SPILL CONTROL AND CONTAINMENT:
  HOUSEHOLD SPILL: WIPE UP WITH AN ABSORBENT CLOTH. RINSE OUT IN THE SINK. WIPE UP RESIDUE WITH A DAMP CLOTH.
  LARGE SPILL: PICK UP WITH ABSORBENT, PUT IN A DISPOSAL CONTAINER. RINSE AREA WITH WATER, PICK UP WITH ABSORBENT.
  PUT IN A DISPOSAL CONTAINER. SEAL AND REMOVE FROM THE WORK AREA.

HEALTH HAZARD INFORMATION:
  WHEN USING THIS PRODUCT WEAR: NO SPECIAL REQUIREMENT. Eye contact will cause stinging, tearing, itching,
  swelling, and redness.

  A physician should be contacted if anyone develops any signs or symptoms and suspects that they are caused by
  exposure to this product.

FIRST AID:
  Eye Exposure:
  Flush with water for 15 minutes while lifting the upper and lower eye lids. Contact lenses should not be worn
  when working with this product. Get medical attention.

  Skin Exposure:
  Rinse off with water.

  Breathing:
  Not expected to be a problem.

  Swallowing:
  Immediately contact the local poison control center for advice. Keep the victim warm and at rest. Get medical
  attention.

PRODUCT: MURPHY OIL SOAP LIQUID

INCOMPATIBILITY: STRONG OXIDIZERS; e.g. Chlorine bleach, Swimming pool acid etc.

INGREDIENTS: POTASSIUM SOAP

OF VEGETABLE OIL	20-30%	TLV:		CAS#:
TRIETHANOLAMINE SOAP				
OF VEGETABLE OIL	0-5%	TLV:		CAS#:
ETHANOL	0-1%	TLV: 1900 mg/m3		CAS#: 64-17-5
WATER	50-80%	TLV:		CAS#: 7732-18-5
SODIUM EDTA	0-5%			
SURFACTANTS	0-5%	FRAGRANCE	0-1%	

IGNI TEMP: NA FP: 250F LEL: NA UEL: NA VP: NA VD: NA SG: 1 PS: LIQUID APPEAR: AMBER ODOR: CITRUS
   PH FACTOR: 11

HAZARD RATINGS: H: F: R: 1 1 0

FIRE FIGHTING:
  HIGH EXPANSION FOAM, ALCOHOL FOAM; DRY CHEMICAL, CARBON DIOXIDE, WATER FOG.
  PROTECTIVE CLOTHING, RUBBER GLOVES, AND BREATHING APPARATUS.
WARNING: 1] STRUCTURAL PROTECTIVE CLOTHING IS PERMEABLE, REMAIN CLEAR OF SMOKE, WATER FALL OUT AND WATER RUN OFF.
         2] KEEP OUT OF THE REACH OF CHILDREN.
         3] CAUTION IN SPILL AREA. LIQUID IS A SLIPPING HAZARD.
         4] THERMAL DECOMPOSITION MAY YIELD CARBON DIOXIDE, CARBON MONOXIDE.
         5] AVOID PROLONGED CONTACT WITH MORE REACTIVE ALLOYS (e.g. ALUMINUM, BRASS, BRONZE, etc.)
         6] MOVE CONTAINERS FROM AREA IF WITHOUT RISK, COOL EXPOSED CONTAINERS.
         7] DIKE AREA FOR CONTROL AND CONTAINMENT TO PREVENT ENTRY INTO SEWERS, DRAIN, AND WATER WAYS.

LARGE SPILL/NO FIRE/RESCUE: WEAR RUBBER OR NEOPRENE BOOTS, GLOVES.

SPILL CONTROL AND CONTAINMENT:
  HOUSEHOLD SPILL: WIPE UP WITH AN ABSORBENT CLOTH. RINSE OUT IN THE SINK. WIPE UP RESIDUE WITH A DAMP CLOTH,
  RINSE OUT IN THE SINK.
  LARGE SPILL: PICK UP WITH ABSORBENT. PUT IN A DISPOSAL CONTAINER. RINSE AREA WITH WATER. PICK UP WITH ABSORBENT.
  PUT IN A DISPOSAL CONTAINER. SEAL AND REMOVE CONTAINER FROM THE WORK AREA.

HEALTH HAZARD INFORMATION:
  WHEN USING THIS PRODUCT WEAR: NO SPECIAL REQUIREMENT. Eye contact may cause stinging, tearing, itching,
  swelling, and redness. Ingestion may cause abdominal pain, nausea, vomiting, and diarrhea.

  A physician should be contacted if anyone develops any signs or symptoms and suspects that they are caused by
  exposure to this product.

FIRST AID:
  Eye Exposure:
  Flush with water for 15 minutes while lifting the upper and lower eye lids. Contact lenses should not be worn
  when working with this product. Get medical attention.

  Skin Exposure:
  Wash off with water.

  Breathing:
  Not expected to be a problem.

  Swallowing:
  Immediately contact the local poison control center for advice. Keep the victim warm and at rest. Get medical
  attention.

PRODUCT: MURPHY OIL SOAP PASTE

INCOMPATIBILITY: STRONG OXIDIZERS; e.g. Chlorine bleach, Swimming pool acid etc.

INGREDIENTS: POTASSIUM SOAP

OF VEGETABLE OIL	35-45%	TLV:	CAS#:
WATER	40-60%	TLV:	CAS#: 7732-18-5
SODIUM EDTA	0-5%	TLV:	CAS#:
PROPYLENE GLYCOL	0-5%	TLV:	CAS#: 4254-15-3
SURFACTANT	0-5%	TLV:	CAS#:
FRAGRANCE	0-5%	TLV:	CAS#:

IGNI TEMP: NA FP: 250F LEL: NA UEL: NA VP: NA VD: NA SG: 1 PS: PASTE APPEAR: AMBER ODOR: CITRUS
   PH FACTOR: 11

HAZARD RATINGS: H: F: R: 1 1 0

FIRE FIGHTING:
   HIGH EXPANSION FOAM, LOW EXPANSION FOAM, ALCOHOL FOAM; DRY CHEMICAL, CARBON DIOXIDE, WATER FOG.
   PROTECTIVE CLOTHING, RUBBER GLOVES, AND BREATHING APPARATUS.
WARNING: 1] STRUCTURAL PROTECTIVE CLOTHING IS PERMEABLE, REMAIN CLEAR OF SMOKE, WATER FALL OUT AND WATER RUN OFF.
         2] KEEP OUT OF THE REACH OF CHILDREN.
         3] CAUTION IN THE SPILL AREA. PASTE IS A SLIPPING HAZARD.
         4] THERMAL DECOMPOSITION MAY YIELD CARBON DIOXIDE, CARBON MONOXIDE.
         5] AVOID PROLONGED CONTACT WITH MORE REACTIVE ALLOYS (e.g. ALUMINUM, BRASS, BRONZE, etc.)
         6] MOVE CONTAINERS FROM AREA IF WITHOUT RISK, COOL EXPOSED CONTAINERS.
         7] DIKE AREA FOR CONTROL AND CONTAINMENT TO PREVENT ENTRY INTO SEWERS, DRAIN, AND WATER WAYS.

LARGE SPILL/NO FIRE/RESCUE: WEAR RUBBER OR NEOPRENE BOOTS, GLOVES.

SPILL CONTROL AND CONTAINMENT:
   HOUSEHOLD SPILL: SCRAPE UP AND PUT BACK IN THE CONTAINER. WIPE UP RESIDUE WITH A DAMP CLOTH. RINSE OUT IN THE
   SINK.
   LARGE SPILL: SCOOP UP PASTE AND PUT IN A DISPOSAL CONTAINER. RINSE AREA WITH WATER. PICK UP WITH ABSORBENT. PUT
   IN A DISPOSAL CONTAINER. SEAL AND REMOVE FROM THE WORK AREA.

HEALTH HAZARD INFORMATION:
   WHEN USING THIS PRODUCT WEAR: NO SPECIAL REQUIREMENT. Eye contact may cause stinging, tearing, itching,
   swelling, and redness. Ingestion may cause abdominal pain, nausea, vomiting, and diarrhea.

   A physician should be contacted if anyone develops any signs or symptoms and suspects that they are caused by
   exposure to this product.

FIRST AID:
   Eye Exposure:
   Flush with water for 15 minutes while lifting the upper and lower eye lids. Contact lenses should not be worn
   when working with this product. Get medical attention.

   Skin Exposure:
   Wash off with water.

   Breathing:
   Not expected to be a problem.

   Swallowing:
   Immediately contact the local poison control center for advice. Keep the victim warm and at rest. Get medical
   attention.

PRODUCT: MURPHY OIL SOAP SPRAY

INCOMPATIBILITY: STRONG OXIDIZERS; e.g. Chlorine bleach, Swimming pool acid etc.

INGREDIENTS: TRIETHANOLAMINE   5% TLV: CAS#: 102-71-6
             ETHANOL   2% TLV: CAS#:   64-17-5
             WATER   85% TLV: CAS#: 7732-18-5
             SODIUM EDTA   1% TLV: CAS#:
             SURFACTANT   1% TLV: CAS#:
             DEFOAMER   1%
             FRAGRANCE   1%
             POTASSIUM SOAP OF VEGETABLE OIL   5%

IGNI TEMP: NA FP: 200F LEL: NA UEL: NA VP: NA VD: NA SG: 1 PS: LIQUID APPEAR: STRAW COLORED
      ODOR: CITRUS   PH FACTOR: 9

HAZARD RATINGS: H: F: R: 0 1 0

FIRE FIGHTING:
   HIGH EXPANSION FOAM, LOW EXPANSION FOAM, ALCOHOL FOAM; DRY CHEMICAL, CARBON DIOXIDE, WATER FOG.
   PROTECTIVE CLOTHING, RUBBER GLOVES, AND BREATHING APPARATUS.
WARNING: 1] STRUCTURAL PROTECTIVE CLOTHING IS PERMEABLE, REMAIN CLEAR OF SMOKE, WATER FALL OUT AND WATER RUN OFF.
         2] KEEP OUT OF THE REACH OF CHILDREN.
         3] CAUTION IN SPILL AREA. LIQUID IS A SLIPPING HAZARD.
         4] THERMAL DECOMPOSITION MAY YIELD CARBON DIOXIDE, CARBON MONOXIDE.
         5] MOVE CONTAINERS FROM AREA IF WITHOUT RISK, COOL EXPOSED CONTAINERS.
         6] DIKE AREA FOR CONTROL AND CONTAINMENT TO PREVENT ENTRY INTO SEWERS, DRAIN, AND WATER WAYS.

LARGE SPILL/NO FIRE/RESCUE: WEAR RUBBER OR NEOPRENE BOOTS, GLOVES.

SPILL CONTROL AND CONTAINMENT:
   HOUSEHOLD SPILL: WIPE UP WITH AN ABSORBENT CLOTH. RINSE OUT IN THE SINK. WIPE UP RESIDUE WITH A DAMP CLOTH.
   RINSE OUT IN THE SINK.
   LARGE SPILL: PICK UP WITH ABSORBENT. PUT IN A DISPOSAL CONTAINER. RINSE AREA WITH WATER. PICK UP WITH ABSORBENT.
   PUT IN A DISPOSAL CONTAINER. SEAL AND REMOVE FROM THE WORK AREA

HEALTH HAZARD INFORMATION:
   WHEN USING THIS PRODUCT WEAR: NO SPECIAL REQUIREMENT. Eye contact may cause stinging, tearing, itching,
   swelling, and redness. Ingestion may cause abdominal pain, nausea, vomiting, and diarrhea.

   A physician should be contacted if anyone develops any signs or symptoms and suspects that they are caused by
   exposure to this product.

FIRST AID:
Eye Exposure:
Flush with water for 15 minutes while lifting the upper and lower eye lids. Contact lenses should not be worn
when working with this product. Get medical attention.

Skin Exposure:
Wash off with water.

Breathing:
Not expected to be a problem.

Swallowing:
Immediately contact the local poison control center for advice. Keep the victim warm and at rest. Get medical
attention.

PRODUCT: OFF (AEROSOL)

INCOMPATIBILITY: NONE KNOWN

INGREDIENTS: N,N-DIETHYL-meta-

TOLUAMIDE	14.25%	TLV:	CAS#: 134-62-3
ETHYL ALCOHOL	70-80.00%	TLV: 1000 ppm	CAS#: 64-17-5
PROPANE		TLV: 1000 ppm	CAS#: 74-98-6
ISOBUTANE_____	10-20.00%	TLV:	CAS#: 75-28-5
N-BUTANE____/		TLV: 800 ppm	CAS#: 106-97-8

IGNI TEMP: NA FP: 20F LEL: NA UEL: NA VP: NA VD: NA SG: .78 PS: LIQUID APPEAR: SPRAY   ODOR: PERFUME
  PH FACTOR: NA

HAZARD RATINGS: H: F: R: 1 4 0

FIRE FIGHTING:
  HIGH EXPANSION FOAM, LOW EXPANSION FOAM, ALCOHOL FOAM; DRY CHEMICAL, CARBON DIOXIDE, WATER FOG.
  PROTECTIVE CLOTHING, RUBBER GLOVES, AND BREATHING APPARATUS.
WARNING: 1] STRUCTURAL PROTECTIVE CLOTHING IS PERMEABLE, REMAIN CLEAR OF SMOKE, WATER FALL OUT AND WATER RUN OFF.
  2] KEEP OUT OF THE REACH OF CHILDREN.
  3] REMOVE ALL SOURCES OF IGNITION IF WITHOUT RISK.
  4] DO NOT TREAT HANDS OF CHILDREN TO PREVENT MOUTH AND EYE CONTACT.
  5] DO NOT APPLY TO EXCESSIVELY SUNBURNED OR DAMAGED SKIN.
  6] CONTAINERS WILL EXPLODE IN A FIRE OR IF HEATED ABOVE 120F.
  7] THERMAL DECOMPOSITION YIELDS CARBON DIOXIDE, CARBON MONOXIDE.
  8] MOVE CONTAINERS FROM AREA IF WITHOUT RISK, COOL EXPOSED CONTAINERS.
  9] DIKE AREA FOR CONTROL AND CONTAINMENT TO PREVENT ENTRY INTO SEWERS, DRAIN, AND WATER WAYS.

LARGE SPILL/NO FIRE/RESCUE: WEAR RUBBER OR NEOPRENE BOOTS, GLOVES.

SPILL CONTROL AND CONTAINMENT:
  HOUSEHOLD SPILL: WIPE UP WITH AN ABSORBENT CLOTH. RINSE OUT IN THE SINK. WIPE UP RESIDUE WITH A DAMP CLOTH,
  RINSE OUT IN THE SINK.
  LARGE SPILL: PICK UP WITH ABSORBENT. PUT IN A DISPOSAL CONTAINER. RINSE AREA WITH WATER. PICK UP WITH ABSORBENT.
  PUT IN A DISPOSAL CONTAINER. SEAL AND REMOVE FROM WORK AREA.

HEALTH HAZARD INFORMATION:
  WHEN USING THIS PRODUCT WEAR: NO SPECIAL REQUIREMENT. Eye contact causes stinging, tearing, itching, swelling,
  and redness. Ingestion may cause abdominal pain, nausea, vomiting, diarrhea, and illness.

A physician should be contacted if anyone develops any signs or symptoms and suspects that they are caused by
exposure to this product.

FIRST AID:
Eye Exposure:
Flush with water for 15 minutes while lifting the upper and lower eye lids. Contact lenses should not be worn
when working with this product. Get medical attention.

  Skin Exposure:
  Not expected to be a problem.

  Breathing:
  Not expected to be a problem.

  Swallowing:
  Immediately contact the local poison control center for advice. Keep the victim warm and at rest. Get medical
  attention.

```
PRODUCT: OFF   (TOWELETTE )

INCOMPATIBILITY: NONE KNOWN

INGREDIENTS: N,N-DIETHYL-meta-
             TOLUAMIDE          23.75%   TLV:           CAS#: 134-62-3
             OTHER ISOMERS       1.25%   TLV:           CAS#:
             ETHYL ALCOHOL      25-35%   TLV: 1000 ppm  CAS#:  64-17-5
             PAPER TOWEL                 TLV:           CAS#:

IGNI TEMP: NA  FP: 76F  LEL: NA  UEL: NA  VP: NA  VD: NA  SG: 1  PS: LIQUID/PAPER  APPEAR: PAPER TOWEL
     ODOR: PERFUME      PH FACTOR: NA

HAZARD RATINGS:    H:  F:  R:     0  3  0
```

FIRE FIGHTING:
  HIGH EXPANSION FOAM, LOW EXPANSION FOAM, ALCOHOL FOAM; DRY CHEMICAL, CARBON DIOXIDE, WATER FOG.
  PROTECTIVE CLOTHING, RUBBER GLOVES, AND BREATHING APPARATUS.
WARNING: 1] STRUCTURAL PROTECTIVE CLOTHING IS PERMEABLE, REMAIN CLEAR OF SMOKE, WATER FALL OUT AND WATER RUN OFF.
         2] KEEP OUT OF THE REACH OF CHILDREN.
         3] THERMAL DECOMPOSITION MAY YIELD CARBON DIOXIDE, CARBON MONOXIDE.
         4] REMOVE ALL SOURCES OF IGNITION IF WITHOUT RISK.
         5] MOVE CONTAINERS FROM AREA IF WITHOUT RISK, COOL EXPOSED CONTAINERS.
         6] DIKE AREA FOR CONTROL AND CONTAINMENT TO PREVENT ENTRY INTO SEWERS, DRAIN, AND WATER WAYS.

LARGE SPILL/NO FIRE/RESCUE: WEAR RUBBER OR NEOPRENE BOOTS, GLOVES.

SPILL CONTROL AND CONTAINMENT:
  HOUSEHOLD SPILL: PICK UP TOWELETTES PACKS. PUT BACK IN BOX. IF LEAKING WIPE UP LIQUID WITH AN ABSORBENT CLOTH.
  RINSE OUT IN THE SINK. PUT LEAKING PACKS INTO DISPOSAL CONTAINER.
  LARGE SPILL: SWEEP UP PACKS, PUT IN A DISPOSAL CONTAINER. PIT UP LIQUID WITH ABSORBENT. PUT INTO A DISPOSAL
  CONTAINER. SEAL AND REMOVE FROM THE WORK AREA.

HEALTH HAZARD INFORMATION:
  WHEN USING THIS PRODUCT WEAR: NO SPECIAL REQUIREMENTS. Eye contact causes stinging, tearing, itching, swelling,
  and redness. Ingestion may cause abdominal pain, nausea, vomiting, and diarrhea.

  A physician should be contacted if anyone develops any signs or symptoms and suspects that they are caused by
  exposure to this product.

FIRST AID:
Eye Exposure:
Flush with water for 15 minutes while lifting the upper and lower eye lids. Contact lenses should not be worn
when working with this product. Get medical attention.

Skin Exposure:
Not expected to be a problem.

Breathing:
Not expected to be a problem.

Swallowing:
Immediately contact the local poison control center for advice. Keep the victim warm and at rest. Get medical
attention.

PRODUCT: OFF, PUMP SPRAY

INCOMPATIBILITY: NONE KNOWN

INGREDIENTS: N,N-DIETHYL-m-TOLUAMIDE   14.25% TLV: CAS#: 134-62-3
             OTHER ISOMERS   .75% TLV: CAS#:
             ETHYL ALCOHOL   75-85% TLV: CAS#: 64-17-5

IGNI TEMP: NA FP: 55F LEL: NA UEL: NA VP: NA VD: NA SG: .82 PS: LIQUID APPEAR: PALE YELLOW
   ODOR: PERFUME   PH FACTOR: NA

HAZARD RATINGS: H: F: R:

FIRE FIGHTING:
   HIGH EXPANSION FOAM, LOW EXPANSION FOAM, ALCOHOL FOAM; DRY CHEMICAL, CARBON DIOXIDE, WATER FOG.
   PROTECTIVE CLOTHING, RUBBER GLOVES, AND BREATHING APPARATUS.
WARNING: 1] STRUCTURAL PROTECTIVE CLOTHING IS PERMEABLE, REMAIN CLEAR OF SMOKE, WATER FALL OUT AND WATER RUN OFF.
         2] KEEP OUT OF THE REACH OF CHILDREN.
         3] REMOVE ALL IGNITION SOURCES IF WITHOUT RISK.
         4] THERMAL DECOMPOSITION MAY YIELD CARBON DIOXIDE, CARBON MONOXIDE.
         5] CONTAINERS WILL MELT IN A FIRE AND LEAK.
         6] MOVE CONTAINERS FROM AREA IF WITHOUT RISK, COOL EXPOSED CONTAINERS.
         7] DIKE AREA FOR CONTROL AND CONTAINMENT TO PREVENT ENTRY INTO SEWERS, DRAIN, AND WATER WAYS.

LARGE SPILL/NO FIRE/RESCUE: WEAR RUBBER OR NEOPRENE BOOTS, GLOVES.

SPILL CONTROL AND CONTAINMENT:
   HOUSEHOLD SPILL: WIPE UP DRY WITH AN ABSORBENT CLOTH. RINSE OUT IN THE SINK. WIPE UP RESIDUE WITH A DAMP CLOTH.
   RINSE OUT IN THE SINK.
   LARGE SPILL: USE NON SPARKING TOOLS. PICK UP WITH ABSORBENT. PUT IN A DISPOSAL CONTAINER. RINSE AREA WITH WATER.
   PICK UP WITH ABSORBENT. PUT IN A DISPOSAL CONTAINER. SEAL AND REMOVE FROM THE WORK AREA.

HEALTH HAZARD INFORMATION:
   WHEN USING THIS PRODUCT WEAR: NO SPECIAL REQUIREMENT. Eye contact will cause stinging, tearing, itching,
   swelling, and redness. Ingestion will cause nausea, vomiting, and illness. Do not treat children's hand to
   prevent eye and mouth contact. Do not apply to excessively sunburned or damaged skin.

   A physician should be contacted if anyone develops any signs or symptoms and suspects that they are caused by
   exposure to this product.

FIRST AID:
   Eye Exposure:
   Flush with water for 15 mi minutes while lifting the upper and lower eye lids. Contact lenses should not be worn
   when working with this product. Get medical attention.

   Skin Exposure:
   Not expected to be a problem.

   Breathing:
   Not expected to be a problem.

   Swallowing:
   Do not induce vomiting. Immediately contact the local poison control center for advice. Keep the victim warm and
   at rest. Get medical attention.

PRODUCT: OFF, DEEP WOODS (AEROSOL)

INCOMPATIBILITY: NONE KNOWN

INGREDIENTS: ETHYL ALCOHOL   60-70.0% TLV: 1000 ppm CAS#: 64-17-5
               PROPANE   TLV: 800 ppm CAS#: 106-97-8
               ISOBUTANE   TLV: CAS#: 72-28-5
               N-BUTANE   TLV: 800 ppm CAS#: 74-98-6
               2,3,4,5-BIS(2-BUTYLENE)
               TETRAHYDRO-2-FURALDEHYDE   1.0% TLV: CAS#: 126-15-8
               N,N-DIETHYL-m-TOLUAMIDE   19.0% TLV: CAS#: 134-62-8
               N-OCTYL BICYCLOHEPTENE DICARBOXYIMIDE   4.0%
               OTHER ISOMERS   1.0%

IGNI TEMP: NA FP: 20F LEL: NA UEL: NA VP: NA VD: NA SG: NA PS: LIQUID APPEAR: CLEAR SPRAY ODOR: PINE
    PH FACTOR: NA

HAZARD RATINGS: H: F: R: 1 4 0

FIRE FIGHTING:
    HIGH EXPANSION FOAM, LOW EXPANSION FOAM, ALCOHOL FOAM, DRY CHEMICAL, CARBON DIOXIDE, WATER FOG.
    PROTECTIVE CLOTHING, RUBBER GLOVES, AND BREATHING APPARATUS.
WARNING: 1] DO NOT SPRAY NEAR OPEN FLAME, HEATED SURFACES OR SPARKS.
         2] KEEP OUT OF REACH OF CHILDREN.
         3] CONTAINER WILL EXPLODE IN A FIRE OR IF HEATED ABOVE 120F.
         4] REMOVE ALL SOURCES OF IGNITION IF WITHOUT RISK.
         5] DO NOT SPRAY INTO FACIAL AREA.
         6] MOVE CONTAINERS FROM FIRE AREA. COOL EXPOSED CONTAINERS.
         7] DIKE AREA FOR CONTROL AND CONTAINMENT TO PREVENT ENTRY INTO SEWERS, DRAINS AND WATER WAYS.

LARGE SPILL/NO FIRE/RESCUE: WEAR RUBBER OR NEOPRENE BOOTS, GLOVES AND BREATHING APPARATUS

SPILL CONTROL AND CONTAINMENT:
    HOUSEHOLD SPILL: ALLOW ALCOHOL TO EVAPORATE. WIPE REMAINING RESIDUE WITH AN ABSORBENT CLOTH. WASH AREA WITH SOAP
    AND WATER. PICK UP WITH ABSORBENT CLOTH.
    LARGE SPILL. COVER WITH ABSORBENT, PICK UP AND PUT IN A DISPOSAL CONTAINER. SCRUB AREA WITH SOAP AND WATER. PICK
    UP WITH ABSORBENT, PLACE IN DISPOSAL CONTAINER. SEAL AND REMOVE FROM WORK AREA.

HEALTH HAZARD INFORMATION:
    Contact with the eyes will cause irritation. Ingestion may cause burns of the mouth, throat, and stomach.
    Excessive inhalation of spray mist may cause irritation of the nasal passage.

    A physician should be contacted if anyone develops any signs of symptoms and suspects that they are caused by
    exposure to this product.

FIRST AID:
    Eye Exposure:
    Immediately flush the eyes with water for at least 15 minutes while lifting the upper and lower eye lids.
    Contact lenses should not be worn when working with this product. Get medical attention.

    Skin Exposure:
    Not expected to be a problem.

    Breathing:
    Treat any signs or symptoms if they appear. Keep victim warm and at rest. Get medical attention.

    Swallowing:
    Contact the local poison control center for advice. Keep victim warm and at rest. Get medical attention.

PRODUCT: OFF, DEEP WOODS (LOTION)

INCOMPATIBILITY: NONE KNOWN

INGREDIENTS: N,N-DIETHYL-m-TOLUAMIDE   28.5% TLV: CAS#: 134-62-3
             2.3;4,5-BIS(2-BUTYLENE)-
             TETRA-HYDRO-2-FURALDEHYDE 1.0% TLV: CAS#: 126-15-8
             WATER   50-65% TLV: CAS#: 7732-18-5
             N-OCTYL-BICYCLOHEPTENE DICARBOXYIMIDE   4.0%
             OTHER ISOMERS   1.5%

IGNI TEMP: NA FP: NA LEL: NA UEL: NA VP: NA VD: NA SG: 1.0 PS: LIQUID APPEAR: OFF WHITE ODOR: PERFUME
   PH FACTOR: NA

HAZARD RATINGS: H: F: R:

FIRE FIGHTING:
   HIGH EXPANSION FOAM, LOW EXPANSION FOAM, ALCOHOL FOAM; DRY CHEMICAL, CARBON DIOXIDE, WATER FOG.
   PROTECTIVE CLOTHING, RUBBER GLOVES, AND BREATHING APPARATUS.
WARNING: 1] STRUCTURAL PROTECTIVE CLOTHING IS PERMEABLE, REMAIN CLEAR OF SMOKE, WATER FALL OUT AND WATER RUN OFF.
         2] KEEP OUT OF THE REACH OF CHILDREN.
         3] SPILL WILL CREATE A SLIPPING HAZARD.
         4] MOVE CONTAINERS FROM AREA IF WITHOUT RISK, COOL EXPOSED CONTAINERS.
         5] DIKE AREA FOR CONTROL AND CONTAINMENT TO PREVENT ENTRY INTO SEWERS, DRAIN, AND WATER WAYS.

LARGE SPILL/NO FIRE/RESCUE: WEAR RUBBER OR NEOPRENE BOOTS, GLOVES.

SPILL CONTROL AND CONTAINMENT:
   HOUSEHOLD SPILL: WIPE UP LIQUID WITH ABSORBENT CLOTH. WASH AREA WITH SOAP AND WATER, PICK UP WITH ABSORBENT
   CLOTH, RINSE OUT IN SINK.
   LARGE SPILL: COVER SPILL WITH ABSORBENT. PICK UP AND PLACE IN A DISPOSAL CONTAINER. SCRUB AREA WITH DETERGENT
   AND WATER. PICK UP WITH ABSORBENT, PUT IN A DISPOSAL CONTAINER. SEAL AND REMOVE FROM WORK AREA.

HEALTH HAZARD INFORMATION:
   WHEN USING THIS PRODUCT WEAR: NO SPECIAL REQUIREMENTS. Do not get into eyes, causes stinging, burning sensation,
   tearing, itching, swelling, and redness. Ingestion may cause abdominal pain, nausea, vomiting, and diarrhea.

   A physician should be contacted if anyone develops any signs or symptoms and suspects that they are caused by
   exposure to this product.

FIRST AID:
Eye Exposure:
Flush eyes with water for 15 minutes while lifting the upper and lower eye lids. Contact lenses should not be
worn when working with this product. If irritation persists, get medical attention.

   Skin Exposure:
   Not expected to be a problem.

   Breathing:
   Not expected to be a problem.

   Swallowing:
   Immediately contact the local poison control center for advice. Keep the victim warm and at rest. Get medical
   attention.

PRODUCT: OFF, DEEP WOOD (PUMP SPRAY)

INCOMPATIBILITY: NONE KNOWN

INGREDIENTS: N,N-DIETHYL-m-TOLUAMIDE   19% TLV: CAS#:   134-62-3
             OTHER ISOMERS   1% TLV: CAS#:
             2,3,4,5-bis(2-BUTYLENE)
             TETRAHYDRO-2-FURALDEHYDE   1% TLV: CAS#:   126-15-8
             ETHYL ALCOHOL   60-80% TLV: 1000 ppm   CAS#:   64-17-5
             N-OCTYL-BICYCLOHEPTENE
             DICARBOXIMIDE   4% TLV: CAS#:

IGNI TEMP: NA FP: 55F LEL: NA UEL: NA VP: NA VD: NA SG: .84 PS: LIQUID APPEAR: PALE YELLOW   ODOR: CITRUS
    PH FACTOR: NA

HAZARD RATINGS: H: F: R:

FIRE FIGHTING:
   HIGH EXPANSION FOAM, LOW EXPANSION FOAM, ALCOHOL FOAM; DRY CHEMICAL, CARBON DIOXIDE, WATER FOG.
   PROTECTIVE CLOTHING, RUBBER GLOVES, AND BREATHING APPARATUS.
WARNING: 1] STRUCTURAL PROTECTIVE CLOTHING IS PERMEABLE, REMAIN CLEAR OF SMOKE, WATER FALL OUT AND WATER RUN OFF.
         2] KEEP OUT OF THE REACH OF CHILDREN.
         3] REMOVE ALL IGNITION SOURCES IF WITHOUT RISK.
         4] THERMAL DECOMPOSITION MAY YIELD CARBON DIOXIDE, CARBON MONOXIDE.
         5] CONTAINERS WILL MELT AND LEAK IN A FIRE.
         6] MOVE CONTAINERS FROM AREA IF WITHOUT RISK, COOL EXPOSED CONTAINERS.
         7] DIKE AREA FOR CONTROL AND CONTAINMENT TO PREVENT ENTRY INTO SEWERS, DRAIN, AND WATER WAYS.

LARGE SPILL/NO FIRE/RESCUE: WEAR RUBBER OR NEOPRENE BOOTS, GLOVES.

SPILL CONTROL AND CONTAINMENT:
   HOUSEHOLD SPILL: WIPE UP DRY WITH AN ABSORBENT CLOTH. RINSE OUT IN THE SINK. WIPE UP RESIDUE WITH A DAMP CLOTH.
   RINSE OUT IN THE SINK.
   LARGE SPILL: USE NON SPARKING TOOLS. PICK UP WITH ABSORBENT. PUT IN A DISPOSAL CONTAINER. RINSE AREA WITH WATER.
   PICK UP WITH ABSORBENT. PUT IN A DISPOSAL CONTAINER. SEAL AND REMOVE FROM THE WORK AREA.

HEALTH HAZARD INFORMATION:
   WHEN USING THIS PRODUCT WEAR: NO SPECIAL REQUIREMENT. Eye contact will cause stinging, tearing, itching,
   swelling, and redness. Ingestion will cause nausea, vomiting, and illness. Do not treat children's hands to
   prevent eye and mouth contact. Do not apply to excessively sunburned or damaged skin.

   A physician should be contacted if anyone develops any signs or symptoms and suspects that they are caused by
   exposure to this product.

FIRST AID:
   Eye Exposure:
   Flush with water for 15 minutes while lifting the upper and lower eye lids. Contact lenses should not be worn
   when working with this product. Get medical attention.

   Skin Exposure:
   Not expected to be a problem.

   Breathing:
   Not expected to be a problem.

   Swallowing:
   Do not induce vomiting. Immediately contact the local poison control center for advice. Keep the victim warm and
   at rest. Get medical attention.

```
PRODUCT: OFF, DEEP WOODS (TOWELETTE)

INCOMPATIBILITY: NONE KNOWN

INGREDIENTS: N,N-DIETHYL-m-TOLUAMIDE  30.69% TLV:           CAS#:   134-62-3
             OTHER ISOMERS           1.63% TLV:           CAS#:
             2.3;4,5-BIS(2-BUTYLENE)-
             TETRA-HYDRO-2-FURALDEHYDE 1.08% TLV:           CAS#:   126-15-8
             N-OCTYL-BICYCLOHEPTENE
               DICARBOXIMIDE          4.31% TLV:           CAS#:
             ETHYL ALCOHOL          20-30% TLV: 1000 ppm  CAS#:   64-17-5
             WATER                  50-65% TLV:           CAS#:   7732-18-5

IGNI TEMP: NA  FP: 76F  LEL: NA  UEL: NA  VP: NA  VD: NA  SG: .9  PS: LIQUID  APPEAR: TOWELETTE  ODOR: ALCOHOL
   PH FACTOR: NA

HAZARD RATINGS:    H:  F:  R:   1  3  0
```

FIRE FIGHTING:
   HIGH EXPANSION FOAM, LOW EXPANSION FOAM, ALCOHOL FOAM; DRY CHEMICAL, CARBON DIOXIDE, WATER FOG.
   PROTECTIVE CLOTHING, RUBBER GLOVES, AND BREATHING APPARATUS.
WARNING: 1] STRUCTURAL PROTECTIVE CLOTHING IS PERMEABLE, REMAIN CLEAR OF SMOKE, WATER FALL OUT AND WATER RUN OFF.
         2] KEEP OUT OF THE REACH OF CHILDREN.
         3] REMOVE ALL SOURCES OF IGNITION.
         4] THERMAL DECOMPOSITION YIELDS TOXIC CARBON DIOXIDE, CARBON MONOXIDE.
         5] MOVE CONTAINERS FROM AREA IF WITHOUT RISK, COOL EXPOSED CONTAINERS.
         6] DIKE AREA FOR CONTROL AND CONTAINMENT TO PREVENT ENTRY INTO SEWERS, DRAIN, AND WATER WAYS.

LARGE SPILL/NO FIRE/RESCUE: WEAR RUBBER OR NEOPRENE BOOTS,GLOVES BREATHING APPARATUS.

SPILL CONTROL AND CONTAINMENT:
   HOUSEHOLD SPILL: WIPE UP LIQUID WITH ABSORBENT CLOTH. WASH AREA WITH SOAP AND WATER, PICK UP WITH ABSORBENT
   CLOTH, RINSE OUT IN SINK.
   LARGE SPILL: COVER SPILL WITH ABSORBENT. PICK UP AND PLACE IN A DISPOSAL CONTAINER. SCRUB AREA WITH DETERGENT
   AND WATER. PICK UP WITH ABSORBENT, PUT IN A DISPOSAL CONTAINER. SEAL AND REMOVE FROM WORK AREA.

HEALTH HAZARD INFORMATION:
   WHEN USING THIS PRODUCT WEAR: NO SPECIAL REQUIREMENTS. Do not get into eyes, causes stinging, burning sensation,
   tearing, itching, swelling, and redness. Ingestion may cause abdominal pain, nausea, vomiting, and diarrhea.

   A physician should be contacted if anyone develops any signs or symptoms and suspects that they are caused by
   exposure to this product.

FIRST AID:
   Eye Exposure:
   Flush eyes with water for 15 minutes while lifting the upper and lower eye lids. Contact lenses should not be
   worn when working with this product. If irritation persists, get medical attention.

   Skin Exposure:
   Not expected to be a problem.

   Breathing:
   Not expected to be a problem.

   Swallowing:
   Immediately contact the local poison control center for advice. Keep the victim warm and at rest. Get medical
   attnetion.

PRODUCT: OFF, MAXIMUM PROTECTION

INCOMPATIBILITY: NONE KNOWN.

INGREDIENTS: N,N-DIETHYL-meta-
              TOLUAMIDE   95% TLV: CAS#: 134-62-8
              OTHER ISOMER   5% TLV: CAS#:

IGNI TEMP: NA   FP: 300F LEL: NA UEL: NA VP: 1 mmHg @ 230F VD: 6.7 SG: 1.0 PS: LIQUID APPEAR: COLORLESS
     ODOR: MILD   PH FACTOR:

HAZARD RATINGS: H: F: R: 2 1 0

FIRE FIGHTING:
    HIGH EXPANSION FOAM, LOW EXPANSION FOAM, ALCOHOL FOAM; DRY CHEMICAL, CARBON DIOXIDE, WATER FOG.
    PROTECTIVE CLOTHING, RUBBER GLOVES, AND BREATHING APPARATUS.
WARNING: 1] STRUCTURAL PROTECTIVE CLOTHING IS PERMEABLE, REMAIN CLEAR OF SMOKE, WATER FALL OUT AND WATER RUN OFF.
         2] KEEP OUT OF THE REACH OF CHILDREN.
         3] CONTAINERS MAY EXPLODE, MELT OR LEAK IN A FIRE.
         4] THERMAL DECOMPOSITION YIELDS TOXIC CARBON DIOXIDE, CARBON MONOXIDE.
         5] DO NOT APPLY TO EXCESSIVELY SUN BURNED OR DAMAGED SKIN.
         6] MOVE CONTAINERS FROM AREA IF WITHOUT RISK, COOL EXPOSED CONTAINERS.
         7] DIKE AREA FOR CONTROL AND CONTAINMENT TO PREVENT ENTRY INTO SEWERS, DRAIN, AND WATER WAYS.

LARGE SPILL/NO FIRE/RESCUE: WEAR RUBBER OR NEOPRENE BOOTS, GLOVES, BREATHING APPARATUS.

SPILL CONTROL AND CONTAINMENT:
    HOUSEHOLD SPILL: WIPE UP WITH AN ABSORBENT CLOTH. RINSE OUT IN THE SINK. WIPE UP RESIDUE WITH A DAMP CLOTH.
    RINSE OUT IN THE SINK.
    LARGE SPILL: PICK UP LIQUID WITH AN ABSORBENT, PUT IN A DISPOSAL CONTAINER. RINSE AREA WITH WATER, PICK UP WITH
    ABSORBENT. PUT IN A DISPOSAL CONTAINER, SEAL AND REMOVE FROM THE WORK AREA.

HEALTH HAZARD INFORMATION:
    WHEN USING THIS PRODUCT WEAR: NO SPECIAL REQUIREMENT. Eye contact may cause stinging, tearing, itching,
    swelling, and redness. Ingestion may cause distress and illness.

    A physician should be contacted if anyone develops any signs or symptoms and suspects that they are caused by
    exposure to this product.

FIRST AID:
    Eye Exposure:
    Flush with water for 15 minutes while lifting the upper and lower eye lids. Contact lenses should not be worn
    when working with this product. Get medical attention.

    Skin Exposure:
    Not expected to be a problem. If irritation develops, wash with soap and water.

    Breathing:
    Not expected to be a problem.

    Swallowing:
    Immediately contact the local poison control center for advice. Keep the victim warm and at rest. Get medical
    attention.

PRODUCT: ONE STEP FINE WOOD FLOOR CARE WAX

INCOMPATIBILITY: NONE KNOWN

INGREDIENTS: ALIPHATIC NAPHTHA   85-95%   TLV: 1000 ppm   CAS#: 64741-41-9

IGNI TEMP: NA FP: 97F LEL: NA UEL: NA VP: NA VD: NA SG: .8 PS: LIQUID APPEAR: CLEAR, LIGHT YELLOW
     ODOR: NAPHTHA   PH FACTOR: NA

HAZARD RATINGS: H: F: R: 1 3 0

FIRE FIGHTING:
  HIGH EXPANSION FOAM, LOW EXPANSION FOAM, ALCOHOL FOAM; DRY CHEMICAL, CARBON DIOXIDE, WATER FOG.
  PROTECTIVE CLOTHING, RUBBER GLOVES, AND BREATHING APPARATUS.
WARNING: 1] STRUCTURAL PROTECTIVE CLOTHING IS PERMEABLE, REMAIN CLEAR OF SMOKE, WATER FALL OUT AND WATER RUN OFF.
         2] KEEP OUT OF THE REACH OF CHILDREN.
         3] CONTAINERS MAY BURST IN A FIRE.
         4] RUN OFF INTO SEWERS MAY CREATE A FIRE/EXPLOSION HAZARD.
         5] REMOVE ALL SOURCES OF IGNITION IF WITHOUT RISK.
         6] MOVE CONTAINERS FROM AREA IF WITHOUT RISK, COOL EXPOSED CONTAINERS.
         7] DIKE AREA FOR CONTROL AND CONTAINMENT TO PREVENT ENTRY INTO SEWERS, DRAIN, AND WATER WAYS.

LARGE SPILL/NO FIRE/RESCUE: WEAR RUBBER OR NEOPRENE BOOTS, GLOVES.

SPILL CONTROL AND CONTAINMENT:
  HOUSEHOLD SPILL: WIPE UP WITH AN ABSORBENT CLOTH. RINSE OUT IN THE SINK. WIPE UP RESIDUE WITH A DAMP CLOTH.
  RINSE OUT IN THE SINK.
  LARGE SPILL: PICK UP WITH ABSORBENT. PUT IN A DISPOSAL CONTAINER. RINSE AREA WITH WATER. PICK UP WITH ABSORBENT.
  PUT IN A DISPOSAL CONTAINER. SEAL AND REMOVE FROM THE WORK AREA.

HEALTH HAZARD INFORMATION:
  WHEN USING THIS PRODUCT WEAR: NO SPECIAL REQUIREMENT. Skin contact may cause irritation. Eye contact will cause stinging, tearing, itching, swelling, and redness. Ingestion will cause nausea, and vomiting. If product gets into the lungs, it may cause chemical pneumonia. Inhalation will cause chemical intoxication at high concentrations.

  A physician should be contacted if anyone develops any signs or symptoms and suspects that they are caused by exposure to this product.

FIRST AID:
  Eye Exposure:
  Flush with water for 15 minutes while lifting the upper and lower eye lids. Contact lenses should not be worn when working with this product. Get medical attention.

  Skin Exposure:
  Wash with plenty of soap and water. If irritation develops, get medical attention.

  Breathing:
  If breathing becomes labored move the victim to fresh air. Keep the victim warm and at rest. Get medical attention.

  Swallowing:
  Do not induce vomiting. Immediately contact the local poison control center for advice. Keep the victim warm and at rest. Get medical attention.

PRODUCT: OXYDOL

INCOMPATIBILITY: NONE KNOWN

INGREDIENTS: ANIONIC SURFACTANTS   TLV: CAS#:
             COMPLEX SODIUM PHOSPHATE   TLV: CAS#:
          OR ALUMINO SILICATES   TLV: CAS#:
               SODIUM CARBONATE   TLV: CAS#:   497-19-8
             SODIUM PERBORATE   TLV: CAS#: 10486-00-7
             SODIUM SULFATE   TLV: CAS#: 7757-82-6
             SODIUM SILICATE   TLV: CAS#: 1344-09-8

IGNI TEMP: NA FP: NA LEL: NA UEL: NA VP: NA VD: NA SG: NA PS: POWDER
   APPEAR: WHITE GRANULAR & GREEN SPECKLES   ODOR: PERFUMED   PH FACTOR: NA

HAZARD RATINGS: H: F: R:

FIRE FIGHTING:
  HIGH EXPANSION FOAM, LOW EXPANSION FOAM, ALCOHOL FOAM; DRY CHEMICAL, CARBON DIOXIDE, WATER FOG.
  PROTECTIVE CLOTHING, RUBBER GLOVES, AND BREATHING APPARATUS.
WARNING: 1] STRUCTURAL PROTECTIVE CLOTHING IS PERMEABLE, REMAIN CLEAR OF SMOKE, WATER FALL OUT AND WATER RUN OFF.
         2] KEEP OUT OF THE REACH OF CHILDREN.
         3] CHECK WHITE RECTANGLE ON TOP OR BOTTOM OF BOX. IF NUMBER STARTS WITH L OR P PRODUCT CONTAINS
            TETRASODIUM PYROPHOSPHATE OR SODIUM PHOSPHATE   7722-88-5   TLV: 5 mg/m3.
         4] MOVE CONTAINERS FROM AREA IF WITHOUT RISK, COOL EXPOSED CONTAINERS.
         5] DIKE AREA FOR CONTROL AND CONTAINMENT TO PREVENT ENTRY INTO SEWERS, DRAIN, N, AND WATER WAYS.

LARGE SPILL/NO FIRE/RESCUE: WEAR RUBBER OR NEOPRENE BOOTS, GLOVES.

SPILL CONTROL AND CONTAINMENT:
  HOUSEHOLD SPILL: AVOID CREATING A DUST. SWEEP UP AND PUT POWDER IN A DISPOSAL CONTAINER. WIPE UP WITH A DAMP
  CLOTH. RINSE OUT IN THE SINK.
  LARGE SPILL: AVOID CREATING A DUST. SCOOP UP AND PUT IN A DISPOSAL CONTAINER. RINSE AREA WITH WATER. PICK UP
  WITH ABSORBENT. PUT IN A DISPOSAL CONTAINER. SEAL AND REMOVE FROM THE WORK AREA.

HEALTH HAZARD INFORMATION:
  WHEN USING THIS PRODUCT WEAR: NO SPECIAL REQUIREMENT. Inhalation may cause irritation of the nose, throat, and
  lungs. Eye contact will cause stinging, tearing, itching, swelling, and redness. Ingestion will result in
  nausea, vomiting, and diarrhea.

  A physician should be contacted if anyone develops any signs or symptoms and   suspects that they are caused by
  exposure to this product.

FIRST AID:
  Eye Exposure:
  Flush with water for 15 minutes while lifting the upper and lower eye lids. Contact lenses should not be worn
  when working with this product. Get medical attention.

  Skin Exposure:
  Wash off with water.

  Breathing:
  If breathing becomes labored move to fresh air.

  Swallowing:
  Immediately contact the local poison control center for advice. Keep the victim warm and at rest. Get medical
  attention.

PRODUCT: OXYDOL, ULTRA (LAUNDRY GRANULES)

INCOMPATIBILITY: NONE KNOWN

INGREDIENTS: SODIUM PERBORATE   TLV: CAS#: 10486-00-7
             SODIUM CARBONATE   TLV: CAS#:   497-19-8
             SODIUM SULFATE   TLV: CAS#: 7757-82-6
             ANIONIC SURFACTANT   TLV: CAS#:
             ALUMINOSILICATE   TLV: CAS#:
             PERFUME   TLV: CAS#:

IGNI TEMP: NA FP: NA LEL: NA UEL: NA VP: NA VD: NA SG: NA PS: POWDER
   APPEAR: WHITE WITH GREEN & BLUE SPECKS ODOR: PERFUMED PH FACTOR: NA

HAZARD RATINGS: H: F: R:

FIRE FIGHTING:
  HIGH EXPANSION FOAM, LOW EXPANSION FOAM, ALCOHOL FOAM; DRY CHEMICAL, CARBON DIOXIDE, WATER FOG.
  PROTECTIVE CLOTHING, RUBBER GLOVES, AND BREATHING APPARATUS.
WARNING: 1] STRUCTURAL PROTECTIVE CLOTHING IS PERMEABLE, REMAIN CLEAR OF SMOKE, WATER FALL OUT AND WATER RUN OFF.
         2] KEEP OUT OF THE REACH OF CHILDREN.
         3] THERMAL DECOMPOSITION MAY YIELD CARBON DIOXIDE, CARBON MONOXIDE.
         4] MOVE CONTAINERS FROM AREA IF WITHOUT RISK, COOL EXPOSED CONTAINERS.
         5] DIKE AREA FOR CONTROL AND CONTAINMENT TO PREVENT ENTRY INTO SEWERS, DRAIN, AND WATER WAYS.

LARGE SPILL/NO FIRE/RESCUE: WEAR RUBBER OR NEOPRENE BOOTS, GLOVES.

SPILL CONTROL AND CONTAINMENT:
  HOUSEHOLD SPILL: SWEEP UP POWDER. PUT IN A DISPOSAL CONTAINER. DAMP MOP UP RESIDUE. RINSE OUT IN THE SINK.
  LARGE SPILL: AVOID CREATING A DUST. SCOOP UP POWDER. PUT IN A DISPOSAL CONTAINER. RINSE AREA WITH WATER. PICK UP
  WITH ABSORBENT. PUT ABSORBENT IS A DISPOSAL CONTAINER. SEAL AND REMOVE FROM THE WORK AREA.

HEALTH HAZARD INFORMATION:
  WHEN USING THIS PRODUCT WEAR: NO SPECIAL REQUIREMENT. Inhalation may cause coughing, sore throat, wheezing, and
  shortness of breath. Eye contact may cause stinging, tearing, itching, swelling, and redness. Ingestion will
  result in nausea, vomiting, and diarrhea.

  A physician should be contacted if anyone develops any signs or symptoms and suspects that they are caused by
  exposure to this product.

FIRST AID:
Eye Exposure:
Flush with water for 15 minutes while lifting the upper and lower eye lids. Contact lenses should not be worn
when working with this product. Get medical attention.

  Skin Exposure:
Wash with water. If irritation develops, get medical attention.

  Breathing:
If irritation develops, move victim to fresh air.

  Swallowing:
If the victim is conscious, give 2-3 glasses of milk or water to drink. Immediately contact the local poison
control center for advice. Keep the victim warm and at rest. Get medical attention.

PRODUCT: PASTE WAX

INCOMPATIBILITY: NONE KNOWN

INGREDIENTS: STODDARD SOLVENT   75-90% TLV: 100 ppm   CAS#: 8052-41-3
             WAX MIXTURE   TLV: CAS#:

IGNI TEMP: NA FP: 88F LEL: NA UEL: NA VP: NA VD: NA SG: .8 PS: PASTE APPEAR: UNIFORM   ODOR: SOLVENT
  PH FACTOR: NA

HAZARD RATINGS: H: F: R: 0 3 0

FIRE FIGHTING:
  HIGH EXPANSION FOAM, LOW EXPANSION FOAM, ALCOHOL FOAM; DRY CHEMICAL, CARBON DIOXIDE, WATER FOG.
  PROTECTIVE CLOTHING, RUBBER GLOVES, AND BREATHING APPARATUS.
WARNING: 1] STRUCTURAL PROTECTIVE CLOTHING IS PERMEABLE, REMAIN CLEAR OF SMOKE, WATER FALL OUT AND WATER RUN OFF.
         2] KEEP OUT OF THE REACH OF CHILDREN.
         3] THERMAL DECOMPOSITION MAY YIELD CARBON DIOXIDE, CARBON MONOXIDE.
         4] MOVE CONTAINERS FROM AREA IF WITHOUT RISK, COOL EXPOSED CONTAINERS.
         5] DIKE AREA FOR CONTROL AND CONTAINMENT TO PREVENT ENTRY INTO SEWERS, DRAIN, AND WATER WAYS.

LARGE SPILL/NO FIRE/RESCUE: WEAR RUBBER OR NEOPRENE BOOTS, GLOVES.

SPILL CONTROL AND CONTAINMENT:
  HOUSEHOLD SPILL: SWEEP OR SCRAPE UP AND RETURN TO CAN OR PUT IN A DISPOSAL CONTAINER. WIPE UP RESIDUE WITH A
  DAMP CLOTH. RINSE OUT IN THE SINK.
  LARGE SPILL: USE NON-SPARKING TOOLS. SWEEP OR SCRAPE UP AND PUT IN A DISPOSAL CONTAINER. SCRUB AREA WITH
  DETERGENT AND WATER. PICK UP WITH ABSORBENT. PUT IN A DISPOSAL CONTAINER. SEAL AND REMOVE FROM THE WORK AREA.

HEALTH HAZARD INFORMATION:
  WHEN USING THIS PRODUCT WEAR: NO SPECIAL REQUIREMENT. Eye contact will cause stinging, tearing, itching,
  swelling, and redness. Prolonged contact will cause skin irritation. Ingestion will cause mild abdominal
  distress, with nausea, vomiting, and diarrhea.

  A physician should be contacted if anyone develops any signs or symptoms and suspects that they are caused by
  exposure to this product.

FIRST AID:
  Eye Exposure:
  Flush with water for 15 minutes while lifting the upper and lower eye lids. Contact lenses should not be worn
  when working with this product. Get medical attention.

  Skin Exposure:
  Wash with soap and water after use.

  Breathing:
  Not expected to be a problem.

  Swallowing:
  Do not induce vomiting. If the victim is conscious give 1-2 glasses of milk or water to drink. Immediately
  contact the local poison control center for advice. Keep the victim warm and at rest. Get medical attention.

PRODUCT: PINE-SOL BROAD SPECTRUM FORMULA

INCOMPATIBILITY: NONE KNOWN

INGREDIENTS:	PINE OIL		TLV:		CAS#: 8002-09-3
	ISOPROPANOL	8%	TLV: 400 ppm		CAS#: 67-63-0
	ANIONIC SURFACTANTS		TLV:		CAS#:
	NONIONIC SURFACTANTS		TLV:		CAS#:

IGNI TEMP: NA FP: 101F LEL: NA UEL: NA VP: NA VD: NA SG: .987 PS: LIQUID APPEAR: CLEAR YELLOW
     ODOR: PINE   PH FACTOR: 3-4

HAZARD RATINGS: H: F: R:

FIRE FIGHTING:
  HIGH EXPANSION FOAM, LOW EXPANSION FOAM, ALCOHOL FOAM; DRY CHEMICAL, CARBON DIOXIDE, WATER FOG.
  PROTECTIVE CLOTHING, RUBBER GLOVES, AND BREATHING APPARATUS.
WARNING: 1] STRUCTURAL PROTECTIVE CLOTHING IS PERMEABLE, REMAIN CLEAR OF SMOKE, WATER FALL OUT AND WATER RUN OFF.
         2] KEEP OUT OF THE REACH OF CHILDREN.
         3] THERMAL DECOMPOSITION MAY YIELD CARBON DIOXIDE, CARBON MONOXIDE.
         4] AVOID CONTACT WITH FOOD.
         5] MOVE CONTAINERS FROM AREA IF WITHOUT RISK, COOL EXPOSED CONTAINERS.
         6] DIKE AREA FOR CONTROL AND CONTAINMENT TO PREVENT ENTRY INTO SEWERS, DRAIN, AND WATER WAYS.

LARGE SPILL/NO FIRE/RESCUE: WEAR RUBBER OR NEOPRENE BOOTS, GLOVES.

SPILL CONTROL AND CONTAINMENT:
  HOUSEHOLD SPILL: WIPE UP WITH AN ABSORBENT CLOTH. RINSE OUT IN THE SINK. WIPE UP RESIDUE WITH A DAMP CLOTH.
  RINSE OUT IN THE SINK.
  LARGE SPILL: USE NON SPARKING TOOLS. PICK UP WITH ABSORBENT. PUT IN A DISPOSAL CONTAINER. RINSE AREA WIPE WATER.
  PICK UP WITH ABSORBENT. PUT IN A DISPOSAL CONTAINER. SEAL AND REMOVE FOR THE WORK AREA.

HEALTH HAZARD INFORMATION:
  WHEN USING THIS PRODUCT WEAR: NO SPECIAL REQUIREMENT. Eye contact will cause stinging, tearing, itching,
  swelling, and redness. Will cause skin irritation. Ingestion will cause abdominal pain, nausea, vomiting, and
  diarrhea.

  A physician should be contacted if anyone develops any signs or symptoms and suspects that they are caused by
  exposure to this product.

FIRST AID:
  Eye Exposure:
  Flush with water for 15 minutes while lifting the upper and lower eye lids. Contact lenses should not be worn
  when working with this product. Get medical attention.

  Skin Exposure:
  Wash with soap and water. If irritation develops, get medical attention.

  Breathing:
  Not expected to be a problem.

  Swallowing:
  Do not induce vomiting. If the victim is conscious give 2-3 glasses of milk or water to drink. Immediately
  contact the local poison control center for advice. Keep the victim warm and at rest. Get medical attention.

```
PRODUCT: PINE-SOL MULTI-ACTION SPRAY CLEANER

INCOMPATIBILITY: NONE KNOWN

INGREDIENTS: ISOPROPANOL            1%   TLV: 400 ppm      CAS#:  67-63-0
             CATIONIC SURFACTANTS        TLV:              CAS#:
             NONIONIC SURFACTANTS        TLV:              CAS#:
             BUTYL CELLOSOLVE       5%   TLV:  25 ppm      CAS#: 111-76-2

IGNI TEMP: NA  FP: 150F  LEL: NA  UEL: NA  VP: NA  VD: NA  SG: 1.003  PS: LIQUID  APPEAR: CLEAR AQUEOUS
   ODOR: PINE  PH FACTOR: 11

HAZARD RATINGS:   H:  F:  R:
```

FIRE FIGHTING:
HIGH EXPANSION FOAM, LOW EXPANSION FOAM, ALCOHOL FOAM; DRY CHEMICAL, CARBON DIOXIDE, WATER FOG.
PROTECTIVE CLOTHING, RUBBER GLOVES, AND BREATHING APPARATUS.
WARNING: 1] STRUCTURAL PROTECTIVE CLOTHING IS PERMEABLE, REMAIN CLEAR OF SMOKE, WATER FALL OUT AND WATER RUN OFF.
         2] KEEP OUT OF THE REACH OF CHILDREN.
         3] THERMAL DECOMPOSITION MAY YIELD CARBON DIOXIDE, CARBON MONOXIDE.
         4] AVOID CONTACT WITH FOOD.
         5] MOVE CONTAINERS FROM AREA IF WITHOUT RISK, COOL EXPOSED CONTAINERS.
         6] DIKE AREA FOR CONTROL AND CONTAINMENT TO PREVENT ENTRY INTO SEWERS, DRAIN, AND WATER WAYS.

LARGE SPILL/NO FIRE/RESCUE: WEAR RUBBER OR NEOPRENE BOOTS, GLOVES.

SPILL CONTROL AND CONTAINMENT:
HOUSEHOLD SPILL: WIPE UP DRY WITH AN ABSORBENT CLOTH. RINSE OUT IN THE SINK. WIPE UP RESIDUE WITH A DAMP CLOTH.
RINSE OUT IN THE SINK.
LARGE SPILL: PICK UP WITH ABSORBENT. PUT IN A DISPOSAL CONTAINER. RINSE AREA WITH WATER. PICK UP WITH ABSORBENT.
PUT IN A DISPOSAL CONTAINER. SEAL AND REMOVE FROM THE WORK AREA.

HEALTH HAZARD INFORMATION:
WHEN USING THIS PRODUCT WEAR: NO SPECIAL REQUIREMENT. Eye contact may cause stinging, tearing, itching,
swelling, and redness. Avoid eye and skin contact. Ingestion may cause abdominal distress with nausea and
vomiting. Avoid inhalation may irritate the nose, throat, and lungs.

A physician should be contacted if anyone develops any signs or symptoms and suspects that they are caused by
exposure to this product.

FIRST AID:
Eye Exposure:
Flush with water for 15 minutes while lifting the upper and lower eye lids. Contact lenses should not be worn
when working with this product. Get medical attention.

Skin Exposure:
Wash with soap and water. If irritation develops, get medical attention.

Breathing:
If breathing becomes labored, move victim to fresh air. If condition persists get medical attention.

Swallowing:
Do not induce vomiting. If victim is conscious, give 2-3 glasses of milk or water to drink. Immediately contact
the local poison control center for advice. Keep the victim warm and at rest. Get medical attention.

PRODUCT: PINE - SOL, SPRING PINE - LIGHT SCENT CLEANER

INCOMPATIBILITY: NONE KNOWN

INGREDIENTS: ISOPROPYL ALCOHOL   5-10% TLV: 400 ppm   CAS#:   67-63-0
             PINE OIL   5-10% TLV: CAS#: 8002-09-3
             NONIONIC SURFACTANTS_   TLV: CAS#:
             ANIONIC SURFACTANTS_____ 15-25% TLV: CAS#:

IGNI TEMP: NA FP: 105F LEL: NA UEL: NA VP: NA VD: NA SG: 1.0 PS: LIQUID APPEAR: AMBER ODOR: LIGHT PINE
    PH FACTOR: 10.5

HAZARD RATINGS: H: F: R: 2 2 0

FIRE FIGHTING:
    HIGH EXPANSION FOAM, LOW EXPANSION FOAM, ALCOHOL FOAM; DRY CHEMICAL, CARBON DIOXIDE, WATER FOG.
    PROTECTIVE CLOTHING, RUBBER GLOVES, AND BREATHING APPARATUS.
WARNING: 1] STRUCTURAL PROTECTIVE CLOTHING IS PERMEABLE, REMAIN CLEAR OF SMOKE, WATER FALL OUT AND WATER RUN OFF.
         2] KEEP OUT OF THE REACH OF CHILDREN.
         3] THERMAL DECOMPOSITION MAY YIELD CARBON DIOXIDE, CARBON MONOXIDE.
         4] MOVE CONTAINERS FROM AREA IF WITHOUT RISK, COOL EXPOSED CONTAINERS.
         5] DIKE AREA FOR CONTROL AND CONTAINMENT TO PREVENT ENTRY INTO SEWERS, DRAIN, AND WATER WAYS.

LARGE SPILL/NO FIRE/RESCUE: WEAR RUBBER OR NEOPRENE BOOTS, GLOVES.

SPILL CONTROL AND CONTAINMENT:
    HOUSEHOLD SPILL: WIPE UP WITH AN ABSORBENT CLOTH. RINSE OUT IN THE SINK. WIPE UP RESIDUE WITH A DAMP CLOTH.
    RINSE OUT IN THE SINK.
    LARGE SPILL: PICK UP WITH ABSORBENT. RINSE AREA WITH WATER. PICK UP WITH ABSORBENT. PUT ALL ABSORBENT IN A
    DISPOSAL CONTAINER. SEAL AND REMOVE FROM THE WORK AREA.

HEALTH HAZARD INFORMATION:
    WHEN USING THIS PRODUCT WEAR: NO SPECIAL REQUIREMENT. Eye contact may cause stinging, tearing, itching,
    swelling, and redness. Skin contact may cause irritation. Ingestion will cause abdominal cramps, nausea,
    vomiting, and diarrhea.

    A physician should be contacted if anyone develops any signs or symptoms and suspects that they are caused by
    exposure to this product.

FIRST AID:
    Eye Exposure:
    Flush with water for 15 minutes while lifting the upper and lower eye lids. Contact lenses should not be worn
    when working with this product. Get medical attention.

    Skin Exposure:
    Wash with soap and water. If irritation develops, get medical attention.

    Breathing:
    Not expected to be a problem.

    Swallowing:
    Do not induce vomiting. If the victim is conscious, give 2-3 glasses of milk or water to drink. Immediately
    contact the local poison control center for advice. Keep the victim warm and at rest. Get medical attention.

PRODUCT: PINE-SOL   SPRUCE UPS

INCOMPATIBILITY: NONE KNOWN

INGREDIENTS: ISOPROPANOL   5%   TLV: 400 ppm   CAS#: 67-63-0
             CATIONIC SURFACTANTS   TLV: CAS#:
             NONIONIC SURFACTANTS   TLV: CAS#:

IGNI TEMP: NA FP: 119F LEL: NA UEL: NA VP: NA VD: NA SG: .99 PS: LIQUID APPEAR: CLEAR, AQUEOUS ODOR: PINE
    PH FACTOR: 5.0-6.0

HAZARD RATINGS: H: F: R:

FIRE FIGHTING:
    HIGH EXPANSION FOAM, LOW EXPANSION FOAM, ALCOHOL FOAM; DRY CHEMICAL, CARBON DIOXIDE, WATER FOG.
    PROTECTIVE CLOTHING, RUBBER GLOVES, AND BREATHING APPARATUS.
WARNING: 1] STRUCTURAL PROTECTIVE CLOTHING IS PERMEABLE, REMAIN CLEAR OF SMOKE, WATER FALL OUT AND WATER RUN OFF.
         2] KEEP OUT OF THE REACH OF CHILDREN.
         3] THERMAL DECOMPOSITION MAY YIELD CARBON DIOXIDE, CARBON MONOXIDE.
         4] AVOID CONTACT WITH FOOD.
         5] MOVE CONTAINERS FROM AREA IF WITHOUT RISK, COOL EXPOSED CONTAINERS.
         6] DIKE AREA FOR CONTROL AND CONTAINMENT TO PREVENT ENTRY INTO SEWERS, DRAIN, AND WATER WAYS.

LARGE SPILL/NO FIRE/RESCUE: WEAR RUBBER OR NEOPRENE BOOTS, GLOVES.

SPILL CONTROL AND CONTAINMENT:
    HOUSEHOLD SPILL: WIPE UP WITH AN ABSORBENT CLOTH. RINSE OUT IN THE SINK. WIPE UP RESIDUE WITH A DAMP CLOTH.
    RINSE OUT IN THE SINK.
    LARGE SPILL: PICK UP WITH ABSORBENT. PUT IN A DISPOSAL CONTAINER. RINSE AREA WITH WATER. PICK UP WITH ABSORBENT.
    PUT IN A DISPOSAL CONTAINER. SEAL AND REMOVE FROM THE WORK AREA.

HEALTH HAZARD INFORMATION:
    WHEN USING THIS PRODUCT WEAR: NO SPECIAL REQUIREMENT. Eye contact will cause stinging, tearing, itching,
    swelling, and redness. Prolonged contact may cause skin irritation. Ingestion may cause abdominal pain, nausea,
    vomiting, and diarrhea.

    A physician should be contacted if anyone develops any signs or symptoms and suspects that they are caused by
    exposure to this product.

FIRST AID:
Eye Exposure:
Flush with water for 15 minutes while lifting the upper and lower eye lids. Contact lenses should not be worn
when working with this product. Get medical attention.

Skin Exposure:
Wash with soap and water. If irritation develops, get medical attention.

Breathing:
Not expected to be a problem.

Swallowing:
Do not induce vomiting. If victim is conscious, give 1-2 glasses of milk or water to drink. Immediately contact
the local poison control center for advice. Keep the victim warm and at rest. Get medical attention.

PRODUCT: PLEDGE

INCOMPATIBILITY: NONE KNOWN

INGREDIENTS: ISOPARAFFINIC   RECOMMENDED
            HYDROCARBON SOLVENT   10-20% TLV: 400 ppm   CAS#: 64742-48-9
            WATER   60-70% TLV: CAS#: 7732-18-5
            PROPANE_____   TLV: 1000 ppm   CAS#: 74-98-6
            ISOBUTANE_____   10-20% TLV: CAS#: 75-28-5
            N-BUTANE_____/   TLV: 800 ppm   CAS#: 106-97-8

IGNI TEMP: NA FP: 20F LEL: NA UEL: NA VP: NA VD: NA SG: .9 PS: LIQUID APPEAR: SPRAY MIST ODOR: PLEASANT
   PH FACTOR: NA

HAZARD RATINGS: H: F: R: 0 4 0

FIRE FIGHTING:
   HIGH EXPANSION FOAM, LOW EXPANSION FOAM, ALCOHOL FOAM; DRY CHEMICAL, CARBON DIOXIDE, WATER FOG.
   PROTECTIVE CLOTHING, RUBBER GLOVES, AND BREATHING APPARATUS.
WARNING: 1] STRUCTURAL PROTECTIVE CLOTHING IS PERMEABLE, REMAIN CLEAR OF SMOKE, WATER FALL OUT AND WATER RUN OFF.
        2] KEEP OUT OF THE REACH OF CHILDREN.
        3] REMOVE ALL IGNITION SOURCES IF WITHOUT RISK.
        4] CONTAINERS WILL EXPLODE IN A FIRE OR IF HEATED ABOVE 120F.
        5] THERMAL DECOMPOSITION MAY YIELD CARBON DIOXIDE, CARBON MONOXIDE.
        6] MOVE CONTAINERS FROM AREA IF WITHOUT RISK, COOL EXPOSED CONTAINERS.
        7] DIKE AREA FOR CONTROL AND CONTAINMENT TO PREVENT ENTRY INTO SEWERS, DRAIN, AND WATER WAYS.

LARGE SPILL/NO FIRE/RESCUE: WEAR RUBBER OR NEOPRENE BOOTS, GLOVES.

SPILL CONTROL AND CONTAINMENT:
   HOUSEHOLD SPILL: WIPE UP DRY WITH AN ABSORBENT CLOTH. PUT IN A DISPOSAL CONTAINER.
   LARGE SPILL: USE NON-SPARKING TOOLS. PICK UP WITH ABSORBENT. PUT IN A DISPOSAL CONTAINER. RINSE AREA WITH WATER.
   PICK UP WITH ABSORBENT. PUT IN A DISPOSAL CONTAINER. SEAL AND REMOVE FROM THE WORK AREA.

HEALTH HAZARD INFORMATION:
   WHEN USING THIS PRODUCT WEAR: NO SPECIAL REQUIREMENT. Inhalation of gas may cause light headedness. Eye contact
will cause stinging, tearing, itching, swelling, and redness.

A physician should be contacted if anyone develops any signs or symptoms and suspects that they are caused by
exposure to this product.

FIRST AID:
Eye Exposure:
Flush with water for 15 minutes while lifting the upper and lower eye lids. Contact lenses should not be worn
when working with this product. Get medical attention.

Skin Exposure:
Not expected to be a problem.

Breathing:
If symptoms appear. move victim to fresh air.

Swallowing:
Immediately contact the local poison control center for advice. Keep the victim warm and at rest. Get medical
attention.

PRODUCT: PLEDGE, LEMON

INCOMPATIBILITY: OPEN FLAME.

INGREDIENTS: ISOPARAFFINIC   Recommended
             HYDROCARBON SOLVENT   10-20% TLV: 400 ppm   CAS#: 64742-48-9
             ISOBUTANE___   TLV: CAS#: 75-28-5
             PROPANE_____   10-20% TLV: 1000 ppm   CAS#: 74-98-6
             N-BUTANE___/   TLV: 800 ppm   CAS#: 106-97-8
             WATER   60-70% TLV: CAS#: 7732-18-5

IGNI TEMP: NA FP: 20F LEL: NA UEL: NA VP: NA VD: NA SG: .9 PS: LIQUID APPEAR: SPRAY MIST   ODOR: LEMON
    PH FACTOR: NA

HAZARD RATINGS: H: F: R: 0 4 0

FIRE FIGHTING:
  HIGH EXPANSION FOAM, LOW EXPANSION FOAM, ALCOHOL FOAM; DRY CHEMICAL, CARBON DIOXIDE, WATER FOG.
  PROTECTIVE CLOTHING, RUBBER GLOVES, AND BREATHING APPARATUS.
WARNING: 1] STRUCTURAL PROTECTIVE CLOTHING IS PERMEABLE, REMAIN CLEAR OF SMOKE, WATER FALL OUT AND WATER RUN OFF.
         2] KEEP OUT OF THE REACH OF CHILDREN.
         3] THERMAL DECOMPOSITION YIELDS TOXIC CARBON DIOXIDE, CARBON MONOXIDE.
         4] CONTAINER WILL EXPLODE IN A FIRE OR IF HEATED ABOVE 120F.
         5] REMOVE ALL SOURCES OF IGNITION IF WITHOUT RISK.
         6] MOVE CONTAINERS FROM AREA IF WITHOUT RISK, COOL EXPOSED CONTAINERS.
         7] DIKE AREA FOR CONTROL AND CONTAINMENT TO PREVENT ENTRY INTO SEWERS, DRAIN, AND WATER WAYS.

LARGE SPILL/NO FIRE/RESCUE: WEAR RUBBER OR NEOPRENE BOOTS, GLOVES.

SPILL CONTROL AND CONTAINMENT:
  HOUSEHOLD SPILL: WIPE UP LIQUID WITH AN ABSORBENT CLOTH, RINSE OUT IN THE SINK. WIPE UP RESIDUE WITH A DRY
  CLOTH.
  LARGE SPILL: COVER SPILL WITH ABSORBENT. SWEEP UP AND PUT IN A DISPOSAL CONTAINER. SCRUB AREA WITH DETERGENT AND
  WATER, PICK UP WITH ABSORBENT. PUT IN A DISPOSAL CONTAINER. SEAL AND REMOVE FROM THE WORK AREA.

HEALTH HAZARD INFORMATION:
  WHEN USING THIS PRODUCT WEAR: NO SPECIAL REQUIREMENT. Contact with the eyes will cause stinging, tearing,
  itching and, swelling with redness. Ingestion will cause nausea and vomiting.

  A physician should be contacted if anyone develops any signs or symptoms and suspects that they are caused by
  exposure to this product.

FIRST AID:
  Eye Exposure:
  Flush with water for 15 minutes while lifting the upper and lower eye lids. Contact lenses should not be worn
  when working with this product. If irritation persists, get medical attention.

  Skin Exposure:
  Wash with soap and water.

  Breathing:
  Not expected to be a problem.

  Swallowing:
  Do not induce vomiting. Immediately contact the local poison control center for advice. Keep the victim warm
  and at rest. Get medical attention.

PRODUCT: PLEDGE, LEMON (PUMP SPRAY)

INCOMPATIBILITY: NONE KNOWN

INGREDIENTS: SILICONE_____   TLV: CAS#:
             WAXES_____ 3-6% TLV: CAS#:
             SURFACTANTS___/   TLV: CAS#:
             ISOPARAFFINIC   Recommended
             HYDROCARBON SOLVENT 1% TLV: 400 ppm   CAS#: 64742-48-9
             WATER   TLV: CAS#: 7732-18-5
             FRAGRANCE   under .5%

IGNI TEMP: NA FP: NA LEL: NA UEL: NA VP: NA VD: NA SG: 1.0 PS: LIQUID APPEAR: WHITE OPAQUE
    ODOR: LEMON   PH FACTOR: 6.3

HAZARD RATINGS: H: F: R:

FIRE FIGHTING:
  HIGH EXPANSION FOAM, LOW EXPANSION FOAM, ALCOHOL FOAM; DRY CHEMICAL, CARBON DIOXIDE, WATER FOG.
  PROTECTIVE CLOTHING, RUBBER GLOVES, AND BREATHING APPARATUS.
WARNING: 1] STRUCTURAL PROTECTIVE CLOTHING IS PERMEABLE, REMAIN CLEAR OF SMOKE, WATER FALL OUT AND WATER RUN OFF.
         2] KEEP OUT OF THE REACH OF CHILDREN.
         3] THERMAL DECOMPOSITION YIELDS TOXIC CARBON DIOXIDE, CARBON MONOXIDE.
         4] MOVE CONTAINERS FROM AREA IF WITHOUT RISK, COOL EXPOSED CONTAINERS.
         5] DIKE AREA FOR CONTROL AND CONTAINMENT TO PREVENT ENTRY INTO SEWERS, DRAIN, AND WATER WAYS.

LARGE SPILL/NO FIRE/RESCUE: WEAR RUBBER OR NEOPRENE BOOTS, GLOVES.

SPILL CONTROL AND CONTAINMENT:
  HOUSEHOLD SPILL: WIPE UP LIQUID WITH ABSORBENT CLOTH.
  LARGE SPILL: COVER SPILL WITH ABSORBENT. PICK UP AND PUT IN A DISPOSAL CONTAINER. RINSE AREA WITH WATER. PICK UP
  WITH ABSORBENT. PUT IN A DISPOSAL CONTAINER. SEAL AND REMOVE FROM THE WORK AREA.

HEALTH HAZARD INFORMATION:
  WHEN USING THIS PRODUCT WEAR: NO SPECIAL REQUIREMENT. Contact with the eyes may cause stinging, tearing,
  itching, swelling, and redness. Ingestion may cause nausea and vomiting. Skin contact may cause redness and
  itching.

  A physician should be contacted if anyone develops any signs or symptoms and suspects that they are caused by
  exposure to this product.

FIRST AID:
  Eye Exposure:
  Flush with water for 15 minutes while lifting the upper and lower eye lids. Contact lenses should not be worn
  when working with this product. If irritation persists, get medical attention.

  Skin Exposure:
  Wash with soap and water.

  Breathing:
  Not expected to be a problem.

  Swallowing:
  Do not induce vomiting. Immediately contact the local poison control center for advice. Keep the victim warm and
  at rest. Get medical attention.

PRODUCT: PLEDGE, SPRING FRESH (PRESSURIZED UNIT)

INCOMPATIBILITY: NONE KNOWN

INGREDIENTS: ISOPARAFFINIC   Recommended
            HYDROCARBON SOLVENT   10-20% TLV: 400 ppm   CAS#: 64742-48-9
            PROPANE   TLV: 1000 ppm   CAS#:   74-98-6
            ISOBUTANE   TLV: CAS#:   75-28-5
            N-BUTANE   TLV: 800 ppm   CAS#: 106-97-8

IGNI TEMP: NA FP: 20F LEL: NA UEL: NA VP: NA VD: NA SG: .9 PS: LIQUID APPEAR: SPRAY MIST
    ODOR: FRESH FLORAL   PH FACTOR: NA

HAZARD RATINGS: H: F: R: 0 4 0

FIRE FIGHTING:
   HIGH EXPANSION FOAM, LOW EXPANSION FOAM, ALCOHOL FOAM; DRY CHEMICAL, CARBON DIOXIDE, WATER FOG.
   PROTECTIVE CLOTHING, RUBBER GLOVES, AND BREATHING APPARATUS.
WARNING: 1] STRUCTURAL PROTECTIVE CLOTHING IS PERMEABLE, REMAIN CLEAR OF SMOKE, WATER FALL OUT AND WATER RUN OFF.
         2] KEEP OUT OF THE REACH OF CHILDREN.
         3] REMOVE ALL IGNITION SOURCES IF WITHOUT RISK.
         4] CONTAINERS WILL EXPLODE IN A FIRE OR IF HEATED ABOVE 120F.
         5] THERMAL DECOMPOSITION MAY YIELD CARBON DIOXIDE, CARBON MONOXIDE.
         6] MOVE CONTAINERS FROM AREA IF WITHOUT RISK, COOL EXPOSED CONTAINERS.
         7] DIKE AREA FOR CONTROL AND CONTAINMENT TO PREVENT ENTRY INTO SEWERS, DRAIN, AND WATER WAYS.

LARGE SPILL/NO FIRE/RESCUE: WEAR RUBBER OR NEOPRENE BOOTS, GLOVES.

SPILL CONTROL AND CONTAINMENT:
   HOUSEHOLD SPILL: WIPE UP DRY WITH AN ABSORBENT CLOTH. RINSE OUT IN THE SINK.
   LARGE SPILL: PICK UP WITH ABSORBENT. RINSE AREA WITH WATER. PICK UP WITH ABSORBENT. PUT ALL ABSORBENT IN A
   DISPOSAL CONTAINER. SEAL AND REMOVE FROM THE WORK AREA.

HEALTH HAZARD INFORMATION:
   WHEN USING THIS PRODUCT WEAR: NO SPECIAL REQUIREMENT. Eye contact may cause stinging, tearing, itching,
   swelling, and redness. Skin contact may cause irritation.

   A physician should be contacted if anyone develops any signs or symptoms and suspects that they are caused by
   exposure to this product.

FIRST AID:
   Eye Exposure:
   Flush with water for 15 minutes while lifting the upper and lower eye lids. Contact lenses should not be worn
   when working with this product. Get medical attention.

   Skin Exposure:
   Wash with soap and water. If irritation develops, get medical attention.

   Breathing:
   Not expected to be a problem.

   Swallowing:
   Do not induce vomiting. Immediately contact the local poison control center for advice. Keep the victim warm and
   at rest. Get medical attention.

PRODUCT: PLEDGE, WOOD SCENT

INCOMPATIBILITY: NONE KNOWN

INGREDIENTS: ISOPARAFFINIC   Recommended

HYDROCARBON SOLVENT	10-20%	TLV: 400 ppm	CAS#: 64742-48-9
PROPANE_____		TLV: 1000 ppm	CAS#: 74-98-6
ISOBUTANE_____	10-20%	TLV:	CAS#: 75-28-5
BUTANE_____/		TLV: 800 ppm	CAS#: 106-97-8
WATER	60-70%	TLV:	CAS#: 7732-18-5

IGNI TEMP: NA FP: 20F LEL: NA UEL: NA VP: NA VD: NA SG: .9 PS: LIQUID APPEAR: SPRAY MIST ODOR: WOODY
 PH FACTOR: NA

HAZARD RATINGS: H: F: R: 0 4 0

FIRE FIGHTING:
 HIGH EXPANSION FOAM, LOW EXPANSION FOAM, ALCOHOL FOAM; DRY CHEMICAL, CARBON DIOXIDE, WATER FOG.
 PROTECTIVE CLOTHING, RUBBER GLOVES, AND BREATHING APPARATUS.
WARNING: 1] STRUCTURAL PROTECTIVE CLOTHING IS PERMEABLE, REMAIN CLEAR OF SMOKE, WATER FALL OUT AND WATER RUN OFF.
   2] KEEP OUT OF THE REACH OF CHILDREN.
   3] REMOVE ALL IGNITION SOURCES IF WITHOUT RISK.
   4] CONTAINERS WILL EXPLODE IN A FIRE OR IF HEATED ABOVE 120F.
   5] THERMAL DECOMPOSITION MAY YIELD CARBON DIOXIDE, CARBON MONOXIDE.
   6] MOVE CONTAINERS FROM AREA IF WITHOUT RISK, COOL EXPOSED CONTAINERS.
   7] DIKE AREA FOR CONTROL AND CONTAINMENT TO PREVENT ENTRY INTO SEWERS, DRAIN, AND WATER WAYS.

LARGE SPILL/NO FIRE/RESCUE: WEAR RUBBER OR NEOPRENE BOOTS, GLOVES.

SPILL CONTROL AND CONTAINMENT:
 HOUSEHOLD SPILL: WIPE UP WITH AN ABSORBENT CLOTH. RINSE OUT IN THE SINK. WIPE UP RESIDUE WITH A DAMP CLOTH.
 RINSE OUT IN THE SINK.
 LARGE SPILL: PICK UP WITH ABSORBENT. RINSE AREA WITH WATER. PICK UP WITH ABSORBENT. PUT ALL ABSORBENT IN A
 DISPOSAL CONTAINER. SEAL AND REMOVE FROM THE WORK AREA.

HEALTH HAZARD INFORMATION:
 WHEN USING THIS PRODUCT WEAR: NO SPECIAL REQUIREMENT. Eye contact may cause stinging, tearing, itching, and
 redness. Skin contact may cause irritation.

 A physician should be contacted if anyone develops any signs or symptoms and suspects that they are caused by
 exposure to this product.

FIRST AID:
 Eye Exposure:
 Flush with water for 15 minutes while lifting the upper and lower eye lids. Contact lenses should not be worn
 when working with this product. Get medical attention.

 Skin Exposure:
 Wash with soap and water. If irritation develops, get medical attention.

 Breathing:
 If irritation develops, move victim to fresh air.

 Swallowing:
 Immediately contact the local poison control center for advice. Keep the victim warm and at rest. Get medical
 attention.

PRODUCT: PLEDGE LEMON OIL

INCOMPATIBILITY: NONE KNOWN

INGREDIENTS: ISOPARAFFINIC   RECOMMENDED
             HYDROCARBON SOLVENT   5-15%   TLV: 300 ppm   CAS#: 64742-47-8

IGNI TEMP: NA   FP: 210F   LEL: NA   UEL: NA   VP: NA   VD: NA   SG: .8   PS: LIQUID   APPEAR: YELLOW   ODOR: LEMON
   PH FACTOR: NA

HAZARD RATINGS: H: F: R: 1   0   0

FIRE FIGHTING:
   HIGH EXPANSION FOAM, LOW EXPANSION FOAM, ALCOHOL FOAM; DRY CHEMICAL, CARBON DIOXIDE, WATER FOG.
   PROTECTIVE CLOTHING, RUBBER GLOVES, AND BREATHING APPARATUS.
WARNING: 1] STRUCTURAL PROTECTIVE CLOTHING IS PERMEABLE, REMAIN CLEAR OF SMOKE, WATER FALL OUT AND WATER RUN OFF.
         2] KEEP OUT OF THE REACH OF CHILDREN.
         3] THERMAL DECOMPOSITION MAY YIELD CARBON DIOXIDE, CARBON MONOXIDE.
         4] RUN OFF INTO SEWERS MAY CREATE A FIRE/EXPLOSION HAZARD.
         5] MOVE CONTAINERS FROM AREA IF WITHOUT RISK, COOL EXPOSED CONTAINERS.
         6] DIKE AREA FOR CONTROL AND CONTAINMENT TO PREVENT ENTRY INTO SEWERS, DRAIN, AND WATER WAYS.

LARGE SPILL/NO FIRE/RESCUE: WEAR RUBBER OR NEOPRENE BOOTS, GLOVES.

SPILL CONTROL AND CONTAINMENT:
   HOUSEHOLD SPILL: WIPE UP DRY WITH AN ABSORBENT CLOTH. PUT IN A DISPOSAL CONTAINER.
   LARGE SPILL: PICK UP WITH ABSORBENT. PUT IN A DISPOSAL CONTAINER. RINSE AREA WITH WATER. PICK UP WITH ABSORBENT.
   PUT IN A DISPOSAL CONTAINER. SEAL AND REMOVE FROM THE WORK AREA.

HEALTH HAZARD INFORMATION:
   WHEN USING THIS PRODUCT WEAR: NO SPECIAL REQUIREMENT. Eye contact will cause stinging, tearing, itching,
   swelling, and redness. Prolonged contact may cause skin irritation. Ingestion will cause nausea and vomiting. If
   product gets into the lungs, it may cause chemical pneumonia.

   A physician should be contacted if anyone develops any signs or symptoms and suspects that they are caused by
   exposure to this product.

FIRST AID:
Eye Exposure:
Flush with water for 15 minutes while lifting the upper and lower eye lids. Contact lenses should not be worn
when working with this product. Get medical attention.

Skin Exposure:
Wash with soap and water. If irritation develops, get medical attention.

Breathing:
Not expected to be a problem.

Swallowing:
Do not induce vomiting. Immediately contact the local poison control center for advice. Keep the victim warm and
at rest. Get medical attention.

PRODUCT: PLUNGE   LIQUID DRAIN OPENER

INCOMPATIBILITY: PRODUCTS WITH AMMONIA; HOUSEHOLD/TOILET BOWL/DRAIN CLEANERS; AMMONIA; ACIDS; VINEGAR.

INGREDIENTS: SODIUM HYPOCHLORITE   under 5% TLV: CAS#: 7681-52-9
             SODIUM HYDROXIDE   under 2% TLV: 2 mg/m3   CAS#: 1310-73-2

IGNI TEMP: NA FP: NA LEL: NA UEL: NA VP: NA VD: NA SG: 1.086 PS: LIQUID APPEAR: CLEAR ODOR: BLEACH
    PH FACTOR: 12

HAZARD RATINGS: H: F: R: 3 0 1   HAZARD CLASS   8

FIRE FIGHTING:
   HIGH EXPANSION FOAM, LOW EXPANSION FOAM, ALCOHOL FOAM; DRY CHEMICAL, CARBON DIOXIDE, WATER FOG.
   PROTECTIVE CLOTHING, RUBBER GLOVES, AND BREATHING APPARATUS.
WARNING: 1] STRUCTURAL PROTECTIVE CLOTHING IS PERMEABLE, REMAIN CLEAR OF SMOKE, WATER FALL OUT AND WATER RUN OFF.
         2] KEEP OUT OF THE REACH OF CHILDREN.
         3] DO NOT MIX WITH INCOMPATIBLE PRODUCTS LISTED ABOVE. HAZARDOUS/TOXIC GASSES WILL BE PRODUCED.
         4] MOVE CONTAINERS FROM AREA IF WITHOUT RISK, COOL EXPOSED CONTAINERS.
         5] DIKE AREA FOR CONTROL AND CONTAINMENT TO PREVENT ENTRY INTO SEWERS, DRAIN, AND WATER WAYS.

LARGE SPILL/NO FIRE/RESCUE: WEAR RUBBER OR NEOPRENE BOOTS, GLOVES.

SPILL CONTROL AND CONTAINMENT:
   HOUSEHOLD SPILL: WEAR RUBBER GLOVES. WIPE UP DRY WITH AN ABSORBENT CLOTH. RINSE AREA WITH WATER. WIPE UP WITH
   ABSORBENT CLOTH. RINSE OUT IN THE SINK. WASH HANDS WITH SOAP AND WATER.
   LARGE SPILL: PICK UP WITH ABSORBENT. RINSE AREA WITH WATER. PICK UP WITH ABSORBENT. PUT ALL ABSORBENT IN A
   DISPOSAL CONTAINER. SEAL AND REMOVE FROM THE WORK AREA.

HEALTH HAZARD INFORMATION:
   WHEN USING THIS PRODUCT WEAR: NO SPECIAL REQUIREMENT. Eye contact will cause severe burns and tissue damage.
   Ingestion will cause burns to the mouth, throat, and stomach with difficult breathing. Inhalation may cause
   irritation. Skin contact may cause burns.

   A physician should be contacted if anyone develops any signs or symptoms and suspects that they are caused by
   exposure to this product.

FIRST AID:
   Eye Exposure:
   Flush with water for 15 minutes while lifting the upper and lower eye lids. Contact lenses should not be worn
   when working with this product. Get medical attention.

   Skin Exposure:
   Remove contaminated clothing. Flush skin with water, wash with soap and water. If irritation develops, get
   medical attention.

   Breathing:
   If irritation develops, move victim to fresh air.

   Swallowing:
   Do not induce vomiting. If victim is conscious, rinse out the mouth, give 2-3 glasses of milk or water to drink.
   Immediately contact the local poison control center for advice. Keep the victim warm and at rest. Get medical
   attention.

PRODUCT: RAID ANT TRAPS

INCOMPATIBILITY: NONE KNOWN

INGREDIENTS: 2-(1-METHYLETHOXY)PHENOL
              METHYL CARBAMATE   .25% TLV: CAS#: 114-26-1
              FOOD   99.75% TLV: CAS#:

IGNI TEMP: NA FP: NA LEL: NA UEL: NA VP: NA VD: NA SG: 1 PS: PASTE APPEAR: BROWN ODOR: PEANUT BUTTER
  PH FACTOR: NA

HAZARD RATINGS: H: F: R: 0 0 0

FIRE FIGHTING:
  HIGH EXPANSION FOAM, LOW EXPANSION FOAM, ALCOHOL FOAM; DRY CHEMICAL, CARBON DIOXIDE, WATER FOG.
  PROTECTIVE CLOTHING, RUBBER GLOVES, AND BREATHING APPARATUS.
WARNING: 1] STRUCTURAL PROTECTIVE CLOTHING IS PERMEABLE, REMAIN CLEAR OF SMOKE, WATER FALL OUT AND WATER RUN OFF.
       2] KEEP OUT OF THE REACH OF CHILDREN.
       3] PRODUCT MAY BE ABSORBED THROUGH THE SKIN.
       4] THERMAL DECOMPOSITION MAY YIELD CARBON DIOXIDE, CARBON MONOXIDE.
       5] MOVE CONTAINERS FROM AREA IF WITHOUT RISK, COOL EXPOSED CONTAINERS.
       6] DIKE AREA FOR CONTROL AND CONTAINMENT TO PREVENT ENTRY INTO SEWERS, DRAIN, AND WATER WAYS.

LARGE SPILL/NO FIRE/RESCUE: WEAR RUBBER OR NEOPRENE BOOTS, GLOVES.

SPILL CONTROL AND CONTAINMENT:
  HOUSEHOLD SPILL: WIPE UP WITH AN ABSORBENT CLOTH. PUT IN DISPOSAL CONTAINER. WASH AREA WITH SOAP AND WATER. WIPE
  UP WITH AN ABSORBENT CLOTH. RINSE OUT IN THE SINK. WASH HANDS WITH SOAP AND WATER.
  LARGE SPILL: SCOOP UP AND PUT IN A DISPOSAL CONTAINER. SCRUB AREA WITH DETERGENT AND WATER. PICK UP WITH
  ABSORBENT. PUT IN A DISPOSAL CONTAINER. SEAL AND REMOVE FROM THE WORK AREA.

HEALTH HAZARD INFORMATION:
  WHEN USING THIS PRODUCT WEAR: NO SPECIAL REQUIREMENT. Eye contact will cause stinging, tearing, itching,
  swelling, and redness. Prolonged contact may cause skin irritation. Ingestion may cause nausea, vomiting, and
  illness.

  A physician should be contacted if anyone develops any signs or symptoms and suspects that they are caused by
  exposure to this product.

FIRST AID:
Eye Exposure:
Flush with water for 15 minutes while lifting the upper and lower eye lids. Contact lenses should not be worn
when working with this product. Get medical attention.

Skin Exposure:
Wash with soap and water. If irritation develops, get medical attention.

Breathing:
Not expected to be a problem.

Swallowing:
Immediately contact the local poison control center for advice. Keep the victim warm and at rest. Get medical
attention.

PRODUCT: RAID ANT TRAP II

INCOMPATIBILITY: NONE KNOWN

INGREDIENTS: (2,2-DIMETHYL-1,3-BENZODIOXOL

        METHYL CARBAMATE   .03% TLV: CAS#: 22781-23-3

        FOOD   99.97% TLV: CAS#:

IGNI TEMP: NA FP: NA LEL: NA UEL: NA VP: NA VD: NA SG: 1 PS: PASTE APPEAR: BROWN ODOR: PEANUT BUTTER
PH FACTOR: NA

HAZARD RATINGS: H: F: R: 0 0 0

FIRE FIGHTING:
  HIGH EXPANSION FOAM, LOW EXPANSION FOAM, ALCOHOL FOAM; DRY CHEMICAL, CARBON DIOXIDE, WATER FOG.
  PROTECTIVE CLOTHING, RUBBER GLOVES, AND BREATHING APPARATUS.
WARNING: 1] STRUCTURAL PROTECTIVE CLOTHING IS PERMEABLE, REMAIN CLEAR OF SMOKE, WATER FALL OUT AND WATER RUN OFF.
      2] KEEP OUT OF THE REACH OF CHILDREN.
      3] PRODUCT MAY BE ABSORBED THROUGH THE SKIN.
      4] THERMAL DECOMPOSITION MAY YIELD CARBON DIOXIDE, CARBON MONOXIDE.
      5] MOVE CONTAINERS FROM AREA IF WITHOUT RISK, COOL EXPOSED CONTAINERS.
      6] DIKE AREA FOR CONTROL AND CONTAINMENT TO PREVENT ENTRY INTO SEWERS, DRAIN, AND WATER WAYS.

LARGE SPILL/NO FIRE/RESCUE: WEAR RUBBER OR NEOPRENE BOOTS, GLOVES.

SPILL CONTROL AND CONTAINMENT:
  HOUSEHOLD SPILL: WIPE UP WITH AN ABSORBENT CLOTH. PUT IN DISPOSAL CONTAINER. WASH AREA WITH SOAP AND WATER. WIPE
UP WITH AN ABSORBENT CLOTH. RINSE OUT IN THE SINK. WASH HANDS WITH SOAP AND WATER.
  LARGE SPILL: SCOOP UP AND PUT IN A DISPOSAL CONTAINER. SCRUB AREA WITH DETERGENT AND WATER. PICK UP WITH
ABSORBENT. PUT IN A DISPOSAL CONTAINER. SEAL AND REMOVE FROM THE WORK AREA.

HEALTH HAZARD INFORMATION:
  WHEN USING THIS PRODUCT WEAR: NO SPECIAL REQUIREMENT. Eye contact will cause stinging, tearing, itching,
swelling, and redness. Prolonged contact may cause skin irritation. Ingestion may cause nausea, vomiting, and
illness.

A physician should be contacted if anyone develops any signs or symptoms and suspects that they are caused by
exposure to this product.

FIRST AID:
Eye Exposure:
Flush with water for 15 minutes while lifting the upper and lower eye lids. Contact lenses should not be worn
when working with this product. Get medical attention.

Skin Exposure:
Wash with soap and water. If irritation develops, get medical attention.

Breathing:
Not expected to be a problem.

Swallowing:
Immediately contact the local poison control center for advice. Keep the victim warm and at rest. Get medical
attention.

PRODUCT: RAID ANT & ROACH KILLER - LIQUID FORMULA II

INCOMPATIBILITY: NONE KNOWN

INGREDIENTS: ALIPHATIC NAPHTHA   80-90%   TLV: 100 ppm   CAS#: 64742-47-8
             ISOPROPYL ALCOHOL   10-15%   TLV: 400 ppm   CAS#:   67-63-0

IGNI TEMP: NA FP: 64F LEL: NA UEL: NA VP: NA VD: NA SG: .79 PS: LIQUID APPEAR: SPRAY MIST ODOR: NAPHTHA
   PH FACTOR: NA

HAZARD RATINGS: H: F: R: 1 3 0

FIRE FIGHTING:
  HIGH EXPANSION FOAM, LOW EXPANSION FOAM, ALCOHOL FOAM; DRY CHEMICAL, CARBON DIOXIDE, WATER FOG.
  PROTECTIVE CLOTHING, RUBBER GLOVES, AND BREATHING APPARATUS.
WARNING: 1] STRUCTURAL PROTECTIVE CLOTHING IS PERMEABLE, REMAIN CLEAR OF SMOKE, WATER FALL OUT AND WATER RUN OFF.
         2] KEEP OUT OF THE REACH OF CHILDREN.
         3] LIQUID RUN OFF INTO SEWER MAY CREATE A FIRE/EXPLOSION HAZARD.
         4] REMOVE ALL IGNITION SOURCES IF WITHOUT RISK.
         5] THERMAL DECOMPOSITION MAY YIELD CARBON DIOXIDE, CARBON MONOXIDE.
         6] CONTAINERS WILL EXPLODE IN A FIRE OR IF HEATED ABOVE 120F.
         7] MOVE CONTAINERS FROM AREA IF WITHOUT RISK, COOL EXPOSED CONTAINERS.
         8] DIKE AREA FOR CONTROL AND CONTAINMENT TO PREVENT ENTRY INTO SEWERS, DRAIN, AND WATER WAYS.

LARGE SPILL/NO FIRE/RESCUE: WEAR RUBBER OR NEOPRENE BOOTS, GLOVES.

SPILL CONTROL AND CONTAINMENT:
  HOUSEHOLD SPILL: WIPE UP DRY WITH AN ABSORBENT CLOTH. WASH AREA WITH SOAP AND WATER. WIPE UP WITH AN ABSORBENT
  CLOTH. RINSE OUT IN THE SINK.
  LARGE SPILL: USE NON-SPARKING TOOLS. PICK UP WITH ABSORBENT. PUT IN A DISPOSAL CONTAINER. RINSE AREA WITH WATER.
  PICK UP WITH ABSORBENT. PUT IN A DISPOSAL CONTAINER, SEAL AND REMOVE FROM WORK AREA.

HEALTH HAZARD INFORMATION:
  WHEN USING THIS PRODUCT WEAR: NO SPECIAL REQUIREMENT. Eye contact will cause stinging, tearing, itching,
  swelling, and redness. Prolonged contact may cause skin irritation. Ingestion may cause nausea and vomiting. If
  product gets into the lungs may cause chemical pneumonia. Spray mist may cause irritation of the nose, throat,
  and lungs. Normal consumer usage not expected to cause adverse effects.

  A physician should be contacted if anyone develops any signs or symptoms and suspects that they are caused by
  exposure to this product.

FIRST AID:
  Eye Exposure:
  Flush with water for 15 minutes while lifting the upper and lower eye lids. Contact lenses should not be worn
  when working with this product. Get medical attention.

  Skin Exposure:
  Wash with soap and water. If irritation develops, get medical attention.

  Breathing:
  If irritation develops, move victim to fresh air. If persistent get medical attention.

  Swallowing:
  If the victim is conscious, give 2-3 glasses of milk or water to drink. Immediately contact the local poison
  control center for advice. Keep the victim warm and at rest. Get medical attention.

PRODUCT: RAID ANT & ROACH KILLER (PRESSURIZED)

INCOMPATIBILITY: NONE KNOWN

INGREDIENTS: HYDROCARBON SOLVENT   85-95% TLV: 100 ppm   CAS#: 64742-47-8
           CARBON DIOXIDE   1-4% TLV: 5000 ppm   CAS#: 124-38-9

IGNI TEMP: NA FP: 140F LEL: NA UEL: NA VP: NA VD: NA SG: .79 PS: LIQUID APPEAR: SPRAY MIST   ODOR: NAPHTHA
  PH FACTOR: NA

HAZARD RATINGS: H: F: R: 1 4 0

FIRE FIGHTING:
  HIGH EXPANSION FOAM, ALCOHOL FOAM; DRY CHEMICAL, CARBON DIOXIDE, WATER FOG.
  PROTECTIVE CLOTHING, RUBBER GLOVES, AND BREATHING APPARATUS.
WARNING: 1] STRUCTURAL PROTECTIVE CLOTHING IS PERMEABLE, REMAIN CLEAR OF SMOKE, WATER FALL OUT AND WATER RUN OFF.
        2] KEEP OUT OF THE REACH OF CHILDREN.
        3] REMOVE ALL IGNITION SOURCES IF WITHOUT RISK.
        4] LIQUID RUN OFF TO SEWER MAY CREATE A FIRE/EXPLOSION HAZARD.
        5] CONTAINERS WILL EXPLODE IN A FIRE OR IF HEATED ABOVE 120F.
        6] THERMAL DECOMPOSITION MAY YIELD CARBON DIOXIDE, CARBON MONOXIDE.
        7] MOVE CONTAINERS FROM AREA IF WITHOUT RISK, COOL EXPOSED CONTAINERS.
        8] DIKE AREA FOR CONTROL AND CONTAINMENT TO PREVENT ENTRY INTO SEWERS, DRAIN, AND WATER WAYS.

LARGE SPILL/NO FIRE/RESCUE: WEAR RUBBER OR NEOPRENE BOOTS, GLOVES.

SPILL CONTROL AND CONTAINMENT:
  <u>HOUSEHOLD</u> <u>SPILL:</u> VENTILATE AREA. WIPE UP WITH ABSORBENT CLOTH. RINSE OUT IN THE SINK. WASH AREA WITH SOAP AND
  WATER. WIPE UP DRY WITH ABSORBENT CLOTH. RINSE OUT IN THE SINK.
  <u>LARGE</u> <u>SPILL:</u> PICK UP WITH ABSORBENT. PUT IN A DISPOSAL CONTAINER. RINSE AREA WITH WATER. PICK UP WITH ABSORBENT.
  PUT IN A DISPOSAL CONTAINER. SEAL AND REMOVE FROM THE WORK AREA.

HEALTH HAZARD INFORMATION:
  WHEN USING THIS PRODUCT WEAR: NO SPECIAL REQUIREMENT. Eye contact will cause stinging, tearing, itching,
  swelling, and redness. Prolonged contact may cause skin irritation. Ingestion may cause nausea and vomiting. If
  product gets into the lungs may, cause chemical pneumonia. Spray mist may cause irritation of the nose, throat,
  and lungs. <u>Normal</u> <u>consumer</u> <u>usage</u> <u>not</u> <u>expected</u> <u>to</u> <u>cause</u> <u>adverse</u> <u>effects.</u>

  A physician should be seen if anyone develops any signs or symptoms and suspects that they are caused by
  exposure to this product.

FIRST AID:
  Eye Exposure:
  Flush with water for 15 minutes while lifting the upper and lower eye lids. Contact lenses should not be worn
  when working with this product. Get medical attention.

  Skin Exposure:
  Wash with soap and water. If irritation develops, get medical attention.

  Breathing:
  If irritation develops, move victim to fresh air.

  Swallowing:
  If victim is conscious, give 2-3 glasses of milk or water to drink. Immediately contact the local poison control
  center for advice. Keep the victim warm and at rest. Get medical attention.

PRODUCT: RAID CRACK & CREVICE

INCOMPATIBILITY: NONE KNOWN

INGREDIENTS: CHLOROPYRIFOS   .5%   TLV: .2 mg/m3   CAS#: 2921-88-2
             HYDROCARBON SOLVENT under 1%   TLV: CAS#:
             SURFACTANT   1-3%   TLV: CAS#:
             WATER   75-85% TLV: CAS#: 7732-18-5
             ISOBUTANE   15-25% TLV: CAS#:   75-28-5

IGNI TEMP: NA FP: 20F LEL: NA UEL: NA VP: NA VD: NA SG: .9 PS: LIQUID APPEAR: FOAM   ODOR:
     PH FACTOR:

HAZARD RATINGS: H: F: R: 1 4 0

FIRE FIGHTING:
   HIGH EXPANSION FOAM, ALCOHOL FOAM; DRY CHEMICAL, CARBON DIOXIDE, WATER FOG.
   PROTECTIVE CLOTHING, RUBBER GLOVES, AND BREATHING APPARATUS.
WARNING: 1] STRUCTURAL PROTECTIVE CLOTHING IS PERMEABLE, REMAIN CLEAR OF SMOKE, WATER FALL OUT AND WATER RUN OFF.
         2] KEEP OUT OF THE REACH OF CHILDREN.
         3] CONTAINERS WILL EXPLODE IN A FIRE OR IF HEATED ABOVE 120F.
         4] REMOVE ALL IGNITION SOURCES IF WITHOUT RISK.
         5] FOAM CAN IGNITE. DO NOT USE NEAR A SOURCE OF IGNITION.
         6] DO NOT ALLOW CHILDREN OR PETS IN TREATED AREA UNTIL FOAM DRIES.
         7] THERMAL DECOMPOSITION MAY YIELD CARBON DIOXIDE, CARBON MONOXIDE.
         8] MOVE CONTAINERS FROM AREA IF WITHOUT RISK, COOL EXPOSED CONTAINERS.
         9] DIKE AREA FOR CONTROL AND CONTAINMENT TO PREVENT ENTRY INTO SEWERS, DRAIN, AND WATER WAYS.

LARGE SPILL/NO FIRE/RESCUE: WEAR NON-SEALED CHEMICAL PROTECTIVE CLOTHING, RUBBER BOOTS, GLOVES AND BREATHING
                           APPARATUS.

SPILL CONTROL AND CONTAINMENT:
   HOUSEHOLD SPILL: WEAR RUBBER GLOVES, WIPE UP FOAM WITH ABSORBENT CLOTH. PUT CLOTH IN A DISPOSAL CONTAINER. WASH
   AREA WITH SOAP AND WATER. WIPE UP DRY WITH ABSORBENT CLOTH. PUT CLOTH IN A DISPOSAL CONTAINER.
   LARGE SPILL: PICK UP WITH ABSORBENT. PUT IN DISPOSAL CONTAINER. SCRUB AREA WITH DETERGENT AND WATER. PICK UP
   WITH ABSORBENT. PUT IN A DISPOSAL CONTAINER. SEAL AND REMOVE FROM THE WORK AREA.

HEALTH HAZARD INFORMATION:
   WHEN USING THIS PRODUCT WEAR: WEAR RUBBER GLOVES. Skin contact can cause irritation. Eye contact can cause
   stinging, tearing, itching, swelling, and redness. Ingestion may cause nausea, vomiting, and illness. Mist may
   cause irritation of the nose, throat, and lungs. Not expected to cause adverse health effects during normal
   consumer usage.

   A physician should be contacted if any one develops any signs or symptoms and suspects that they are caused by
   exposure to this product.

FIRST AID:
Eye Exposure:
Flush with water for 15 minutes while lifting the upper and lower eye lids. Contact lenses should not be worn if
using this product. Get medical attention.

Skin Exposure:
Wash with soap and water. If irritation develops, get medical attention.

Breathing:
If irritation develops, move victim to fresh air.

Swallowing:
Immediately contact the local poison control center for advice. Keep the victim warm and at rest. Get medical
attention.

PRODUCT: RAID FLYING INSECT KILLER - FORMULA V

INCOMPATIBILITY: NONE KNOWN

INGREDIENTS: ISOPARAFFINIC   RECOMMENDED
             HYDROCARBON SOLVENT   1-2% TLV: 300 ppm   CAS#: 64742-47-8
             PROPANE_____   TLV: 1000 ppm   CAS#:   74-98-6
             ISOBUTANE_____   25-35% TLV: CAS#:   75-28-5
             N-BUTANE____/   TLV: 800 ppm   CAS#: 106-97-8
             WATER   60-75% TLV: CAS#: 7732-18-5

IGNI TEMP: NA FP: 20F LEL: NA UEL: NA VP: NA VD: NA SG: .8 PS: LIQUID APPEAR: SPRAY MIST ODOR: MILD
   PH FACTOR: NA

HAZARD RATINGS: H: F: R: 0 4 0

FIRE FIGHTING:
   HIGH EXPANSION FOAM, LOW EXPANSION FOAM, ALCOHOL FOAM; DRY CHEMICAL, CARBON DIOXIDE, WATER FOG.
   PROTECTIVE CLOTHING, RUBBER GLOVES, AND BREATHING APPARATUS.
WARNING: 1] STRUCTURAL PROTECTIVE CLOTHING IS PERMEABLE, REMAIN CLEAR OF SMOKE, WATER FALL OUT AND WATER RUN OFF.
         2] KEEP OUT OF THE REACH OF CHILDREN.
         3] CONTAINERS WILL EXPLODE IN A FIRE OR IF HEATED ABOVE 120F.
         4] REMOVE ALL IGNITION SOURCES IF WITHOUT RISK.
         5] THERMAL DECOMPOSITION MAY YIELD CARBON DIOXIDE, CARBON MONOXIDE.
         6] DO NOT REMAIN IN AN ENCLOSED AREA AFTER SPRAYING THIS PRODUCT.
         7] MOVE CONTAINERS FROM AREA IF WITHOUT RISK, COOL EXPOSED CONTAINERS.
         8] DIKE AREA FOR CONTROL AND CONTAINMENT TO PREVENT ENTRY INTO SEWERS, DRAIN, AND WATER WAYS.

LARGE SPILL/NO FIRE/RESCUE: WEAR NON-SEALED CHEMICAL PROTECTIVE CLOTHING, RUBBER BOOTS, GLOVES AND BREATHING
                           APPARATUS.

SPILL CONTROL AND CONTAINMENT:
   HOUSEHOLD SPILL: WEAR RUBBER GLOVES. WIPE UP WITH ABSORBENT CLOTH. PUT IN A DISPOSAL CONTAINER. WASH AREA WITH
   SOAP AND WATER, WIPE UP DRY WITH ABSORBENT CLOTH. PUT CLOTH IN A DISPOSAL CONTAINER.
   LARGE SPILL: PICK UP WITH ABSORBENT. PUT IN A DISPOSAL CONTAINER. SCRUB AREA WITH DETERGENT AND WATER. PICK UP
   WITH ABSORBENT. PUT IN DISPOSAL CONTAINER. SEAL AND REMOVE FROM THE WORK AREA.

HEALTH HAZARD INFORMATION:
   WHEN USING THIS PRODUCT WEAR: RUBBER GLOVES. Skin contact may cause irritation. Inhalation may cause irritation
   of the nose, throat, and lungs.

   A physician should be contacted if anyone develops any signs or symptoms and suspects that they are caused by
   exposure to this product.

FIRST AID:
   Eye Exposure:
   Flush with water for 15 minutes while lifting the upper and lower eye lids. Contact lenses should not be worn
   when working with this product. Get medical attention.

   Skin Exposure:
   Wash with soap and water. If irritation develops, get medical attention.

   Breathing:
   If irritation develops, move victim to fresh air.

   Swallowing:
   Immediately contact the local poison control center for advice. Keep the victim warm and at rest. Get medical
   attention.

PRODUCT: RAID FIRE ANT KILLER - FORMULA 2

INCOMPATIBILITY: NONE KNOWN

INGREDIENTS: CHLOROPYRIFOS   2.0%   TLV: .2 mg/m3   CAS#: 2921-88-2
             XYLENE   1-2%   TLV: 100 ppm   CAS#: 1330-20-7
             WATER   90-97% TLV: CAS#: 7732-18-5

IGNI TEMP: NA FP: NA LEL: NA UEL: NA VP: NA VD: NA SG: 1.0 PS: LIQUID APPEAR: ODOR: MILD
    PH FACTOR: 6.3

HAZARD RATINGS: H: F: R:

FIRE FIGHTING:
  HIGH EXPANSION FOAM, LOW EXPANSION FOAM, ALCOHOL FOAM; DRY CHEMICAL, CARBON DIOXIDE, WATER FOG.
  PROTECTIVE CLOTHING, RUBBER GLOVES, AND BREATHING APPARATUS.
WARNING: 1] STRUCTURAL PROTECTIVE CLOTHING IS PERMEABLE, REMAIN CLEAR OF SMOKE, WATER FALL OUT AND WATER RUN OFF.
         2] KEEP OUT OF THE REACH OF CHILDREN.
         3] CONTAINERS WILL MELT AND LEAK IN A FIRE.
         4] THERMAL DECOMPOSITION MAY YIELD CARBON DIOXIDE, CARBON MONOXIDE.
         5] DO NOT ALLOW CHILDREN OR PETS IN TREATED AREA UNTIL DRY.
         6] MOVE CONTAINERS FROM AREA IF WITHOUT RISK, COOL EXPOSED CONTAINERS.
         7] DIKE AREA FOR CONTROL AND CONTAINMENT TO PREVENT ENTRY INTO SEWERS, DRAIN, AND WATER WAYS.

LARGE SPILL/NO FIRE/RESCUE: WEAR RUBBER OR NEOPRENE BOOTS, GLOVES.

SPILL CONTROL AND CONTAINMENT:
  HOUSEHOLD SPILL: WEAR RUBBER GLOVES. WIPE UP WITH ABSORBENT CLOTH. PUT CLOTH IN DISPOSAL CONTAINER. WASH WITH
  SOAP AND WATER. WIPE UP DRY WITH ABSORBENT CLOTH. PUT CLOTH IN DISPOSAL CONTAINER.
  LARGE SPILL: PICK UP WITH ABSORBENT. PUT IN A DISPOSAL CONTAINER. SCRUB AREA WITH DETERGENT AND WATER. PICK UP
  WITH ABSORBENT. PUT IN DISPOSAL CONTAINER. SEAL AND REMOVE FROM THE WORK AREA.

HEALTH HAZARD INFORMATION:
  WHEN USING THIS PRODUCT WEAR: RUBBER GLOVES. Prolonged contact may cause skin irritation. Eye contact will cause
  stinging, tearing, itching, swelling, and redness. Ingestion may cause nausea, vomiting, and illness. Inhalation
  of spray mist may cause irritation of the nose, throat, and lungs.

  A physician should be contacted if anyone develops any signs or symptoms and suspects that they are caused by
  exposure to this product.

FIRST AID:
  Eye Exposure:
  Flush with water for 15 minutes while lifting the upper and lower eye lids. Contact lenses should not be worn
  when working with this product. Get medical attention.

  Skin Exposure:
  Wash with soap and water. If irritation develops, get medical attention.

  Breathing:
  If irritation develops, move victim to fresh air.

  Swallowing:
  If victim is conscious, give 2-3 glasses of milk or water to drink. Immediately contact the local poison control
  center for advice. Keep the victim warm and at rest. Get medical attention.

PRODUCT: RAID FLEA FUMIGATION CARTRIDGE

INCOMPATIBILITY: STRONG ACIDS (e.g. SWIMMING POOL ACID) WATER

INGREDIENTS: PERMETHRIN	20%	TLV:	CAS#: 52645-53-1
PYNAMIN FORTE	5%	TLV:	CAS#: 42534-61-2
INERT INGREDIENTS	75%	TLV:	CAS#:

IGNI TEMP: NA FP: NA LEL: NA UEL: NA VP: NA VD: NA SG: NA PS: SOLID APPEAR: CYLINDRICAL CONTAINER
   ODOR: NONE   PH FACTOR: NA

HAZARD RATINGS: H: F: R:

FIRE FIGHTING:
  HIGH EXPANSION FOAM, LOW EXPANSION FOAM, ALCOHOL FOAM; DRY CHEMICAL, CARBON DIOXIDE, WATER FOG.
  PROTECTIVE CLOTHING, RUBBER GLOVES, AND BREATHING APPARATUS.
WARNING: 1] STRUCTURAL PROTECTIVE CLOTHING IS PERMEABLE, REMAIN CLEAR OF SMOKE, WATER FALL OUT AND WATER RUN OFF.
         2] KEEP OUT OF THE REACH OF CHILDREN.
         3] DO NOT REMAIN IN ENCLOSED AREA AFTER USE.
         4] REMOVE ALL PLANTS, PETS, BIRDS AND FOOD BEFORE USING.
         5] VENTILATE AREA BEFORE REOCCUPYING.
         6] THERMAL DECOMPOSITION MAY YIELD CARBON DIOXIDE, CARBON MONOXIDE.
         7] MOVE CONTAINERS FROM AREA IF WITHOUT RISK, COOL EXPOSED CONTAINERS.
         8] DIKE AREA FOR CONTROL AND CONTAINMENT TO PREVENT ENTRY INTO SEWERS, DRAIN, AND WATER WAYS.

LARGE SPILL/NO FIRE/RESCUE: WEAR TYVEK NON-SEALED SUITE RUBBER BOOTS, GLOVES AND BREATHING APPARATUS.

SPILL CONTROL AND CONTAINMENT:
  HOUSEHOLD SPILL: VENTILATE AREA. WEAR RUBBER GLOVES. SWEEP OR SCRAPE UP MATERIAL. PUT IN A DISPOSAL CONTAINER.
  SCRUB AREA WITH SOAP AND WATER. WIPE UP DRY WITH AN ABSORBENT CLOTH. PUT CLOTH IN DISPOSAL CONTAINER. WASH HANDS
  OR EXPOSED SKIN WITH SOAP AND WATER.
  LARGE SPILL: SWEEP OR SCRAPE UP AND PUT IN A DISPOSAL CONTAINER. SCRUB AREA WITH DETERGENT AND WATER. PICK UP
  WITH ABSORBENT. PUT IN DISPOSAL CONTAINER. SEAL AND REMOVE CONTAINER FROM THE WORK AREA.

HEALTH HAZARD INFORMATION:
  WHEN USING THIS PRODUCT WEAR: NO SPECIAL REQUIREMENT. Eye contact can cause stinging, tearing, itching,
  swelling, and redness. Ingestion can cause nausea, vomiting, and illness. Fog may cause irritation of the nose,
  throat, and lungs.

  A physician should be contacted if anyone develops any signs or symptoms and suspects that they are caused by
  exposure to this product.

FIRST AID:
  Eye Exposure:
  Flush with water for 15 minutes while lifting the upper and lower eye lids. Contact lenses should not be worn
  when working with this product. Get medical attention.

  Skin Exposure:
  Wash with soap and water. If irritation develops, get medical attention.

  Breathing:
  If irritation develops, move victim to fresh air. Get medical attention if breathing becomes labored.

  Swallowing:
  If the victim is conscious, give 2-3 glasses of milk or water to drink. Immediately contact the local poison
  control center for advice. Keep the victim warm and at rest. Get medical attention.

PRODUCT: RAID FLEA KILLER

INCOMPATIBILITY: NONE KNOWN

INGREDIENTS: PYRETHRINS   .140% TLV: CAS#:
             PIPERONYL BUTOXIDE   1.000% TLV: CAS#:   31-03-6
             TETRAMETHRIN   .063% TLV: CAS#:
             WATER   65-80.000% TLV: CAS#: 7732-18-5
             ISOBUTANE_____   TLV: CAS#:   75-28-5
             PROPANE_____/   TLV: 1000 ppm   CAS#:   74-98-6
N-OCTYL BICYCLOHEPTENE DICARBOXIMIDE .98% TLV: CAS#:

IGNI TEMP: NA FP: 20F LEL: NA UEL: NA VP: NA VD: NA SG: .8 PS: LIQUID APPEAR: SPRAY MIST   ODOR: MILD
    PH FACTOR: NA

HAZARD RATINGS: H: F: R: 0 4 0

FIRE FIGHTING:
  LOW EXPANSION FOAM, ALCOHOL FOAM; DRY CHEMICAL, CARBON DIOXIDE, WATER FOG.
  PROTECTIVE CLOTHING, RUBBER GLOVES, AND BREATHING APPARATUS.
WARNING: 1] STRUCTURAL PROTECTIVE CLOTHING IS PERMEABLE, REMAIN CLEAR OF SMOKE, WATER FALL OUT AND WATER RUN OFF.
         2] KEEP OUT OF THE REACH OF CHILDREN.
         3] DO NOT REMAIN IN ENCLOSED AREA AFTER USE.
         4] VENTILATE AREAS BEFORE REOCCUPYING.
         5] CONTAINERS WILL EXPLODE IN A FIRE OR IF HEATED ABOVE 120F.
         6] REMOVE ALL IGNITION SOURCES BEFORE USE.
         7] MOVE CONTAINERS FROM AREA IF WITHOUT RISK, COOL EXPOSED CONTAINERS.
         8] DIKE AREA FOR CONTROL AND CONTAINMENT TO PREVENT ENTRY INTO SEWERS, DRAIN, AND WATER WAYS.

LARGE SPILL/NO FIRE/RESCUE: WEAR NON-SEALED CHEMICAL PROTECTIVE CLOTHING, BOOTS, GLOVES, AND BREATHING APPARATUS.

SPILL CONTROL AND CONTAINMENT:
  HOUSEHOLD SPILL: WEAR RUBBER GLOVES. WIPE UP WITH ABSORBENT CLOTH. PUT IN A DISPOSAL CONTAINER. WASH AREA WITH
  SOAP AND WATER. WIPE UP DRY WITH AN ABSORBENT CLOTH. PUT IN A DISPOSAL CONTAINER.
  LARGE SPILL: PICK UP WITH ABSORBENT. PUT IN A DISPOSAL CONTAINER. SCRUB AREA WITH DETERGENT AND WATER. PICK UP
  WITH ABSORBENT. PUT IN DISPOSAL CONTAINER. SEAL AND REMOVE FROM THE WORK AREA.

HEALTH HAZARD INFORMATION:
  WHEN USING THIS PRODUCT WEAR: NO SPECIAL REQUIREMENT. Eye contact may cause stinging, tearing, itching,
  swelling, and redness. Ingestion may result in nausea, vomiting, and diarrhea.

  A physician should be contacted if anyone develops any signs or symptoms and suspects that they are caused by
  exposure to this product.

FIRST AID:
Eye Exposure:
Flush with water for 15 minutes while lifting the upper and lower eye lids. Contact lenses should not be worn
when working with this product. Get medical attention.

Skin Exposure:
Wash with soap and water. If irritation develops, get medical attention.

Breathing:
If irritation develops, move victim to fresh air.

Swallowing:
Immediately contact the local poison control center for advice. Keep the victim warm and at rest. Get medical
attention.

```
PRODUCT: RAID  FLEA KILLER PLUS


INCOMPATIBILITY: NONE KNOWN


INGREDIENTS: METHOPRENE            .03% TLV:              CAS#: 40596-69-8
             PIPERONYL BUTOXIDE   1.00% TLV:              CAS#:    51-03-6
             METHYLENE CHLORIDE   .1-.2% TLV:             CAS#:    75-09-2
             PROPANE_____              TLV: 1000 ppm     CAS#:    74-98-6
             ISOBUTANE_____     20-35% TLV:             CAS#:    75-28-5
             N-BUTANE___/              TLV:  800 ppm     CAS#:   106-97-8
             PYRETHRINS           .14% TLV:              CAS#:
             TETRAMETHRIN         .063% TLV:             CAS#:  7696-12-0
       N-OCTYL BICYCLOHEPTENE DICARBOXIMIDE        1.00%


IGNI TEMP: NA  FP: 20F  LEL: NA  UEL: NA  VP: NA  VD: NA  SG: .83  PS: LIQUID  APPEAR: SPRAY  ODOR: MILD
   PH FACTOR: NA


HAZARD RATINGS:   H: F: R:    1  4  0
```

FIRE FIGHTING:
  HIGH EXPANSION FOAM, ALCOHOL FOAM; DRY CHEMICAL, CARBON DIOXIDE, WATER FOG.
  PROTECTIVE CLOTHING, RUBBER GLOVES, AND BREATHING APPARATUS.
WARNING: 1] STRUCTURAL PROTECTIVE CLOTHING IS PERMEABLE, REMAIN CLEAR OF SMOKE, WATER FALL OUT AND WATER RUN OFF.
         2] KEEP OUT OF THE REACH OF CHILDREN.
         3] REMOVE ALL PLANTS, PETS, BIRDS FROM AREA. DO NOT STAY AFTER USE.
         4] VENTILATE BEFORE REOCCUPYING.
         5] CONTAINERS WILL EXPLODE IN A FIRE OR IF HEATED ABOVE 130F.
         6] THERMAL DECOMPOSITION MAY YIELD CARBON DIOXIDE, CARBON MONOXIDE.
         7] MOVE CONTAINERS FROM AREA IF WITHOUT RISK, COOL EXPOSED CONTAINERS.
         8] DIKE AREA FOR CONTROL AND CONTAINMENT TO PREVENT ENTRY INTO SEWERS, DRAIN, AND WATER WAYS.

LARGE SPILL/NO FIRE/RESCUE: WEAR NON-SEALED CHEMICAL PROTECTIVE CLOTHING, BOOTS, GLOVES AND BREATHING APPARATUS.

SPILL CONTROL AND CONTAINMENT:
  HOUSEHOLD SPILL: WEAR RUBBER GLOVES. WIPE UP DRY WITH AN ABSORBENT CLOTH. WASH WITH SOAP AND WATER. WIPE UP DRY
  AND PUT CLOTHS IN DISPOSAL CONTAINER.
  LARGE SPILL: PICK UP WITH ABSORBENT. RINSE AREA WITH WATER. PICK UP WITH ABSORBENT. PUT ALL IN A DISPOSAL
  CONTAINER. SEAL AND REMOVE FROM THE WORK AREA.

HEALTH HAZARD INFORMATION:
  WHEN USING THIS PRODUCT WEAR: NO SPECIAL REQUIREMENTS. Eye contact may cause stinging, tearing, itching,
  swelling, and redness. Inhalation may cause irritation of the nose, throat, and lungs. Ingestion may cause
  nausea, vomiting, and illness.

  A physician should be seen if anyone develops any signs or symptoms and suspects that they are caused by
  exposure to this product.

FIRST AID:
  Eye Exposure:
  Flush with water for 15 minutes while lifting the upper and lower eye lids. Contact lenses should not be worn
  when working with this product. Get medical attention.

  Skin Exposure:
  Wash with soap and water. If irritation develops, get medical attention.

  Breathing:
  If symptoms develop, move victim to fresh air.

  Swallowing:
  Immediately contact the local poison control center for advice. Keep the victim warm and at rest. Get medical
  attention.

PRODUCT: RAID FUMIGATOR

INCOMPATIBILITY: STRONG ACIDS (e.g. SWIMMING POOL ACID)

INGREDIENTS: PERMETHRIN   12.6% TLV: CAS#: 52645-53-1
             CALCIUM OXIDE   TLV: 2 mg/m3   CAS#: 1305-78-8

IGNI TEMP: NA FP: 400F LEL: NA UEL: NA VP: NA VD: NA SG: NA PS: SOLID APPEAR: CYLINDRICAL CONTAINER
   ODOR: NONE   PH FACTOR: NA

HAZARD RATINGS: H: F: R: 1 0 1

FIRE FIGHTING:
   HIGH EXPANSION FOAM, LOW EXPANSION FOAM, ALCOHOL FOAM; DRY CHEMICAL, CARBON DIOXIDE, WATER FOG.
   PROTECTIVE CLOTHING, RUBBER GLOVES, AND BREATHING APPARATUS.
WARNING: 1] STRUCTURAL PROTECTIVE CLOTHING IS PERMEABLE, REMAIN CLEAR OF SMOKE, WATER FALL OUT AND WATER RUN OFF.
         2] KEEP OUT OF THE REACH OF CHILDREN.
         3] REMOVE PLANTS, PETS, BIRDS FROM AREA. COVER & TURN OFF FISH TANKS.
         4] DO NOT REMAIN ON TREATED BUILDING.
         5] VENTILATE BEFORE REOCCUPYING.
         6] THERMAL DECOMPOSITION MAY YIELD CARBON DIOXIDE, CARBON MONOXIDE.
         7] MOVE CONTAINERS FROM AREA IF WITHOUT RISK, COOL EXPOSED CONTAINERS.
         8] DIKE AREA FOR CONTROL AND CONTAINMENT TO PREVENT ENTRY INTO SEWERS, DRAIN, AND WATER WAYS.

LARGE SPILL/NO FIRE/RESCUE: IF ACTIVATED WEAR NON-SEALED CHEMICAL PROTECTIVE CLOTHING, BOOTS, GLOVES, BREATHING
                           APPARATUS.

SPILL CONTROL AND CONTAINMENT:
   HOUSEHOLD SPILL: SWEEP UP AND PUT IN DISPOSAL CONTAINER.
   LARGE SPILL: SCOOP UP AND PUT IN A DISPOSAL CONTAINER. SWEEP UP RESIDUE AND PUT IN A DISPOSAL CONTAINER.

HEALTH HAZARD INFORMATION:
   WHEN USING THIS PRODUCT: REMOVE ALL PLANTS, PETS AND BIRDS. LEAVE TREATMENT AREA FOR SPECIFIED TIME. VENTILATE
   BEFORE REOCCUPYING. Eye contact will cause irritation. Inhaling fog will irritate the nose, throat, and lungs.
   Ingestion will cause abdominal pain, nausea, vomiting, and illness. Harmful if absorbed through the skin.

   A physician should be contacted if anyone develops any signs or symptoms and suspects that they are caused by
   exposure to this product.

FIRST AID:
   Eye Exposure:
   Flush with water for 15 minutes while lifting the upper and lower eye lids. Contact lenses should not be worn
   when working with this product. Get medical attention.

   Skin Exposure:
   Wash with soap and water. If irritation develops, get medical attention.

   Breathing:
   If symptoms develop, move victim to fresh air. Get medical attention.

   Swallowing:
   Immediately contact the local poison control center for advice. Keep the victim warm and at rest. Get medical
   attention.

PRODUCT: RAID GYPSY MOTH & JAPANESE BEETLE KILLER

INCOMPATIBILITY: NONE KNOWN

```
INGREDIENTS: (5-BENZYL-3-FURYL) METHYL
             2,2-DIMETHYL-3-(2-METHYLPROPENYL)
             CYCLOPROPANE CARBOXYLATE    .37% TLV:          CAS#: 10453-86-8
             SURFACTANT            25-35.00% TLV:          CAS#:
             ISOBUTANE              5-15.00% TLV:          CAS#:   75-28-5
             WATER                 55-65.00% TLV:          CAS#:  7732-18-5
```

IGNI TEMP: NA FP: 20F LEL: NA UEL: NA VP: NA VD: NA SG: .9 PS: LIQUID APPEAR: SPRAY ODOR: MILD
   PH FACTOR: NA

HAZARD RATINGS: H: F: R: 1 4 0

FIRE FIGHTING:
  HIGH EXPANSION FOAM, LOW EXPANSION FOAM, ALCOHOL FOAM; DRY CHEMICAL, CARBON DIOXIDE, WATER FOG.
  PROTECTIVE CLOTHING, RUBBER GLOVES, AND BREATHING APPARATUS.
WARNING: 1] STRUCTURAL PROTECTIVE CLOTHING IS PERMEABLE, REMAIN CLEAR OF SMOKE, WATER FALL OUT AND WATER RUN OFF.
         2] KEEP OUT OF THE REACH OF CHILDREN.
         3] REMOVE ALL IGNITION SOURCES IF WITHOUT RISK.
         4] CONTAINERS WILL EXPLODE IN A FIRE OR IF HEATED ABOVE 120F.
         5] THERMAL DECOMPOSITION MAY YIELD CARBON DIOXIDE, CARBON MONOXIDE.
         6] MOVE CONTAINERS FROM AREA IF WITHOUT RISK, COOL EXPOSED CONTAINERS.
         7] DIKE AREA FOR CONTROL AND CONTAINMENT TO PREVENT ENTRY INTO SEWERS, DRAIN, AND WATER WAYS.

LARGE SPILL/NO FIRE/RESCUE: WEAR RUBBER OR NEOPRENE BOOTS, GLOVES.

SPILL CONTROL AND CONTAINMENT:
  HOUSEHOLD SPILL: WIPE UP WITH AN ABSORBENT CLOTH. RINSE OUT IN THE SINK. WASH AREA WITH SOAP AND WATER. WIPE UP
  DRY WITH ABSORBENT CLOTH. RINSE OUT IN THE SINK.
  LARGE SPILL: PICK UP WITH ABSORBENT. PUT IN A DISPOSAL CONTAINER. RINSE AREA WITH WATER. PICK UP WITH ABSORBENT.
  PUT IN DISPOSAL CONTAINER. SEAL AND REMOVE FROM THE WORK AREA.

HEALTH HAZARD INFORMATION:
  WHEN USING THIS PRODUCT WEAR: NO SPECIAL REQUIREMENT. Eye contact will cause stinging, tearing, itching,
  swelling, and redness. Ingestion may cause nausea, vomiting, and illness. Inhalation may cause irritation of the
  nose, throat, and lungs.

  A physician should be contacted if anyone develops any signs or symptoms and suspects that they are caused by
  exposure to this product.

FIRST AID:
  Eye Exposure:
  Flush with water for 15 minutes while lifting the upper and lower eye lids. Contact lenses should not be worn
  when working with this product. Get medical attention.

  Skin Exposure:
  Wash with soap and water. If irritation develops, get medical attention.

  Breathing:
  If symptoms develop, move victim to fresh air.

  Swallowing:
  Immediately contact the local poison control center for advice. Keep the victim warm and at rest. Get medical
  attention.

PRODUCT: RAID HOUSE & GARDEN BUG KILLER

INCOMPATIBILITY: NONE KNOWN

INGREDIENTS: ISOPARAFFINIC   Recommended
             ·HYDROCARBON SOLVENT   1-3% TLV: 300 ppm   CAS#: 64742-47-8
              PROPANE_____   TLV: 1000 ppm   CAS#:   74-98-6
              ISOBUTANE_____ 25-35% TLV: CAS#:   75-28-5
              N-BUTANE____/   TLV: CAS#: 106-97-8
              WATER   65-75% TLV: CAS#: 7732-18-5

IGNI TEMP: NA FP: 20F LEL: NA UEL: NA VP: NA VD: NA SG: .79 PS: LIQUID APPEAR: SPRAY ODOR: MILD
    PH FACTOR: NA

HAZARD RATINGS: H: F: R: 1 4 0

FIRE FIGHTING:
    HIGH EXPANSION FOAM, LOW EXPANSION FOAM, ALCOHOL FOAM; DRY CHEMICAL, CARBON DIOXIDE, WATER FOG.
    PROTECTIVE CLOTHING, RUBBER GLOVES, AND BREATHING APPARATUS.
WARNING: 1] STRUCTURAL PROTECTIVE CLOTHING IS PERMEABLE, REMAIN CLEAR OF SMOKE, WATER FALL OUT AND WATER RUN OFF.
         2] KEEP OUT OF THE REACH OF CHILDREN.
         3] REMOVE ALL IGNITION SOURCES IF WITHOUT RISK.
         4] CONTAINERS WILL EXPLODE IN A FIRE OR IF HEATED ABOVE 120F.
         5] THERMAL DECOMPOSITION MAY YIELD CARBON DIOXIDE, CARBON MONOXIDE.
         6] MOVE CONTAINERS FROM AREA IF WITHOUT RISK, COOL EXPOSED CONTAINERS.
         7] DIKE AREA FOR CONTROL AND CONTAINMENT TO PREVENT ENTRY INTO SEWERS, DRAIN, AND WATER WAYS.

LARGE SPILL/NO FIRE/RESCUE: WEAR RUBBER OR NEOPRENE BOOTS, GLOVES.

SPILL CONTROL AND CONTAINMENT:
    HOUSEHOLD SPILL: WIPE UP DRY WITH AN ABSORBENT CLOTH. RINSE OUT IN THE SINK. WASH HANDS WITH SOAP AND WATER.
    LARGE SPILL: PICK UP WITH ABSORBENT. PUT IN A DISPOSAL CONTAINER. RINSE AREA WITH WATER. PICK UP WITH ABSORBENT.
    PUT IN A DISPOSAL CONTAINER. SEAL AND REMOVE FROM THE WORK AREA.

HEALTH HAZARD INFORMATION:
    WHEN USING THIS PRODUCT WEAR: NO SPECIAL REQUIREMENT. Eye contact will cause stinging, tearing, itching,
    swelling, and redness. Inhalation may cause irritation of the nose and throat.

    A physician should be contacted if anyone develops any signs or symptoms and suspects that they are caused by
    exposure to this product.

FIRST AID:
    Eye Exposure:
    Flush with water for 15 minutes while lifting the upper and lower eye lids. Contact lenses should not be worn
    when working with this product. Get medical attention.

    Skin Exposure:
    Wash with soap and water. If irritation develops, get medical attention.

    Breathing:
    If irritation develops, move victim to fresh air.

    Swallowing:
    Immediately contact the local poison control center for advice. Keep the victim warm and at rest. Get medical
    attention.

PRODUCT: RAID INDOOR FOGGER PLUS

INCOMPATIBILITY: STRONG ACIDS (e.g. SWIMMING POOL ACID; OXIDIZERS LIKE CHLORINE.)

INGREDIENTS: ISOPARAFFINIC   Recommended

HYDROCARBON SOLVENT	8-12%	TLV: 300 ppm	CAS#: 64742-47-8
PIPERONYL BUTOXIDE	1-3%	TLV:	CAS#: 51-03-6
1,1,1-TRICHLOROETHANE	60-75%	TLV: 350 ppm	CAS#: 71-55-6
PROPANE	15-20%	TLV: 1000 ppm	CAS#: 74-98-6
ISOBUTANE	5-10%	TLV:	CAS#: 75-28-5
N-OCTYL BICYCLOHEPTENE DICARBOXIMIDE	1-3%	TLV:	CAS#: 113-48-4

IGNI TEMP: NA FP: 20F LEL: NA UEL: NA VP: NA VD: NA SG: 1.17 PS: LIQUID APPEAR: SPRAY MIST
     ODOR: SOLVENT   PH FACTOR: NA

HAZARD RATINGS: H: F: R: 1 4 0

FIRE FIGHTING:
  HIGH EXPANSION FOAM, ALCOHOL FOAM; DRY CHEMICAL, CARBON DIOXIDE, WATER FOG.
  PROTECTIVE CLOTHING, RUBBER GLOVES, AND BREATHING APPARATUS.
WARNING: 1] STRUCTURAL PROTECTIVE CLOTHING IS PERMEABLE, REMAIN CLEAR OF SMOKE, WATER FALL OUT AND WATER RUN OFF.
        2] KEEP OUT OF THE REACH OF CHILDREN.
        3] REMOVE ALL IGNITION SOURCES.
        4] REMOVE PLANTS, PETS, BIRDS FROM AREA. COVER AND TURN OFF FISH TANKS.
        5] VENTILATE AREA BEFORE REOCCUPYING. DO NOT STAY DURING USE.
        6] CONTAINERS WILL EXPLODE IN A FIRE OR IF HEATED ABOVE 120F.
        7] THERMAL DECOMPOSITION MAY YIELD CARBON DIOXIDE, CARBON MONOXIDE.
        8] MOVE CONTAINERS FROM AREA IF WITHOUT RISK, COOL EXPOSED CONTAINERS.
        9] DIKE AREA FOR CONTROL AND CONTAINMENT TO PREVENT ENTRY INTO SEWERS, DRAIN, AND WATER WAYS.

LARGE SPILL/NO FIRE/RESCUE: WEAR NON-SEALED CHEMICAL PROTECTIVE CLOTHING, BOOTS, GLOVES AND BREATHING APPARATUS.

SPILL CONTROL AND CONTAINMENT:
  HOUSEHOLD SPILL: WEAR RUBBER GLOVES. WIPE UP DRY WITH AN ABSORBENT CLOTH. WASH AREA WITH SOAP AND WATER. WIPE UP
  DRY. PUT CLOTHS IN DISPOSAL CONTAINER.
  LARGE SPILL: PICK UP WITH ABSORBENT. PUT IN A DISPOSAL CONTAINER. RINSE AREA WITH WATER. PICK UP WITH ABSORBENT,
  PUT IN A DISPOSAL CONTAINER. SEAL AND REMOVE FROM THE WORK AREA.

HEALTH HAZARD INFORMATION:
  WHEN USING THIS PRODUCT WEAR: NO SPECIAL REQUIREMENT. Eye contact will cause irritation. Inhaled vapors may
  cause chemical intoxication at high levels. Ingestion may cause nausea and vomiting. If product gets into the
  lungs it may cause chemical pneumonia. May cause skin irritation.

  A physician should be contacted if anyone develops any signs or symptoms and suspects that they are caused by
  exposure to this product.

FIRST AID:
  Eye Exposure:
  Flush with water for 15 minutes while lifting the upper and lower eye lids. Contact lenses should not be worn
  when working with this product. Get medical attention.

  Skin Exposure:
  Wash with soap and water. If irritation develops, get medical attention.

  Breathing:
  If irritation develops, move victim to fresh air.

  Swallowing:
  Immediately contact the local poison control center for advice. Keep the victim warm and at rest. Get medical
  attention.

PRODUCT: RAID INDOOR FOGGER PLUS

INCOMPATIBILITY: STRONG ACIDS (e.g. SWIMMING POOL ACID; OXIDIZERS LIKE CHLORINE.)

INGREDIENTS: ALIPHATIC NAPHTHA   8-12% TLV: 300 ppm RECOMMENDED   CAS#: 64742-47-8
             PIPERONYL BUTOXIDE TECH.   1% TLV: CAS#:   51-03-6
             1,1,1-TRICHLOROETHANE   55-70% TLV: 350 ppm   CAS#:   71-55-6
             PROPANE_____   20-30% TLV: 1000 ppm   CAS#:   74-98-6
             ISOBUTANE____/   TLV: CAS#:   75-28-5
N-OCTYL BICYCLOHEPTENE DICARBOXIMIDE 1.67% TLV: CAS#: 113-48-4

IGNI TEMP: NA FP: 20F LEL: NA UEL: NA VP: NA VD: NA SG: .9 PS: LIQUID APPEAR: SPRAY MIST
     ODOR: CHLORINATED SOLVENT   PH FACTOR: NA

HAZARD RATINGS: H: F: R: 1 4 0

FIRE FIGHTING:
  HIGH EXPANSION FOAM, ALCOHOL FOAM; DRY CHEMICAL, CARBON DIOXIDE, WATER FOG.
  PROTECTIVE CLOTHING, RUBBER GLOVES, AND BREATHING APPARATUS.
WARNING: 1] STRUCTURAL PROTECTIVE CLOTHING IS PERMEABLE, REMAIN CLEAR OF SMOKE, WATER FALL OUT AND WATER RUN OFF.
         2] KEEP OUT OF THE REACH OF CHILDREN.
         3] REMOVE ALL IGNITION SOURCES
         4] REMOVE PLANTS, PETS, BIRDS FROM AREA. COVER & TURN OFF FISH TANKS.
         5] VENTILATE AREA BEFORE REOCCUPYING. DO NOT STAY DURING USE
         6] CONTAINERS WILL EXPLODE IN A FIRE OR IF HEATED ABOVE 120F.
         7] THERMAL DECOMPOSITION MAY YIELD CARBON DIOXIDE, CARBON MONOXIDE.
         8] MOVE CONTAINERS FROM AREA IF WITHOUT RISK, COOL EXPOSED CONTAINERS.
         9] DIKE AREA FOR CONTROL AND CONTAINMENT TO PREVENT ENTRY INTO SEWERS, DRAIN, AND WATER WAYS.

LARGE SPILL/NO FIRE/RESCUE: WEAR NON-SEALED CHEMICAL PROTECTIVE CLOTHING, BOOTS, GLOVES AND BREATHING APPARATUS.

SPILL CONTROL AND CONTAINMENT:
  HOUSEHOLD SPILL: WEAR RUBBER GLOVES. WIPE UP DRY WITH AN ABSORBENT CLOTH. WASH AREA WITH SOAP AND WATER. WIPE UP
  DRY. PUT CLOTHS IN DISPOSAL CONTAINER.
  LARGE SPILL: PICK UP WITH ABSORBENT. PUT IN A DISPOSAL CONTAINER. RINSE AREA WITH WATER. PICK UP WITH ABSORBENT,
  PUT IN A DISPOSAL CONTAINER. SEAL AND REMOVE FROM THE WORK AREA.

HEALTH HAZARD INFORMATION:
  WHEN USING THIS PRODUCT WEAR: NO SPECIAL REQUIREMENT. Eye contact will cause irritation. Inhaled vapors may
  cause chemical intoxication at high levels. Ingestion may cause nausea and vomiting. If product gets into the
  lungs it may cause chemical pneumonia. May cause skin irritation.

  A physician should be contacted if anyone develops any signs or symptoms and suspects that they are caused by
  exposure to this product.

FIRST AID:
  Eye Exposure:
  Flush with water for 15 minutes while lifting the upper and lower eye lids. Contact lenses should not be worn
  when working with this product. Get medical attention.

  Skin Exposure:
  Wash with soap and water. If irritation develops, get medical attention.

  Breathing:
  If irritation develops, move victim to fresh air.

  Swallowing:
  Immediately contact the local poison control center for advice. Keep the victim warm and at rest. Get medical
  attention.

PRODUCT: RAID   MOSQUITO REPELLENT COIL

INCOMPATIBILITY: NONE KNOWN

INGREDIENTS: ALLETHRIN   .35% TLV: CAS#: 584-79-2
             VEGETABLE MATTER_____ 98.04% TLV: CAS#:
             WOOD FIBER_____/   TLV: CAS#:
             COLOR & PRESERVATIVE 1.61% TLV: CAS#:

IGNI TEMP: NA FP: NA LEL: NA UEL: NA VP: NA VD: NA SG: NA PS: COIL SHAPED SOLID APPEAR: GREEN ODOR: NONE
  PH FACTOR: NA

HAZARD RATINGS: H: F: R:

FIRE FIGHTING:
  HIGH EXPANSION FOAM, LOW EXPANSION FOAM, ALCOHOL FOAM; DRY CHEMICAL, CARBON DIOXIDE, WATER FOG.
  PROTECTIVE CLOTHING, RUBBER GLOVES, AND BREATHING APPARATUS.
WARNING: 1] STRUCTURAL PROTECTIVE CLOTHING IS PERMEABLE, REMAIN CLEAR OF SMOKE, WATER FALL OUT AND WATER RUN OFF.
         2] KEEP OUT OF THE REACH OF CHILDREN.
         3] COIL MAY RE-IGNITE AFTER EXTINGUISHMENT
         4] THERMAL DECOMPOSITION MAY YIELD CARBON DIOXIDE, CARBON MONOXIDE.
         5] MOVE CONTAINERS FROM AREA IF WITHOUT RISK, COOL EXPOSED CONTAINERS.
         6] DIKE AREA FOR CONTROL AND CONTAINMENT TO PREVENT ENTRY INTO SEWERS, DRAIN, AND WATER WAYS.

LARGE SPILL/NO FIRE/RESCUE: WEAR RUBBER OR NEOPRENE BOOTS, GLOVES.

SPILL CONTROL AND CONTAINMENT:
  HOUSEHOLD SPILL: SWEEP UP AND PUT IN A DISPOSAL CONTAINER.
  LARGE SPILL: SWEEP OR SCOOP UP AND PUT IN A CONTAINER FOR RECYCLE OR DISPOSAL.

HEALTH HAZARD INFORMATION:
  WHEN USING THIS PRODUCT WEAR: NO SPECIAL REQUIREMENT. Inhalation of dust or fumes may irritate the nose, throat,
  and lungs. If so, remove coil, extinguish, and ventilate area.

  A physician should be contacted if anyone develops any signs or symptoms and suspects that they are caused by
  exposure to this product.

FIRST AID:
  Eye Exposure:
  Flush with water for 15 minutes while lifting the upper and lower eye lids. Contact lenses should not be worn
  when working with this product. Get medical attention.

  Skin Exposure:
  Wash with soap and water.

  Breathing:
  If irritation develops, move victim to fresh air.

  Swallowing:
  Immediately contact the local poison control center for advice. Keep the victim warm and at rest. Get medical
  attention.

PRODUCT: RAID   OUTDOOR FLEA KILLER

INCOMPATIBILITY: NONE KNOWN

INGREDIENTS: DURSBAN   3.81%   TLV: .2 mg/m3   CAS#: 2921-88-2
             XYLENE   2.30%   TLV: 100 ppm   CAS#: 1330-20-7
             ISOBUTANE   5-20%   TLV: CAS#:   75-28-5
             WATER   70-85%   TLV: CAS#: 7732-18-4

IGNI TEMP: NA FP: 20F LEL: NA UEL: NA VP: NA VD: NA SG: 1 PS: LIQUID APPEAR: SPRAY ODOR: MILD
    PH FACTOR: NA

HAZARD RATINGS: H: F: R: 1 4 0

FIRE FIGHTING:
    HIGH EXPANSION FOAM, LOW EXPANSION FOAM, ALCOHOL FOAM; DRY CHEMICAL, CARBON DIOXIDE, WATER FOG.
    PROTECTIVE CLOTHING, RUBBER GLOVES, AND BREATHING APPARATUS.
WARNING: 1] STRUCTURAL PROTECTIVE CLOTHING IS PERMEABLE, REMAIN CLEAR OF SMOKE, WATER FALL OUT AND WATER RUN OFF.
         2] KEEP OUT OF THE REACH OF CHILDREN.
         3] REMOVE ALL IGNITION SOURCES IF WITHOUT RISK.
         4] CONTAINERS WILL EXPLODE IN A FIRE OR IF HEATED ABOVE 120F.
         5] THERMAL DECOMPOSITION MAY YIELD CARBON DIOXIDE, CARBON MONOXIDE.
         6] MOVE CONTAINERS FROM AREA IF WITHOUT RISK, COOL EXPOSED CONTAINERS.
         7] DIKE AREA FOR CONTROL AND CONTAINMENT TO PREVENT ENTRY INTO SEWERS, DRAIN, AND WATER WAYS.

LARGE SPILL/NO FIRE/RESCUE: WEAR RUBBER OR NEOPRENE BOOTS, GLOVES.

SPILL CONTROL AND CONTAINMENT:
    HOUSEHOLD SPILL: WIPE UP DRY WITH AN ABSORBENT CLOTH. PUT IN A DISPOSAL CONTAINER.
    LARGE SPILL: PICK UP WITH ABSORBENT. RINSE AREA WITH WATER. PICK UP WITH ABSORBENT. PUT ALL ABSORBENT IN A
    DISPOSAL CONTAINER. SEAL AND REMOVE FROM THE WORK AREA.

HEALTH HAZARD INFORMATION:
    WHEN USING THIS PRODUCT WEAR: NO SPECIAL REQUIREMENT. Eye contact will cause stinging, tearing, itching,
    swelling, and redness. Prolonged contact with skin may cause irritation.

    A physician should be contacted if anyone develops any signs or symptoms and suspects that they are caused by
    exposure to this product.

FIRST AID:
Eye Exposure:
Flush with water for 15 minutes while lifting the upper and lower eye lids. Contact lenses should not be worn
when working with this product. Get medical attention.

Skin Exposure:
Wash with soap and water. If irritation develops, get medical attention.

Breathing:
Not expected to be a problem.

Swallowing:
Do not induce vomiting. If victim is conscious, give 2-3 glasses of milk or water to drink. Immediately contact
the local poison control center for advice. Keep the victim warm and at rest. Get medical attention.

PRODUCT: RAID   PROFESSIONAL STRENGTH ANT & ROACH KILLER (LIQUID)

INCOMPATIBILITY: NONE KNOWN

INGREDIENTS: BAYGON   .95% TLV: CAS#:   114-26-1
             ISOPROPYL ALCOHOL   10-20% TLV: 400 ppm   CAS#:   67-63-0
             ALIPHATIC NAPHTHA   80-90% TLV: 200 ppm   CAS#: 64742-47-8

IGNI TEMP: NA FP: 59F LEL: NA UEL: NA VP: NA VD: NA SG: .78 PS: LIQUID APPEAR: CLEAR   ODOR: NAPHTHA
   PH FACTOR: NA

HAZARD RATINGS: H: F: R: 1 3 0

FIRE FIGHTING:
   HIGH EXPANSION FOAM, ALCOHOL FOAM; DRY CHEMICAL, CARBON DIOXIDE, WATER FOG.
   PROTECTIVE CLOTHING, RUBBER GLOVES, AND BREATHING APPARATUS.
WARNING: 1] STRUCTURAL PROTECTIVE CLOTHING IS PERMEABLE, REMAIN CLEAR OF SMOKE, WATER FALL OUT AND WATER RUN OFF.
         2] KEEP OUT OF THE REACH OF CHILDREN.
         3] CONTAINERS MAY BURST IN A FIRE.
         4] THERMAL DECOMPOSITION MAY YIELD CARBON DIOXIDE, CARBON MONOXIDE.
         5] PRODUCT IS A CHOLINESTERASE INHIBITOR.
         6] MOVE CONTAINERS FROM AREA IF WITHOUT RISK, COOL EXPOSED CONTAINERS.
         7] DIKE AREA FOR CONTROL AND CONTAINMENT TO PREVENT ENTRY INTO SEWERS, DRAIN, AND WATER WAYS.

LARGE SPILL/NO FIRE/RESCUE: WEAR RUBBER OR NEOPRENE BOOTS, GLOVES, BREATHING APPARATUS.

SPILL CONTROL AND CONTAINMENT:
   HOUSEHOLD SPILL: WEAR RUBBER GLOVES. WIPE UP DRY WITH AN ABSORBENT CLOTH. PUT IN A DISPOSAL CONTAINER. WASH
   HANDS WITH SOAP AND WATER.
   LARGE SPILL: PICK UP WITH ABSORBENT. RINSE AREA WITH WATER. PICK UP WITH ABSORBENT. PUT ALL ABSORBENT IN A
   DISPOSAL CONTAINER. SEAL AND REMOVE FROM THE WORK AREA.

HEALTH HAZARD INFORMATION:
   WHEN USING THIS PRODUCT WEAR: NO SPECIAL REQUIREMENT. Eye contact will cause stinging, tearing, itching,
   swelling, and redness. Prolonged contact with the skin may cause irritation. Ingestion may cause nausea and
   vomiting. If product gets into the lungs, it may cause chemical pneumonia.

   A physician should be contacted if anyone develops any signs or symptoms and suspects that they are caused by
   exposure to this product.

FIRST AID:
   Eye Exposure:
   Flush with water for 15 minutes while lifting the upper and lower eye lids. Contact lenses should not be worn
   when working with this product. Get medical attention.

   Skin Exposure:
   Remove contaminated clothing. Wash with soap and water. If irritation develops, get medical attention.

   Breathing:
   If vapors cause discomfort, move victim to fresh air. If irritation persists, get medical attention.

   Swallowing:
   Immediately contact the local poison control center for advice. Keep the victim warm and at rest. Get medical
   attention.

PRODUCT: RAID   PROFESSIONAL STRENGTH ANT & ROACH KILLER (PRESSURIZED)

INCOMPATIBILITY: NONE KNOWN

INGREDIENTS: ISOPROPYL ALCOHOL   12-20% TLV: 400 ppm   CAS#:   67-63-0
             PETROLEUM SOLVENT   75-85% TLV: 100 ppm   CAS#: 64742-47-8
             CARBON DIOXIDE   2-4% TLV: 5000 ppm   CAS#: 124-38-9

IGNI TEMP: NA FP: 59F LEL: NA UEL: NA VP: NA VD: NA SG: .79 PS: LIQUID APPEAR: SPRAY ODOR: NAPHTHA
  PH FACTOR: NA

HAZARD RATINGS: H: F: R: 1 4 0

FIRE FIGHTING:
  HIGH EXPANSION FOAM, LOW EXPANSION FOAM, ALCOHOL FOAM; DRY CHEMICAL, CARBON DIOXIDE, WATER FOG.
  PROTECTIVE CLOTHING, RUBBER GLOVES, AND BREATHING APPARATUS.
WARNING: 1] STRUCTURAL PROTECTIVE CLOTHING IS PERMEABLE, REMAIN CLEAR OF SMOKE, WATER FALL OUT AND WATER RUN OFF.
         2] KEEP OUT OF THE REACH OF CHILDREN.
         3] CONTAINERS WILL EXPLODE IN A FIRE OR IF HEATED ABOVE 120F.
         4] REMOVE ALL IGNITION SOURCES IF WITHOUT RISK.
         5] THERMAL DECOMPOSITION MAY YIELD CARBON DIOXIDE, CARBON MONOXIDE.
         6] MOVE CONTAINERS FROM AREA IF WITHOUT RISK, COOL EXPOSED CONTAINERS.
         7] DIKE AREA FOR CONTROL AND CONTAINMENT TO PREVENT ENTRY INTO SEWERS, DRAIN, AND WATER WAYS.

LARGE SPILL/NO FIRE/RESCUE: WEAR RUBBER OR NEOPRENE BOOTS, GLOVES, BREATHING APPARATUS.

SPILL CONTROL AND CONTAINMENT:
  HOUSEHOLD SPILL: WIPE UP DRY WITH ABSORBENT CLOTH. PUT IN DISPOSAL CONTAINER.
  LARGE SPILL: USE NON-SPARKING TOOLS. PICK UP WITH ABSORBENT. RINSE AREA WITH WATER. PICK UP WITH ABSORBENT. PUT
  ALL ABSORBENT IN A DISPOSAL CONTAINER. SEAL AND REMOVE FROM THE WORK AREA.

HEALTH HAZARD INFORMATION:
  WHEN USING THIS PRODUCT WEAR: NO SPECIAL REQUIREMENT. Eye contact will cause stinging, tearing, itching,
  swelling, and redness. Ingestion can cause nausea, vomiting, and illness. If product gets into the lungs, it can
  cause chemical pneumonia. Prolonged contact with skin can cause irritation. Vapor may cause irritation of the
  nose, throat, and lungs.

  A physician should be contacted if anyone develops any signs or symptoms and suspects that they are caused by
  exposure to this product. Antidote is atropine use only after symptoms appear.

FIRST AID:
  Eye Exposure:
  Flush with water for 15 minutes while lifting the upper and lower eye lids. Contact lenses should not be worn
  when working with this product. Get medical attention.

  Skin Exposure:
  Wash with soap and water. If irritation develops, get medical attention.

  Breathing:
  If symptoms develop, move victim to fresh air.

  Swallowing:
  Immediately contact the local poison control center for advice. Keep the victim warm and at rest. Get medical
  attention.

PRODUCT: RAID   PROFESSIONAL STRENGTH FLYING INSECT KILLER

INCOMPATIBILITY: NONE KNOWN

INGREDIENTS: ISOPARAFFINIC   Recommended
             HYDROCARBON SOLVENT   5-10%   TLV: 300 ppm   CAS#: 64742-48-9
             PROPANE_____   TLV: 1000 ppm   CAS#:   74-98-6
             ISOBUTANE_____   20-35% TLV: CAS#:   75-28-5
             N-BUTANE_____/   TLV: 800 ppm   CAS#: 106-97-8
             WATER   55-65% TLV: CAS#: 7732-18-5

IGNI TEMP: NA FP: 20F LEL: NA UEL: NA VP: NA VD: NA SG: .80 PS: LIQUID APPEAR: SPRAY ODOR: NAPHTHA
    PH FACTOR: NA

HAZARD RATINGS: H: F: R: 0 4 0

FIRE FIGHTING:
    HIGH EXPANSION FOAM, LOW EXPANSION FOAM, ALCOHOL FOAM; DRY CHEMICAL, CARBON DIOXIDE, WATER FOG.
    PROTECTIVE CLOTHING, RUBBER GLOVES, AND BREATHING APPARATUS.
WARNING: 1] STRUCTURAL PROTECTIVE CLOTHING IS PERMEABLE, REMAIN CLEAR OF SMOKE, WATER FALL OUT AND WATER RUN OFF.
         2] KEEP OUT OF THE REACH OF CHILDREN.
         3] CONTAINERS WILL EXPLODE IN A FIRE OR IF HEATED ABOVE 120F.
         4] REMOVE ALL IGNITION SOURCES IF WITHOUT RISK.
         5] THERMAL DECOMPOSITION MAY YIELD CARBON DIOXIDE, CARBON MONOXIDE.
         6] MOVE CONTAINERS FROM AREA IF WITHOUT RISK, COOL EXPOSED CONTAINERS.
         7] DIKE AREA FOR CONTROL AND CONTAINMENT TO PREVENT ENTRY INTO SEWERS, DRAIN, AND WATER WAYS.

LARGE SPILL/NO FIRE/RESCUE: WEAR RUBBER OR NEOPRENE BOOTS, GLOVES.

SPILL CONTROL AND CONTAINMENT:
    HOUSEHOLD SPILL: WIPE UP DRY WITH AN ABSORBENT CLOTH. RINSE OUT IN THE SINK. WIPE UP RESIDUE WITH A DAMP CLOTH.
    RINSE OUT IN THE SINK.
    LARGE SPILL: PICK UP WITH ABSORBENT. RINSE AREA WITH WATER. PICK UP WITH ABSORBENT. PUT ALL ABSORBENT IN A
    DISPOSAL CONTAINER. SEAL AND REMOVE FROM THE WORK AREA.

HEALTH HAZARD INFORMATION:
    WHEN USING THIS PRODUCT WEAR: NO SPECIAL REQUIREMENT. Eye contact will cause stinging, tearing, itching,
    swelling, and redness. Inhalation of spray mist may cause irritation of the nose, throat, and lungs.

    A physician should be contacted if anyone develops any signs or symptoms and suspects that they are caused by
    exposure to this product.

FIRST AID:
    Eye Exposure:
    Flush with water for 15 minutes while lifting the upper and lower eye lids. Contact lenses should not be worn
    when working with this product. Get medical attention.

    Skin Exposure:
    Remove contaminated clothing. Wash with soap and water.

    Breathing:
    If irritation develops, move victim to fresh air. If it persists get medical attention.

    Swallowing:
    Do not induce vomiting. Immediately contact the local poison control center for advice. Keep the victim warm and
    at rest. Get medical attention.

PRODUCT: RAID   ROACH CONTROLLER   (BAIT)

INCOMPATIBILITY: NONE KNOWN

INGREDIENTS: CHLOROPYRIFOS   .5% TLV: .2 mg/m3   CAS#: 2921-88-2
             WAX; FOOD   99.5% TLV: CAS#:

IGNI TEMP: NA FP: NA LEL: NA UEL: NA VP: NA VD: NA SG: NA PS: SOLID APPEAR: YELLOW   ODOR: NA
    PH FACTOR: NA

HAZARD RATINGS: H: F: R: 1 0 0

FIRE FIGHTING:
   HIGH EXPANSION FOAM, LOW EXPANSION FOAM, ALCOHOL FOAM; DRY CHEMICAL, CARBON DIOXIDE, WATER FOG.
   PROTECTIVE CLOTHING, RUBBER GLOVES, AND BREATHING APPARATUS.
WARNING: 1] STRUCTURAL PROTECTIVE CLOTHING IS PERMEABLE, REMAIN CLEAR OF SMOKE, WATER FALL OUT AND WATER RUN OFF.
         2] KEEP OUT OF THE REACH OF CHILDREN.
         3] THERMAL DECOMPOSITION MAY YIELD CARBON DIOXIDE, CARBON MONOXIDE.
         4] PRODUCT IS A CHOLINESTERASE INHIBITOR.
         5] MOVE CONTAINERS FROM AREA IF WITHOUT RISK, COOL EXPOSED CONTAINERS.
         6] DIKE AREA FOR CONTROL AND CONTAINMENT TO PREVENT ENTRY INTO SEWERS, DRAIN, AND WATER WAYS.

LARGE SPILL/NO FIRE/RESCUE: WEAR RUBBER OR NEOPRENE BOOTS, GLOVES.

SPILL CONTROL AND CONTAINMENT:
   HOUSEHOLD SPILL: SWEEP UP WAXY SUBSTANCE AND PUT IN A DISPOSAL CONTAINER.
   LARGE SPILL: SWEEP UP WAXY SUBSTANCE. SCRUB AREA WITH DETERGENT AND WATER. PICK UP WITH ABSORBENT. PUT ALL
   ABSORBENT IN A DISPOSAL CONTAINER. SEAL AND REMOVE FROM THE WORK AREA.

HEALTH HAZARD INFORMATION:
   WHEN USING THIS PRODUCT WEAR: NO SPECIAL REQUIREMENT. Prolonged skin contact may cause irritation. Ingestion may
   cause nausea and vomiting.

   A physician should be contacted if anyone develops any signs or symptoms and suspects that they are caused by
   exposure to this product.

FIRST AID:
   Eye Exposure:
   Flush with water for 15 minutes while lifting the upper and lower eye lids. Contact lenses should not be worn
   when working with this product. Get medical attention.

   Skin Exposure:
   Wash with soap and water. If irritation develops, get medical attention.

   Breathing:
   Not expected to be a problem.

   Swallowing:
   Immediately contact the local poison control center for advice. Keep the victim warm and at rest. Get medical
   attention.

```
PRODUCT: RAID   ROACH & FLEA KILLER

INCOMPATIBILITY: NONE KNOWN

INGREDIENTS: CHLOROPYRIFOS        .25% TLV:          CAS#: 2921-88-2
             PYRETHRINS           .08% TLV:          CAS#:
             PIPERONYL BUTOXIDE   .40% TLV:          CAS#:    51-03-6
             ALIPHATIC NAPHTHA under 1.00% TLV:      CAS#:
             XYLENE          under 1.00% TLV:        CAS#: 1330-20-7
             WATER                     TLV:          CAS#: 7732-18-5

IGNI TEMP: NA  FP: NA  LEL: NA  UEL: NA  VP: NA  VD: NA  SG: 1.0  PS: LIQUID  APPEAR: SPRAY  ODOR: MILD
    PH FACTOR: 6.4

HAZARD RATINGS:    H:  F:  R:
```

FIRE FIGHTING:
  HIGH EXPANSION FOAM, LOW EXPANSION FOAM, ALCOHOL FOAM; DRY CHEMICAL, CARBON DIOXIDE, WATER FOG.
  PROTECTIVE CLOTHING, RUBBER GLOVES, AND BREATHING APPARATUS.
WARNING: 1] STRUCTURAL PROTECTIVE CLOTHING IS PERMEABLE, REMAIN CLEAR OF SMOKE, WATER FALL OUT AND WATER RUN OFF.
         2] KEEP OUT OF THE REACH OF CHILDREN.
         3] CONTAINER WILL MELT AND LEAK IN A FIRE.
         4] THERMAL DECOMPOSITION MAY YIELD CARBON DIOXIDE, CARBON MONOXIDE.
         5] PRODUCT IS A CHOLINESTERASE INHIBITOR.
         6] MOVE CONTAINERS FROM AREA IF WITHOUT RISK, COOL EXPOSED CONTAINERS.
         7] DIKE AREA FOR CONTROL AND CONTAINMENT TO PREVENT ENTRY INTO SEWERS, DRAIN, AND WATER WAYS.

LARGE SPILL/NO FIRE/RESCUE: WEAR RUBBER OR NEOPRENE BOOTS, GLOVES.

SPILL CONTROL AND CONTAINMENT:
  HOUSEHOLD SPILL: WEAR RUBBER GLOVES. WIPE UP WITH AN ABSORBENT CLOTH. RINSE OUT IN THE SINK. WIPE UP RESIDUE
  WITH A DAMP CLOTH. RINSE OUT IN THE SINK.
  LARGE SPILL: PICK UP WITH ABSORBENT. SCRUB AREA WITH DETERGENT AND WATER. PICK UP WITH ABSORBENT. PUT IN A
  DISPOSAL CONTAINER. SEAL AND REMOVE FROM THE WORK AREA.

HEALTH HAZARD INFORMATION:
  WHEN USING THIS PRODUCT WEAR: NO SPECIAL REQUIREMENT. Eye contact will cause stinging, tearing, itching,
  swelling, and redness. Prolonged skin contact may cause irritation. Ingestion may cause nausea and vomiting.

  A physician should be contacted if anyone develops any signs or symptoms and suspects that they are caused by
  exposure to this product.

FIRST AID:
  Eye Exposure:
  Flush with water for 15 minutes while lifting the upper and lower eye lids. Contact lenses should not be worn
  when working with this product. Get medical attention.

  Skin Exposure:
  Wash with soap and water. If irritation develops, get medical attention.

  Breathing:
  Not expected to be a problem.

  Swallowing:
  Immediately contact the local poison control center for advice. Keep the victim warm and at rest. Get medical
  attention.

PRODUCT: RAID   ROACH FUMIGATION CARTRIDGE

INCOMPATIBILITY: STRONG ACIDS (e.g. SWIMMING POOL ACID) WATER

INGREDIENTS: PERMETHRIN   20%   TLV: CAS#: 52645-53-1
             PYNAMIN FORTE   5%   TLV: CAS#: 42534-61-2
             INERT INGREDIENTS 75%   TLV: CAS#:

IGNI TEMP: NA FP: NA LEL: NA UEL: NA VP: NA VD: NA SG: NA PS: SOLID APPEAR: TUBULAR   ODOR: NONE
    PH FACTOR: NA

HAZARD RATINGS: H: F: R:

FIRE FIGHTING:
    HIGH EXPANSION FOAM, LOW EXPANSION FOAM, ALCOHOL FOAM; DRY CHEMICAL, CARBON DIOXIDE, WATER FOG.
    PROTECTIVE CLOTHING, RUBBER GLOVES, AND BREATHING APPARATUS.
WARNING: 1] STRUCTURAL PROTECTIVE CLOTHING IS PERMEABLE, REMAIN CLEAR OF SMOKE, WATER FALL OUT AND WATER RUN OFF.
         2] KEEP OUT OF THE REACH OF CHILDREN.
         3] REMOVE PLANTS, PETS, BIRD, TURN OFF & COVER FISH TANKS.
         4] DO NOT REMAIN IN AREA DURING USE.
         5] VENTILATE AREA BEFORE REOCCUPYING.
         6] THERMAL DECOMPOSITION MAY YIELD CARBON DIOXIDE, CARBON MONOXIDE.
         7] MOVE CONTAINERS FROM AREA IF WITHOUT RISK, COOL EXPOSED CONTAINERS.
         8] DIKE AREA FOR CONTROL AND CONTAINMENT TO PREVENT ENTRY INTO SEWERS, DRAIN, AND WATER WAYS.

LARGE SPILL/NO FIRE/RESCUE: WEAR RUBBER OR NEOPRENE BOOTS, GLOVES.

SPILL CONTROL AND CONTAINMENT:
    HOUSEHOLD SPILL: SWEEP OR SCRAPE UP AND PUT IN A DISPOSAL CONTAINER.
    LARGE SPILL: SWEEP OR SCRAPE UP MATERIAL. PUT IN A DISPOSAL CONTAINER.

HEALTH HAZARD INFORMATION:
    WHEN USING THIS PRODUCT WEAR: NO SPECIAL REQUIREMENT. Eye contact will cause stinging, tearing, itching,
    swelling, and redness. Ingestion may cause nausea, vomiting, and illness. Inhalation of fog or mist may cause
    irritation to the nose, throat, and lungs.

    A physician should be contacted if anyone develops any signs or symptoms and suspects that they are caused by
    exposure to this product.

FIRST AID:
    Eye Exposure:
    Flush with water for 15 minutes while lifting the upper and lower eye lids. Contact lenses should not be worn
    when working with this product. Get medical attention.

    Skin Exposure:
    Wash with soap and water. If irritation develops, get medical attention.

    Breathing:
    If irritation develops, move victim to fresh air. If it persists get medical attention.

    Swallowing:
    Immediately contact the local poison control center for advice. Keep the victim warm and at rest. Get medical
    attention.

PRODUCT: RAID   ROACH TRAPS

INCOMPATIBILITY: NONE KNOWN

INGREDIENTS: POLYMERS_____   90-95% TLV: CAS#:
             RESINS_____/   TLV: CAS#:
             FOOD   5-10% TLV: CAS#:
                PRODUCT IS ONLY A TRAP NO INSECTICIDES

IGNI TEMP: NA FP: NA LEL: NA UEL: NA VP: NA VD: NA SG: NA PS: SOLID IN A OPEN ENDED CARDBOARD BOX
   ODOR: NA   PH FACTOR: NA

HAZARD RATINGS: H: F: R:

FIRE FIGHTING:
  HIGH EXPANSION FOAM, LOW EXPANSION FOAM, ALCOHOL FOAM; DRY CHEMICAL, CARBON DIOXIDE, WATER FOG.
  PROTECTIVE CLOTHING, RUBBER GLOVES, AND BREATHING APPARATUS.
WARNING: 1] STRUCTURAL PROTECTIVE CLOTHING IS PERMEABLE, REMAIN CLEAR OF SMOKE, WATER FALL OUT AND WATER RUN OFF.
         2] KEEP OUT OF THE REACH OF CHILDREN.
         3] THERMAL DECOMPOSITION MAY YIELD CARBON DIOXIDE, CARBON MONOXIDE.
         4] MOVE CONTAINERS FROM AREA IF WITHOUT RISK, COOL EXPOSED CONTAINERS.
         5] DIKE AREA FOR CONTROL AND CONTAINMENT TO PREVENT ENTRY INTO SEWERS, DRAIN, AND WATER WAYS.

LARGE SPILL/NO FIRE/RESCUE: WEAR RUBBER OR NEOPRENE BOOTS, GLOVES.

SPILL CONTROL AND CONTAINMENT:
  HOUSEHOLD SPILL: PICK UP AND PUT IN A DISPOSAL CONTAINER.
  LARGE SPILL: SCOOP UP CONTAINERS AND PUT IN A DISPOSAL CONTAINER.

HEALTH HAZARD INFORMATION:
  WHEN USING THIS PRODUCT WEAR: NO SPECIAL REQUIREMENT. Not expected to be a health problem.

  A physician should be contacted if anyone develops any signs or symptoms and suspects that they are caused by exposure to this product.

FIRST AID:
  Eye Exposure:
  Flush with water for 15 minutes while lifting the upper and lower eye lids. Contact lenses should not be worn when working with this product. Get medical attention.

  Skin Exposure:
  Wash hand with soap and water.

  Breathing:
  Not expected to be a problem.

  Swallowing:
  Immediately contact the local poison control center for advice. Keep the victim warm and at rest. Get medical attention.

PRODUCT: RAID   STRIP FLYING INSECT KILLER

INCOMPATIBILITY: NONE KNOWN.

INGREDIENTS: DICHLORVOS   18.6% TLV: .1 ppm (skin) CAS#: 62-73-7
             RELATED COMPOUNDS   1.4% TLV: CAS#:
             POLYMER BASED MATERIAL 80.0% TLV: CAS#:

IGNI TEMP: NA FP: NA LEL: NA UEL: NA VP: NA VD: NA SG: NA PS: SOLID APPEAR: YELLOW   ODOR: MILD
    PH FACTOR: NA

HAZARD RATINGS: H: F: R:

FIRE FIGHTING:
   HIGH EXPANSION FOAM, LOW EXPANSION FOAM, ALCOHOL FOAM; DRY CHEMICAL, CARBON DIOXIDE, WATER FOG.
   PROTECTIVE CLOTHING, RUBBER GLOVES, AND BREATHING APPARATUS.
WARNING: 1] STRUCTURAL PROTECTIVE CLOTHING IS PERMEABLE, REMAIN CLEAR OF SMOKE, WATER FALL OUT AND WATER RUN OFF.
         2] KEEP OUT OF THE REACH OF CHILDREN.
         3] PRODUCT IS A CHOLINESTERASE INHIBITOR.
         4] DUE NOT USE IN A BABY NURSERY OR BED RIDDEN PATIENT ROOMS.
         5] THERMAL DECOMPOSITION MAY YIELD CARBON DIOXIDE, CARBON MONOXIDE.
         6] MOVE CONTAINERS FROM AREA IF WITHOUT RISK, COOL EXPOSED CONTAINERS.
         7] DIKE AREA FOR CONTROL AND CONTAINMENT TO PREVENT ENTRY INTO SEWERS, DRAIN, AND WATER WAYS.

LARGE SPILL/NO FIRE/RESCUE: WEAR RUBBER OR NEOPRENE BOOTS, GLOVES.

SPILL CONTROL AND CONTAINMENT:
   HOUSEHOLD SPILL: PICK UP STRIP AND PUT IN A DISPOSAL CONTAINER. WASH HANDS WITH SOAP AND WATER.
   LARGE SPILL: SCOOP UP STRIPS AND PUT IN A DISPOSAL CONTAINER. RINSE AREA WITH WATER. PICK UP WITH ABSORBENT. PUT
   IN A DISPOSAL CONTAINER. SEAL AND REMOVE FROM THE WORK AREA.

HEALTH HAZARD INFORMATION:
   WHEN USING THIS PRODUCT WEAR: NO SPECIAL REQUIREMENT. Eye contact may cause irritation. Prolonged skin contact
   may cause irritation. Ingestion may cause nausea, and vomiting.

   A physician should be contacted if anyone develops any signs or symptoms and suspects that they are caused by
   exposure to this product.

FIRST AID:
   Eye Exposure:
   Flush with water for 15 minutes while lifting the upper and lower eye lids. Contact lenses should not be worn
   when working with this product. Get medical attention.

   Skin Exposure:
   Wash hands with soap and water.

   Breathing:
   Not expected to be a problem.

   Swallowing:
   Immediately contact the local poison control center for advice. Keep the victim warm and at rest. Get medical
   attention.

```
PRODUCT: RAID   TOMATO & VEGETABLE FOGGER


INCOMPATIBILITY: NONE KNOWN.


INGREDIENTS: PYRETHRINS          .2%  TLV:            CAS#:
             PROPANE____             TLV: 1000 ppm   CAS#:   74-96-8
             ISOBUTANE_____  20-40% TLV:            CAS#:   75-28-5
             N-BUTANE___/            TLV:  800 ppm   CAS#:  106-97-8
             WATER           60-75%  TLV:            CAS#: 7732-18-5


IGNI TEMP: NA  FP: 20F  LEL: NA  UEL: NA  VP: NA  VD: NA  SG: .8  PS: LIQUID  APPEAR: SPRAY  ODOR: MILD
   PH FACTOR: NA


HAZARD RATINGS:    H:  F:  R:
```

FIRE FIGHTING:

```
  HIGH EXPANSION FOAM, LOW EXPANSION FOAM, ALCOHOL FOAM; DRY CHEMICAL, CARBON DIOXIDE, WATER FOG.
  PROTECTIVE CLOTHING, RUBBER GLOVES, AND BREATHING APPARATUS.
WARNING: 1] STRUCTURAL PROTECTIVE CLOTHING IS PERMEABLE, REMAIN CLEAR OF SMOKE, WATER FALL OUT AND WATER RUN OFF.
         2] KEEP OUT OF THE REACH OF CHILDREN.
         3] REMOVE ALL IGNITION SOURCES IF WITHOUT RISK.
         4] CONTAINERS WILL EXPLODE IN A FIRE OR IF HEATED ABOVE 120F.
         5] THERMAL DECOMPOSITION MAY YIELD CARBON DIOXIDE, CARBON MONOXIDE.
         6] MOVE CONTAINERS FROM AREA IF WITHOUT RISK, COOL EXPOSED CONTAINERS.
         7] DIKE AREA FOR CONTROL AND CONTAINMENT TO PREVENT ENTRY INTO SEWERS, DRAIN, AND WATER WAYS.


LARGE SPILL/NO FIRE/RESCUE: WEAR RUBBER OR NEOPRENE BOOTS, GLOVES.


SPILL CONTROL AND CONTAINMENT:
  HOUSEHOLD SPILL: WIPE UP DRY WITH AN ABSORBENT CLOTH. RINSE OUT IN THE SINK. WIPE UP RESIDUE WITH A DAMP CLOTH.
  RINSE OUT IN THE SINK.
  LARGE SPILL: PICK UP WITH ABSORBENT. RINSE AREA WITH WATER. PICK UP WITH ABSORBENT. PUT ALL ABSORBENT IN A
  DISPOSAL CONTAINER SEAL AND REMOVE FROM THE WORK AREA.
```

HEALTH HAZARD INFORMATION:
  WHEN USING THIS PRODUCT WEAR: NO SPECIAL REQUIREMENT. Ingestion may cause illness. Eye contact will cause
  stinging, tearing, itching, swelling, and redness. Skin contact may cause irritation.

  A physician should be contacted if anyone develops any signs or symptoms and suspects that they are caused by
  exposure to this product.

FIRST AID:
  Eye Exposure:
  Flush with water for 15 minutes while lifting the upper and lower eye lids. Contact lenses should not be worn
  when working with this product. Get medical attention.

  Skin Exposure:
  Remove contaminated clothing. Wash with soap and water.

  Breathing:
  Not expected to be a problem.

  Swallowing:
  Immediately contact the local poison control center for advice. Keep the victim warm and at rest. Get medical
  attention.

PRODUCT: RAID WASP & HORNET KILLER

INCOMPATIBILITY: STRONG OXIDIZERS; ALKALIS; BASES; CAUSTICS; STRONG ACIDS; ALUMINUM; MAGNESIUM; ZINC.

INGREDIENTS: PETROLEUM SOLVENT   15-20% TLV: 100 ppm   CAS#: 64742-47-8
             1,1,1-TRICHLOROETHANE   75-85% TLV: 350 ppm   CAS#:   71-55-6
             CARBON DIOXIDE   1-5%   TLV: 5000 ppm   CAS#: 124-38-0

IGNI TEMP: NA FP: NA LEL: NA UEL: NA VP: NA VD: NA SG: 1.07 PS: LIQUID APPEAR: CLEAR SPRAY STREAM
    ODOR: SOLVENT   PH FACTOR: NA

HAZARD RATINGS: H: F: R: 1 2 0

FIRE FIGHTING:
  HIGH EXPANSION FOAM, LOW EXPANSION FOAM, ALCOHOL FOAM; DRY CHEMICAL, CARBON DIOXIDE, WATER FOG.
  PROTECTIVE CLOTHING, RUBBER GLOVES, AND BREATHING APPARATUS.
WARNING: 1] STRUCTURAL PROTECTIVE CLOTHING IS PERMEABLE, REMAIN CLEAR OF SMOKE, WATER FALL OUT AND WATER RUN OFF.
         2] KEEP OUT OF THE REACH OF CHILDREN.
         3] THERMAL DECOMPOSITION MAY FORM ACID MIST AND TOXIC FUMES.
         4] PRODUCT IS A CHOLINESTERASE INHIBITOR.
         5] MOVE CONTAINERS FROM AREA IF WITHOUT RISK, COOL EXPOSED CONTAINERS.
         6] DIKE AREA FOR CONTROL AND CONTAINMENT TO PREVENT ENTRY INTO SEWERS, DRAIN, AND WATER WAYS.

LARGE SPILL/NO FIRE/RESCUE: WEAR ORGANIC VAPOR MASK, RUBBER BOOTS, GLOVES.

SPILL CONTROL AND CONTAINMENT:
  HOUSEHOLD SPILL: WIPE UP DRY WITH AN ABSORBENT CLOTH. RINSE OUT IN THE SINK. RINSE WITH WATER. WIPE UP DRY WITH
  AN ABSORBENT CLOTH. RINSE OUT IN THE SINK.
  LARGE SPILL: PICK UP WITH ABSORBENT. RINSE AREA WITH WATER. PICK UP WITH ABSORBENT. PUT ALL ABSORBENT IN A
  DISPOSAL CONTAINER. SEAL AND REMOVE FROM THE WORK AREA.

HEALTH HAZARD INFORMATION:
  WHEN USING THIS PRODUCT WEAR: NO SPECIAL REQUIREMENT. Eye contact will cause severe stinging, tearing, itching,
  swelling, and redness. Prolonged contact with skin may cause irritation. Ingestion may cause nausea and
  vomiting. If product gets into the lungs, it may cause chemical pneumonia.

  A physician should be contacted if anyone develops any signs or symptoms and suspects that they are caused by
  exposure to this product. ANTIDOTE IS ATROPINE USE ONLY AFTER SYMPTOMS APPEAR

FIRST AID:
  Eye Exposure:
  Flush with water for 15 minutes while lifting the upper and lower eye lids. Contact lenses should not be worn
  when working with this product. Get medical attention.

  Skin Exposure:
  Wash with soap and water. If irritation develops, get medical attention.

  Breathing:
  Not expected to be a problem.

  Swallowing:
  Do not induce vomiting. Immediately contact the local poison control center for advice. Keep the victim warm and
  at rest. Get medical attention.

PRODUCT: RAID   YARD GUARD

INCOMPATIBILITY: NONE KNOWN

INGREDIENTS: PROPANE_____   TLV: 1000 ppm   CAS#:   74-98-6
             ISOBUTANE_____ 60-75% TLV: CAS#:   75-28-5
             N-BUTANE____/   TLV: 800 ppm   CAS#:   106-97-8
             WATER   20-40% TLV: CAS#: 7732-18-5

IGNI TEMP: NA FP: 20F LEL: NA UEL: NA VP: NA VD: NA SG: .83 PS: LIQUID APPEAR: SPRAY FOG   ODOR: MILD
   PH FACTOR: NA

HAZARD RATINGS: H: F: R: 1 4 0

FIRE FIGHTING:
   HIGH EXPANSION FOAM, LOW EXPANSION FOAM, ALCOHOL FOAM; DRY CHEMICAL, CARBON DIOXIDE, WATER FOG.
   PROTECTIVE CLOTHING, RUBBER GLOVES, AND BREATHING APPARATUS.
WARNING: 1] STRUCTURAL PROTECTIVE CLOTHING IS PERMEABLE, REMAIN CLEAR OF SMOKE, WATER FALL OUT AND WATER RUN OFF.
         2] KEEP OUT OF THE REACH OF CHILDREN.
         3] FOR OUT DOOR USE ONLY
         4] REMOVE ALL IGNITION SOURCES IF WITHOUT RISK.
         5] CONTAINERS WILL EXPLODE IN A FIRE OR IF HEATED ABOVE 120F.
         6] THERMAL DECOMPOSITION MAY YIELD CARBON DIOXIDE, CARBON MONOXIDE.
         7] MOVE CONTAINERS FROM AREA IF WITHOUT RISK, COOL EXPOSED CONTAINERS.
         8] DIKE AREA FOR CONTROL AND CONTAINMENT TO PREVENT ENTRY INTO SEWERS, DRAIN, AND WATER WAYS.

LARGE SPILL/NO FIRE/RESCUE: VENTILATE ENCLOSED SPACES. WEAR RUBBER BOOTS, GLOVES, BREATHING APPARATUS.

SPILL CONTROL AND CONTAINMENT:
   HOUSEHOLD SPILL: VENTILATE ENCLOSED SPACES. WIPE UP DRY WITH AN ABSORBENT CLOTH. RINSE OUT IN THE SINK.
   LARGE SPILL: VENTILATE ENCLOSED SPACES. PICK UP WITH ABSORBENT. PUT IN A DISPOSAL CONTAINER. SEAL AND REMOVE
   FROM THE WORK PLACE.

HEALTH HAZARD INFORMATION:
   WHEN USING THIS PRODUCT WEAR: NO SPECIAL REQUIREMENT. Eye contact may cause irritation. Skin contact may cause
   irritation. Ingestion may cause nausea and vomiting. Inhalation may cause irritation to the nose, throat, and
   lungs.

   A physician should be contacted if anyone develops any signs or symptoms and suspects that they are caused by
   exposure to this product.

FIRST AID:
Eye Exposure:
Flush with water for 15 minutes while lifting the upper and lower eye lids. Contact lenses should not be worn
when working with this product. Get medical attention.

Skin Exposure:
Wash with soap and water. If irritation develops, get medical attention.

Breathing:
If irritation develops, move the victim to fresh air.

Swallowing:
Immediately contact the local poison control center for advice. Keep the victim warm and at rest. Get medical
attention.

PRODUCT: RAIN BARREL

INCOMPATIBILITY: NONE KNOWN

INGREDIENTS: QUARTERNARY SOFTENERS   9-15% TLV: CAS#:
             ISOPROPYL ALCOHOL   2-4% TLV: 400 ppm   CAS#: 67-63-0
             WATER   80-90% TLV: CAS#: 7732-18-5

IGNI TEMP: NA FP: 200F LEL: NA UEL: NA VP: NA VD: NA SG: 1.0 PS: LIQUID APPEAR: BLUE   ODOR: PLEASANT
   PH FACTOR: 4.7

HAZARD RATINGS: H: F: R:

FIRE FIGHTING:
   HIGH EXPANSION FOAM, LOW EXPANSION FOAM, ALCOHOL FOAM; DRY CHEMICAL, CARBON DIOXIDE, WATER FOG.
   PROTECTIVE CLOTHING, RUBBER GLOVES, AND BREATHING APPARATUS.
WARNING: 1] STRUCTURAL PROTECTIVE CLOTHING IS PERMEABLE, REMAIN CLEAR OF SMOKE, WATER FALL OUT AND WATER RUN OFF.
         2] KEEP OUT OF THE REACH OF CHILDREN.
         3] CONTAINERS WILL MELT AND LEAK IN A FIRE.
         4] THERMAL DECOMPOSITION MAY YIELD CARBON DIOXIDE, CARBON MONOXIDE.
         5] MOVE CONTAINERS FROM AREA IF WITHOUT RISK, COOL EXPOSED CONTAINERS.
         6] DIKE AREA FOR CONTROL AND CONTAINMENT TO PREVENT ENTRY INTO SEWERS, DRAIN, AND WATER WAYS.

LARGE SPILL/NO FIRE/RESCUE: WEAR RUBBER OR NEOPRENE BOOTS, GLOVES.

SPILL CONTROL AND CONTAINMENT:
   HOUSEHOLD SPILL: WIPE UP DRY WITH AN ABSORBENT CLOTH. RINSE OUT IN THE SINK.
   LARGE SPILL: PICK UP WITH ABSORBENT. RINSE AREA WITH WATER. PICK UP WITH ABSORBENT. PUT ALL ABSORBENT IN A
   DISPOSAL CONTAINER. SEAL AND REMOVE FROM THE WORK AREA.

HEALTH HAZARD INFORMATION:
   WHEN USING THIS PRODUCT WEAR: NO SPECIAL REQUIREMENT. Eye contact may cause stinging, tearing, itching,
   swelling, and redness. Ingestion may cause nausea and vomiting.

   A physician should be contacted if anyone develops any signs or symptoms and suspects that they are caused by
   exposure to this product.

FIRST AID:
   Eye Exposure:
   Flush with water for 15 minutes while lifting the upper and lower eye lids. Contact lenses should not be worn
   when working with this product. Get medical attention.

   Skin Exposure:
   Not expected to be a problem.

   Breathing:
   Not expected to be a problem.

   Swallowing:
   Immediately contact the local poison control center for advice. Keep the victim warm and at rest. Get medical
   attention.

PRODUCT: RENUZIT ADJUSTABLE AIR FRESHENER (SOLID)

INCOMPATIBILITY: NONE KNOWN

INGREDIENTS: ALL ARE NON-HAZARDOUS BY OSHA 1910.120 STANDARD

IGNI TEMP: NA FP: NA LEL: NA UEL: NA VP: NA VD: NA SG: 1 PS: GEL APPEAR: COLORED   ODOR: FRAGRANCE
    PH FACTOR: NA

HAZARD RATINGS: H: F: R: 0 0 0

FIRE FIGHTING:
    HIGH EXPANSION FOAM, LOW EXPANSION FOAM, ALCOHOL FOAM; DRY CHEMICAL, CARBON DIOXIDE, WATER FOG.
    PROTECTIVE CLOTHING, RUBBER GLOVES, AND BREATHING APPARATUS.
WARNING: 1] STRUCTURAL PROTECTIVE CLOTHING IS PERMEABLE, REMAIN CLEAR OF SMOKE, WATER FALL OUT AND WATER RUN OFF.
         2] KEEP OUT OF THE REACH OF CHILDREN.
         3] THERMAL DECOMPOSITION MAY YIELD CARBON DIOXIDE, CARBON MONOXIDE.
         5] MOVE CONTAINERS FROM AREA IF WITHOUT RISK, COOL EXPOSED CONTAINERS.
         6] DIKE AREA FOR CONTROL AND CONTAINMENT TO PREVENT ENTRY INTO SEWERS, DRAIN, AND WATER WAYS.

LARGE SPILL/NO FIRE/RESCUE: WEAR RUBBER OR NEOPRENE BOOTS, GLOVES.

SPILL CONTROL AND CONTAINMENT:
    HOUSEHOLD SPILL: WIPE UP WITH AN ABSORBENT CLOTH. RINSE OUT IN THE SINK. WIPE UP RESIDUE WITH A DAMP CLOTH.
    RINSE OUT IN THE SINK.
    LARGE SPILL: SCOOP UP SOLID GEL AND PUT IN A DISPOSAL CONTAINER. RINSE AREA WITH WATER. PICK UP WITH ABSORBENT.
    PUT IN DISPOSAL CONTAINER. SEAL AND REMOVE FROM WORK AREA.

HEALTH HAZARD INFORMATION:
    WHEN USING THIS PRODUCT WEAR: NO SPECIAL REQUIREMENT. Eye contact may cause irritation. Skin contact may cause
    irritation. Ingestion may result in nausea, vomiting, and diarrhea.

    A physician should be contacted if anyone develops any signs or symptoms and suspects that they are caused by
    exposure to this product.

FIRST AID:
    Eye Exposure:
    Flush with water for 15 minutes while lifting the upper and lower eye lids. Contact lenses should not be worn
    when working with this product. Get medical attention.

    Skin Exposure:
    Wash with soap and water. If irritation develops, get medical attention.

    Breathing:
    Not expected to be a problem.

    Swallowing:
    If victim is conscious, give 2-3 glasses of milk or water to drink. Immediately contact the local poison control
    center for advice. Keep the victim warm and at rest. Get medical attention.

PRODUCT: RENUZIT AIR DEODORIZER - AEROSOL

INCOMPATIBILITY: NONE KNOWN

INGREDIENTS: MINERAL SPIRITS less than 10% TLV: 100 ppm   CAS#: 8030-30-6
             PROPANE_____   TLV: 1000 ppm   CAS#: 74-98-6
             BUTANE_____ under 30% TLV: 800 ppm   CAS#: 106-97-8

IGNI TEMP: NA FP: 128F LEL: 1.8 UEL: 9.5 VP: NA VD: 2.0 SG: .97 PS: LIQUID   APPEAR: CLEAR TO YELLOWISH
     ODOR: FRAGRANCE PH FACTOR: NA

HAZARD RATINGS: H: F: R: 0 4 0

FIRE FIGHTING:
   HIGH EXPANSION FOAM, LOW EXPANSION FOAM, ALCOHOL FOAM; DRY CHEMICAL, CARBON DIOXIDE, WATER FOG.
   PROTECTIVE CLOTHING, RUBBER GLOVES, AND BREATHING APPARATUS.
WARNING: 1] STRUCTURAL PROTECTIVE CLOTHING IS PERMEABLE, REMAIN CLEAR OF SMOKE, WATER FALL OUT AND WATER RUN OFF.
         2] KEEP OUT OF THE REACH OF CHILDREN.
         3] REMOVE ALL IGNITION SOURCES IF WITHOUT RISK.
         4] CONTAINERS WILL EXPLODE IN A FIRE OR IF HEATED ABOVE 120F.
         5] MOVE CONTAINERS FROM AREA IF WITHOUT RISK, COOL EXPOSED CONTAINERS.
         6] DIKE AREA FOR CONTROL AND CONTAINMENT TO PREVENT ENTRY INTO SEWERS, DRAIN, AND WATER WAYS.

LARGE SPILL/NO FIRE/RESCUE: WEAR RUBBER OR NEOPRENE BOOTS, GLOVES.

SPILL CONTROL AND CONTAINMENT:
   HOUSEHOLD SPILL: WIPE UP DRY WITH AN ABSORBENT CLOTH. RINSE OUT IN THE SINK.
   LARGE SPILL: PICK UP WITH ABSORBENT. RINSE AREA WITH WATER. PICK UP WITH ABSORBENT. PUT ALL ABSORBENT IN A
   DISPOSAL CONTAINER. SEAL AND REMOVE FROM THE WORK AREA.

HEALTH HAZARD INFORMATION:
   WHEN USING THIS PRODUCT WEAR: NO SPECIAL REQUIREMENT. Eye contact may cause stinging, tearing, itching,
   swelling, and redness. Skin contact may cause irritation. Ingestion may result in nausea and vomiting.

   A physician should be contacted if anyone develops any signs or symptoms and suspects that they are caused by
   exposure to this product.

FIRST AID:
   Eye Exposure:
   Flush with water for 15 minutes while lifting the upper and lower eye lids. Contact lenses should not be worn
   when working with this product. Get medical attention.

   Skin Exposure:
   Wash with soap and water. If irritation develops, get medical attention.

   Breathing:
   Not expected to be a problem.

   Swallowing:
   If the victim is conscious, give 2-3 glasses of milk or water to drink. Immediately contact the local poison
   control center for advice. Keep the victim warm and at rest. Get medical attention.

PRODUCT: RENUZIT FRAGRANCE JAR - BAYBERRY

INCOMPATIBILITY: STRONG OXIDIZERS.

INGREDIENTS: CONCENTRATED PERFUME OIL 100% TLV: CAS#:

IGNI TEMP: NA FP: 192F LEL: NA UEL: NA VP: under 1 VD: over 1 SG: .98 PS: LIQUID APPEAR: CLEAR TO YELLOW
     ODOR: FRAGRANT   PH FACTOR: NA

HAZARD RATINGS: H: F: R: 1 2 0

FIRE FIGHTING:
  HIGH EXPANSION FOAM, LOW EXPANSION FOAM, ALCOHOL FOAM; DRY CHEMICAL, CARBON DIOXIDE, WATER FOG.
  PROTECTIVE CLOTHING, RUBBER GLOVES, AND BREATHING APPARATUS.
WARNING: 1] STRUCTURAL PROTECTIVE CLOTHING IS PERMEABLE, REMAIN CLEAR OF SMOKE, WATER FALL OUT AND WATER RUN OFF.
         2] KEEP OUT OF THE REACH OF CHILDREN.
         3] KEEP AWAY FROM HEAT AND FLAME.
         4] THERMAL DECOMPOSITION MAY YIELD CARBON DIOXIDE, CARBON MONOXIDE.
         5] MOVE CONTAINERS FROM AREA IF WITHOUT RISK, COOL EXPOSED CONTAINERS.
         6] DIKE AREA FOR CONTROL AND CONTAINMENT TO PREVENT ENTRY INTO SEWERS, DRAIN, AND WATER WAYS.

LARGE SPILL/NO FIRE/RESCUE: WEAR RUBBER OR NEOPRENE BOOTS, GLOVES.

SPILL CONTROL AND CONTAINMENT:
  HOUSEHOLD SPILL: WIPE UP WITH ABSORBENT CLOTH. RINSE OUT IN THE SINK. WIPE UP RESIDUE WITH A DRY CLOTH.
  LARGE SPILL: PICK UP WITH ABSORBENT. WASH AREA WITH DETERGENT AND WATER. PICK UP WITH ABSORBENT. PUT ALL
  ABSORBENT IN A DISPOSAL CONTAINER. SEAL AND REMOVE FROM THE WORK AREA.

HEALTH HAZARD INFORMATION:
  WHEN USING THIS PRODUCT WEAR: NO SPECIAL REQUIREMENT. Eye contact may cause stinging, tearing, itching,
  swelling, and redness. Skin contact may cause irritation. Ingestion may cause nausea and vomiting.

  A physician should be contacted if anyone develops any signs or symptoms and suspects that they are caused by
  exposure to this product.

FIRST AID:
  Eye Exposure:
  Flush with water for 15 minutes while lifting the upper and lower eye lids. Contact lenses should not be worn
  when working with this product. Get medical attention.

  Skin Exposure:
  Wash with soap and water. If irritation develops, get medical attention.

  Breathing:
  Not expected to be a problem.

  Swallowing:
  If the victim is conscious, give 2-3 glasses of milk or water to drink. Immediately contact the local poison
  control center for advice. Keep the victim warm and at rest. Get medical attention.

PRODUCT: RENUZIT FRAGRANCE JAR - CINNAMON

INCOMPATIBILITY: STRONG OXIDIZERS.

INGREDIENTS: CONCENTRATED PERFUME OIL 100% TLV: CAS#:

IGNI TEMP: NA FP: 192F LEL: NA UEL: NA VP: under 1 VD: over 1 SG: 1.02 PS: LIQUID APPEAR: CLEAR
    ODOR: FRAGRANT   PH FACTOR: NA

HAZARD RATINGS: H: F: R: 1 2 0

FIRE FIGHTING:
  HIGH EXPANSION FOAM, LOW EXPANSION FOAM, ALCOHOL FOAM; DRY CHEMICAL, CARBON DIOXIDE, WATER FOG.
  PROTECTIVE CLOTHING, RUBBER GLOVES, AND BREATHING APPARATUS.
WARNING: 1] STRUCTURAL PROTECTIVE CLOTHING IS PERMEABLE, REMAIN CLEAR OF SMOKE, WATER FALL OUT AND WATER RUN OFF.
         2] KEEP OUT OF THE REACH OF CHILDREN.
         3] KEEP AWAY FROM HEAT AND FLAME.
         4] THERMAL DECOMPOSITION MAY YIELD CARBON DIOXIDE, CARBON MONOXIDE.
         5] MOVE CONTAINERS FROM AREA IF WITHOUT RISK, COOL EXPOSED CONTAINERS.
         6] DIKE AREA FOR CONTROL AND CONTAINMENT TO PREVENT ENTRY INTO SEWERS, DRAIN, AND WATER WAYS.

LARGE SPILL/NO FIRE/RESCUE: WEAR RUBBER OR NEOPRENE BOOTS, GLOVES.

SPILL CONTROL AND CONTAINMENT:
  HOUSEHOLD SPILL: WIPE UP WITH ABSORBENT CLOTH. RINSE OUT IN THE SINK. WIPE UP RESIDUE WITH A DRY CLOTH.
  LARGE SPILL: PICK UP WITH ABSORBENT. WASH AREA WITH DETERGENT AND WATER. PICK UP WITH ABSORBENT. PUT ALL
  ABSORBENT IN A DISPOSAL CONTAINER. SEAL AND REMOVE FROM THE WORK AREA.

HEALTH HAZARD INFORMATION:
  WHEN USING THIS PRODUCT WEAR: NO SPECIAL REQUIREMENT. Eye contact may cause stinging, tearing, itching,
  swelling, and redness. Skin contact may cause irritation. Ingestion may cause nausea and vomiting.

  A physician should be contacted if anyone develops any signs or symptoms and suspects that they are caused by
  exposure to this product.

FIRST AID:
Eye Exposure:
Flush with water for 15 minutes while lifting the upper and lower eye lids. Contact lenses should not be worn
when working with this product. Get medical attention.

Skin Exposure:
Wash with soap and water. If irritation develops, get medical attention.

Breathing:
Not expected to be a problem.

Swallowing:
If the victim is conscious, give 2-3 glasses of milk or water to drink. Immediately contact the local poison
control center for advice. Keep the victim warm and at rest. Get medical attention.

PRODUCT: RENUZIT FRAGRANCE JAR - FLORAL POTPOURRI

INCOMPATIBILITY: STRONG OXIDIZERS.

INGREDIENTS: CONCENTRATED PERFUME OIL   91% TLV: CAS#:
        ISOPARAFFINIC   Recommended
        PETROLEUM SOLVENT   9% TLV: 300 ppm   CAS#: 64742-47-8

IGNI TEMP: NA FP: 171F LEL: NA UEL: NA VP: under 1 VD: over 1 SG: .96 PS: LIQUID APPEAR: CLEAR TO YELLOW
   ODOR: FRAGRANT PH FACTOR: NA

HAZARD RATINGS: H: F: R: 1 2 0

FIRE FIGHTING:
  HIGH EXPANSION FOAM, LOW EXPANSION FOAM, ALCOHOL FOAM; DRY CHEMICAL, CARBON DIOXIDE, WATER FOG.
  PROTECTIVE CLOTHING, RUBBER GLOVES, AND BREATHING APPARATUS.
WARNING: 1] STRUCTURAL PROTECTIVE CLOTHING IS PERMEABLE, REMAIN CLEAR OF SMOKE, WATER FALL OUT AND WATER RUN OFF.
       2] KEEP OUT OF THE REACH OF CHILDREN.
       3] KEEP AWAY FROM HEAT AND FLAME.
       4] THERMAL DECOMPOSITION MAY YIELD CARBON DIOXIDE, CARBON MONOXIDE.
       5] MOVE CONTAINERS FROM AREA IF WITHOUT RISK, COOL EXPOSED CONTAINERS.
       6] DIKE AREA FOR CONTROL AND CONTAINMENT TO PREVENT ENTRY INTO SEWERS, DRAIN, AND WATER WAYS.

LARGE SPILL/NO FIRE/RESCUE: WEAR RUBBER OR NEOPRENE BOOTS, GLOVES.

SPILL CONTROL AND CONTAINMENT:
  HOUSEHOLD SPILL: WIPE UP WITH ABSORBENT CLOTH. RINSE OUT IN THE SINK. WIPE UP RESIDUE WITH A DRY CLOTH.
  LARGE SPILL: PICK UP WITH ABSORBENT. WASH AREA WITH DETERGENT AND WATER. PICK UP WITH ABSORBENT. PUT ALL
  ABSORBENT IN A DISPOSAL CONTAINER. SEAL AND REMOVE FROM THE WORK AREA.

HEALTH HAZARD INFORMATION:
  WHEN USING THIS PRODUCT WEAR: NO SPECIAL REQUIREMENT. Eye contact may cause stinging, tearing, itching,
  swelling, and redness. Skin contact may cause irritation. Ingestion may cause nausea and vomiting.

  A physician should be contacted if anyone develops any signs or symptoms and suspects that they are caused by
  exposure to this product.

FIRST AID:
  Eye Exposure:
  Flush with water for 15 minutes while lifting the upper and lower eye lids. Contact lenses should not be worn
  when working with this product. Get medical attention.

  Skin Exposure:
  Wash with soap and water. If irritation develops, get medical attention.

  Breathing:
  Not expected to be a problem.

  Swallowing:
  If the victim is conscious, give 2-3 glasses of milk or water to drink. Immediately contact the local poison
  control center for advice. Keep the victim warm and at rest. Get medical attention.

PRODUCT: RENUZIT FRAGRANCE JAR - FLORAL POTPOURRI (27%)

INCOMPATIBILITY: STRONG OXIDIZERS.

INGREDIENTS: CONCENTRATED PERFUME OIL   43% TLV: CAS#:
             DIETHYLENE GLYCOL
             MONO ETHYL ETHER   30% TLV: CAS#:   111-90-0
             ISOPARAFFINIC   Recommended
             PETROLEUM SOLVENT   27% TLV: 300 ppm   CAS#: 64742-47-8

IGNI TEMP: NA FP: 171F LEL: NA UEL: NA VP: under 1 VD: over 1 SG: .91 PS: LIQUID APPEAR: CLEAR TO YELLOW
     ODOR: FRAGRANT   PH FACTOR: NA

HAZARD RATINGS: H: F: R: 1 2 0

FIRE FIGHTING:
  HIGH EXPANSION FOAM, LOW EXPANSION FOAM, ALCOHOL FOAM; DRY CHEMICAL, CARBON DIOXIDE, WATER FOG.
  PROTECTIVE CLOTHING, RUBBER GLOVES, AND BREATHING APPARATUS.
WARNING: 1] STRUCTURAL PROTECTIVE CLOTHING IS PERMEABLE, REMAIN CLEAR OF SMOKE, WATER FALL OUT AND WATER RUN OFF.
         2] KEEP OUT OF THE REACH OF CHILDREN.
         3] KEEP AWAY FROM HEAT AND FLAME.
         4] THERMAL DECOMPOSITION MAY YIELD CARBON DIOXIDE, CARBON MONOXIDE.
         5] MOVE CONTAINERS FROM AREA IF WITHOUT RISK, COOL EXPOSED CONTAINERS.
         6] DIKE AREA FOR CONTROL AND CONTAINMENT TO PREVENT ENTRY INTO SEWERS, DRAIN, AND WATER WAYS.

LARGE SPILL/NO FIRE/RESCUE: WEAR RUBBER OR NEOPRENE BOOTS, GLOVES.

SPILL CONTROL AND CONTAINMENT:
  HOUSEHOLD SPILL: WIPE UP WITH ABSORBENT CLOTH. RINSE OUT IN THE SINK. WIPE UP RESIDUE WITH A DRY CLOTH.
  LARGE SPILL: PICK UP WITH ABSORBENT. WASH AREA WITH DETERGENT AND WATER. PICK UP WITH ABSORBENT. PUT ALL
  ABSORBENT IN A DISPOSAL CONTAINER. SEAL AND REMOVE FROM THE WORK AREA.

HEALTH HAZARD INFORMATION:
  WHEN USING THIS PRODUCT WEAR: NO SPECIAL REQUIREMENT. Eye contact may cause stinging, tearing, itching,
  swelling, and redness. Skin contact may cause irritation. Ingestion may cause nausea and vomiting. If product
  gets into the lungs, it may cause chemical pneumonia.

  A physician should be contacted if anyone develops any signs or symptoms and suspects that they are caused by
  exposure to this product.

FIRST AID:
  Eye Exposure:
  Flush with water for 15 minutes while lifting the upper and lower eye lids. Contact lenses should not be worn
  when working with this product. Get medical attention.

  Skin Exposure:
  Wash with soap and water. If irritation develops, get medical attention.

  Breathing:
  If breathing becomes labored give oxygen. Get medical attention.

  Swallowing:
  Do not induce vomiting. If the victim is conscious, give 2-3 glasses of milk or water to drink. Immediately
  contact the local poison control center for advice. Keep the victim warm and at rest. Get medical attention.

PRODUCT: RENUZIT FRAGRANCE JAR - FRESH CUT FLOWERS

INCOMPATIBILITY: STRONG OXIDIZERS.

INGREDIENTS: CONCENTRATED PERFUME OIL 100% TLV: CAS#:

IGNI TEMP: NA FP: 161F LEL: NA UEL: NA VP: under 1 VD: over 1 SG: 1 PS: LIQUID APPEAR: CLEAR TO YELLOW
    ODOR: FRAGRANT   PH FACTOR: NA

HAZARD RATINGS: H: F: R: 1 2 0

FIRE FIGHTING:
  HIGH EXPANSION FOAM, LOW EXPANSION FOAM, ALCOHOL FOAM; DRY CHEMICAL, CARBON DIOXIDE, WATER FOG.
  PROTECTIVE CLOTHING, RUBBER GLOVES, AND BREATHING APPARATUS.
WARNING: 1] STRUCTURAL PROTECTIVE CLOTHING IS PERMEABLE, REMAIN CLEAR OF SMOKE, WATER FALL OUT AND WATER RUN OFF.
         2] KEEP OUT OF THE REACH OF CHILDREN.
         3] KEEP AWAY FROM HEAT AND FLAME.
         4] THERMAL DECOMPOSITION MAY YIELD CARBON DIOXIDE, CARBON MONOXIDE.
         5] MOVE CONTAINERS FROM AREA IF WITHOUT RISK, COOL EXPOSED CONTAINERS.
         6] DIKE AREA FOR CONTROL AND CONTAINMENT TO PREVENT ENTRY INTO SEWERS, DRAIN, AND WATER WAYS.

LARGE SPILL/NO FIRE/RESCUE: WEAR RUBBER OR NEOPRENE BOOTS, GLOVES.

SPILL CONTROL AND CONTAINMENT:
  HOUSEHOLD SPILL: WIPE UP WITH ABSORBENT CLOTH. RINSE OUT IN THE SINK. WIPE UP RESIDUE WITH A DRY CLOTH.
  LARGE SPILL: PICK UP WITH ABSORBENT. WASH AREA WITH DETERGENT AND WATER. PICK UP WITH ABSORBENT. PUT ALL
  ABSORBENT IN A DISPOSAL CONTAINER. SEAL AND REMOVE FROM THE WORK AREA.

HEALTH HAZARD INFORMATION:
  WHEN USING THIS PRODUCT WEAR: NO SPECIAL REQUIREMENT. Eye contact may cause stinging, tearing, itching,
  swelling, and redness. Skin contact may cause irritation. Ingestion may cause nausea and vomiting.

  A physician should be contacted if anyone develops any signs or symptoms and suspects that they are caused by
  exposure to this product.

FIRST AID:
  Eye Exposure:
  Flush with water for 15 minutes while lifting the upper and lower eye lids. Contact lenses should not be worn
  when working with this product. Get medical attention.

  Skin Exposure:
  Wash with soap and water. If irritation develops, get medical attention.

  Breathing:
  Not expected to be a problem.

  Swallowing:
  If the victim is conscious, give 2-3 glasses of milk or water to drink. Immediately contact the local poison
  control center for advice. Keep the victim warm and at rest. Get medical attention.

PRODUCT: RENUZIT FRAGRANCE JAR - FRESH POTPOURRI

INCOMPATIBILITY: STRONG OXIDIZERS.

INGREDIENTS: CONCENTRATED PERFUME OIL 100% TLV: CAS#:

IGNI TEMP: NA FP: 184F LEL: NA UEL: NA VP: .3 mmHg VD: over 1 SG: .989 PS: LIQUID APPEAR: CLEAR TO YELLOW
     ODOR: FRAGRANT   PH FACTOR: NA

HAZARD RATINGS: H: F: R: 1 2 0

FIRE FIGHTING:
   HIGH EXPANSION FOAM, LOW EXPANSION FOAM, ALCOHOL FOAM; DRY CHEMICAL, CARBON DIOXIDE, WATER FOG.
   PROTECTIVE CLOTHING, RUBBER GLOVES, AND BREATHING APPARATUS.
WARNING: 1] STRUCTURAL PROTECTIVE CLOTHING IS PERMEABLE, REMAIN CLEAR OF SMOKE, WATER FALL OUT AND WATER RUN OFF.
         2] KEEP OUT OF THE REACH OF CHILDREN.
         3] KEEP AWAY FROM HEAT AND FLAME.
         4] THERMAL DECOMPOSITION MAY YIELD CARBON DIOXIDE, CARBON MONOXIDE.
         5] MOVE CONTAINERS FROM AREA IF WITHOUT RISK, COOL EXPOSED CONTAINERS.
         6] DIKE AREA FOR CONTROL AND CONTAINMENT TO PREVENT ENTRY INTO SEWERS, DRAIN, AND WATER WAYS.

LARGE SPILL/NO FIRE/RESCUE: WEAR RUBBER OR NEOPRENE BOOTS, GLOVES.

SPILL CONTROL AND CONTAINMENT:
   HOUSEHOLD SPILL: WIPE UP WITH ABSORBENT CLOTH. RINSE OUT IN THE SINK. WIPE UP RESIDUE WITH A DRY CLOTH.
   LARGE SPILL: PICK UP WITH ABSORBENT. WASH AREA WITH DETERGENT AND WATER. PICK UP WITH ABSORBENT. PUT ALL
   ABSORBENT IN A DISPOSAL CONTAINER. SEAL AND REMOVE FROM THE WORK AREA.

HEALTH HAZARD INFORMATION:
   WHEN USING THIS PRODUCT WEAR: NO SPECIAL REQUIREMENT. Eye contact may cause stinging, tearing, itching,
   swelling, and redness. Skin contact may cause irritation. Ingestion may cause nausea and vomiting.

   A physician should be contacted if anyone develops any signs or symptoms and suspects that they are caused by
   exposure to this product.

FIRST AID:
Eye Exposure:
Flush with water for 15 minutes while lifting the upper and lower eye lids. Contact lenses should not be worn
when working with this product. Get medical attention.

   Skin Exposure:
   Wash with soap and water. If irritation develops, get medical attention.

   Breathing:
   Not expected to be a problem.

   Swallowing:
   If the victim is conscious, give 2-3 glasses of milk or water to drink. Immediately contact the local poison
   control center for advice. Keep the victim warm and at rest. Get medical attention.

PRODUCT: RENUZIT FRAGRANCE JAR - JADE BREEZES

INCOMPATIBILITY: STRONG OXIDIZERS.

INGREDIENTS: CONCENTRATED PERFUME OIL 100% TLV: CAS#:

IGNI TEMP: NA FP: 198F LEL: NA UEL: NA VP: .07 mmHg VD: over 1 SG: .94 PS: LIQUID APPEAR: CLEAR TO YELLOW
      ODOR: FRAGRANT   PH FACTOR: NA

HAZARD RATINGS: H: F: R: 1 2 0

FIRE FIGHTING:
    HIGH EXPANSION FOAM, LOW EXPANSION FOAM, ALCOHOL FOAM; DRY CHEMICAL, CARBON DIOXIDE, WATER FOG.
    PROTECTIVE CLOTHING, RUBBER GLOVES, AND BREATHING APPARATUS.
WARNING: 1] STRUCTURAL PROTECTIVE CLOTHING IS PERMEABLE, REMAIN CLEAR OF SMOKE, WATER FALL OUT AND WATER RUN OFF.
         2] KEEP OUT OF THE REACH OF CHILDREN.
         3] KEEP AWAY FROM HEAT AND FLAME.
         4] THERMAL DECOMPOSITION MAY YIELD CARBON DIOXIDE, CARBON MONOXIDE.
         5] MOVE CONTAINERS FROM AREA IF WITHOUT RISK, COOL EXPOSED CONTAINERS.
         6] DIKE AREA FOR CONTROL AND CONTAINMENT TO PREVENT ENTRY INTO SEWERS, DRAIN, AND WATER WAYS.

LARGE SPILL/NO FIRE/RESCUE: WEAR RUBBER OR NEOPRENE BOOTS, GLOVES.

SPILL CONTROL AND CONTAINMENT:
    HOUSEHOLD SPILL: WIPE UP WITH ABSORBENT CLOTH. RINSE OUT IN THE SINK. WIPE UP RESIDUE WITH A DRY CLOTH.
    LARGE SPILL: PICK UP WITH ABSORBENT. WASH AREA WITH DETERGENT AND WATER. PICK UP WITH ABSORBENT. PUT ALL
    ABSORBENT IN A DISPOSAL CONTAINER. SEAL AND REMOVE FROM THE WORK AREA.

HEALTH HAZARD INFORMATION:
    WHEN USING THIS PRODUCT WEAR: NO SPECIAL REQUIREMENT. Eye contact may cause stinging, tearing, itching,
    swelling, and redness. Skin contact may cause irritation. Ingestion may cause nausea and vomiting.

    A physician should be contacted if anyone develops any signs or symptoms and suspects that they are caused by
    exposure to this product.

FIRST AID:
    Eye Exposure:
    Flush with water for 15 minutes while lifting the upper and lower eye lids. Contact lenses should not be worn
    when working with this product. Get medical attention.

    Skin Exposure:
    Wash with soap and water. If irritation develops, get medical attention.

    Breathing:
    Not expected to be a problem.

    Swallowing:
    If the victim is conscious, give 2-3 glasses of milk or water to drink. Immediately contact the local poison
    control center for advice. Keep the victim warm and at rest. Get medical attention.

PRODUCT: RENUZIT FRAGRANCE JAR - ORCHARD POTPOURRI

INCOMPATIBILITY: STRONG OXIDIZERS.

INGREDIENTS: CONCENTRATED PERFUME OIL 100% TLV: CAS#:

IGNI TEMP: NA FP: 161F LEL: NA UEL: NA VP: under 1 VD: over 1 SG: .94 PS: LIQUID APPEAR: CLEAR TO YELLOW
ODOR: FRAGRANT PH FACTOR: NA

HAZARD RATINGS: H: F: R: 1 2 0

FIRE FIGHTING:
HIGH EXPANSION FOAM, LOW EXPANSION FOAM, ALCOHOL FOAM; DRY CHEMICAL, CARBON DIOXIDE, WATER FOG.
PROTECTIVE CLOTHING, RUBBER GLOVES, AND BREATHING APPARATUS.
WARNING: 1] STRUCTURAL PROTECTIVE CLOTHING IS PERMEABLE, REMAIN CLEAR OF SMOKE, WATER FALL OUT AND WATER RUN OFF.
2] KEEP OUT OF THE REACH OF CHILDREN.
3] KEEP AWAY FROM HEAT AND FLAME.
4] THERMAL DECOMPOSITION MAY YIELD CARBON DIOXIDE, CARBON MONOXIDE.
5] MOVE CONTAINERS FROM AREA IF WITHOUT RISK, COOL EXPOSED CONTAINERS.
6] DIKE AREA FOR CONTROL AND CONTAINMENT TO PREVENT ENTRY INTO SEWERS, DRAIN, AND WATER WAYS.

LARGE SPILL/NO FIRE/RESCUE: WEAR RUBBER OR NEOPRENE BOOTS, GLOVES.

SPILL CONTROL AND CONTAINMENT:
HOUSEHOLD SPILL: WIPE UP WITH ABSORBENT CLOTH. RINSE OUT IN THE SINK. WIPE UP RESIDUE WITH A DRY CLOTH.
LARGE SPILL: PICK UP WITH ABSORBENT. WASH AREA WITH DETERGENT AND WATER. PICK UP WITH ABSORBENT. PUT ALL
ABSORBENT IN A DISPOSAL CONTAINER. SEAL AND REMOVE FROM THE WORK AREA.

HEALTH HAZARD INFORMATION:
WHEN USING THIS PRODUCT WEAR: NO SPECIAL REQUIREMENT. Eye contact may cause stinging, tearing, itching,
swelling, and redness. Skin contact may cause irritation. Ingestion may cause nausea and vomiting.

A physician should be contacted if anyone develops any signs or symptoms and suspects that they are caused by
exposure to this product.

FIRST AID:
Eye Exposure:
Flush with water for 15 minutes while lifting the upper and lower eye lids. Contact lenses should not be worn
when working with this product. Get medical attention.

Skin Exposure:
Wash with soap and water. If irritation develops, get medical attention.

Breathing:
Not expected to be a problem.

Swallowing:
If the victim is conscious, give 2-3 glasses of milk or water to drink. Immediately contact the local poison
control center for advice. Keep the victim warm and at rest. Get medical attention.

PRODUCT: RENUZIT FRAGRANCE JAR - POWDER POTPOURRI

INCOMPATIBILITY: STRONG OXIDIZERS.

INGREDIENTS: CONCENTRATED PERFUME OIL 100% TLV: CAS#:

IGNI TEMP: NA FP: 161F LEL: NA UEL: NA VP: .05 VD: over 1 SG: .98 PS: LIQUID APPEAR: CLEAR TO YELLOW
    ODOR: FRAGRANT   PH FACTOR: NA

HAZARD RATINGS: H: F: R: 1 2 0

FIRE FIGHTING:
  HIGH EXPANSION FOAM, LOW EXPANSION FOAM, ALCOHOL FOAM; DRY CHEMICAL, CARBON DIOXIDE, WATER FOG.
  PROTECTIVE CLOTHING, RUBBER GLOVES, AND BREATHING APPARATUS.
WARNING: 1] STRUCTURAL PROTECTIVE CLOTHING IS PERMEABLE, REMAIN CLEAR OF SMOKE, WATER FALL OUT AND WATER RUN OFF.
         2] KEEP OUT OF THE REACH OF CHILDREN.
         3] KEEP AWAY FROM HEAT AND FLAME.
         4] THERMAL DECOMPOSITION MAY YIELD CARBON DIOXIDE, CARBON MONOXIDE.
         5] MOVE CONTAINERS FROM AREA IF WITHOUT RISK, COOL EXPOSED CONTAINERS.
         6] DIKE AREA FOR CONTROL AND CONTAINMENT TO PREVENT ENTRY INTO SEWERS, DRAIN, AND WATER WAYS.

LARGE SPILL/NO FIRE/RESCUE: WEAR RUBBER OR NEOPRENE BOOTS, GLOVES.

SPILL CONTROL AND CONTAINMENT:
  HOUSEHOLD SPILL: WIPE UP WITH ABSORBENT CLOTH. RINSE OUT IN THE SINK. WIPE UP RESIDUE WITH A DRY CLOTH.
  LARGE SPILL: PICK UP WITH ABSORBENT. WASH AREA WITH DETERGENT AND WATER. PICK UP WITH ABSORBENT. PUT ALL
  ABSORBENT IN A DISPOSAL CONTAINER. SEAL AND REMOVE FROM THE WORK AREA.

HEALTH HAZARD INFORMATION:
  WHEN USING THIS PRODUCT WEAR: NO SPECIAL REQUIREMENT. Eye contact may cause stinging, tearing, itching,
  swelling, and redness. Skin contact may cause irritation. Ingestion may cause nausea and vomiting.

  A physician should be contacted if anyone develops any signs or symptoms and suspects that they are caused by
  exposure to this product.

FIRST AID:
  Eye Exposure:
  Flush with water for 15 minutes while lifting the upper and lower eye lids. Contact lenses should not be worn
  when working with this product. Get medical attention.

  Skin Exposure:
  Wash with soap and water. If irritation develops, get medical attention.

  Breathing:
  Not expected to be a problem.

  Swallowing:
  If the victim is conscious, give 2-3 glasses of milk or water to drink. Immediately contact the local poison
  control center for advice. Keep the victim warm and at rest. Get medical attention.

PRODUCT: RENUZIT FRESH 'n DRY SPRAY (AEROSOL)

INCOMPATIBILITY: NONE KNOWN

INGREDIENTS: PROPANE_____ 99%   TLV: 1000 ppm   CAS#: 74-98-6
             BUTANE_____/   TLV: 800 ppm   CAS#: 106-97-8

IGNI TEMP: NA FP: 20F LEL: 1.8 UEL: 9.5 VP: NA VD: 1.9 SG: .6 PS: GAS APPEAR: CLEAR
    ODOR: VARIOUS FRAGRANTS   PH FACTOR: 7

HAZARD RATINGS: H: F: R: 0   4   0

FIRE FIGHTING:
    HIGH EXPANSION FOAM, LOW EXPANSION FOAM, ALCOHOL FOAM; DRY CHEMICAL, CARBON DIOXIDE, WATER FOG.
    PROTECTIVE CLOTHING, RUBBER GLOVES, AND BREATHING APPARATUS.
WARNING: 1] STRUCTURAL PROTECTIVE CLOTHING IS PERMEABLE, REMAIN CLEAR OF SMOKE, WATER FALL OUT AND WATER RUN OFF.
         2] KEEP OUT OF THE REACH OF CHILDREN.
         3] KEEP AWAY FROM HEAT AND FLAME.
         4] REMOVE ALL IGNITION SOURCES IF WITHOUT RISK.
         5] CONTAINERS WILL EXPLODE IN A FIRE OR IF HEATED ABOVE 120F.
         6] THERMAL DECOMPOSITION MAY YIELD CARBON DIOXIDE, CARBON MONOXIDE.
         7] MOVE CONTAINERS FROM AREA IF WITHOUT RISK, COOL EXPOSED CONTAINERS.
         8] DIKE AREA FOR CONTROL AND CONTAINMENT TO PREVENT ENTRY INTO SEWERS, DRAIN, AND WATER WAYS.

LARGE SPILL/NO FIRE/RESCUE: WEAR RUBBER OR NEOPRENE BOOTS, GLOVES, BREATHING APPARATUS.

SPILL CONTROL AND CONTAINMENT:
    HOUSEHOLD SPILL: OPEN WINDOWS AND DOORS. VENTILATE ENCLOSED SPACES. DO NOT TURN ON OR OFF ANY ELECTRICAL
    SWITCHES, BLOW OUT PILOT LIGHTS IN HEATERS, STOVES AND OTHER. LEAVE AREA UNTIL GAS IS CLEAR.
    LARGE SPILL: VENTILATE ALL ENCLOSED SPACES. REMOVE ALL IGNITION SOURCES IF WITHOUT RISK.

HEALTH HAZARD INFORMATION:
    WHEN USING THIS PRODUCT WEAR: NO SPECIAL REQUIREMENT. Inhalation may cause dizziness, loss of consciousness, and
    death by asphyxiation (suffocation). Contact spray at a very close range may freeze skin tissue or the eyes.
    Symptoms are a whitish coloring of the skin and redness or loss of feeling.

    A physician should be contacted if anyone develops any signs or symptoms and suspects that they are caused by
    exposure to this product.

FIRST AID:
    Eye Exposure:
    Apply a luke warm compress. Contact lenses should not be worn when working with this product. Get medical
    attention.

    Skin Exposure:
    Apply a luke war, compress. Keep victim warm and at rest. Get medical attention.

    Breathing:
    If symptoms develop, move victim to fresh air. Give oxygen. Get medical attention.

    Swallowing:
    Unlikely since product is a gas.

PRODUCT: RENUZIT FRESHELL LONG LASTING AIR FRESHENER - MULTIPLE FRAGRANCES

INCOMPATIBILITY: NONE KNOWN

INGREDIENTS: ISOPROPYL ALCOHOL   8% TLV: 400 ppm   CAS#: 67-63-0

IGNI TEMP: NA FP: 102F LEL: NA UEL: NA VP: 21 mmHg VD: 1 SG: .99 PS: LIQUID APPEAR: VARIOUS COLORS
    ODOR: FRAGRANCE   PH FACTOR: 5.5

HAZARD RATINGS: H: F: R: 0 2 0

FIRE FIGHTING:
    HIGH EXPANSION FOAM, LOW EXPANSION FOAM, ALCOHOL FOAM; DRY CHEMICAL, CARBON DIOXIDE, WATER FOG.
    PROTECTIVE CLOTHING, RUBBER GLOVES, AND BREATHING APPARATUS.
WARNING: 1] STRUCTURAL PROTECTIVE CLOTHING IS PERMEABLE, REMAIN CLEAR OF SMOKE, WATER FALL OUT AND WATER RUN OFF.
         2] KEEP OUT OF THE REACH OF CHILDREN.
         3] KEEP AWAY FROM HEAT AND FLAME.
         4] THERMAL DECOMPOSITION MAY YIELD CARBON DIOXIDE, CARBON MONOXIDE.
         5] MOVE CONTAINERS FROM AREA IF WITHOUT RISK, COOL EXPOSED CONTAINERS.
         6] DIKE AREA FOR CONTROL AND CONTAINMENT TO PREVENT ENTRY INTO SEWERS, DRAIN, AND WATER WAYS.

LARGE SPILL/NO FIRE/RESCUE: WEAR RUBBER OR NEOPRENE BOOTS, GLOVES.

SPILL CONTROL AND CONTAINMENT:
    HOUSEHOLD SPILL: WIPE UP WITH ABSORBENT CLOTH. RINSE OUT IN THE SINK. WIPE UP RESIDUE WITH A DRY CLOTH.
    LARGE SPILL: PICK UP WITH ABSORBENT. WASH AREA WITH DETERGENT AND WATER. PICK UP WITH ABSORBENT. PUT ALL
    ABSORBENT IN A DISPOSAL CONTAINER. SEAL AND REMOVE FROM THE WORK AREA.

HEALTH HAZARD INFORMATION:
    WHEN USING THIS PRODUCT WEAR: NO SPECIAL REQUIREMENT. Eye contact may cause stinging, tearing, itching,
    swelling, and redness. Skin contact may cause irritation. Ingestion may cause nausea, vomiting, and diarrhea.

    A physician should be contacted if anyone develops any signs or symptoms and suspects that they are caused by
    exposure to this product.

FIRST AID:
    Eye Exposure:
    Flush with water for 15 minutes while lifting the upper and lower eye lids. Contact lenses should not be worn
    when working with this product. Get medical attention.

    Skin Exposure:
    Wash with soap and water. If irritation develops, get medical attention.

    Breathing:
    Not expected to be a problem.

    Swallowing:
    If the victim is conscious, give 2-3 glasses of milk or water to drink. Immediately contact the local poison
    control center for advice. Keep the victim warm and at rest. Get medical attention.

PRODUCT: RENUZIT ROOMMATE LIQUID AIR FRESHENER - MULTIPLE FRAGRANCES

INCOMPATIBILITY: NONE KNOWN

INGREDIENTS: ISOPROPYL ALCOHOL   8% TLV: 400 ppm   CAS#: 67-63-0

IGNI TEMP: NA FP: 106F LEL: NA UEL: NA VP: 21 mmHg VD: 1.0 SG: 1.0 PS: LIQUID APPEAR: CLEAR MOBILE
    ODOR: FRAGRANCE   PH FACTOR: 5.5

HAZARD RATINGS: H: F: R: 0 2 0

FIRE FIGHTING:
    HIGH EXPANSION FOAM, LOW EXPANSION FOAM, ALCOHOL FOAM; DRY CHEMICAL, CARBON DIOXIDE, WATER FOG.
    PROTECTIVE CLOTHING, RUBBER GLOVES, AND BREATHING APPARATUS.
WARNING: 1] STRUCTURAL PROTECTIVE CLOTHING IS PERMEABLE, REMAIN CLEAR OF SMOKE, WATER FALL OUT AND WATER RUN OFF.
         2] KEEP OUT OF THE REACH OF CHILDREN.
         3] KEEP AWAY FROM HEAT AND FLAME.
         4] THERMAL DECOMPOSITION MAY YIELD CARBON DIOXIDE, CARBON MONOXIDE.
         5] MOVE CONTAINERS FROM AREA IF WITHOUT RISK, COOL EXPOSED CONTAINERS.
         6] DIKE AREA FOR CONTROL AND CONTAINMENT TO PREVENT ENTRY INTO SEWERS, DRAIN, AND WATER WAYS.

LARGE SPILL/NO FIRE/RESCUE: WEAR RUBBER OR NEOPRENE BOOTS, GLOVES.

SPILL CONTROL AND CONTAINMENT:
    HOUSEHOLD SPILL: WIPE UP WITH ABSORBENT CLOTH. RINSE OUT IN THE SINK. WIPE UP RESIDUE WITH A DRY CLOTH.
    LARGE SPILL: PICK UP WITH ABSORBENT. WASH AREA WITH DETERGENT AND WATER. PICK UP WITH ABSORBENT. PUT ALL
    ABSORBENT IN A DISPOSAL CONTAINER. SEAL AND REMOVE FROM THE WORK AREA.

HEALTH HAZARD INFORMATION:
    WHEN USING THIS PRODUCT WEAR: NO SPECIAL REQUIREMENT. Eye contact may cause stinging, tearing, itching,
    swelling, and redness. Skin contact may cause irritation. Ingestion may cause nausea and vomiting.

    A physician should be contacted if anyone develops any signs or symptoms and suspects that they are caused by
    exposure to this product.

FIRST AID:
Eye Exposure:
Flush with water for 15 minutes while lifting the upper and lower eye lids. Contact lenses should not be worn
when working with this product. Get medical attention.

Skin Exposure:
Wash with soap and water. If irritation develops, get medical attention.

Breathing:
Not expected to be a problem.

Swallowing:
If the victim is conscious, give 2-3 glasses of milk or water to drink. Immediately contact the local poison
control center for advice. Keep the victim warm and at rest. Get medical attention.

PRODUCT: RINSO NON-PHOSPHATE POWDER DETERGENT

INCOMPATIBILITY: NONE KNOWN

INGREDIENTS: SODIUM SILICATE   1% TLV: CAS#: 1344-09-8
             SODIUM CARBONATE   1% TLV: CAS#: 497-19-8

IGNI TEMP: NA FP: NA LEL: NA UEL: NA VP: NA VD: NA SG: NA PS: POWDER APPEAR: BLUE SPECKS   ODOR:
   PH FACTOR:

HAZARD RATINGS: H: F: R: 2 0 0

FIRE FIGHTING:
   HIGH EXPANSION FOAM, LOW EXPANSION FOAM, ALCOHOL FOAM; DRY CHEMICAL, CARBON DIOXIDE, WATER FOG.
   PROTECTIVE CLOTHING, RUBBER GLOVES, AND BREATHING APPARATUS.
WARNING: 1] STRUCTURAL PROTECTIVE CLOTHING IS PERMEABLE, REMAIN CLEAR OF SMOKE, WATER FALL OUT AND WATER RUN OFF.
         2] KEEP OUT OF THE REACH OF CHILDREN.
         3] MOVE CONTAINERS FROM AREA IF WITHOUT RISK, COOL EXPOSED CONTAINERS.
         4] DIKE AREA FOR CONTROL AND CONTAINMENT TO PREVENT ENTRY INTO SEWERS, DRAIN, AND WATER WAYS.

LARGE SPILL/NO FIRE/RESCUE: WEAR RUBBER OR NEOPRENE BOOTS, GLOVES.

SPILL CONTROL AND CONTAINMENT:
   HOUSEHOLD SPILL: AVOID CREATING A DUST. SWEEP UP POWDER. DAMP MOP UP RESIDUE. RINSE OUT IN THE SINK.
   LARGE SPILL: AVOID CREATING A DUST. SWEEP UP POWDER. PUT IN A DISPOSAL CONTAINER. RINSE AREA WITH WATER. PICK UP
   WITH ABSORBENT. PUT IN A DISPOSAL CONTAINER. SEAL AND REMOVE FROM THE WORK AREA.

HEALTH HAZARD INFORMATION:
   WHEN USING THIS PRODUCT WEAR: NO SPECIAL REQUIREMENT. Eye contact will cause stinging, tearing, itching,
   swelling, and redness. Prolonged contact with the skin may cause drying and irritation. Ingestion may cause
   nausea and vomiting. Inhalation may cause irritation of the nose, throat, and lungs.

   A physician should be contacted if anyone develops any signs or symptoms and suspects that they are caused by
   exposure to this product.

FIRST AID:
   Eye Exposure:
   Flush with water for 15 minutes while lifting the upper and lower eye lids. Contact lenses should not be worn
   when working with this product. Get medical attention.

   Skin Exposure:
   Wash with water. If irritation develops, get medical attention.

   Breathing:
   If symptoms develop, move the victim to fresh air.

   Swallowing:
   If the victim is conscious, give 2-3 glasses of milk or water to drink. Immediately contact the local poison
   control center for advice. Keep the victim warm and at rest. Get medical attention.

PRODUCT: RHULI CREAM

INCOMPATIBILITY: NONE KNOWN

INGREDIENTS: ISOPROPYL ALCOHOL   8.8% TLV: 400 ppm   CAS#:   64-17-5
             WATER   TLV: CAS#: 7732-18-5
             ZIRCONIUM OXIDE   1%   TLV: CAS#: 1314-23-4
             BENZOCAINE   1%   TLV: CAS#:   94-09-7
             MENTHOL   .7% TLV: CAS#:
             CAMPHOR   .3% TLV: CAS#:
             ALUMINUM SILICATE; CELLULOSE; MAGNESIUM; PRESERVATIVES; PIGMENT.

IGNI TEMP: NA FP: NA LEL: NA UEL: NA VP: NA VD: NA SG: NA PS: PASTE APPEAR: PINK   ODOR: CAMPHOR
    PH FACTOR: NA

HAZARD RATINGS: H: F: R:

FIRE FIGHTING:
    HIGH EXPANSION FOAM, LOW EXPANSION FOAM, ALCOHOL FOAM; DRY CHEMICAL, CARBON DIOXIDE, WATER FOG.
    PROTECTIVE CLOTHING, RUBBER GLOVES, AND BREATHING APPARATUS.
WARNING: 1] STRUCTURAL PROTECTIVE CLOTHING IS PERMEABLE, REMAIN CLEAR OF SMOKE, WATER FALL OUT AND WATER RUN OFF.
         2] KEEP OUT OF THE REACH OF CHILDREN.
         3] THERMAL DECOMPOSITION MAY YIELD CARBON DIOXIDE, CARBON MONOXIDE.
         4] MOVE CONTAINERS FROM AREA IF WITHOUT RISK, COOL EXPOSED CONTAINERS.
         5] DIKE AREA FOR CONTROL AND CONTAINMENT TO PREVENT ENTRY INTO SEWERS, DRAIN, AND WATER WAYS.

LARGE SPILL/NO FIRE/RESCUE: WEAR RUBBER OR NEOPRENE BOOTS, GLOVES.

SPILL CONTROL AND CONTAINMENT:
    HOUSEHOLD SPILL: SCOOP UP AND PUT IN DISPOSAL CONTAINER. WIPE UP RESIDUE WITH A DAMP CLOTH.
    LARGE SPILL: SCOOP UP MATERIAL, PUT IN DISPOSAL CONTAINER. RINSE AREA WITH WATER. PICK UP WITH ABSORBENT. PUT IN
    A DISPOSAL CONTAINER. SEAL AND REMOVE FROM THE WORK AREA.

HEALTH HAZARD INFORMATION:
    WHEN USING THIS PRODUCT WEAR: NO SPECIAL REQUIREMENT. Eye contact may cause stinging, tearing, itching,
    swelling, and redness. Ingestion may cause nausea and vomiting.

    A physician should be contacted if anyone develops any signs or symptoms and suspects that they are caused by
    exposure to this product.

FIRST AID:
    Eye Exposure:
    Flush with water for 15 minutes while lifting the upper and lower eye lids. Contact lenses should not be worn
    when working with this product. Get medical attention.

    Skin Exposure:
    Not expected to be a problem.

    Breathing:
    Not expected to be a problem.

    Swallowing:
    Immediately contact the local poison control center for advice. Keep the victim warm and at rest. Get medical
    attention.

```
PRODUCT: RHULI GEL

INCOMPATIBILITY: NONE KNOWN

INGREDIENTS: ETHYL ALCOHOL        31.0%   TLV: 1000 ppm    CAS#:   64-17-5
             WATER              55-65.0%  TLV:             CAS#: 7732-18-5
             PHENYLCARBINOL        2.0%   TLV:             CAS#:  100-51-6
             MENTHOL                .3%   TLV:             CAS#:
             CAMPHOR                .3%   TLV:             CAS#:

IGNI TEMP: NA  FP: NA  LEL: NA  UEL: NA  VP: NA  VD: NA  SG: NA  PS: GEL  APPEAR: CLEAR, COLORLESS
    ODOR: CAMPHOR/MENTHOL      PH FACTOR: NA

HAZARD RATINGS:   H:  F:  R:
```

FIRE FIGHTING:

HIGH EXPANSION FOAM, LOW EXPANSION FOAM, ALCOHOL FOAM; DRY CHEMICAL, CARBON DIOXIDE, WATER FOG.
PROTECTIVE CLOTHING, RUBBER GLOVES, AND BREATHING APPARATUS.
WARNING: 1] STRUCTURAL PROTECTIVE CLOTHING IS PERMEABLE, REMAIN CLEAR OF SMOKE, WATER FALL OUT AND WATER RUN OFF.
         2] KEEP OUT OF THE REACH OF CHILDREN.
         3] THERMAL DECOMPOSITION MAY YIELD CARBON DIOXIDE, CARBON MONOXIDE.
         4] MOVE CONTAINERS FROM AREA IF WITHOUT RISK, COOL EXPOSED CONTAINERS.
         5] DIKE AREA FOR CONTROL AND CONTAINMENT TO PREVENT ENTRY INTO SEWERS, DRAIN, AND WATER WAYS.

LARGE SPILL/NO FIRE/RESCUE: WEAR RUBBER OR NEOPRENE BOOTS, GLOVES.

SPILL CONTROL AND CONTAINMENT:

HOUSEHOLD SPILL: SCOOP UP GEL AND PUT IN A DISPOSAL CONTAINER. WIPE UP RESIDUE WITH A DAMP CLOTH. RINSE OUT IN
THE SINK.
LARGE SPILL: SCOOP UP GEL AND PUT IN A DISPOSAL CONTAINER. RINSE AREA WITH WATER. PICK UP WITH ABSORBENT. PUT IN
A DISPOSAL CONTAINER. SEAL AND REMOVE FROM THE WORK AREA.

HEALTH HAZARD INFORMATION:

WHEN USING THIS PRODUCT WEAR: NO SPECIAL REQUIREMENT. Eye contact may cause
irritation. Ingestion may cause nausea and vomiting.

A physician should be contacted if anyone develops any signs or symptoms and suspects that they are caused by
exposure to this product.

FIRST AID:

Eye Exposure:
Flush with water for 15 minutes while lifting the upper and lower eye lids. Contact lenses should not be worn
when working with this product. Get medical attention.

Skin Exposure:
Not expected to be a problem.

Breathing:
Not expected to be a problem.

Swallowing:
Immediately contact the local poison control center for advice. Keep the victim warm and at rest. Get medical
attention.

PRODUCT: RHULI SPRAY

INCOMPATIBILITY: NONE KNOWN

INGREDIENTS: ISOBUTANE   60-65% TLV: CAS#:   75-28-5
             ISOPROPYL ALCOHOL   28.767% TLV: 400 ppm   CAS#:   67-63-0
             BENZOCAINE   1.153% TLV: CAS#:   94-09-7
             CALAMINE   4.710% TLV: CAS#: 8011-96-9
             PHENYLCARBINOL   .674% TLV: CAS#: 100-51-6
             CAMPHOR; MENTHOL

IGNI TEMP: NA FP: 20F LEL: NA UEL: NA VP: NA VD: NA SG: .64 PS: LIQUID APPEAR: SPRAY MIST ODOR:
   PH FACTOR: NA

HAZARD RATINGS: H: F: R:

FIRE FIGHTING:
   HIGH EXPANSION FOAM, LOW EXPANSION FOAM, ALCOHOL FOAM; DRY CHEMICAL, CARBON DIOXIDE, WATER FOG.
   PROTECTIVE CLOTHING, RUBBER GLOVES, AND BREATHING APPARATUS.
WARNING: 1] STRUCTURAL PROTECTIVE CLOTHING IS PERMEABLE, REMAIN CLEAR OF SMOKE, WATER FALL OUT AND WATER RUN OFF.
         2] KEEP OUT OF THE REACH OF CHILDREN.
         3] REMOVE ALL IGNITION SOURCES IF WITHOUT RISK.
         4] CONTAINERS WILL EXPLODE IN A FIRE OR IF HEATED ABOVE 120F.
         5] THERMAL DECOMPOSITION MAY YIELD CARBON DIOXIDE, CARBON MONOXIDE.
         6] MOVE CONTAINERS FROM AREA IF WITHOUT RISK, COOL EXPOSED CONTAINERS.
         7] DIKE AREA FOR CONTROL AND CONTAINMENT TO PREVENT ENTRY INTO SEWERS, DRAIN, AND WATER WAYS.

LARGE SPILL/NO FIRE/RESCUE: WEAR RUBBER OR NEOPRENE BOOTS, GLOVES.

SPILL CONTROL AND CONTAINMENT:
   HOUSEHOLD SPILL: WIPE UP WITH AN ABSORBENT CLOTH. RINSE OUT IN THE SINK. WIPE UP RESIDUE WITH A DAMP CLOTH.
   RINSE OUT IN THE SINK.
   LARGE SPILL: PICK UP WITH ABSORBENT. RINSE AREA WITH WATER. PICK UP WITH ABSORBENT. PUT ABSORBENT IN A DISPOSAL
   CONTAINER. SEAL AND REMOVE FROM THE WORK AREA.

HEALTH HAZARD INFORMATION:
   WHEN USING THIS PRODUCT WEAR: NO SPECIAL REQUIREMENT. Eye contact may cause stinging, tearing, itching, and
   redness.

   A physician should be contacted if anyone develops any signs or symptoms and suspects that they are caused by
   exposure to this product.

FIRST AID:
   Eye Exposure:
   Flush with water for 15 minutes while lifting the upper and lower eye lids. Contact lenses should not be worn
   when working with this product. Get medical attention.

   Skin Exposure:
   Not expected to be a problem.

   Breathing:
   Not expected to be a problem.

   Swallowing:
   Immediately contact the local poison control center for advice. Keep the victim warm and at rest. Get medical
   attention.

PRODUCT: SAFEGUARD DEODORANT TOILET BAR SOAP

INCOMPATIBILITY: NONE KNOWN

INGREDIENTS:			
SODIUM CHLORIDE	TLV:	CAS#:	6747-14-5
TITANIUM DIOXIDE	TLV:	CAS#:	13463-67-7
TETRASODIUM EDTA	TLV:	CAS#:	10378-23-1
TRICLOCARBAN	TLV:	CAS#:	
SODIUM TALLOWATE	TLV:	CAS#:	
WATER	TLV:	CAS#:	7732-18-5
BUTYLATED HYDROXYTOLUENE	TLV:	CAS#:	128-37-0

IGNI TEMP: NA FP: NA LEL: NA UEL: NA VP: NA VD: NA SG: 1.04 PS: SOLID APPEAR: WHITE, BEIGE OR GOLD BARS
   ODOR: SPICY   PH FACTOR: NA

HAZARD RATINGS: H: F: R:

FIRE FIGHTING:
  HIGH EXPANSION FOAM, LOW EXPANSION FOAM, ALCOHOL FOAM; DRY CHEMICAL, CARBON DIOXIDE, WATER FOG.
  PROTECTIVE CLOTHING, RUBBER GLOVES, AND BREATHING APPARATUS.
WARNING: 1] STRUCTURAL PROTECTIVE CLOTHING IS PERMEABLE, REMAIN CLEAR OF SMOKE, WATER FALL OUT AND WATER RUN OFF.
         2] KEEP OUT OF THE REACH OF CHILDREN.
         3] MOVE CONTAINERS FROM AREA IF WITHOUT RISK, COOL EXPOSED CONTAINERS.
         4] DIKE AREA FOR CONTROL AND CONTAINMENT TO PREVENT ENTRY INTO SEWERS, DRAIN, AND WATER WAYS.

LARGE SPILL/NO FIRE/RESCUE: WEAR RUBBER OR NEOPRENE BOOTS, GLOVES.

SPILL CONTROL AND CONTAINMENT:
  HOUSEHOLD SPILL: PICK UP BARS. WIPE UP ANY RESIDUE WITH A DAMP CLOTH.
  LARGE SPILL: SCOOP UP BARS AND PUT IN A DISPOSAL CONTAINER. RINSE WITH WATER. PICK UP WITH ABSORBENT. PUT
  ABSORBENT IN A DISPOSAL CONTAINER.

HEALTH HAZARD INFORMATION:
  WHEN USING THIS PRODUCT WEAR: NO SPECIAL REQUIREMENT. Eye contact may cause irritation. May cause skin
  irritation if allergic to product. Ingestion may cause mild abdominal cramps, nausea, vomiting, and diarrhea.

  A physician should be contacted if anyone develops any signs or symptoms and suspects that they are caused by
  exposure to this product.

FIRST AID:
  Eye Exposure:
  Flush with water for 15 minutes while lifting the upper and lower eye lids. Contact lenses should not be worn
  when working with this product. Get medical attention.

  Skin Exposure:
  If irritation develops, rinse off with water. Discontinue use of product.

  Breathing:
  Not expected to be a problem.

  Swallowing:
  If the victim is conscious, give 1-2 glasses of water to drink. Immediately contact the local poison control
  center for advice. Keep the victim warm and at rest. Get medical attention.

PRODUCT: SAFEGUARD DS FOR DRY SKIN PROTECTION - DEODORANT TOILET BAR SOAP

INCOMPATIBILITY: NONE KNOWN

INGREDIENTS: GLYCERINE   TLV: CAS#:   56-81-5
             SODIUM CHLORIDE   TLV: CAS#: 7647-14-5
             TETRASODIUM EDTA   TLV: CAS#: 10378-23-1
             BHT   TLV: CAS#:   128-37-0
             TITANIUM DIOXIDE   TLV: CAS#: 13463-67-7
             WATER   TLV: CAS#: 7732-18-5
             SODIUM TALLOWATE

IGNI TEMP: NA FP: NA LEL: NA UEL: NA VP: NA VD: NA SG: NA PS: SOLID APPEAR: WHITE OR BEIGE BARS
     ODOR: SPICY   PH FACTOR: NA

HAZARD RATINGS: H: F: R:

FIRE FIGHTING:
  HIGH EXPANSION FOAM, LOW EXPANSION FOAM, ALCOHOL FOAM; DRY CHEMICAL, CARBON DIOXIDE, WATER FOG.
  PROTECTIVE CLOTHING, RUBBER GLOVES, AND BREATHING APPARATUS.
WARNING: 1] STRUCTURAL PROTECTIVE CLOTHING IS PERMEABLE, REMAIN CLEAR OF SMOKE, WATER FALL OUT AND WATER RUN OFF.
         2] KEEP OUT OF THE REACH OF CHILDREN.
         3] MOVE CONTAINERS FROM AREA IF WITHOUT RISK, COOL EXPOSED CONTAINERS.
         4] DIKE AREA FOR CONTROL AND CONTAINMENT TO PREVENT ENTRY INTO SEWERS, DRAIN, AND WATER WAYS.

LARGE SPILL/NO FIRE/RESCUE: WEAR RUBBER OR NEOPRENE BOOTS, GLOVES.

SPILL CONTROL AND CONTAINMENT:
  HOUSEHOLD SPILL: PICK UP BAR. WIPE UP RESIDUE WITH A DAMP CLOTH.
  LARGE SPILL: SCOOP UP MATERIAL AND PUT IN A DISPOSAL CONTAINER. RINSE AREA WITH WATER. PICK UP WITH ABSORBENT.
  PUT IN A DISPOSAL CONTAINER. SEAL AND REMOVE FROM THE WORK AREA.

HEALTH HAZARD INFORMATION:
  WHEN USING THIS PRODUCT WEAR: NO SPECIAL REQUIREMENT. Eye contact may cause irritation. May cause skin
  irritation if allergic to product. Ingestion may cause mild abdominal cramps, nausea, vomiting, and diarrhea.

  A physician should be contacted if anyone develops any signs or symptoms and suspects that they are caused by
  exposure to this product.

FIRST AID:
  Eye Exposure:
  Flush with water for 15 minutes while lifting the upper and lower eye lids. Contact lenses should not be worn
  when working with this product. Get medical attention.

  Skin Exposure:
  If irritation develops, rinse off with water. Discontinue use of product.

  Breathing:
  Not expected to be a problem.

  Swallowing:
  If the victim is conscious, give 1-2 glasses of water to drink. Immediately contact the local poison control
  center for advice. Keep the victim warm and at rest. Get medical attention.

PRODUCT: SHIELD DEODORANT BAR

INCOMPATIBILITY: NONE KNOWN

INGREDIENTS: INGREDIENTS LISTED AS NON HAZARDOUS BY OSHA 1910.1200 STANDARD

IGNI TEMP: NA FP: NA LEL: NA UEL: NA VP: NA VD: NA SG: 1.04 PS: SOLID APPEAR: GREEN BAR ODOR:
 PH FACTOR: 10

HAZARD RATINGS: H: F: R:

FIRE FIGHTING:
 HIGH EXPANSION FOAM, LOW EXPANSION FOAM, ALCOHOL FOAM; DRY CHEMICAL, CARBON DIOXIDE, WATER FOG.
 PROTECTIVE CLOTHING, RUBBER GLOVES, AND BREATHING APPARATUS.
WARNING: 1] STRUCTURAL PROTECTIVE CLOTHING IS PERMEABLE, REMAIN CLEAR OF SMOKE, WATER FALL OUT AND WATER RUN OFF.
     2] KEEP OUT OF THE REACH OF CHILDREN.
     3] MOVE CONTAINERS FROM AREA IF WITHOUT RISK, COOL EXPOSED CONTAINERS.
     4] DIKE AREA FOR CONTROL AND CONTAINMENT TO PREVENT ENTRY INTO SEWERS, DRAIN, AND WATER WAYS.

LARGE SPILL/NO FIRE/RESCUE: WEAR RUBBER OR NEOPRENE BOOTS, GLOVES.

SPILL CONTROL AND CONTAINMENT:
 HOUSEHOLD SPILL: PICK UP BAR. WIPE UP ANY RESIDUE WITH A DAMP CLOTH.
 LARGE SPILL: SCOOP UP MATERIAL AND PUT IN A DISPOSAL CONTAINER. RINSE AREA WITH WATER. PICK UP WITH ABSORBENT.
 PUT ABSORBENT IN A DISPOSAL CONTAINER. SEAL AND REMOVE FROM THE WORK AREA.

HEALTH HAZARD INFORMATION:
 WHEN USING THIS PRODUCT WEAR: NO SPECIAL REQUIREMENT. Eye contact will cause stinging, tearing, itching,
 swelling, and redness. Prolonged skin exposure may cause irritation. Ingestion may cause mild abdominal cramps,
 nausea, vomiting, and diarrhea.

 A physician should be contacted if anyone develops any signs or symptoms and suspects that they are caused by
 exposure to this product.

FIRST AID:
 Eye Exposure:
 Flush with water for 15 minutes while lifting the upper and lower eye lids. Contact lenses should not be worn
 when working with this product. Get medical attention.

 Skin Exposure:
 If irritation develops, rinse off with water. Discontinue use of product.

 Breathing:
 Not expected to be a problem.

 Swallowing:
 If the victim is conscious, give 2-3 glasses of milk or water to drink. Immediately contact the local poison
 control center for advice. Keep the victim warm and at rest. Get medical attention.

PRODUCT: SHOUT AEROSOL

INCOMPATIBILITY: NONE KNOWN

INGREDIENTS: MINERAL SPIRITS   5-15% TLV: 100 ppm   CAS#: 64741-65-7
             WATER   65-80% TLV: CAS#: 7732-18-5
             PROPANE_____   TLV: 1000 ppm   CAS#:   74-98-6
             ISOBUTANE_____   6-15% TLV: CAS#:   75-28-5
             N-BUTANE____/   TLV: 800 ppm   CAS#: 1-6-97-8
             SURFACTANTS   6-12% TLV: CAS#:
             BUILDERS, CORROSION INHIBITORS.

IGNI TEMP: NA FP: 20F LEL: NA UEL: NA VP: NA VD: NA SG: 1.0 PS: LIQUID APPEAR: SPRAY ODOR: SOLVENT
   PH FACTOR: NA

HAZARD RATINGS: H: F: R: 2 4 0

FIRE FIGHTING:
   HIGH EXPANSION FOAM, LOW EXPANSION FOAM, ALCOHOL FOAM; DRY CHEMICAL, CARBON DIOXIDE, WATER FOG.
   PROTECTIVE CLOTHING, RUBBER GLOVES, AND BREATHING APPARATUS.
WARNING: 1] STRUCTURAL PROTECTIVE CLOTHING IS PERMEABLE, REMAIN CLEAR OF SMOKE, WATER FALL OUT AND WATER RUN OFF.
         2] KEEP OUT OF THE REACH OF CHILDREN.
         3] REMOVE ALL IGNITION SOURCES IF WITHOUT RISK.
         4] CONTAINERS WILL EXPLODE IN A FIRE OR IF HEATED ABOVE 120F.
         5] THERMAL DECOMPOSITION MAY YIELD CARBON DIOXIDE, CARBON MONOXIDE.
         6] MOVE CONTAINERS FROM AREA IF WITHOUT RISK, COOL EXPOSED CONTAINERS.
         7] DIKE AREA FOR CONTROL AND CONTAINMENT TO PREVENT ENTRY INTO SEWERS, DRAIN, AND WATER WAYS.

LARGE SPILL/NO FIRE/RESCUE: WEAR RUBBER OR NEOPRENE BOOTS, GLOVES.

SPILL CONTROL AND CONTAINMENT:
   HOUSEHOLD SPILL: WIPE UP WITH AN ABSORBENT CLOTH. RINSE OUT IN THE SINK. WIPE UP RESIDUE WITH A DAMP CLOTH.
   RINSE OUT IN THE SINK.
   LARGE SPILL: PICK UP WITH ABSORBENT. RINSE AREA WITH WATER. PICK UP WITH ABSORBENT. PUT ALL ABSORBENT IN A
   DISPOSAL CONTAINER. SEAL AND REMOVE FROM THE WORK AREA.

HEALTH HAZARD INFORMATION:
   WHEN USING THIS PRODUCT WEAR: NO SPECIAL REQUIREMENT. Eye contact will cause stinging, tearing, itching,
   swelling, and redness. May cause skin irritation. Ingestion may cause mild abdominal cramps, nausea, vomiting,
   and diarrhea.

   A physician should be contacted if anyone develops any signs or symptoms and suspects that they are caused by
   exposure to this product.

FIRST AID:
Eye Exposure:
Flush with water for 15 minutes while lifting the upper and lower eye lids. Contact lenses should not be worn
when working with this product. Get medical attention.

Skin Exposure:
Wash with soap and water. If irritation develops, get medical attention.

Breathing:
If any breathing problem occurs, move victim to fresh air.

Swallowing:
Do not induce vomiting. If victim is conscious, give 2-3 glasses of water to drink. Immediately contact the
local poison control center for advice. Keep the victim warm and at rest. Get medical attention.

PRODUCT: SHOUT LIQUID

INCOMPATIBILITY: NONE KNOWN

INGREDIENTS: POLYOXYETHYLENE

NONYL PHENYL ETHER	5-12%	TLV:	CAS#: 26027-38-3
SODIUM CITRATE	1-3%	TLV:	CAS#: 6132-04-3
WATER	85-88%	TLV:	CAS#: 7732-18-5

IGNI TEMP: NA FP: NA LEL: NA UEL: NA VP: NA VD: NA SG: 1.02 PS: LIQUID APPEAR: OPAQUE, BLUE
   ODOR: PLEASANT   PH FACTOR: 7.0

HAZARD RATINGS: H: F: R: 1 0 0

FIRE FIGHTING:
   HIGH EXPANSION FOAM, LOW EXPANSION FOAM, ALCOHOL FOAM; DRY CHEMICAL, CARBON DIOXIDE, WATER FOG.
   PROTECTIVE CLOTHING, RUBBER GLOVES, AND BREATHING APPARATUS.
WARNING: 1] STRUCTURAL PROTECTIVE CLOTHING IS PERMEABLE, REMAIN CLEAR OF SMOKE, WATER FALL OUT AND WATER RUN OFF.
         2] KEEP OUT OF THE REACH OF CHILDREN.
         3] CONTAINERS MAY MELT IN A FIRE AND LEAK.
         4] THERMAL DECOMPOSITION MAY YIELD CARBON DIOXIDE, CARBON MONOXIDE.
         5] MOVE CONTAINERS FROM AREA IF WITHOUT RISK, COOL EXPOSED CONTAINERS.
         6] DIKE AREA FOR CONTROL AND CONTAINMENT TO PREVENT ENTRY INTO SEWERS, DRAIN, AND WATER WAYS.

LARGE SPILL/NO FIRE/RESCUE: WEAR RUBBER OR NEOPRENE BOOTS, GLOVES.

SPILL CONTROL AND CONTAINMENT:
   HOUSEHOLD SPILL: WIPE UP WITH AN ABSORBENT CLOTH. RINSE OUT IN THE SINK. WIPE UP RESIDUE WITH A DAMP CLOTH.
   RINSE OUT IN THE SINK.
   LARGE SPILL: PICK UP WITH ABSORBENT. RINSE AREA WITH WATER. PICK UP WITH ABSORBENT. PUT ALL ABSORBENT IN A
   DISPOSAL CONTAINER. SEAL AND REMOVE FROM THE WORK AREA.

HEALTH HAZARD INFORMATION:
   WHEN USING THIS PRODUCT WEAR: NO SPECIAL REQUIREMENT. Eye contact may cause stinging, tearing, itching,
   swelling, and redness. Skin contact may cause irritation. Ingestion may cause mild abdominal cramps, nausea,
   vomiting, and diarrhea.

   A physician should be contacted if anyone develops any signs or symptoms and suspects that they are caused by
   exposure to this product.

FIRST AID:
   Eye Exposure:
   Flush with water for 15 minutes while lifting the upper and lower eye lids. Contact lenses should not be worn
   when working with this product. Get medical attention.

   Skin Exposure:
   Wash with soap and water. If irritation develops, get medical attention.

   Breathing:
   Not expected to be a problem.

   Swallowing:
   Immediately contact the local poison control center for advice. Keep the victim warm and at rest. Get medical
   attention.

PRODUCT: SKINTASTIC (ALL FRAGRANCES)

INCOMPATIBILITY: NONE KNOWN

INGREDIENTS: ETHYL ALCOHOL   45-55% TLV: 1000 ppm   CAS#: 64-17-5
             WATER   TLV: CAS#: 7732-18-5
             GLYCERINE   TLV: CAS#: 56-81-5
             DIMETHICONE   TLV: CAS#:
             TRIETHANOLAMINE   TLV: CAS#: 102-71-6
             BENZO-PHENONE   TLV: CAS#: 119-61-9

IGNI TEMP: NA FP: 70F LEL: NA UEL: NA VP: NA VD: NA SG: .9 PS: GEL APPEAR: TRANSLUCENT, PEACH
     ODOR: FRAGRANT   PH FACTOR:6-7

HAZARD RATINGS: H: F: R:

FIRE FIGHTING:
   HIGH EXPANSION FOAM, LOW EXPANSION FOAM, ALCOHOL FOAM; DRY CHEMICAL, CARBON DIOXIDE, WATER FOG.
   PROTECTIVE CLOTHING, RUBBER GLOVES, AND BREATHING APPARATUS.
WARNING: 1] STRUCTURAL PROTECTIVE CLOTHING IS PERMEABLE, REMAIN CLEAR OF SMOKE, WATER FALL OUT AND WATER RUN OFF.
         2] KEEP OUT OF THE REACH OF CHILDREN.
         3] THERMAL DECOMPOSITION MAY YIELD CARBON DIOXIDE, CARBON MONOXIDE.
         4] CONTAINERS MAY MELT IN A FIRE AND CONTENTS WILL BURN.
         5] MOVE CONTAINERS FROM AREA IF WITHOUT RISK, COOL EXPOSED CONTAINERS.
         6] DIKE AREA FOR CONTROL AND CONTAINMENT TO PREVENT ENTRY INTO SEWERS, DRAIN, AND WATER WAYS.

LARGE SPILL/NO FIRE/RESCUE: WEAR RUBBER OR NEOPRENE BOOTS, GLOVES.

SPILL CONTROL AND CONTAINMENT:
   HOUSEHOLD SPILL: WIPE UP WITH AN ABSORBENT CLOTH. RINSE OUT IN THE SINK. WIPE UP RESIDUE WITH DAMP CLOTH.
   RINSE OUT IN THE SINK.
   LARGE SPILL: SCOOP UP THE GEL AND PUT IN A DISPOSAL CONTAINER. RINSE AREA WITH WATER. PICK UP WITH ABSORBENT.
   PUT ABSORBENT IN A DISPOSAL CONTAINER. SEAL AND REMOVE FROM THE AREA.

HEALTH HAZARD INFORMATION:
   WHEN USING THIS PRODUCT WEAR: NO SPECIAL REQUIREMENT. Eye contact may cause stinging, tearing, swelling, and
   redness. Use on freshly shaved skin may cause irritation.

   A physician should be contacted if anyone develops any signs or symptoms and suspects that they are caused by
   exposure to this product.

FIRST AID:
   Eye Exposure:
   Flush with water for 15 minutes while lifting the upper and lower eye lids. Contact lenses should not be worn
   when working with this product. Get medical attention.

   Skin Exposure:
   If irritation develops, wash off with soap and water. If it persists, get medical attention.

   Breathing:
   Not expected to be a problem.

   Swallowing:
   Immediately contact the local poison control center for advice. Keep the victim warm and at rest. Get medical
   attention.

PRODUCT: SLEEK LIQUID CAR WAX

INCOMPATIBILITY: NONE KNOWN

INGREDIENTS: MORPHOLINE   under   1.0% TLV: 20 ppm   CAS#: 110-91-8
             AMMONIA   under   0.2% TLV: 25 ppm   CAS#: 7664-41-7
             WATER   TLV: CAS#: 7732-18-5
             SILICATE   10-15%   SILICON FLUID   3-6%
             WAXES   3-5%   SURFACTANTS under   1%
             ACRYLIC POLYMERS 1-3%   ALIPHATIC HYDROCARBON SOLVENT 25-35%

IGNI TEMP: NA FP: 133F LEL: NA UEL: NA VP: NA VD: NA SG: 1.0 PS: LIQUID APPEAR: OPAQUE, TAN
    ODOR: SOLVENT   PH FACTOR: 8.4 - 9.4

HAZARD RATINGS: H: F: R:

FIRE FIGHTING:
   HIGH EXPANSION FOAM, LOW EXPANSION FOAM, ALCOHOL FOAM; DRY CHEMICAL, CARBON DIOXIDE, WATER FOG.
   PROTECTIVE CLOTHING, RUBBER GLOVES, AND BREATHING APPARATUS.
WARNING: 1] STRUCTURAL PROTECTIVE CLOTHING IS PERMEABLE, REMAIN CLEAR OF SMOKE, WATER FALL OUT AND WATER RUN OFF.
         2] KEEP OUT OF THE REACH OF CHILDREN.
         3] REMOVE ALL IGNITION SOURCES IF WITHOUT RISK.
         4] CONTAINERS MAY BURST IN THE HEAT OF A FIRE.
         5] LIQUID RUN-OFF TO SEWERS MAY CREATE A FIRE/EXPLOSION HAZARD.
         6] THERMAL DECOMPOSITION MAY YIELD CARBON DIOXIDE, CARBON MONOXIDE,
         7] MOVE CONTAINERS FROM AREA IF WITHOUT RISK, COOL EXPOSED CONTAINERS.
         8] DIKE AREA FOR CONTROL AND CONTAINMENT TO PREVENT ENTRY INTO SEWERS, DRAIN, AND WATER WAYS.

LARGE SPILL/NO FIRE/RESCUE: WEAR RUBBER OR NEOPRENE BOOTS, GLOVES.

SPILL CONTROL AND CONTAINMENT:
   HOUSEHOLD SPILL: WIPE UP WITH AN ABSORBENT CLOTH. RINSE OUT IN THE SINK. WIPE UP RESIDUE WITH A DAMP CLOTH.
   RINSE OUT IN THE SINK.
   LARGE SPILL: PICK UP WITH ABSORBENT. RINSE AREA WITH WATER. PICK UP WITH ABSORBENT. PUT ALL ABSORBENT IN A
   DISPOSAL CONTAINER. SEAL AND REMOVE FROM THE WORK AREA.

HEALTH HAZARD INFORMATION:
   WHEN USING THIS PRODUCT WEAR: NO SPECIAL REQUIREMENT. Eye contact may cause stinging, tearing, itching, and
   redness. Skin contact may cause irritation.

   A physician should be contacted if anyone develops any signs or symptoms and suspects that they are caused by
   exposure to this product.

FIRST AID:
   Eye Exposure:
   Flush with water for 15 minutes while lifting the upper and lower eye lids. Contact lenses should not be worn
   when working with this product. Get medical attention.

   Skin Exposure:
   Wash off with soap and water. If irritation develops, get medical attention.

   Breathing:
   Not expected to be a problem.

   Swallowing:
   Immediately contact the local poison control center for advice. Keep the victim warm and at rest. Get medical
   attention.

PRODUCT: SLEEK SOFT PASTE

INCOMPATIBILITY: NONE KNOWN

INGREDIENTS: MORPHOLINE   under   1.0% TLV: CAS#: 110-91-8
             AMMONIA   under   .5% TLV: CAS#: 7664-41-7
             ALIPHATIC PETROLEUM SOLVENT 30-40% TLV: CAS#:
             ALUMINUM SILICATE   10-15% TLV: 10 mg/m3   CAS#: 1302-76-7
             SURFACTANTS   about   1.0% TLV: CAS#:
             SILICONE__   TLV: CAS#:
             WAXES_____   10-15% TLV: CAS#:
             ACRYLIC POLYMERS___   TLV: CAS#:
             RESINS_____   1-3% TLV: CAS#:

IGNI TEMP: NA FP: 109F LEL: NA UEL: NA VP: NA VD: NA SG: 1.0 PS: PASTE APPEAR: ODOR: SOLVENT
     PH FACTOR: 8.6 - 9.5

HAZARD RATINGS: H: F: R:

FIRE FIGHTING:
   HIGH EXPANSION FOAM, LOW EXPANSION FOAM, ALCOHOL FOAM; DRY CHEMICAL, CARBON DIOXIDE, WATER FOG.
   PROTECTIVE CLOTHING, RUBBER GLOVES, AND BREATHING APPARATUS.
WARNING: 1] STRUCTURAL PROTECTIVE CLOTHING IS PERMEABLE, REMAIN CLEAR OF SMOKE, WATER FALL OUT AND WATER RUN OFF.
         2] KEEP OUT OF THE REACH OF CHILDREN.
         3] REMOVE ALL IGNITION SOURCES IF WITHOUT RISK.
         4] CONTAINERS MAY EXPLODE IN A FIRE.
         5] THERMAL DECOMPOSITION MAY YIELD CARBON DIOXIDE, CARBON MONOXIDE.
         6] MOVE CONTAINERS FROM AREA IF WITHOUT RISK, COOL EXPOSED CONTAINERS.
         7] DIKE AREA FOR CONTROL AND CONTAINMENT TO PREVENT ENTRY INTO SEWERS, DRAIN, AND WATER WAYS.

LARGE SPILL/NO FIRE/RESCUE: WEAR RUBBER OR NEOPRENE BOOTS, GLOVES.

SPILL CONTROL AND CONTAINMENT:
   HOUSEHOLD SPILL: SWEEP OR SCRAPE UP PASTE. PUT IN A DISPOSAL CONTAINER. WIPE UP RESIDUE WITH AN ABSORBENT CLOTH.
   LARGE SPILL: USE ABSORBENT MATERIAL. SWEEP UP AND PUT IN DISPOSAL CONTAINER. SOLVENT MAY HAVE TO BE USED TO
   DISSOLVE PASTE. IF SO PICK UP WITH ABSORBENT. PUT IN A DISPOSAL CONTAINER. SEAL AND REMOVE FROM THE WORK AREA.

HEALTH HAZARD INFORMATION:
   WHEN USING THIS PRODUCT WEAR: NO SPECIAL REQUIREMENT. Eye contact may cause irritation. Skin contact may cause
   irritation.

   A physician should be contacted if anyone develops any signs or symptoms and suspects that they are caused by
   exposure to this product.

FIRST AID:
Eye Exposure:
Flush with water for 15 minutes while lifting the upper and lower eye lids. Contact lenses should not be worn
when working with this product. Get medical attention.

Skin Exposure:
Scrub with soap and water. If irritation develops, get medical attention.

Breathing:
Not expected to be a problem.

Swallowing:
Immediately contact the local poison control center for advice. Keep the victim warm and at rest. Get medical
attention.

PRODUCT: SNUGGLE DRYER FABRIC SOFTENER SHEETS

INCOMPATIBILITY: NONE KNOWN

INGREDIENTS: NOT LISTED BY MANUFACTURER NON HAZARDOUS IN ACCORDANCE WITH OSHA 1910.1200 STANDARD

IGNI TEMP: NA FP: NA LEL: NA UEL: NA VP: NA VD: NA SG: NA PS: SHEETS APPEAR: POLYESTER FABRIC   ODOR:
  PH FACTOR:

HAZARD RATINGS: H: F: R: 1 0 0

FIRE FIGHTING:
  HIGH EXPANSION FOAM, LOW EXPANSION FOAM, ALCOHOL FOAM; DRY CHEMICAL, CARBON DIOXIDE, WATER FOG.
  PROTECTIVE CLOTHING, RUBBER GLOVES, AND BREATHING APPARATUS.
WARNING: 1] STRUCTURAL PROTECTIVE CLOTHING IS PERMEABLE, REMAIN CLEAR OF SMOKE, WATER FALL OUT AND WATER RUN OFF.
         2] KEEP OUT OF THE REACH OF CHILDREN.
         3] PRODUCT MAY CONTAIN SURFACTANTS WHICH ARE MODERATELY TOXIC IF THEY ARE INGESTED.
         4] THERMAL DECOMPOSITION MAY YIELD CARBON DIOXIDE, CARBON MONOXIDE.
         5] MOVE CONTAINERS FROM AREA IF WITHOUT RISK, COOL EXPOSED CONTAINERS.
         6] DIKE AREA FOR CONTROL AND CONTAINMENT TO PREVENT ENTRY INTO SEWERS, DRAIN, AND WATER WAYS.

LARGE SPILL/NO FIRE/RESCUE: WEAR RUBBER OR NEOPRENE BOOTS, GLOVES.

SPILL CONTROL AND CONTAINMENT:
  HOUSEHOLD SPILL: PICK UP SHEETS. PUT IN A DISPOSAL CONTAINER. WIPE UP ANY RESIDUE WITH A DAMP CLOTH.
  LARGE SPILL: SWEEP UP SHEETS. PUT IN A DISPOSAL CONTAINER. SEAL AND REMOVE FROM THE WORK AREA.

HEALTH HAZARD INFORMATION:
  WHEN USING THIS PRODUCT WEAR: NO SPECIAL REQUIREMENT. Eye contact may cause stinging, tearing, itching,
  swelling, and redness. Prolonged contact with the skin may cause irritation. Ingestion may cause mild abdominal
  cramps, nausea, vomiting, and diarrhea. Inhalation may cause irritation to the nose and throat.

  A physician should be contacted if anyone develops any signs or symptoms and suspects that they are caused by
  exposure to this product.

FIRST AID:
  Eye Exposure:
  Flush with water for 15 minutes while lifting the upper and lower eye lids. Contact lenses should not be worn
  when working with this product. Get medical attention.

  Skin Exposure:
  Wash with soap and water. If irritation develops and persists, get medical attention.

  Breathing:
  If irritation develops, move victim to fresh air.

  Swallowing:
  If the victim is conscious, give 2-3 glasses of milk or water to drink. Immediately contact the local poison
  control center for advice. Keep the victim warm and at rest. Get medical attention.

PRODUCT: SNUGGLE LIQUID FABRIC SOFTENER

INCOMPATIBILITY: NONE KNOWN

INGREDIENTS: NOT LISTED BY MANUFACTURER NON HAZARDOUS IN ACCORDANCE WITH OSHA 1910.1200 STANDARD

IGNI TEMP: NA FP: NA LEL: NA UEL: NA VP: NA VD: NA SG: .99 PS: LIQUID APPEAR: ODOR:
   PH FACTOR: 4.0 - 6.0

HAZARD RATINGS: H: F: R:

FIRE FIGHTING:
   HIGH EXPANSION FOAM, LOW EXPANSION FOAM, ALCOHOL FOAM; DRY CHEMICAL, CARBON DIOXIDE, WATER FOG.
   PROTECTIVE CLOTHING, RUBBER GLOVES, AND BREATHING APPARATUS.
WARNING: 1] STRUCTURAL PROTECTIVE CLOTHING IS PERMEABLE, REMAIN CLEAR OF SMOKE, WATER FALL OUT AND WATER RUN OFF.
         2] KEEP OUT OF THE REACH OF CHILDREN.
         3] PRODUCT MAY CONTAIN SURFACTANTS WHICH ARE MODERATELY TOXIC IF THEY ARE INGESTED.
         4] THERMAL DECOMPOSITION MAY YIELD CARBON DIOXIDE, CARBON MONOXIDE.
         5] MOVE CONTAINERS FROM AREA IF WITHOUT RISK, COOL EXPOSED CONTAINERS.
         6] DIKE AREA FOR CONTROL AND CONTAINMENT TO PREVENT ENTRY INTO SEWERS, DRAIN, AND WATER WAYS.

LARGE SPILL/NO FIRE/RESCUE: WEAR RUBBER OR NEOPRENE BOOTS, GLOVES.

SPILL CONTROL AND CONTAINMENT:
   HOUSEHOLD SPILL: WIPE UP WITH AN ABSORBENT CLOTH. RINSE OUT IN THE SINK. WIPE UP RESIDUE WITH A DAMP CLOTH.
   RINSE OUT IN THE SINK.
   LARGE SPILL: PICK UP WITH ABSORBENT. RINSE AREA WITH WATER. PICK UP WITH ABSORBENT. PUT ALL ABSORBENT IN A
   DISPOSAL CONTAINER. SEAL AND REMOVE FROM THE WORK AREA.

HEALTH HAZARD INFORMATION:
   WHEN USING THIS PRODUCT WEAR: NO SPECIAL REQUIREMENT. Eye contact may cause stinging, tearing, itching, and
   redness. Prolonged skin contact may cause irritation. Ingestion will cause mild abdominal cramps, nausea, and
   vomiting. Inhalation may cause irritation of the nose and throat.

   A physician should be contacted if anyone develops any signs or symptoms and suspects that they are caused by
   exposure to this product.

FIRST AID:
   Eye Exposure:
   Flush with water for 15 minutes while lifting the upper and lower eye lids. Contact lenses should not be worn
   when working with this product. Get medical attention.

   Skin Exposure:
   Wash with soap and water. If irritation develops, get medical attention.

   Breathing:
   If irritation develops, move victim to fresh air.

   Swallowing:
   If the victim is conscious, give 2-3 glasses of milk or water to drink. Immediately contact the local poison
   control center for advice. Keep the victim warm and at rest. Get medical attention.

PRODUCT: SNUGGLE LIQUID FABRIC SOFTENER, YELLOW

INCOMPATIBILITY: NONE KNOWN

INGREDIENTS: NON LISTED BY THE MANUFACTURER

IGNI TEMP: NA   FP: NA   LEL: NA   UEL: NA   VP: NA   VD: NA   SG: .99   PS: LIQUID   APPEAR: YELLOW   ODOR: NA
   PH FACTOR: 4.0 - 6.0

HAZARD RATINGS: H: F: R:

FIRE FIGHTING:
   HIGH EXPANSION FOAM, LOW EXPANSION FOAM, ALCOHOL FOAM; DRY CHEMICAL, CARBON DIOXIDE, WATER FOG.
   PROTECTIVE CLOTHING, RUBBER GLOVES, AND BREATHING APPARATUS.
WARNING: 1] STRUCTURAL PROTECTIVE CLOTHING IS PERMEABLE, REMAIN CLEAR OF SMOKE, WATER FALL OUT AND WATER RUN OFF.
         2] KEEP OUT OF THE REACH OF CHILDREN.
         3] THERMAL DECOMPOSITION PRODUCTS ARE NOT KNOWN.
         4] MOVE CONTAINERS FROM AREA IF WITHOUT RISK, COOL EXPOSED CONTAINERS.
         5] DIKE AREA FOR CONTROL AND CONTAINMENT TO PREVENT ENTRY INTO SEWERS, DRAIN, AND WATER WAYS.

LARGE SPILL/NO FIRE/RESCUE: WEAR RUBBER OR NEOPRENE BOOTS, GLOVES.

SPILL CONTROL AND CONTAINMENT:
   HOUSEHOLD SPILL: WIPE UP WITH AN ABSORBENT CLOTH. RINSE OUT IN THE SINK. WIPE UP RESIDUE WITH A DAMP CLOTH.
   RINSE OUT IN THE SINK.
   LARGE SPILL: PICK UP WITH ABSORBENT. RINSE AREA WITH WATER. PICK UP WITH ABSORBENT. PUT ALL ABSORBENT IN A
   DISPOSAL CONTAINER. SEAL AND REMOVE FROM THE WORK AREA.

HEALTH HAZARD INFORMATION:
   WHEN USING THIS PRODUCT WEAR: NO SPECIAL REQUIREMENT. Eye contact will cause stinging, tearing, itching, and
   redness. Prolonged skin contact may cause irritation. Ingestion may result in nausea and vomiting and diarrhea.
   Inhalation may result in irritation to the nasal passage.

   A physician should be contacted if anyone develops any signs or symptoms and suspects that they are caused by
   exposure to this product.

FIRST AID:
   Eye Exposure:
   Flush with water for 15 minutes while lifting the upper and lower eye lids. Contact lenses should not be worn
   when working with this product. Get medical attention.

   Skin Exposure:
   Wash with soap and water. If irritation develops, get medical attention.

   Breathing:
   If irritation develops, move victim to fresh air.

   Swallowing:
   If the victim is conscious, give 2-3 glasses of milk or water to drink. Immediately contact the local poison
   control center for advice. Keep the victim warm and at rest. Get medical attention.

PRODUCT: SOFT SCRUB

INCOMPATIBILITY: NONE KNOWN

INGREDIENTS: CALCIUM CARBONATE   50-75% TLV: 10 mg/m3(dust) CAS#: 1317-65-3

IGNI TEMP: NA FP: NA LEL: NA UEL: NA VP: NA VD: NA SG: NA PS: LIQUID APPEAR: CREAMY WHITE
    ODOR: FRAGRANCE   PH FACTOR: 10.2

HAZARD RATINGS: H: F: R: 3 0 0

FIRE FIGHTING:
  HIGH EXPANSION FOAM, LOW EXPANSION FOAM, ALCOHOL FOAM; DRY CHEMICAL, CARBON DIOXIDE, WATER FOG.
  PROTECTIVE CLOTHING, RUBBER GLOVES, AND BREATHING APPARATUS.
WARNING: 1] STRUCTURAL PROTECTIVE CLOTHING IS PERMEABLE, REMAIN CLEAR OF SMOKE, WATER FALL OUT AND WATER RUN OFF.
         2] KEEP OUT OF THE REACH OF CHILDREN.
         3] PRODUCT MAY CONTAIN SODIUM HYPOCHLORITE.
         4] THERMAL DECOMPOSITION MAY YIELD CARBON DIOXIDE, CARBON MONOXIDE.
         5] MOVE CONTAINERS FROM AREA IF WITHOUT RISK, COOL EXPOSED CONTAINERS.
         6] DIKE AREA FOR CONTROL AND CONTAINMENT TO PREVENT ENTRY INTO SEWERS, DRAIN, AND WATER WAYS.

LARGE SPILL/NO FIRE/RESCUE: WEAR RUBBER OR NEOPRENE BOOTS, GLOVES.

SPILL CONTROL AND CONTAINMENT:
  HOUSEHOLD SPILL: WIPE UP WITH AN ABSORBENT CLOTH. RINSE OUT IN THE SINK. WIPE UP RESIDUE WITH A DAMP CLOTH.
  RINSE OUT IN THE SINK.
  LARGE SPILL: PICK UP WITH ABSORBENT. RINSE AREA WITH WATER. PICK UP WITH ABSORBENT. PUT ALL ABSORBENT IN A
  DISPOSAL CONTAINER. SEAL AND REMOVE FROM THE WORK AREA.

HEALTH HAZARD INFORMATION:
  WHEN USING THIS PRODUCT WEAR: RUBBER GLOVES FOR PROLONGED SKIN CONTACT. Eye contact may cause stinging, tearing,
  itching, swelling, and redness. Prolonged skin contact may cause irritation.

  A physician should be contacted if anyone develops any signs or symptoms and suspects that they are caused by
  exposure to this product.

FIRST AID:
  Eye Exposure:
  Flush with water for 15 minutes while lifting the upper and lower eye lids. Contact lenses should not be worn
  when working with this product. Get medical attention.

  Skin Exposure:
  Wash with soap and water. If irritation develops, get medical attention.

  Breathing:
  Not expected to be a problem.

  Swallowing:
  Do not induce vomiting. Immediately contact the local poison control center for advice. Keep the victim warm and
  at rest. Get medical attention.

PRODUCT: SOFT SCRUB WITH BLEACH

INCOMPATIBILITY: AMMONIA; PRODUCTS WITH AMMONIA; TOILET BOWL CLEANERS; RUST REMOVERS; ACIDS; VINEGAR.

INGREDIENTS: CALCIUM CARBONATE   25-50% TLV: 10mg/m3(dust) CAS#: 1317-65-3
             SODIUM HYPOCHLORITE   .5-2% TLV: CAS#: 7168-52-9

IGNI TEMP: NA FP: NA LEL: NA UEL: NA VP: NA VD: NA SG: NA PS: LIQUID APPEAR: WHITE, THICK ODOR: FRAGRANCE
   PH FACTOR: 12.5

HAZARD RATINGS: H: F: R: 2 0 1

FIRE FIGHTING:
   HIGH EXPANSION FOAM, LOW EXPANSION FOAM, ALCOHOL FOAM; DRY CHEMICAL, CARBON DIOXIDE, WATER FOG.
   PROTECTIVE CLOTHING, RUBBER GLOVES, AND BREATHING APPARATUS.
WARNING: 1] STRUCTURAL PROTECTIVE CLOTHING IS PERMEABLE, REMAIN CLEAR OF SMOKE, WATER FALL OUT AND WATER RUN OFF.
         2] KEEP OUT OF THE REACH OF CHILDREN.
         3] DO NOT MIX WITH PRODUCT LISTED ABOVE IN THE INCOMPATIBLE LIST. VERY HAZARDOUS GASES WILL BE RELEASED.
         4] MOVE CONTAINERS FROM AREA IF WITHOUT RISK, COOL EXPOSED CONTAINERS.
         5] DIKE AREA FOR CONTROL AND CONTAINMENT TO PREVENT ENTRY INTO SEWERS, DRAIN, AND WATER WAYS.

LARGE SPILL/NO FIRE/RESCUE: WEAR RUBBER OR NEOPRENE BOOTS, GLOVES.

SPILL CONTROL AND CONTAINMENT:
   HOUSEHOLD SPILL: WIPE UP WITH AN ABSORBENT CLOTH. RINSE OUT IN THE SINK. WIPE UP RESIDUE WITH A DAMP CLOTH.
   LARGE SPILL: PICK UP WITH ABSORBENT. RINSE AREA WITH WATER. PICK UP WITH ABSORBENT. PUT ALL ABSORBENT IN A
   DISPOSAL CONTAINER. SEAL AND REMOVE FROM THE WORK AREA.

HEALTH HAZARD INFORMATION:
   WHEN USING THIS PRODUCT WEAR: WEAR RUBBER GLOVES FOR PROLONGED SKIN EXPOSURE. Eye contact will cause stinging,
   tearing, itching, swelling, and redness. Prolonged contact with the skin may cause irritation.

   A physician should be contacted if anyone develops any signs or symptoms and suspects that they are caused by
   exposure to this product.

FIRST AID:
   Eye Exposure:
   Flush with water for 15 minutes while lifting the upper and lower eye lids. Contact lenses should not be worn
   when working with this product. Get medical attention.

   Skin Exposure:
   Wash with soap and water. If irritation develops, get medical attention.

   Breathing:
   Not expected to be a problem.

   Swallowing:
   Immediately contact the local poison control center for advice. Keep the victim warm and at rest. Get medical
   attention.

PRODUCT: SOFT SENSE ALOE FORMULA

INCOMPATIBILITY: NONE KNOWN

INGREDIENTS: NOT LISTED BY MANUFACTURER NON HAZARDOUS IN ACCORDANCE WITH OSHA 1910.1200 STANDARD

IGNI TEMP: NA FP: NA LEL: NA UEL: NA VP: NA VD: NA SG: .99 PS: LIQUID APPEAR: CREAMY LIGHT COLORED
   ODOR: MILD   PH FACTOR: 4.5 - 6.0

HAZARD RATINGS: H: F: R:

FIRE FIGHTING:
   HIGH EXPANSION FOAM, LOW EXPANSION FOAM, ALCOHOL FOAM; DRY CHEMICAL, CARBON DIOXIDE, WATER FOG.
   PROTECTIVE CLOTHING, RUBBER GLOVES, AND BREATHING APPARATUS.
WARNING: 1] STRUCTURAL PROTECTIVE CLOTHING IS PERMEABLE, REMAIN CLEAR OF SMOKE, WATER FALL OUT AND WATER RUN OFF.
         2] KEEP OUT OF THE REACH OF CHILDREN.
         3] THERMAL DECOMPOSITION MAY YIELD CARBON DIOXIDE, CARBON MONOXIDE.
         4] MOVE CONTAINERS FROM AREA IF WITHOUT RISK, COOL EXPOSED CONTAINERS.
         5] DIKE AREA FOR CONTROL AND CONTAINMENT TO PREVENT ENTRY INTO SEWERS, DRAIN, AND WATER WAYS.

LARGE SPILL/NO FIRE/RESCUE: WEAR RUBBER OR NEOPRENE BOOTS, GLOVES.

SPILL CONTROL AND CONTAINMENT:
   HOUSEHOLD SPILL: WIPE UP WITH AN ABSORBENT CLOTH, RINSE OUT IN THE SINK. WIPE UP RESIDUE WITH A DAMP CLOTH.
   RINSE OUT IN THE SINK.
   LARGE SPILL: PICK UP WITH ABSORBENT. RINSE AREA WITH WATER. PICK UP WITH ABSORBENT. PUT ALL ABSORBENT IN A
   DISPOSAL CONTAINER. SEAL AND REMOVE FROM THE WORK AREA.

HEALTH HAZARD INFORMATION:
   WHEN USING THIS PRODUCT WEAR: RUBBER GLOVES FOR PROLONGED SKIN CONTACT. Eye contact may cause stinging, tearing,
   itching, swelling, and redness. Prolonged skin contact may cause irritation.

   A physician should be contacted if anyone develops any signs or symptoms and suspects that they are caused by
   exposure to this product.

FIRST AID:
   Eye Exposure:
   Flush with water for 15 minutes while lifting the upper and lower eye lids. Contact lenses should not be worn
   when working with this product. Get medical attention.

   Skin Exposure:
   Wash with soap and water. If irritation develops, get medical attention.

   Breathing:
   Not expected to be a problem.

   Swallowing:
   Immediately contact the local poison control center for advice. Keep the victim warm and at rest. Get medical
   attention.

PRODUCT: SOFT SENSE   ALOE SHAVE GEL

INCOMPATIBILITY: NONE KNOWN

INGREDIENTS: PENTENE   1-4%   TLV: 600 ppm   CAS#: 109-66-0
             ISOBUTANE   3-6%   TLV: CAS#: 75-28-5

IGNI TEMP: NA FP: 68F LEL: NA UEL: NA VP: NA VD: NA SG: NA PS: GEL APPEAR: ODOR: MILD
    PH FACTOR: NA

HAZARD RATINGS: H: F: R: 1 4 0

FIRE FIGHTING:
   HIGH EXPANSION FOAM, LOW EXPANSION FOAM, ALCOHOL FOAM; DRY CHEMICAL, CARBON DIOXIDE, WATER FOG.
   PROTECTIVE CLOTHING, RUBBER GLOVES, AND BREATHING APPARATUS.
WARNING: 1] STRUCTURAL PROTECTIVE CLOTHING IS PERMEABLE, REMAIN CLEAR OF SMOKE, WATER FALL OUT AND WATER RUN OFF.
         2] KEEP OUT OF THE REACH OF CHILDREN.
         3] REMOVE ALL IGNITION SOURCES IF WITHOUT RISK.
         4] CONTAINERS WILL EXPLODE ON A FIRE OR IF HEATED ABOVE 120F.
         5] THERMAL DECOMPOSITION MAY YIELD CARBON DIOXIDE, CARBON MONOXIDE.
         6] MOVE CONTAINERS FROM AREA IF WITHOUT RISK, COOL EXPOSED CONTAINERS.
         7] DIKE AREA FOR CONTROL AND CONTAINMENT TO PREVENT ENTRY INTO SEWERS, DRAIN, AND WATER WAYS.

LARGE SPILL/NO FIRE/RESCUE: WEAR RUBBER OR NEOPRENE BOOTS, GLOVES.

SPILL CONTROL AND CONTAINMENT:
   HOUSEHOLD SPILL: WIPE UP WITH AN ABSORBENT CLOTH. RINSE OUT IN TH SINK. WIPE UP RESIDUE WITH A DAMP CLOTH. RINSE
   OUT IN THE SINK.
   LARGE SPILL: PICK UP WITH ABSORBENT. RINSE AREA WITH WATER, PICK UP WITH ABSORBENT. PUT ALL ABSORBENT IN A
   DISPOSAL CONTAINER. SEAL AND REMOVE FROM THE WORK AREA.

HEALTH HAZARD INFORMATION:
   WHEN USING THIS PRODUCT WEAR: NO SPECIAL REQUIREMENT. Eye contact may cause stinging, tearing, itching,
   swelling, and redness. Ingestion may cause nausea and vomiting.

   A physician should be contacted if anyone develops any signs or symptoms and suspects that they are caused by
   exposure to this product.

FIRST AID:
   Eye Exposure:
   Flush with water for 15 minutes while lifting the upper and lower eye lids. Contact lenses should not be worn
   when working with this product. Get medical attention.

   Skin Exposure:
   Not expected to be a problem.

   Breathing:
   Not expected to be a problem.

   Swallowing:
   Immediately contact the local poison control center for advice. Keep the victim warm and at rest. Get medical
   attention.

PRODUCT: SOFT SENSE BODY MOUSSE

INCOMPATIBILITY: NONE KNOWN

INGREDIENTS: DIFLUOROETHANE   1-3% TLV: 1000 ppm   CAS#: 75-37-6
             ISOBUTANE____   TLV: CAS#: 75-28-5
             PROPANE_____   1-4% TLV: 1000 ppm   CAS#: 74-98-6

IGNI TEMP: NA FP: 20F LEL: NA UEL: NA VP: NA VD: NA SG: .98 PS: LIQUID   APPEAR: FOAM   ODOR: FRAGRANCE
   PH FACTOR: 7.5 - 8.5

HAZARD RATINGS: H: F: R: 0 4 0

FIRE FIGHTING:
   HIGH EXPANSION FOAM, LOW EXPANSION FOAM, ALCOHOL FOAM; DRY CHEMICAL, CARBON DIOXIDE, WATER FOG.
   PROTECTIVE CLOTHING, RUBBER GLOVES, AND BREATHING APPARATUS.
WARNING: 1] STRUCTURAL PROTECTIVE CLOTHING IS PERMEABLE, REMAIN CLEAR OF SMOKE, WATER FALL OUT AND WATER RUN OFF.
         2] KEEP OUT OF THE REACH OF CHILDREN.
         3] REMOVE ALL IGNITION SOURCES IF WITHOUT RISK.
         4] CONTAINERS WILL EXPLODE IN A FIRE OR IF HEATED ABOVE 120F.
         5] THERMAL DECOMPOSITION MAY YIELD CARBON DIOXIDE, CARBON MONOXIDE.
         6] MOVE CONTAINERS FROM AREA IF WITHOUT RISK, COOL EXPOSED CONTAINERS.
         7] DIKE AREA FOR CONTROL AND CONTAINMENT TO PREVENT ENTRY INTO SEWERS, DRAIN, AND WATER WAYS.

LARGE SPILL/NO FIRE/RESCUE: WEAR RUBBER OR NEOPRENE BOOTS, GLOVES.

SPILL CONTROL AND CONTAINMENT:
   HOUSEHOLD SPILL: WIPE UP WITH AN ABSORBENT CLOTH. RINSE OUT IN THE SINK. WIPE UP RESIDUE WITH A DAMP CLOTH.
   RINSE OUT IN THE SINK.
   LARGE SPILL: PICK UP WITH ABSORBENT. RINSE AREA WITH WATER. PICK UP WITH ABSORBENT. PUT ALL ABSORBENT IN A
   DISPOSAL CONTAINER. SEAL AND REMOVE FROM THE WORK AREA.

HEALTH HAZARD INFORMATION:
   WHEN USING THIS PRODUCT WEAR: NO SPECIAL REQUIREMENT. Eye contact may cause stinging, tearing, itching,
   swelling, and redness.

   A physician should be contacted if anyone develops any signs or symptoms and suspects that they are caused by
   exposure to this product.

FIRST AID:
   Eye Exposure:
   Flush with water for 15 minutes while lifting the upper and lower eye lids. Contact lenses should not be worn
   when working with this product. Get medical attention.

   Skin Exposure:
   Not expected to be a problem. If irritation develops, get medical attention.

   Breathing:
   Not expected to be a problem.

   Swallowing:
   Immediately contact the local poison control center for advice. Keep the victim warm and at rest. Get medical
   attention.

PRODUCT: SOFT SENSE LANOLIN SHAVE GEL

INCOMPATIBILITY: NONE KNOWN

INGREDIENTS: PENTENE   1-4% TLV: 600 ppm   CAS#: 109-66-0
             ISOBUTANE   3-6% TLV: CAS#: 75-28-5

IGNI TEMP: NA FP: 20F LEL: NA UEL: NA VP: NA VD: NA SG: NA PS: LIQUID APPEAR: GEL   ODOR: MILD
  PH FACTOR: NA

HAZARD RATINGS: H: F: R: 1 4 0

FIRE FIGHTING:
  HIGH EXPANSION FOAM, LOW EXPANSION FOAM, ALCOHOL FOAM; DRY CHEMICAL, CARBON DIOXIDE, WATER FOG.
  PROTECTIVE CLOTHING, RUBBER GLOVES, AND BREATHING APPARATUS.
WARNING: 1] STRUCTURAL PROTECTIVE CLOTHING IS PERMEABLE, REMAIN CLEAR OF SMOKE, WATER FALL OUT AND WATER RUN OFF.
         2] KEEP OUT OF THE REACH OF CHILDREN.
         3] REMOVE ALL IGNITION SOURCES IF WITHOUT RISK.
         4] CONTAINERS WILL EXPLODE IN A FIRE OR IF HEATED ABOVE 120F.
         5] THERMAL DECOMPOSITION MAY YIELD CARBON DIOXIDE, CARBON MONOXIDE.
         6] MOVE CONTAINERS FROM AREA IF WITHOUT RISK, COOL EXPOSED CONTAINERS.
         7] DIKE AREA FOR CONTROL AND CONTAINMENT TO PREVENT ENTRY INTO SEWERS, DRAIN, AND WATER WAYS.

LARGE SPILL/NO FIRE/RESCUE: WEAR RUBBER OR NEOPRENE BOOTS, GLOVES.

SPILL CONTROL AND CONTAINMENT:
  HOUSEHOLD SPILL: WIPE UP WITH AN ABSORBENT CLOTH. RINSE OUT IN THE SINK. WIPE UP RESIDUE WITH A DAMP CLOTH.
  LARGE SPILL: PICK UP WITH ABSORBENT. RINSE AREA WITH WATER. PICK UP WITH ABSORBENT. PUT ALL ABSORBENT IN A
  DISPOSAL CONTAINER. SEAL AND REMOVE FROM THE WORK AREA.

HEALTH HAZARD INFORMATION:
  WHEN USING THIS PRODUCT WEAR: NO SPECIAL REQUIREMENT. Eye contact may cause stinging, tearing, itching,
  swelling, and redness.

  A physician should be contacted if anyone develops any signs or symptoms and suspects that they are caused by
  exposure to this product.

FIRST AID:
  Eye Exposure:
  Flush with water for 15 minutes while lifting the upper and lower eye lids. Contact lenses should not be worn
  when working with this product. Get medical attention.

  Skin Exposure:
  Not expected to be a problem.

  Breathing:
  Not expected to be a problem.

  Swallowing:
  Immediately contact the local poison control center for advice. Keep the victim warm and at rest. Get medical
  attention.

PRODUCT: SOFT SENSE SKIN LOTION

INCOMPATIBILITY: NONE KNOWN

INGREDIENTS: SODIUM CHLORIDE   TLV: CAS#: 7647-14-5
             GLYCERINE   TLV: CAS#:   56-81-5
             PETROLATUM   TLV: CAS#: 8009-03-8
             CETYL ALCOHOL   TLV: CAS#: 36653-82-4
             METHYL PARABEN   TLV: CAS#:   99-76-3
             PROPYL PARABEN   TLV: CAS#:   94-13-3
             ISOPROPYL PALMITATE   TLV: CAS#:   142-91-6
             QUATERNIUM-5, DEIONIZED WATER, DIMETHICONE, FRAGRANCE

IGNI TEMP: NA FP: NA LEL: NA UEL: NA VP: NA VD: NA SG: 1.0 PS: LIQUID APPEAR: CREAMY, LIGHT
    ODOR: FRAGRANT   PH FACTOR: 5.3

HAZARD RATINGS: H: F: R:

FIRE FIGHTING:
    HIGH EXPANSION FOAM, LOW EXPANSION FOAM, ALCOHOL FOAM; DRY CHEMICAL, CARBON DIOXIDE, WATER FOG.
    PROTECTIVE CLOTHING, RUBBER GLOVES, AND BREATHING APPARATUS.
WARNING: 1] STRUCTURAL PROTECTIVE CLOTHING IS PERMEABLE, REMAIN CLEAR OF SMOKE, WATER FALL OUT AND WATER RUN OFF.
         2] KEEP OUT OF THE REACH OF CHILDREN.
         3] CONTAINERS MAY MELT AND LEAK IN THE HEAT OF A FIRE.
         4] THERMAL DECOMPOSITION MAY YIELD CARBON DIOXIDE, CARBON MONOXIDE.
         5] MOVE CONTAINERS FROM AREA IF WITHOUT RISK, COOL EXPOSED CONTAINERS.
         6] DIKE AREA FOR CONTROL AND CONTAINMENT TO PREVENT ENTRY INTO SEWERS, DRAIN, AND WATER WAYS.

LARGE SPILL/NO FIRE/RESCUE: WEAR RUBBER OR NEOPRENE BOOTS, GLOVES.

SPILL CONTROL AND CONTAINMENT:
    HOUSEHOLD SPILL: WIPE UP WITH AN ABSORBENT CLOTH. RINSE OUT IN THE SINK. WIPE UP RESIDUE WITH A DAMP CLOTH.
    RINSE OUT IN THE SINK.
    LARGE SPILL: PICK UP WITH ABSORBENT. RINSE AREA WITH WATER. PICK UP WITH ABSORBENT. PUT ALL ABSORBENT IN A
    DISPOSAL CONTAINER. SEAL AND REMOVE FROM THE WORK AREA.

HEALTH HAZARD INFORMATION:
    WHEN USING THIS PRODUCT WEAR: NO SPECIAL REQUIREMENT. Eye contact may cause stinging, tearing, itching,
    swelling, and redness.

    A physician should be contacted if anyone develops any signs or symptoms and suspects that they are caused by
    exposure to this product.

FIRST AID:
Eye Exposure:
Flush with water for 15 minutes while lifting the upper and lower eye lids. Contact lenses should not be worn
when working with this product. Get medical attention.

Skin Exposure:
Not expected to be a problem.

Breathing:
Not expected to be a problem.

Swallowing:
Immediately contact the local poison control center for advice. Keep the victim warm and at rest. Get medical
attention.

PRODUCT: SOILOVE SOIL AND STAIN REMOVER

INCOMPATIBILITY: CHLORINE OR PRODUCT WITH CHLORINE (Sodium Hypochlorite)

```
INGREDIENTS: ETHYLENE GLYCOL MONOBUTYL ETHER  TLV:        CAS#:  111-76-2
             SODIUM METASILICATE              TLV:        CAS#: 6834-92-0
             PENTA HYDRATE                    TLV:        CAS#:
             PHOSPHORUS                       TLV:        CAS#:
             AMMONIUM HYDROXIDE               TLV:        CAS#: 1336-21-6
```

IGNI TEMP: NA FP: NA LEL: NA UEL: NA VP: NA VD: NA SG: 1.0 PS: LIQUID APPEAR: WHITE ODOR: AMMONIA
    PH FACTOR: NA

HAZARD RATINGS: H: F: R:

FIRE FIGHTING:
    HIGH EXPANSION FOAM, LOW EXPANSION FOAM, ALCOHOL FOAM; DRY CHEMICAL, CARBON DIOXIDE, WATER FOG.
    PROTECTIVE CLOTHING, RUBBER GLOVES, AND BREATHING APPARATUS.
WARNING: 1] STRUCTURAL PROTECTIVE CLOTHING IS PERMEABLE, REMAIN CLEAR OF SMOKE, WATER FALL OUT AND WATER RUN OFF.
         2] KEEP OUT OF THE REACH OF CHILDREN.
         3] THERMAL DECOMPOSITION MAY YIELD TOXIC FUMES.
         4] MOVE CONTAINERS FROM AREA IF WITHOUT RISK, COOL EXPOSED CONTAINERS.
         5] DIKE AREA FOR CONTROL AND CONTAINMENT TO PREVENT ENTRY INTO SEWERS, DRAIN, AND WATER WAYS.

LARGE SPILL/NO FIRE/RESCUE: WEAR RUBBER OR NEOPRENE BOOTS, GLOVES.

SPILL CONTROL AND CONTAINMENT:
    HOUSEHOLD SPILL: WIPE UP WITH AN ABSORBENT CLOTH. RINSE OUT IN THE SINK. WIPE UP RESIDUE WITH A DAMP CLOTH.
    RINSE OUT IN THE SINK.
    LARGE SPILL: PICK UP WITH ABSORBENT. RINSE AREA WITH WATER. PICK UP WITH ABSORBENT. PUT ALL ABSORBENT IN A
    DISPOSAL CONTAINER. SEAL AND REMOVE FROM THE WORK AREA.

HEALTH HAZARD INFORMATION:
    WHEN USING THIS PRODUCT WEAR: RUBBER GLOVES. Product may cause skin irritation. Eye contact may cause stinging,
    tearing, itching, swelling, and redness. Inhalation may cause irritation of the nose, throat, and lungs.

    A physician should be contacted if anyone develops any signs or symptoms and suspects that they are caused by
    exposure to this product.

FIRST AID:
Eye Exposure:
Flush with water for 15 minutes while lifting the upper and lower eye lids. Contact lenses should not be worn
when working with this product. Get medical attention.

Skin Exposure:
Wash with soap and water. If irritation develops, get medical attention.

Breathing:
If irritation develops, move the victim to fresh air. If irritation persists get medical attention.

Swallowing:
Immediately contact the local poison control center for advice. Keep the victim warm and at rest. Get medical
attention.

PRODUCT: SOLO (LIQUID LAUNDRY DETERGENT)

INCOMPATIBILITY: NONE KNOWN

INGREDIENTS: NONIONIC SURFACTANTS   TLV: CAS#:
             ETHYL ALCOHOL   TLV: 1000 ppm   CAS#:   64-17-5
             WATER   TLV: CAS#: 7732-18-5
             PERFUME   TLV: CAS#:

IGNI TEMP: NA FP: NA LEL: NA UEL: NA VP: NA VD: NA SG: NA PS: LIQUID APPEAR: BLUE   ODOR: PERFUMED
   PH FACTOR: NA

HAZARD RATINGS: H: F: R:

FIRE FIGHTING:
   HIGH EXPANSION FOAM, LOW EXPANSION FOAM, ALCOHOL FOAM; DRY CHEMICAL, CARBON DIOXIDE, WATER FOG.
   PROTECTIVE CLOTHING, RUBBER GLOVES, AND BREATHING APPARATUS.
WARNING: 1] STRUCTURAL PROTECTIVE CLOTHING IS PERMEABLE, REMAIN CLEAR OF SMOKE, WATER FALL OUT AND WATER RUN OFF.
         2] KEEP OUT OF THE REACH OF CHILDREN.
         3] THERMAL DECOMPOSITION MAY YIELD CARBON DIOXIDE, CARBON MONOXIDE.
         4] MOVE CONTAINERS FROM AREA IF WITHOUT RISK, COOL EXPOSED CONTAINERS.
         5] DIKE AREA FOR CONTROL AND CONTAINMENT TO PREVENT ENTRY INTO SEWERS, DRAIN, AND WATER WAYS.

LARGE SPILL/NO FIRE/RESCUE: WEAR RUBBER OR NEOPRENE BOOTS, GLOVES.

SPILL CONTROL AND CONTAINMENT:
   HOUSEHOLD SPILL: WIPE UP WITH AN ABSORBENT CLOTH. RINSE OUT IN THE SINK. WIPE UP RESIDUE WITH A DAMP CLOTH.
   RINSE OUT IN THE SINK.
   LARGE SPILL: PICK UP WITH ABSORBENT. RINSE AREA WITH WATER. PICK UP WITH ABSORBENT. PUT ALL ABSORBENT IN A
   DISPOSAL CONTAINER. SEAL AND REMOVE FROM THE WORK AREA.

HEALTH HAZARD INFORMATION:
   WHEN USING THIS PRODUCT WEAR: NO SPECIAL REQUIREMENT. Eye contact may cause stinging, tearing, itching,
   swelling, and redness. Prolonged contact with skin may cause irritation. Ingestion may result in nausea,
   vomiting, and diarrhea.

   A physician should be contacted if anyone develops any signs or symptoms and suspects that they are caused by
   exposure to this product.

FIRST AID:
Eye Exposure:
Flush with water for 15 minutes while lifting the upper and lower eye lids. Contact lenses should not be worn
when working with this product. Get medical attention.

Skin Exposure:
Wash with soap and water. If irritation develops, get medical attention.

Breathing:
Not expected to be a problem.

Swallowing:
If the victim is conscious, give 2-3 glasses of water to drink. Immediately contact the local poison control
center for advice. Keep the victim warm and at rest. Get medical attention.

PRODUCT: SOL-ZOL HAND CLEANER

INCOMPATIBILITY: STRONG OXIDIZING AGENTS.

INGREDIENTS: ALKYL PHENOL ETHOXYLATE   TLV: CAS#:
             HYDROTREATED NAPHTHENIC
             DISTILLATE   TLV: CAS#:

IGNI TEMP: NA FP: 250F LEL: NA UEL: NA VP: NA VD: NA SG: NA PS: LIQUID APPEAR: WHITE ODOR:
  PH FACTOR:

HAZARD RATINGS: H: F: R: 1 0 0

FIRE FIGHTING:
  HIGH EXPANSION FOAM, LOW EXPANSION FOAM, ALCOHOL FOAM; DRY CHEMICAL, CARBON DIOXIDE, WATER FOG.
  PROTECTIVE CLOTHING, RUBBER GLOVES, AND BREATHING APPARATUS.
WARNING: 1] STRUCTURAL PROTECTIVE CLOTHING IS PERMEABLE, REMAIN CLEAR OF SMOKE, WATER FALL OUT AND WATER RUN OFF.
         2] KEEP OUT OF THE REACH OF CHILDREN.
         3] THERMAL DECOMPOSITION MAY YIELD CARBON DIOXIDE, CARBON MONOXIDE.
         4] MOVE CONTAINERS FROM AREA IF WITHOUT RISK, COOL EXPOSED CONTAINERS.
         5] DIKE AREA FOR CONTROL AND CONTAINMENT TO PREVENT ENTRY INTO SEWERS, DRAIN, AND WATER WAYS.

LARGE SPILL/NO FIRE/RESCUE: WEAR RUBBER OR NEOPRENE BOOTS, GLOVES.

SPILL CONTROL AND CONTAINMENT:
  HOUSEHOLD SPILL: WIPE UP WITH AN ABSORBENT CLOTH. RINSE OUT IN THE SINK. WIPE UP RESIDUE WITH A DAMP CLOTH.
  RINSE OUT IN THE SINK.
  LARGE SPILL: PICK UP WITH ABSORBENT. RINSE AREA WITH WATER. PICK UP WITH ABSORBENT. PUT ALL ABSORBENT IN A
  DISPOSAL CONTAINER. SEAL AND REMOVE FROM THE WORK AREA.

HEALTH HAZARD INFORMATION:
  WHEN USING THIS PRODUCT WEAR: NO SPECIAL REQUIREMENT. Eye contact may cause stinging, tearing, itching,
  swelling, and redness. Prolonged contact with skin may cause irritation. Ingestion may result in nausea,
  vomiting, and diarrhea.

  A physician should be contacted if anyone develops any signs or symptoms and suspects that they are caused by
  exposure to this product.

FIRST AID:
  Eye Exposure:
  Flush with water for 15 minutes while lifting the upper and lower eye lids. Contact lenses should not be worn
  when working with this product. Get medical attention.

  Skin Exposure:
  Wash with soap and water. If irritation develops, get medical attention.

  Breathing:
  Not expected to be a problem.

  Swallowing:
  Do not induce vomiting. Immediately contact the local poison control center for advice. Keep the victim warm and
  at rest. Get medical attention.

PRODUCT: SPIC AND SPAN - BUILT SYNTHETIC DETERGENT (PHOSPHATE)

INCOMPATIBILITY: NONE KNOWN

INGREDIENTS: SODIUM TRIPOLYPHOSPHATE   TLV: CAS#: 7758-29-4
             ANIONIC SURFACTANT   TLV: CAS#:
             SODIUM SILICATE   TLV: CAS#: 1344-09-8
             SODIUM SULFATE   TLV: CAS#: 7757-82-6
             COLORANTS   PERFUME

IGNI TEMP: NA FP: NA LEL: NA UEL: NA VP: NA VD: NA SG: .4 PS: POWDER APPEAR: GREEN ODOR: PINE
    PH FACTOR: NA

HAZARD RATINGS: H: F: R:

FIRE FIGHTING:
    HIGH EXPANSION FOAM, LOW EXPANSION FOAM, ALCOHOL FOAM; DRY CHEMICAL, CARBON DIOXIDE, WATER FOG.
    PROTECTIVE CLOTHING, RUBBER GLOVES, AND BREATHING APPARATUS.
WARNING: 1] STRUCTURAL PROTECTIVE CLOTHING IS PERMEABLE, REMAIN CLEAR OF SMOKE, WATER FALL OUT AND WATER RUN OFF.
         2] KEEP OUT OF THE REACH OF CHILDREN.
         3] THERMAL DECOMPOSITION MAY YIELD CARBON DIOXIDE, CARBON MONOXIDE.
         4] MOVE CONTAINERS FROM AREA IF WITHOUT RISK, COOL EXPOSED CONTAINERS.
         5] DIKE AREA FOR CONTROL AND CONTAINMENT TO PREVENT ENTRY INTO SEWERS, DRAIN, AND WATER WAYS.

LARGE SPILL/NO FIRE/RESCUE: WEAR RUBBER OR NEOPRENE BOOTS, GLOVES.

SPILL CONTROL AND CONTAINMENT:
    HOUSEHOLD SPILL: SWEEP UP POWDER AND PUT IN A DISPOSAL CONTAINER. DAMP MOP AREA TO PICK UP RESIDUE.
    LARGE SPILL: AVOID CREATING A DUST. SWEEP OR VACUUM UP POWDER. RINSE AREA WITH WATER. PICK UP WITH ABSORBENT.
    PUT IN A DISPOSAL CONTAINER. SEAL AND REMOVE FROM THE WORK AREA.

HEALTH HAZARD INFORMATION:
    WHEN USING THIS PRODUCT WEAR: NO SPECIAL REQUIREMENT. Eye contact may cause stinging, tearing, itching,
    swelling, and redness. Prolonged contact with skin may cause drying and irritation. Ingestion may cause
    abdominal cramps, nausea, vomiting, and diarrhea.

    A physician should be contacted if anyone develops any signs or symptoms and suspects that they are caused by
    exposure to this product.

FIRST AID:
Eye Exposure:
Flush with water for 15 minutes while lifting the upper and lower eye lids. Contact lenses should not be worn
when working with this product. Get medical attention.

Skin Exposure:
Wash with water. If irritation develops, get medical attention.

Breathing:
Not expected to be a problem.

Swallowing:
If the victim is conscious, give 1-2 glasses of water to drink. Immediately contact the local poison control
center for advice. Keep the victim warm and at rest. Get medical attention.

PRODUCT: SPIC AND SPAN - BUILT SYNTHETIC DETERGENT (NON-PHOSPHATE)

INCOMPATIBILITY: NONE KNOWN

INGREDIENTS: SODIUM CARBONATE   TLV: CAS#: 497-19-8
             ANIONIC SURFACTANT   TLV: CAS#:
             SODIUM SILICATE   TLV: CAS#: 1344-09-8
             SODIUM SULFATE   TLV: CAS#: 7757-82-6
             COLORANTS   PERFUME

IGNI TEMP: NA FP: NA LEL: NA UEL: NA VP: NA VD: NA SG: .4 PS: POWDER APPEAR: GREEN ODOR: PINE
   PH FACTOR: NA

HAZARD RATINGS: H: F: R:

FIRE FIGHTING:
   HIGH EXPANSION FOAM, LOW EXPANSION FOAM, ALCOHOL FOAM; DRY CHEMICAL, CARBON DIOXIDE, WATER FOG.
   PROTECTIVE CLOTHING, RUBBER GLOVES, AND BREATHING APPARATUS.
WARNING: 1] STRUCTURAL PROTECTIVE CLOTHING IS PERMEABLE, REMAIN CLEAR OF SMOKE, WATER FALL OUT AND WATER RUN OFF.
         2] KEEP OUT OF THE REACH OF CHILDREN.
         3] THERMAL DECOMPOSITION MAY YIELD CARBON DIOXIDE, CARBON MONOXIDE.
         4] MOVE CONTAINERS FROM AREA IF WITHOUT RISK, COOL EXPOSED CONTAINERS.
         5] DIKE AREA FOR CONTROL AND CONTAINMENT TO PREVENT ENTRY INTO SEWERS, DRAIN, AND WATER WAYS.

LARGE SPILL/NO FIRE/RESCUE: WEAR RUBBER OR NEOPRENE BOOTS, GLOVES.

SPILL CONTROL AND CONTAINMENT:
   HOUSEHOLD SPILL: SWEEP UP POWDER AND PUT IN A DISPOSAL CONTAINER. DAMP MOP AREA TO PICK UP THE RESIDUE.
   LARGE SPILL: AVOID CREATING A DUST. SWEEP OR VACUUM UP POWDER. RINSE AREA WITH WATER. PICK UP WITH ABSORBENT.
   PUT ABSORBENT IN A DISPOSAL CONTAINER. SEAL AND REMOVE FROM THE WORK AREA.

HEALTH HAZARD INFORMATION:
   WHEN USING THIS PRODUCT WEAR: NO SPECIAL REQUIREMENT. Eye contact may cause stinging, tearing, itching,
   swelling, and redness. Prolonged skin contact may cause drying and irritation. Ingestion may result in abdominal
   cramps, nausea, vomiting, and diarrhea.

   A physician should be contacted if anyone develops any signs or symptoms and suspects that they are caused by
   exposure to this product.

FIRST AID:
   Eye Exposure:
   Flush with water for 15 minutes while lifting the upper and lower eye lids. Contact lenses should not be worn
   when working with this product. Get medical attention.

   Skin Exposure:
   Wash with water. If irritation develops, get medical attention.

   Breathing:
   Not expected to be a problem.

   Swallowing:
   If the victim is conscious, give 1-2 glasses of water to drink. Immediately contact the local poison control
   center for advice. Keep the victim warm and at rest. Get medical attention.

PRODUCT: SPIC AND SPAN, LIQUID (PHOSPHATE)

INCOMPATIBILITY: NONE KNOWN

INGREDIENTS: ANIONIC SURFACTANT   TLV: CAS#:
             BUTYL DIGLYCOL   TLV: CAS#:
             TETRAPOTASSIUM PYROPHOSPHATE TLV: CAS#: 7320-34-5
             SODIUM CUMENE SULFONATE   TLV: CAS#:
             COLOR   PERFUME

IGNI TEMP: NA FP: 200F LEL: NA UEL: NA VP: NA VD: NA SG: 1.09 PS: LIQUID APPEAR: GREEN ODOR: PINE
   PH FACTOR: NA

HAZARD RATINGS: H: F: R:

FIRE FIGHTING:
   HIGH EXPANSION FOAM, LOW EXPANSION FOAM, ALCOHOL FOAM; DRY CHEMICAL, CARBON DIOXIDE, WATER FOG.
   PROTECTIVE CLOTHING, RUBBER GLOVES, AND BREATHING APPARATUS.
WARNING: 1] STRUCTURAL PROTECTIVE CLOTHING IS PERMEABLE, REMAIN CLEAR OF SMOKE, WATER FALL OUT AND WATER RUN OFF.
         2] KEEP OUT OF THE REACH OF CHILDREN.
         3] THERMAL DECOMPOSITION MAY YIELD CARBON DIOXIDE, CARBON MONOXIDE.
         4] MOVE CONTAINERS FROM AREA IF WITHOUT RISK, COOL EXPOSED CONTAINERS.
         5] DIKE AREA FOR CONTROL AND CONTAINMENT TO PREVENT ENTRY INTO SEWERS, DRAIN, AND WATER WAYS.

LARGE SPILL/NO FIRE/RESCUE: WEAR RUBBER OR NEOPRENE BOOTS, GLOVES.

SPILL CONTROL AND CONTAINMENT:
   HOUSEHOLD SPILL: WIPE UP WITH AN ABSORBENT CLOTH. RINSE OUT IN THE SINK. WIPE UP RESIDUE WITH A DAMP CLOTH.
   RINSE OUT IN THE SINK.
   LARGE SPILL: PICK UP WITH ABSORBENT. RINSE AREA WITH WATER. PICK UP WITH ABSORBENT. PUT ALL ABSORBENT IN A
   DISPOSAL CONTAINER. SEAL AND REMOVE FROM THE WORK AREA.

HEALTH HAZARD INFORMATION:
   WHEN USING THIS PRODUCT WEAR: NO SPECIAL REQUIREMENT. Eye contact may cause stinging, tearing, itching,
   swelling, and redness. Prolonged skin contact may cause irritation. Ingestion may result in abdominal cramps,
   nausea, vomiting, and diarrhea.

   A physician should be contacted if anyone develops any signs or symptoms and suspects that they are caused by
   exposure to this product.

FIRST AID:
Eye Exposure:
Flush with water for 15 minutes while lifting the upper and lower eye lids. Contact lenses should not be worn
when working with this product. Get medical attention.

Skin Exposure:
Wash with water. If irritation develops, get medical attention.

Breathing:
Not expected to be a problem.

Swallowing:
If the victim is conscious, give 1-2 glasses of water to drink. Immediately contact the local poison control
center for advice. Keep the victim warm and at rest. Get medical attention.

PRODUCT: SPIC AND SPAN, LIQUID (NON-PHOSPHATE)

INCOMPATIBILITY: NONE KNOWN

INGREDIENTS: ANIONIC SURFACTANT   TLV: CAS#:
             BUTYL DIGLYCOL   TLV: CAS#:
             SODIUM CITRATE   TLV: CAS#: 6132-04-3
             TETRASODIUM EDTA   TLV: CAS#: 10378-23-1
             SODIUM CUMENE SULFONATE TLV: CAS#:
             PERFUME   TLV: CAS#:

IGNI TEMP: NA FP: 200F LEL: NA UEL: NA VP: NA VD: NA SG: 1.09 PS: LIQUID APPEAR: GREEN ODOR: PINE
   PH FACTOR: NA

HAZARD RATINGS: H: F: R:

FIRE FIGHTING:
   HIGH EXPANSION FOAM, LOW EXPANSION FOAM, ALCOHOL FOAM; DRY CHEMICAL, CARBON DIOXIDE, WATER FOG.
   PROTECTIVE CLOTHING, RUBBER GLOVES, AND BREATHING APPARATUS.
WARNING: 1] STRUCTURAL PROTECTIVE CLOTHING IS PERMEABLE, REMAIN CLEAR OF SMOKE, WATER FALL OUT AND WATER RUN OFF.
         2] KEEP OUT OF THE REACH OF CHILDREN.
         3] THERMAL DECOMPOSITION MAY YIELD CARBON DIOXIDE, CARBON MONOXIDE.
         4] MOVE CONTAINERS FROM AREA IF WITHOUT RISK, COOL EXPOSED CONTAINERS.
         5] DIKE AREA FOR CONTROL AND CONTAINMENT TO PREVENT ENTRY INTO SEWERS, DRAIN, AND WATER WAYS.

LARGE SPILL/NO FIRE/RESCUE: WEAR RUBBER OR NEOPRENE BOOTS, GLOVES.

SPILL CONTROL AND CONTAINMENT:
   HOUSEHOLD SPILL: WIPE UP WITH AN ABSORBENT CLOTH. RINSE OUT IN THE SINK. WIPE UP RESIDUE WITH A DAMP CLOTH.
   RINSE OUT IN THE SINK.
   LARGE SPILL: PICK UP WITH ABSORBENT. RINSE AREA WITH WATER. PICK UP WITH ABSORBENT. PUT ALL ABSORBENT IN A
   DISPOSAL CONTAINER. SEAL AND REMOVE FROM THE WORK AREA.

HEALTH HAZARD INFORMATION:
   WHEN USING THIS PRODUCT WEAR: NO SPECIAL REQUIREMENT. Eye contact may cause stinging, tearing, itching,
   swelling, and redness. Skin contact may cause irritation. Ingestion may result in abdominal cramps, nausea,
   vomiting, and diarrhea.

   A physician should be contacted if anyone develops any signs or symptoms and suspects that they are caused by
   exposure to this product.

FIRST AID:
   Eye Exposure:
   Flush with water for 15 minutes while lifting the upper and lower eye lids. Contact lenses should not be worn
   when working with this product. Get medical attention.

   Skin Exposure:
   Wash with water. If irritation develops, get medical attention.

   Breathing:
   Not expected to be a problem.

   Swallowing:
   If the victim is conscious, give 1-2 glasses of water to drink. Immediately contact the local poison control
   center for advice. Keep the victim warm and at rest. Get medical attention.

PRODUCT: SPIC AND SPAN PINE CLEANER - LIQUID HOUSEHOLD CLEANER

INCOMPATIBILITY: NONE KNOWN

INGREDIENTS:	ANIONIC SURFACTANT	TLV:	CAS#:
	PINE OIL	TLV:	CAS#: 8002-09-3
	ISOPROPYL ALCOHOL	TLV: 400 ppm	CAS#: 67-63-0
	WATER	TLV:	CAS#: 7732-18-5
	PERFUME	TLV:	CAS#:

IGNI TEMP: NA FP: 106F LEL: NA UEL: NA VP: NA VD: NA SG: 1.07 PS: LIQUID APPEAR: AMBER ODOR: PINE
   PH FACTOR: NA

HAZARD RATINGS: H: F: R:

FIRE FIGHTING:
   HIGH EXPANSION FOAM, LOW EXPANSION FOAM, ALCOHOL FOAM; DRY CHEMICAL, CARBON DIOXIDE, WATER FOG.
   PROTECTIVE CLOTHING, RUBBER GLOVES, AND BREATHING APPARATUS.
WARNING: 1] STRUCTURAL PROTECTIVE CLOTHING IS PERMEABLE, REMAIN CLEAR OF SMOKE, WATER FALL OUT AND WATER RUN OFF.
         2] KEEP OUT OF THE REACH OF CHILDREN.
         3] THERMAL DECOMPOSITION MAY YIELD CARBON DIOXIDE, CARBON MONOXIDE.
         4] MOVE CONTAINERS FROM AREA IF WITHOUT RISK, COOL EXPOSED CONTAINERS.
         5] DIKE AREA FOR CONTROL AND CONTAINMENT TO PREVENT ENTRY INTO SEWERS, DRAIN, AND WATER WAYS.

LARGE SPILL/NO FIRE/RESCUE: WEAR RUBBER OR NEOPRENE BOOTS, GLOVES.

SPILL CONTROL AND CONTAINMENT:
   HOUSEHOLD SPILL: WIPE UP WITH AN ABSORBENT CLOTH. RINSE OUT IN THE SINK. WIPE UP RESIDUE WITH A DAMP CLOTH.
   RINSE OUT IN THE SINK.
   LARGE SPILL: PICK UP WITH ABSORBENT. RINSE AREA WITH WATER. PICK UP WITH ABSORBENT. PUT ALL ABSORBENT IN A
   DISPOSAL CONTAINER. SEAL AND REMOVE FROM THE WORK AREA.

HEALTH HAZARD INFORMATION:
   WHEN USING THIS PRODUCT WEAR: NO SPECIAL REQUIREMENT. Eye contact may cause stinging. tearing, itching,
   swelling, and redness. Skin contact may cause irritation. Ingestion may result in abdominal cramps, nausea,
   vomiting, and diarrhea.

   A physician should be contacted if anyone develops any signs or symptoms and suspects that they are caused by
   exposure to this product.

FIRST AID:
   Eye Exposure:
   Flush with water for 15 minutes while lifting the upper and lower eye lids. Contact lenses should not be worn
   when working with this product. Get medical attention.

   Skin Exposure:
   Wash with water. If irritation develops, get medical attention.

   Breathing:
   Not expected to be a problem.

   Swallowing:
   If the victim is conscious, give 1-2 glasses of water to drink. Immediately contact the local poison control
   center for advice. Keep the victim warm and at rest. Get medical attention.

PRODUCT: SPIC AND SPAN BATHROOM CLEANER SPRAY

INCOMPATIBILITY: NONE KNOWN

INGREDIENTS:	TLV:	CAS#:
ANIONIC SURFACTANT	TLV:	CAS#:
BUTYL DIGLYCOL	TLV:	CAS#:
SODIUM CITRATE	TLV:	CAS#: 6132-04-3
SODIUM CARBONATE	TLV:	CAS#: 497-19-8
WATER	TLV:	CAS#: 7732-18-5
PERFUME	TLV:	CAS#:

IGNI TEMP: NA FP: 200F LEL: NA UEL: NA VP: NA VD: NA SG: 1.03 PS: LIQUID APPEAR: COLORLESS   ODOR: PINE
   PH FACTOR: NA

HAZARD RATINGS: H: F: R:

FIRE FIGHTING:
   HIGH EXPANSION FOAM, LOW EXPANSION FOAM, ALCOHOL FOAM; DRY CHEMICAL, CARBON DIOXIDE, WATER FOG.
   PROTECTIVE CLOTHING, RUBBER GLOVES, AND BREATHING APPARATUS.
WARNING: 1] STRUCTURAL PROTECTIVE CLOTHING IS PERMEABLE, REMAIN CLEAR OF SMOKE, WATER FALL OUT AND WATER RUN OFF.
         2] KEEP OUT OF THE REACH OF CHILDREN.
         3] THERMAL DECOMPOSITION MAY YIELD CARBON DIOXIDE, CARBON MONOXIDE.
         4] MOVE CONTAINERS FROM AREA IF WITHOUT RISK, COOL EXPOSED CONTAINERS.
         5] DIKE AREA FOR CONTROL AND CONTAINMENT TO PREVENT ENTRY INTO SEWERS, DRAIN, AND WATER WAYS.

LARGE SPILL/NO FIRE/RESCUE: WEAR RUBBER OR NEOPRENE BOOTS, GLOVES.

SPILL CONTROL AND CONTAINMENT:
   HOUSEHOLD SPILL: WIPE UP WITH AN ABSORBENT CLOTH. RINSE OUT IN THE SINK. WIPE UP RESIDUE WITH A DAMP CLOTH.
   RINSE OUT IN THE SINK.
   LARGE SPILL: PICK UP WITH ABSORBENT. RINSE AREA WITH WATER. PICK UP WITH ABSORBENT. PUT ALL ABSORBENT IN A
   DISPOSAL CONTAINER. SEAL AND REMOVE FROM THE WORK AREA.

HEALTH HAZARD INFORMATION:
   WHEN USING THIS PRODUCT WEAR: NO SPECIAL REQUIREMENT. Eye contact may cause stinging, tearing, itching,
   swelling, and redness. Skin contact may cause irritation. Inhalation may cause irritation of the nose, throat,
   and lungs. Ingestion may result in abdominal cramps, nausea, vomiting, and diarrhea.

   A physician should be contacted if anyone develops any signs or symptoms and suspects that they are caused by
   exposure to this product.

FIRST AID:
   Eye Exposure:
   Flush with water for 15 minutes while lifting the upper and lower eye lids. Contact lenses should not be worn
   when working with this product. Get medical attention.

   Skin Exposure:
   Wash with soap and water. If irritation develops, get medical attention.

   Breathing:
   If irritation develops, move victim to fresh air.

   Swallowing:
   If the victim is conscious, give 1-2 glasses of water to drink. Immediately contact the local poison control
   center for advice. Keep the victim warm and at rest. Get medical attention.

PRODUCT: STEP SAVER

INCOMPATIBILITY: NONE KNOWN

INGREDIENTS: MODIFIED ACRYLIC POLYMERS 10-20% TLV: CAS#:
             DIETHYLENE GLYCOL
             MONO ETHYL ETHER   1-3%   TLV: CAS#:   111-90-0
             WATER   80-90% TLV: CAS#:   7732-18-5

IGNI TEMP: NA FP: NA LEL: NA UEL: NA VP: NA VD: NA SG: 1.02 PS: LIQUID APPEAR: TRANSLUCENT
     ODOR: MILD AMMONIA   PH FACTOR: 9.8

HAZARD RATINGS: H: F: R: 1 0 0

FIRE FIGHTING:
   HIGH EXPANSION FOAM, LOW EXPANSION FOAM, ALCOHOL FOAM; DRY CHEMICAL, CARBON DIOXIDE, WATER FOG.
   PROTECTIVE CLOTHING, RUBBER GLOVES, AND BREATHING APPARATUS.
WARNING: 1] STRUCTURAL PROTECTIVE CLOTHING IS PERMEABLE, REMAIN CLEAR OF SMOKE, WATER FALL OUT AND WATER RUN OFF.
         2] KEEP OUT OF THE REACH OF CHILDREN.
         3] THERMAL DECOMPOSITION MAY YIELD CARBON DIOXIDE, CARBON MONOXIDE.
         4] MOVE CONTAINERS FROM AREA IF WITHOUT RISK, COOL EXPOSED CONTAINERS.
         5] DIKE AREA FOR CONTROL AND CONTAINMENT TO PREVENT ENTRY INTO SEWERS, DRAIN, AND WATER WAYS.

LARGE SPILL/NO FIRE/RESCUE: WEAR RUBBER OR NEOPRENE BOOTS, GLOVES.

SPILL CONTROL AND CONTAINMENT:
   HOUSEHOLD SPILL: WIPE UP WITH AN ABSORBENT CLOTH. RINSE OUT IN THE SINK. WIPE UP RESIDUE WITH A DAMP CLOTH.
   RINSE OUT IN THE SINK.
   LARGE SPILL: PICK UP WITH ABSORBENT. RINSE AREA WITH WATER. PICK UP WITH ABSORBENT. PUT ALL ABSORBENT IN A
   DISPOSAL CONTAINER. SEAL AND REMOVE FROM THE WORK AREA.

HEALTH HAZARD INFORMATION:
   WHEN USING THIS PRODUCT WEAR: NO SPECIAL REQUIREMENT. Eye contact may cause stinging, tearing, itching,
   swelling, and redness. Prolonged skin contact may cause irritation.

   A physician should be contacted if anyone develops any signs or symptoms and suspects that they are caused by
   exposure to this product.

FIRST AID:
   Eye Exposure:
   Flush with water for 15 minutes while lifting the upper and lower eye lids. Contact lenses should not be worn
   when working with this product. Get medical attention.

   Skin Exposure:
   Wash with soap and water. If irritation develops, get medical attention.

   Breathing:
   Not expected to be a problem.

   Swallowing:
   Immediately contact the local poison control center for advice. Keep the victim warm and at rest. Get medical
   attention.

PRODUCT: SUNLIGHT AUTOMATIC DISH WASHING DETERGENT

INCOMPATIBILITY: STRONG ACIDS; AMMONIA; PRODUCTS CONTAINING AMMONIA.

INGREDIENTS: SODIUM SILICATE   1% TLV: CAS#: 1344-09-8
             SODIUM CARBONATE   1% TLV: CAS#: 497-19-8

IGNI TEMP: NA FP: NA LEL: NA UEL: NA VP: NA VD: NA SG: .84-.98 PS: POWDER APPEAR: YELLOW GRANULES
     ODOR: LEMON   PH FACTOR: NA

HAZARD RATINGS: H: F: R: 2 0 0

FIRE FIGHTING:
  HIGH EXPANSION FOAM, LOW EXPANSION FOAM, ALCOHOL FOAM; DRY CHEMICAL, CARBON DIOXIDE, WATER FOG.
  PROTECTIVE CLOTHING, RUBBER GLOVES, AND BREATHING APPARATUS.
WARNING: 1] STRUCTURAL PROTECTIVE CLOTHING IS PERMEABLE, REMAIN CLEAR OF SMOKE, WATER FALL OUT AND WATER RUN OFF.
         2] KEEP OUT OF THE REACH OF CHILDREN.
         3] THERMAL DECOMPOSITION MAY YIELD CHLORINE GAS.
         4] MOVE CONTAINERS FROM AREA IF WITHOUT RISK, COOL EXPOSED CONTAINERS.
         5] DIKE AREA FOR CONTROL AND CONTAINMENT TO PREVENT ENTRY INTO SEWERS, DRAIN, AND WATER WAYS.

LARGE SPILL/NO FIRE/RESCUE: WEAR RUBBER OR NEOPRENE BOOTS, GLOVES.

SPILL CONTROL AND CONTAINMENT:
  HOUSEHOLD SPILL: SWEEP UP POWDER. DAMP MOP UP RESIDUE.
  LARGE SPILL: AVOID CREATING A DUST. SWEEP UP POWDER. RINSE AREA WITH WATER. PICK UP WITH ABSORBENT. PUT ALL
  MATERIAL IN A DISPOSAL CONTAINER. SEAL AND REMOVE FROM THE WORK AREA.

HEALTH HAZARD INFORMATION:
  WHEN USING THIS PRODUCT WEAR: NO SPECIAL REQUIREMENT. Eye contact will cause burning sensation, tearing,
  itching, swelling, and redness. Prolonged skin contact may cause irritation. Ingestion may result in abdominal
  cramps, nausea, vomiting, and diarrhea. Inhalation may cause irritation of the nose, throat, and lungs.

  A physician should be contacted if anyone develops any signs or symptoms and suspects that they are caused by
  exposure to this product.

FIRST AID:
  Eye Exposure:
  Flush with water for 15 minutes while lifting the upper and lower eye lids. Contact lenses should not be worn
  when working with this product. Get medical attention.

  Skin Exposure:
  Wash with water. If irritation develops, get medical attention.

  Breathing:
  If irritation develops, move victim to fresh air. If it persists get medical attention.

  Swallowing:
  If the victim is conscious, give 1-2 glasses of milk or water to drink. Immediately contact the local poison
  control center for advice. Keep the victim warm and at rest. Get medical attention.

PRODUCT: SUNLIGHT AUTOMATIC DISH WASHING DETERGENT, LIQUID

INCOMPATIBILITY: NONE KNOWN.

INGREDIENTS: SODIUM HYPOCHLORITE   1%   TLV: CAS#: 7681-52-9
             SODIUM CARBONATE   1%   TLV: CAS#: 497-19-8
             SODIUM SILICATE   1%   TLV: CAS#: 1344-09-8

IGNI TEMP: NA FP: NA LEL: NA UEL: NA VP: NA VD: NA SG: 1.31 PS: LIQUID APPEAR: ODOR:
    PH FACTOR: 11.8 - 12.5

HAZARD RATINGS: H: F: R: 2 1 1

FIRE FIGHTING:
    HIGH EXPANSION FOAM, LOW EXPANSION FOAM, ALCOHOL FOAM; DRY CHEMICAL, CARBON DIOXIDE, WATER FOG.
    PROTECTIVE CLOTHING, RUBBER GLOVES, AND BREATHING APPARATUS.
WARNING: 1] STRUCTURAL PROTECTIVE CLOTHING IS PERMEABLE, REMAIN CLEAR OF SMOKE, WATER FALL OUT AND WATER RUN OFF.
         2] KEEP OUT OF THE REACH OF CHILDREN.
         3] THERMAL DECOMPOSITION MAY YIELD CHLORINE GAS.
         4] MOVE CONTAINERS FROM AREA IF WITHOUT RISK, COOL EXPOSED CONTAINERS.
         5] DIKE AREA FOR CONTROL AND CONTAINMENT TO PREVENT ENTRY INTO SEWERS, DRAIN, AND WATER WAYS.

LARGE SPILL/NO FIRE/RESCUE: WEAR RUBBER OR NEOPRENE BOOTS, GLOVES.

SPILL CONTROL AND CONTAINMENT:
    HOUSEHOLD SPILL: WIPE UP WITH AN ABSORBENT CLOTH, RINSE OUT IN THE SINK. WIPE UP RESIDUE WITH A DAMP CLOTH,
    RINSE OUT IN THE SINK. WASH HANDS WITH SOAP AND WATER.
    LARGE SPILL: PICK UP WITH ABSORBENT, PUT IN A DISPOSAL CONTAINER. RINSE AREA WITH WATER, PICK UP WITH ABSORBENT.
    PUT IN A DISPOSAL CONTAINER, SEAL AND REMOVE FROM THE WORK AREA.

HEALTH HAZARD INFORMATION:
    WHEN USING THIS PRODUCT WEAR: NO SPECIAL REQUIREMENT. Eye contact will cause stinging, tearing, swelling,
    redness and possible corneal ulceration and permanent eye injury. Skin contact may produce ulceration of nasal
    passage and inside of the mouth. Ingestion may produce ulceration in the mouth, throat, and stomach. If ingested
    in sufficient amount, spontaneous vomiting may occur.

    A physician should be contacted if anyone develops any signs or symptoms and suspects that they are caused by
    exposure to this product.

FIRST AID:
Eye Exposure:
Flush with water for 15 minutes while lifting the upper and lower eye lids. Contact lenses should not be worn
when working with this product. Get medical attention.

Skin Exposure:
Remove contaminated clothing and wash with water. If irritation develops, get medical attention.

Breathing:
Not Expected to be a problem.

Swallowing:
If the victim is conscious, give 1-2 glasses of milk or water to drink. Immediately contact the local poison
control center for advice. Keep the victim warm and at rest. Get medical attention.

PRODUCT: SUNLIGHT DISH WASHING DETERGENT (LIQUID)

INCOMPATIBILITY: ALL PRODUCTS CONTAINING CHLORINE (SODIUM HYPOCHLORITE)

INGREDIENTS: NOT IDENTIFIED BY MANUFACTURER

IGNI TEMP: NA FP: NA LEL: NA UEL: NA VP: NA VD: NA SG: NA PS: LIQUID APPEAR: YELLOW TRANSPARENT
    ODOR: LEMON   PH FACTOR: 6.5 - 6.9

HAZARD RATINGS: H: F: R: 2 0 0

FIRE FIGHTING:
    HIGH EXPANSION FOAM, LOW EXPANSION FOAM, ALCOHOL FOAM; DRY CHEMICAL, CARBON DIOXIDE, WATER FOG.
    PROTECTIVE CLOTHING, RUBBER GLOVES, AND BREATHING APPARATUS.
WARNING: 1] STRUCTURAL PROTECTIVE CLOTHING IS PERMEABLE, REMAIN CLEAR OF SMOKE, WATER FALL OUT AND WATER RUN OFF.
         2] KEEP OUT OF THE REACH OF CHILDREN.
         3] THERMAL DECOMPOSITION MAY YIELD TOXIC OXIDES OF NITROGEN, OXIDES OF SULFUR
         4] MOVE CONTAINERS FROM AREA IF WITHOUT RISK, COOL EXPOSED CONTAINERS.
         5] DIKE AREA FOR CONTROL AND CONTAINMENT TO PREVENT ENTRY INTO SEWERS, DRAIN, AND WATER WAYS.

LARGE SPILL/NO FIRE/RESCUE: WEAR RUBBER OR NEOPRENE BOOTS, GLOVES.

SPILL CONTROL AND CONTAINMENT:
    HOUSEHOLD SPILL: WIPE UP WITH AN ABSORBENT CLOTH. RINSE OUT IN THE SINK. WIPE UP RESIDUE WITH A DAMP CLOTH.
    RINSE OUT IN THE SINK.
    LARGE SPILL: PICK UP WITH ABSORBENT. RINSE AREA WITH WATER. PICK UP WITH ABSORBENT. PUT ALL ABSORBENT IN A
    DISPOSAL CONTAINER. SEAL AND REMOVE FROM THE WORK AREA.

HEALTH HAZARD INFORMATION:
    WHEN USING THIS PRODUCT WEAR: NO SPECIAL REQUIREMENT. Eye contact may cause stinging, tearing, itching,
    swelling, and redness. Prolonged skin contact may cause irritation. Ingestion may result in abdominal cramps,
    nausea, vomiting, and diarrhea. Inhalation may cause irritation of the nose, throat, and lungs.

    A physician should be contacted if anyone develops any signs or symptoms and suspects that they are caused by
    exposure to this product.

FIRST AID:
Eye Exposure:
Flush with water for 15 minutes while lifting the upper and lower eye lids. Contact lenses should not be worn
when working with this product. Get medical attention.

Skin Exposure:
Wash with water. If irritation develops, get medical attention.

Breathing:
If irritation develops, move victim to fresh air. If it persists get medical attention.

Swallowing:
If the victim is conscious, give 2-3 glasses of milk or water to drink. Immediately contact the local poison
control center for advice. Keep the victim warm and at rest. Get medical attention.

PRODUCT: SUNLIGHT GEL AUTOMATIC DISH WASHING DETERGENT

INCOMPATIBILITY: ACIDS; VINEGAR.

INGREDIENTS: SODIUM SILICATE   1% TLV: CAS#: 1344-09-8
             SODIUM HYDROXIDE   1% TLV: 2 mg/m3   CAS#: 1310-73-2
             SODIUM HYPOCHLORITE   1% TLV: CAS#: 7681-52-9

IGNI TEMP: NA FP: NA LEL: NA UEL: NA VP: NA VD: NA SG: 1.25 - 1.4 PS: LIQUID APPEAR: VISCOUS
    ODOR: SWEET CITRUS PH FACTOR: 11.8 - 12.5

HAZARD RATINGS: H: F: R: 2 1 1

FIRE FIGHTING:
    HIGH EXPANSION FOAM, LOW EXPANSION FOAM, ALCOHOL FOAM; DRY CHEMICAL, CARBON DIOXIDE, WATER FOG.
    PROTECTIVE CLOTHING, RUBBER GLOVES, AND BREATHING APPARATUS.
WARNING: 1] STRUCTURAL PROTECTIVE CLOTHING IS PERMEABLE, REMAIN CLEAR OF SMOKE, WATER FALL OUT AND WATER RUN OFF.
         2] KEEP OUT OF THE REACH OF CHILDREN.
         3] THERMAL DECOMPOSITION MAY YIELD CHLORINE GAS.
         4] DO NOT MIX WITH ANY ACIDIC PRODUCT. HAZARDOUS GASES RELEASED.
         5] MOVE CONTAINERS FROM AREA IF WITHOUT RISK, COOL EXPOSED CONTAINERS.
         6] DIKE AREA FOR CONTROL AND CONTAINMENT TO PREVENT ENTRY INTO SEWERS, DRAIN, AND WATER WAYS.

LARGE SPILL/NO FIRE/RESCUE: WEAR RUBBER OR NEOPRENE BOOTS, GLOVES.

SPILL CONTROL AND CONTAINMENT:
    HOUSEHOLD SPILL: WIPE UP WITH AN ABSORBENT CLOTH. RINSE OUT IN THE SINK. WIPE UP RESIDUE WITH A DAMP CLOTH.
    RINSE OUT IN THE SINK.
    LARGE SPILL: PICK UP WITH ABSORBENT. RINSE AREA WITH WATER. PICK UP WITH ABSORBENT. PUT ALL ABSORBENT IN A
    DISPOSAL CONTAINER. SEAL AND REMOVE FROM THE WORK AREA.

HEALTH HAZARD INFORMATION:
    WHEN USING THIS PRODUCT WEAR: NO SPECIAL REQUIREMENT. Eye contact will cause severe burning sensation, tearing,
    swelling, and redness. Skin contact will cause severe irritation, with redness and swelling. Ingestion will
    cause severe irritation, possible burns to the mouth, throat, and stomach, nausea, and vomiting. Inhalation will
    cause irritation of the nose, throat and lungs.

    A physician should be contacted if anyone develops any signs or symptoms and suspects that they are caused by
    exposure to this product.

FIRST AID:
Eye Exposure:
Flush with water for 15 minutes while lifting the upper and lower eye lids. s. Contact lenses should not be worn
when working with this product. Get medical attention.

Skin Exposure:
Wash with soap and water. If irritation develops, get medical attention.

Breathing:
If irritation develops, move victim to fresh air. If it persists, get medical attention.

Swallowing:
Do not induce vomiting. Immediately rinse out the mouth, If the victim is conscious, give 2-3 glasses of milk or
water to drink. Immediately contact the local poison control center for advice. Keep the victim warm and at
rest. Get medical attention.

PRODUCT: SURF LAUNDRY DETERGENT, ULTRA PHOSPHATE

INCOMPATIBILITY: NONE KNOWN

INGREDIENTS: SODIUM SILICATE   1% TLV: CAS#: 1344-09-8
             SODIUM CARBONATE   1% TLV: CAS#: 497-19-8

IGNI TEMP: NA FP: NA LEL: NA UEL: NA VP: NA VD: NA SG: NA PS: POWDER APPEAR: WHITE - GREEN SPECKS
    ODOR: NONE   PH FACTOR: 13.0

HAZARD RATINGS: H: F: R: 2 0 0   HAZARD CLASS 8

FIRE FIGHTING:
  HIGH EXPANSION FOAM, LOW EXPANSION FOAM, ALCOHOL FOAM; DRY CHEMICAL, CARBON DIOXIDE, WATER FOG.
  PROTECTIVE CLOTHING, RUBBER GLOVES, AND BREATHING APPARATUS.
WARNING: 1] STRUCTURAL PROTECTIVE CLOTHING IS PERMEABLE, REMAIN CLEAR OF SMOKE, WATER FALL OUT AND WATER RUN OFF.
         2] KEEP OUT OF THE REACH OF CHILDREN.
         3] THERMAL DECOMPOSITION MAY YIELD OXIDES OF SULFUR.
         4] MOVE CONTAINERS FROM AREA IF WITHOUT RISK, COOL EXPOSED CONTAINERS.
         5] DIKE AREA FOR CONTROL AND CONTAINMENT TO PREVENT ENTRY INTO SEWERS, DRAIN, AND WATER WAYS.

LARGE SPILL/NO FIRE/RESCUE: WEAR RUBBER OR NEOPRENE BOOTS, GLOVES.

SPILL CONTROL AND CONTAINMENT:
  HOUSEHOLD SPILL: SWEEP UP POWDER. PUT IN A DISPOSAL CONTAINER. DAMP MOP UP RESIDUE. RINSE OUT IN THE SINK.
  LARGE SPILL: AVOID CREATING A DUST. SCOOP UP POWDER. PUT IN A DISPOSAL CONTAINER. RINSE AREA WITH WATER. PICK UP
  WITH ABSORBENT. PUT IN A DISPOSAL CONTAINER. SEAL AND REMOVE FROM THE WORK AREA.

HEALTH HAZARD INFORMATION:
  WHEN USING THIS PRODUCT WEAR: NO SPECIAL REQUIREMENT. Inhalation may cause coughing, sore throat, wheezing, and
  shortness of breath. Eye contact will cause stinging, tearing, itching, swelling, and redness. Skin contact may
  cause irritation. Ingestion may cause nausea, vomiting, and diarrhea.

  A physician should be contacted if anyone develops any signs or symptoms and suspects that they are caused by
  exposure to this product.

FIRST AID:
  Eye Exposure:
  Flush with water for 15 minutes while lifting the upper and lower eye lids. Contact lenses should not be worn
  when working with this product. Get medical attention.

  Skin Exposure:
  Wash with water. If irritation develops, get medical attention.

  Breathing:
  If irritation develops, move victim to fresh air.

  Swallowing:
  If the victim is conscious, give 2-3 glasses of milk or water to drink. Immediately contact the local poison
  control center for advice. Keep the victim warm and at rest. Get medical attention.

PRODUCT: SURF LAUNDRY DETERGENT, ULTRA ZEOLITE POWDER

INCOMPATIBILITY: NONE KNOWN

INGREDIENTS: SODIUM SILICATE   1% TLV: CAS#: 1344-09-8
             SODIUM CARBONATE   1% TLV: CAS#: 497-19-8

IGNI TEMP: NA FP: NA LEL: NA UEL: NA VP: NA VD: NA SG: NA PS: POWDER APPEAR: WHITE - GREEN SPECKS
   ODOR: NONE   PH FACTOR: 13.0

HAZARD RATINGS: H: F: R: 2 0 0   HAZARD CLASS   8

FIRE FIGHTING:
  HIGH EXPANSION FOAM, LOW EXPANSION FOAM, ALCOHOL FOAM; DRY CHEMICAL, CARBON DIOXIDE, WATER FOG.
  PROTECTIVE CLOTHING, RUBBER GLOVES, AND BREATHING APPARATUS.
WARNING: 1] STRUCTURAL PROTECTIVE CLOTHING IS PERMEABLE, REMAIN CLEAR OF SMOKE, WATER FALL OUT AND WATER RUN OFF.
         2] KEEP OUT OF THE REACH OF CHILDREN.
         3] THERMAL DECOMPOSITION MAY YIELD OXIDES OF SULFUR
         4] MOVE CONTAINERS FROM AREA IF WITHOUT RISK, COOL EXPOSED CONTAINERS.
         5] DIKE AREA FOR CONTROL AND CONTAINMENT TO PREVENT ENTRY INTO SEWERS, DRAIN, AND WATER WAYS.

LARGE SPILL/NO FIRE/RESCUE: WEAR RUBBER OR NEOPRENE BOOTS, GLOVES.

SPILL CONTROL AND CONTAINMENT:
  HOUSEHOLD SPILL: SWEEP UP POWDER. PUT IN A DISPOSAL CONTAINER. DAMP MOP UP RESIDUE. RINSE OUT IN THE SINK.
  LARGE SPILL: SCOOP UP POWDER. PUT IN A DISPOSAL CONTAINER. RINSE AREA WITH WATER. PICK UP WITH ABSORBENT. PUT IN
  A DISPOSAL CONTAINER. SEAL AND REMOVE FROM THE WORK AREA.

HEALTH HAZARD INFORMATION:
  WHEN USING THIS PRODUCT WEAR: NO SPECIAL REQUIREMENT. Inhalation may cause coughing, sore throat, wheezing, and
  shortness of breath. Eye contact may cause stinging, tearing, itching, swelling, and redness. Skin contact may
  cause irritation. Ingestion may result in nausea, vomiting, and diarrhea.

  A physician should be contacted if anyone develops any signs or symptoms and suspects that they are caused by
  exposure to this product.

FIRST AID:
  Eye Exposure:
  Flush with water for 15 minutes while lifting the upper and lower eye lids. Contact lenses should not be worn
  when working with this product. Get medical attention.

  Skin Exposure:
  Wash with water. If irritation develops, get medical attention.

  Breathing:
  If irritation develops, move the victim to fresh air.

  Swallowing:
  If the victim is conscious, give 2-3 glasses of milk or water to drink. Immediately contact the local poison
  control center for advice. Keep the victim warm and at rest. Get medical attention.

PRODUCT: SURF LIQUID LAUNDRY DETERGENT, NON-PHOSPHATE

INCOMPATIBILITY: STRONG ACIDS AND PRODUCT CONTAINING CHLORINE

INGREDIENTS: ETHANOLAMINE   1.0%   TLV: 3 ppm   CAS#: 141-43-5

IGNI TEMP: NA FP: NA LEL: NA UEL: NA VP: NA VD: NA SG: 1.12 PS: LIQUID APPEAR: GREEN   ODOR: FLORAL
    PH FACTOR: 9

HAZARD RATINGS: H: F: R: 2 0 0

FIRE FIGHTING:
    HIGH EXPANSION FOAM, LOW EXPANSION FOAM, ALCOHOL FOAM; DRY CHEMICAL, CARBON DIOXIDE, WATER FOG.
    PROTECTIVE CLOTHING, RUBBER GLOVES, AND BREATHING APPARATUS.
WARNING: 1] STRUCTURAL PROTECTIVE CLOTHING IS PERMEABLE, REMAIN CLEAR OF SMOKE, WATER FALL OUT AND WATER RUN OFF.
        2] KEEP OUT OF THE REACH OF CHILDREN.
        3] THERMAL DECOMPOSITION MAY YIELD CARBON DIOXIDE, CARBON MONOXIDE.
        4] MOVE CONTAINERS FROM AREA IF WITHOUT RISK, COOL EXPOSED CONTAINERS.
        5] DIKE AREA FOR CONTROL AND CONTAINMENT TO PREVENT ENTRY INTO SEWERS, DRAIN, AND WATER WAYS.

LARGE SPILL/NO FIRE/RESCUE: WEAR RUBBER OR NEOPRENE BOOTS, GLOVES.

SPILL CONTROL AND CONTAINMENT:
    HOUSEHOLD SPILL: WIPE UP WITH AN ABSORBENT CLOTH, RINSE OUT IN THE SINK. WIPE UP RESIDUE WITH A DAMP CLOTH.
    RINSE OUT IN THE SINK.
    LARGE SPILL: PICK UP WITH ABSORBENT. PUT IN A DISPOSAL CONTAINER. RINSE AREA WITH WATER. PICK UP WITH ABSORBENT.
    PUT IN A DISPOSAL CONTAINER. SEAL AND REMOVE FROM THE WORK AREA.

HEALTH HAZARD INFORMATION:
    WHEN USING THIS PRODUCT WEAR: NO SPECIAL REQUIREMENT. Eye contact will cause stinging, tearing, itching,
    swelling, and redness. Prolonged contact may cause drying of the skin and irritation. Ingestion may cause
    abdominal pain, nausea, vomiting, and diarrhea. Inhalation may cause irritation of the upper respiratory tract.

    A physician should be contacted if anyone develops any signs or symptoms and suspects that they are caused by
    exposure to this product.

FIRST AID:
    Eye Exposure:
    Flush with water for 15 minutes while lifting the upper and lower eye lids. Contact lenses should not be worn
    when working with this product. Get medical attention.

    Skin Exposure:
    Remove contaminated clothing and wash with water. If irritation appears, get medical attention.

    Breathing:
    If breathing becomes labored, move victim to fresh air. If difficulty persists, get medical attention.

    Swallowing:
    If victim is conscious, give 1-2 glasses of milk or water to drink. Immediately contact the local poison control
    center for advice. Keep the victim warm and at rest. Get medical attention.

PRODUCT: SURF POWDER LAUNDRY DETERGENT, NON-PHOSPHATE

INCOMPATIBILITY: NONE KNOWN

INGREDIENTS: SODIUM SILICATE   1.0% TLV: CAS#: 1344-09-8
             SODIUM CARBONATE   1.0% TLV: CAS#: 497-19-8

IGNI TEMP: NA FP: NA LEL: NA UEL: NA VP: NA VD: NA SG: NA PS: POWDER APPEAR: WHITE   ODOR:
   PH FACTOR:

HAZARD RATINGS: H: F: R: 2 0 0

FIRE FIGHTING:
  HIGH EXPANSION FOAM, LOW EXPANSION FOAM, ALCOHOL FOAM; DRY CHEMICAL, CARBON DIOXIDE, WATER FOG.
  PROTECTIVE CLOTHING, RUBBER GLOVES, AND BREATHING APPARATUS.
WARNING: 1] STRUCTURAL PROTECTIVE CLOTHING IS PERMEABLE, REMAIN CLEAR OF SMOKE, WATER FALL OUT AND WATER RUN OFF.
         2] KEEP OUT OF THE REACH OF CHILDREN.
         3] THERMAL DECOMPOSITION MAY YIELD OXIDES OF SULFUR.
         4] MOVE CONTAINERS FROM AREA IF WITHOUT RISK, COOL EXPOSED CONTAINERS.
         5] DIKE AREA FOR CONTROL AND CONTAINMENT TO PREVENT ENTRY INTO SEWERS, DRAIN, AND WATER WAYS.

LARGE SPILL/NO FIRE/RESCUE: WEAR RUBBER OR NEOPRENE BOOTS, GLOVES.

SPILL CONTROL AND CONTAINMENT:
  HOUSEHOLD SPILL: AVOID CREATING A DUST. GENTLY SWEEP UP POWDER AND PUT IN A DISPOSAL CONTAINER. WIPE UP RESIDUE
  WITH A DAMP CLOTH, RINSE OUT IN THE SINK.
  LARGE SPILL: AVOID CREATING A DUST. GENTLY SWEEP UP POWDER. PUT IN A DISPOSAL CONTAINER. RINSE AREA WITH WATER.
  PICK UP WITH AN ABSORBENT. PUT IN A DISPOSAL CONTAINER. SEAL AND REMOVE FROM THE WORK AREA.

HEALTH HAZARD INFORMATION:
  WHEN USING THIS PRODUCT WEAR: NO SPECIAL REQUIREMENT. Eye contact will cause stinging, tearing, itching,
  swelling, and redness. Prolonged contact may cause drying of the skin and irritation. Ingestion of sufficient
  amounts will cause abdominal cramps, nausea, and vomiting. Inhalation may cause irritation of the respiratory
  tract.

  A physician should be contacted if anyone develops any signs or symptoms and suspects that they are caused by
  exposure to this product.

FIRST AID:
  Eye Exposure:
  Flush with water for 15 minutes while lifting the upper and lower eye lids. Contact lenses should not be worn
  when working with this product. Get medical attention.

  Skin Exposure:
  Wash with water. If irritation develops, get medical attention.

  Breathing:
  If breathing becomes labored, move victim to fresh air. If persistent get medical attention.

  Swallowing:
  If the victim is conscious, give 1-2 glasses of milk or water to drink. Immediately contact the local poison
  control center for advice. Keep the victim warm and at rest. Get medical attention.

PRODUCT: TACKLE

INCOMPATIBILITY: TOILET BOWL CLEANERS; AMMONIA; RUST REMOVERS; VINEGAR; ACIDS; PRODUCTS CONTAINING AMMONIA.

INGREDIENTS: SODIUM HYDROXIDE   .5-2% TLV: 2 mg/m3   CAS#: 1310-73-2
             SODIUM HYPOCHLORITE   2-5% TLV: CAS#: 7681-52-9

IGNI TEMP: NA FP: NA LEL: NA UEL: NA VP: NA VD: NA SG: 1.034 PS: LIQUID APPEAR: CLEAR YELLOW
   ODOR: CHLORINE   PH FACTOR: 12.4-12.8

HAZARD RATINGS: H: F: R: 2 0 1   HAZARD CLASS 8

FIRE FIGHTING:
   HIGH EXPANSION FOAM, LOW EXPANSION FOAM, ALCOHOL FOAM; DRY CHEMICAL, CARBON DIOXIDE, WATER FOG.
   PROTECTIVE CLOTHING, RUBBER GLOVES, AND BREATHING APPARATUS.
WARNING: 1] STRUCTURAL PROTECTIVE CLOTHING IS PERMEABLE, REMAIN CLEAR OF SMOKE, WATER FALL OUT AND WATER RUN OFF.
         2] KEEP OUT OF THE REACH OF CHILDREN.
         3] THERMAL DECOMPOSITION RELEASES SODIUM CHLORATE A CORROSIVE/OXIDIZER
         4] DO NOT MIX WITH PRODUCTS LISTED ABOVE AS INCOMPATIBLE. HAZARDOUS GASES WILL BE GENERATED.
         5] MOVE CONTAINERS FROM AREA IF WITHOUT RISK, COOL EXPOSED CONTAINERS.
         6] DIKE AREA FOR CONTROL AND CONTAINMENT TO PREVENT ENTRY INTO SEWERS, DRAIN, AND WATER WAYS.

LARGE SPILL/NO FIRE/RESCUE: WEAR RUBBER OR NEOPRENE BOOTS, GLOVES.

SPILL CONTROL AND CONTAINMENT:
   HOUSEHOLD SPILL: WIPE UP WITH AN ABSORBENT CLOTH. RINSE OUT IN THE SINK. WIPE UP RESIDUE WITH A DAMP CLOTH.
   LARGE SPILL: PICK UP WITH ABSORBENT. RINSE AREA WITH WATER. PICK UP WITH ABSORBENT. PUT ALL ABSORBENT IN A
   DISPOSAL CONTAINER. SEAL AND REMOVE FROM THE WORK AREA.

HEALTH HAZARD INFORMATION:
   WHEN USING THIS PRODUCT WEAR: RUBBER GLOVES. Skin contact causes irritation. Eye contact causes moderate
   stinging, tearing, swelling, and redness. Ingestion may cause abdominal cramps, nausea, and vomiting. Inhalation
   of vapors or mist may cause irritation of the nose, throat, and lungs.

   A physician should be contacted if anyone develops any signs or symptoms and suspects that they are caused by
   exposure to this product.

FIRST AID:
   Eye Exposure:
   Flush with water for 15 minutes while lifting the upper and lower eye lids. Contact lenses should not be worn
   when working with this product. Get medical attention.

   Skin Exposure:
   Wash with soap and water. If irritation develops, get medical attention.

   Breathing:
   If irritation develops, move the victim to fresh air.

   Swallowing:
   If the victim is conscious, give 2-3 glasses of water to drink. Immediately contact the local poison control
   center for advice. Keep the victim warm and at rest. Get medical attention.

PRODUCT: TIDE (LAUNDRY GRANULES)

INCOMPATIBILITY: NONE KNOWN

INGREDIENTS: SODIUM PYROPHOSPHATE   TLV: 5 mg/m3   CAS#: 7722-88-5
             SODIUM PHOSPHATE   TLV: CAS#:
             SODIUM CARBONATE   TLV: CAS#:   497-19-8
             SODIUM SULFATE   TLV: CAS#: 7757-82-6
             SODIUM SILICATES   TLV: CAS#: 1344-09-8
             ANIONIC SURFACTANTS, ENZYME, PERFUME

IGNI TEMP: NA FP: NA LEL: NA UEL: NA VP: NA VD: NA SG: NA PS: POWDER APPEAR: WHITE GRANULES
    ODOR: WITH OR WITHOUT PERFUME   PH FACTOR: NA

HAZARD RATINGS: H: F: R:

FIRE FIGHTING:
    HIGH EXPANSION FOAM, LOW EXPANSION FOAM, ALCOHOL FOAM; DRY CHEMICAL, CARBON DIOXIDE, WATER FOG.
    PROTECTIVE CLOTHING, RUBBER GLOVES, AND BREATHING APPARATUS.
WARNING: 1] STRUCTURAL PROTECTIVE CLOTHING IS PERMEABLE, REMAIN CLEAR OF SMOKE, WATER FALL OUT AND WATER RUN OFF.
         2] KEEP OUT OF THE REACH OF CHILDREN.
         3] THERMAL DECOMPOSITION MAY YIELD CARBON DIOXIDE, CARBON MONOXIDE.
         4] MOVE CONTAINERS FROM AREA IF WITHOUT RISK, COOL EXPOSED CONTAINERS.
         5] DIKE AREA FOR CONTROL AND CONTAINMENT TO PREVENT ENTRY INTO SEWERS, DRAIN, AND WATER WAYS.

LARGE SPILL/NO FIRE/RESCUE: WEAR RUBBER OR NEOPRENE BOOTS, GLOVES.

SPILL CONTROL AND CONTAINMENT:
    HOUSEHOLD SPILL: SWEEP UP POWDER. DAMP MOP UP RESIDUE.
    LARGE SPILL: AVOID CREATING A DUST. GENTLY SWEEP UP POWDER AND PUT IN A DISPOSAL CONTAINER. RINSE AREA WITH
    WATER. PICK UP WITH ABSORBENT. PUT IN A DISPOSAL CONTAINER. SEAL AND REMOVE FROM THE WORK AREA.

HEALTH HAZARD INFORMATION:
    WHEN USING THIS PRODUCT WEAR: NO SPECIAL REQUIREMENT. Inhalation may cause coughing, sore throat, wheezing, and
    shortness of breath. Eye contact will cause stinging, tearing, swelling, and redness. Ingestion may result in
    abdominal cramps, nausea, and vomiting.

    A physician should be contacted if anyone develops any signs or symptoms and suspects that they are caused by
    exposure to this product.

FIRST AID:
Eye Exposure:
Flush with water for 15 minutes while lifting the upper and lower eye lids. Contact lenses should not be worn
when working with this product. Get medical attention.

Skin Exposure:
Wash with soap and water.If irritation develops, get medical attention.

Breathing:
If irritation develops, move victim to fresh air.

Swallowing:
If the victim is conscious, give 2-3 glasses of water to drink. Immediately contact the local poison control
center for advice. Keep the victim warm and at rest. Get medical attention.

PRODUCT: TIDE, LIQUID (LAUNDRY DETERGENT)

INCOMPATIBILITY: ACIDS AND ACID FUMES

INGREDIENTS: ANIONIC SURFACTANTS   TLV: CAS#:
             NONIONIC SURFACTANTS   TLV: CAS#:
             CATIONIC SURFACTANTS   TLV: CAS#:
             ETHYL ALCOHOL   TLV: 1900 mg/m3   CAS#:   64-17-5
             WATER   TLV: CAS#: 7732-18-5

IGNI TEMP: NA FP: 200F LEL: NA UEL: NA VP: NA VD: NA SG: 1.1 PS: LIQUID APPEAR: DARK AMBER
     ODOR: MAY OR MAY NOT HAVE PERFUME   PH FACTOR: NA

HAZARD RATINGS: H: F: R:

FIRE FIGHTING:
  HIGH EXPANSION FOAM, LOW EXPANSION FOAM, ALCOHOL FOAM; DRY CHEMICAL, CARBON DIOXIDE, WATER FOG.
  PROTECTIVE CLOTHING, RUBBER GLOVES, AND BREATHING APPARATUS.
WARNING: 1] STRUCTURAL PROTECTIVE CLOTHING IS PERMEABLE, REMAIN CLEAR OF SMOKE, WATER FALL OUT AND WATER RUN OFF.
         2] KEEP OUT OF THE REACH OF CHILDREN.
         3] THERMAL DECOMPOSITION MAY YIELD CARBON DIOXIDE, CARBON MONOXIDE.
         4] PRODUCT IS AN AQUEOUS SOLUTION CONTAINING ETHYL ALCOHOL AND DOES NOT SUSTAIN COMBUSTION.
         5] MOVE CONTAINERS FROM AREA IF WITHOUT RISK, COOL EXPOSED CONTAINERS.
         6] DIKE AREA FOR CONTROL AND CONTAINMENT TO PREVENT ENTRY INTO SEWERS, DRAIN, AND WATER WAYS.

LARGE SPILL/NO FIRE/RESCUE: WEAR RUBBER OR NEOPRENE BOOTS, GLOVES.

SPILL CONTROL AND CONTAINMENT:
  HOUSEHOLD SPILL: WIPE UP WITH AN ABSORBENT CLOTH, RINSE OUT IN THE SINK. WIPE UP RESIDUE WITH A DAMP CLOTH,
  RINSE OUT IN THE SINK.
  LARGE SPILL: PICK UP WITH ABSORBENT, PUT IN A DISPOSAL CONTAINER. RINSE AREA WITH WATER, PICK UP WITH ABSORBENT.
  PUT IN A DISPOSAL CONTAINER, SEAL AND REMOVE FROM THE WORK AREA.

HEALTH HAZARD INFORMATION:
  WHEN USING THIS PRODUCT WEAR: NO SPECIAL REQUIREMENT. Eye contact may cause stinging, tearing, itching,
  swelling, and redness. Prolonged contact with concentrated liquid may cause drying of the skin and irritation.
  Ingestion may result in abdominal pain, nausea, vomiting, and diarrhea.

  A physician should be contacted if anyone develops any signs or symptoms and suspects that they are caused by
  exposure to this product.

FIRST AID:
  Eye Exposure:
  Flush with water for 15 minutes while lifting the upper and lower eye lids. Contact lenses should not be worn
  when working with this product. Get medical attention.

  Skin Exposure:
  Wash off with water. If irritation develops, get medical attention.

  Breathing:
  Not expected to be a problem.

  Swallowing:
  If victim is conscious, give 2-3 glasses of milk or water to drink. Immediately contact the local poison control
  center for advice. Keep the victim warm and at rest. Get medical attention.

PRODUCT: TIDE, ULTRA (LAUNDRY GRANULES)

INCOMPATIBILITY: NONE KNOWN

INGREDIENTS: SODIUM CARBONATE   TLV: CAS#: 497-19-8
             SODIUM SILICATES   TLV: CAS#: 1344-09-8
             SODIUM SULFATE   TLV: CAS#: 7757-82-6
             ANIONIC SURFACTANT, SODIUM PHOSPHATES, ALUMINOSILICATE, ENZYME, PERFUME

IGNI TEMP: NA FP: NA LEL: NA UEL: NA VP: NA VD: NA SG: NA PS: POWDER APPEAR: WHITE
     ODOR: WITH OR WITHOUT PERFUME   PH FACTOR: NA

HAZARD RATINGS: H: F: R:

FIRE FIGHTING:
   HIGH EXPANSION FOAM, LOW EXPANSION FOAM, ALCOHOL FOAM; DRY CHEMICAL, CARBON DIOXIDE, WATER FOG.
   PROTECTIVE CLOTHING, RUBBER GLOVES, AND BREATHING APPARATUS.
WARNING: 1] STRUCTURAL PROTECTIVE CLOTHING IS PERMEABLE, REMAIN CLEAR OF SMOKE, WATER FALL OUT AND WATER RUN OFF.
         2] KEEP OUT OF THE REACH OF CHILDREN.
         3] THERMAL DECOMPOSITION MAY YIELD CARBON DIOXIDE, CARBON MONOXIDE.
         4] MOVE CONTAINERS FROM AREA IF WITHOUT RISK, COOL EXPOSED CONTAINERS.
         5] DIKE AREA FOR CONTROL AND CONTAINMENT TO PREVENT ENTRY INTO SEWERS, DRAIN, AND WATER WAYS.

LARGE SPILL/NO FIRE/RESCUE: WEAR RUBBER OR NEOPRENE BOOTS, GLOVES.

SPILL CONTROL AND CONTAINMENT:
   HOUSEHOLD SPILL: SWEEP UP POWDER. PUT IN A DISPOSAL CONTAINER. DAMP MOP UP RESIDUE. RINSE OUT IN THE SINK.
   LARGE SPILL: AVOID CREATING A DUST. SCOOP UP POWDER. PUT IN A DISPOSAL CONTAINER. RINSE AREA WITH WATER. PICK UP
   WITH ABSORBENT. PUT IN A DISPOSAL CONTAINER. SEAL AND REMOVE FROM THE WORK AREA.

HEALTH HAZARD INFORMATION:
   WHEN USING THIS PRODUCT WEAR: NO SPECIAL REQUIREMENT. Inhalation may cause coughing, sore throat, wheezing, and
   shortness of breath. Eye contact will cause stinging, tearing, itching, swelling, and redness. Ingestion may
   result in nausea, vomiting, and diarrhea. Skin contact may cause irritation.

   A physician should be contacted if anyone develops any signs or symptoms and suspects that they are caused by
   exposure to this product.

FIRST AID:
Eye Exposure:
Flush with water for 15 minutes while lifting the upper and lower eye lids. Contact lenses should not be worn
when working with this product. Get medical attention.

Skin Exposure:
Wash with water. If irritation develops, get medical attention.

Breathing:
If irritation develops, move the victim to fresh air.

Swallowing:
If the victim is conscious, give 2-3 glasses of milk or water to drink. Immediately contact the local poison
control center for advice. Keep the victim warm and at rest. Get medical attention.

PRODUCT: TIDE, ULTRA   (O - P)   (LAUNDRY GRANULES)

INCOMPATIBILITY: NONE KNOWN

INGREDIENTS: SODIUM CARBONATE   TLV: CAS#: 497-19-8
             SODIUM SILICATES   TLV: CAS#: 1344-09-8
             SODIUM SULFATE   TLV: CAS#: 7757-82-6
             ANIONIC SURFACTANT, SODIUM PHOSPHATES, ALUMINOSILICATE, ENZYME, PERFUME

IGNI TEMP: NA FP: NA LEL: NA UEL: NA VP: NA VD: NA SG: NA PS: POWDER APPEAR: WHITE
   ODOR: WITH OR WITHOUT PERFUME   PH FACTOR: NA

HAZARD RATINGS: H: F: R:

FIRE FIGHTING:
  HIGH EXPANSION FOAM, LOW EXPANSION FOAM, ALCOHOL FOAM; DRY CHEMICAL, CARBON DIOXIDE, WATER FOG.
  PROTECTIVE CLOTHING, RUBBER GLOVES, AND BREATHING APPARATUS.
WARNING: 1] STRUCTURAL PROTECTIVE CLOTHING IS PERMEABLE, REMAIN CLEAR OF SMOKE, WATER FALL OUT AND WATER RUN OFF.
         2] KEEP OUT OF THE REACH OF CHILDREN.
         3] THERMAL DECOMPOSITION MAY YIELD CARBON DIOXIDE, CARBON MONOXIDE.
         4] MOVE CONTAINERS FROM AREA IF WITHOUT RISK, COOL EXPOSED CONTAINERS.
         5] DIKE AREA FOR CONTROL AND CONTAINMENT TO PREVENT ENTRY INTO SEWERS, DRAIN, AND WATER WAYS.

LARGE SPILL/NO FIRE/RESCUE: WEAR RUBBER OR NEOPRENE BOOTS, GLOVES.

SPILL CONTROL AND CONTAINMENT:
  HOUSEHOLD SPILL: SWEEP UP POWDER. PUT IN A DISPOSAL CONTAINER. DAMP MOP UP RESIDUE. RINSE OUT IN THE SINK.
  LARGE SPILL: AVOID CREATING A DUST. SCOOP UP POWDER. PUT IN A DISPOSAL CONTAINER. RINSE AREA WITH WATER. PICK UP
  WITH ABSORBENT. PUT IN A DISPOSAL CONTAINER. SEAL AND REMOVE FROM THE WORK AREA.

HEALTH HAZARD INFORMATION:
  WHEN USING THIS PRODUCT WEAR: NO SPECIAL REQUIREMENT. Inhalation may cause coughing, sore throat, wheezing, and
  shortness of breath. Eye contact will cause stinging, tearing, itching, swelling, and redness. Ingestion may
  result in nausea, vomiting, and diarrhea. Skin contact may cause irritation.

  A physician should be contacted if anyone develops any signs or symptoms and suspects that they are caused by
  exposure to this product.

FIRST AID:
  Eye Exposure:
  Flush with water for 15 minutes while lifting the upper and lower eye lids. Contact lenses should not be worn
  when working with this product. Get medical attention.

  Skin Exposure:
  Wash with water. If irritation develops, get medical attention.

  Breathing:
  If irritation develops, move the victim to fresh air.

  Swallowing:
  If the victim is conscious, give 2-3 glasses of milk or water to drink. Immediately contact the local poison
  control center for advice. Keep the victim warm and at rest. Get medical attention.

PRODUCT: TIDE WITH BLEACH ALTERNATIVE, LIQUID (LAUNDRY DETERGENT)

INCOMPATIBILITY: ACIDS AND ACID FUMES.

INGREDIENTS: ANIONIC SURFACTANTS   TLV: CAS#:
             NONIONIC SURFACTANTS   TLV: CAS#:
             CATIONIC SURFACTANTS   TLV: CAS#:
             ETHYL ALCOHOL   TLV: 1900 mg/m3 CAS#:   64-17-5
             WATER   TLV: CAS#: 7732-18-5

IGNI TEMP: NA FP: 200F LEL: NA UEL: NA VP: NA VD: NA SG: 1.1 PS: LIQUID APPEAR: DARK AMBER
   ODOR: PERFUMED   PH FACTOR: NA

HAZARD RATINGS: H: F: R:

FIRE FIGHTING:
   HIGH EXPANSION FOAM, LOW EXPANSION FOAM, ALCOHOL FOAM; DRY CHEMICAL, CARBON DIOXIDE, WATER FOG.
   PROTECTIVE CLOTHING, RUBBER GLOVES, AND BREATHING APPARATUS.
WARNING: 1] STRUCTURAL PROTECTIVE CLOTHING IS PERMEABLE, REMAIN CLEAR OF SMOKE, WATER FALL OUT AND WATER RUN OFF.
         2] KEEP OUT OF THE REACH OF CHILDREN.
         3] THERMAL DECOMPOSITION MAY YIELD CARBON DIOXIDE, CARBON MONOXIDE.
         4] PRODUCT IS AN AQUEOUS SOLUTION CONTAINING ETHYL ALCOHOL AND DOES NOT SUSTAIN COMBUSTION.
         5] MOVE CONTAINERS FROM AREA IF WITHOUT RISK, COOL EXPOSED CONTAINERS.
           6] DIKE AREA FOR CONTROL AND CONTAINMENT TO PREVENT ENTRY INTO SEWERS, DRAIN, AND WATER WAYS.

LARGE SPILL/NO FIRE/RESCUE: WEAR RUBBER OR NEOPRENE BOOTS, GLOVES.

SPILL CONTROL AND CONTAINMENT:
   HOUSEHOLD SPILL: WIPE UP WITH AN ABSORBENT CLOTH, RINSE OUT IN THE SINK. WIPE UP RESIDUE WITH A DAMP CLOTH,
   RINSE OUT IN THE SINK.
   LARGE SPILL: PICK UP WITH ABSORBENT, PUT IN A DISPOSAL CONTAINER. RINSE AREA WITH WATER, PICK UP WITH ABSORBENT.
   PUT IN A DISPOSAL CONTAINER, SEAL AND REMOVE FROM THE WORK AREA.

HEALTH HAZARD INFORMATION:
   WHEN USING THIS PRODUCT WEAR: NO SPECIAL REQUIREMENT. Eye contact will cause stinging, tearing, itching,
   swelling, and redness. Prolonged contact with concentrated liquid may cause drying of the skin and irritation.
   Ingestion may result in abdominal pain, nausea, vomiting, and diarrhea.

   A physician should be contacted if anyone develops any signs or symptoms and suspects that they are caused by
   exposure to this product.

FIRST AID:
   Eye Exposure:
   Flush with water for 15 minutes while lifting the upper and lower eye lids. Contact lenses should not be worn
   when working with this product. Get medical attention.

   Skin Exposure:
   Wash skin with water. If irritation develops, get medical attention.

   Breathing:
   Not expected to be a problem.

   Swallowing:
   If the victim is conscious, give 2-3 glasses of milk or water to drink. Immediately contact the local poison
   control center for advice. Keep the victim warm and at rest. Get medical attention.

PRODUCT: TIDE WITH BLEACH (LAUNDRY GRANULES)

INCOMPATIBILITY: NONE KNOWN

INGREDIENTS: SODIUM PYROPHOSPHATE   TLV: 5 mg/m3   CAS#: 7722-88-5
             SODIUM PHOSPHATE   TLV: CAS#:
             SODIUM CARBONATE   TLV: CAS#:   497-19-8
             SODIUM SULFATE   TLV: CAS#: 7757-82-6
             SODIUM SILICATES   TLV: CAS#: 1344-09-8
             SODIUM PERBORATE   TLV: CAS#: 10486-00-7
             ANIONIC SURFACTANTS, ENZYME, PERFUME

IGNI TEMP: NA FP: NA LEL: NA UEL: NA VP: NA VD: NA SG: NA PS: POWDER APPEAR: WHITE WITH BLUE SPECKS
     ODOR: PERFUME   PH FACTOR: NA

HAZARD RATINGS: H: F: R:

FIRE FIGHTING:
   HIGH EXPANSION FOAM, LOW EXPANSION FOAM, ALCOHOL FOAM; DRY CHEMICAL, CARBON DIOXIDE, WATER FOG.
   PROTECTIVE CLOTHING, RUBBER GLOVES, AND BREATHING APPARATUS.
WARNING: 1] STRUCTURAL PROTECTIVE CLOTHING IS PERMEABLE, REMAIN CLEAR OF SMOKE, WATER FALL OUT AND WATER RUN OFF.
         2] KEEP OUT OF THE REACH OF CHILDREN.
         3] THERMAL DECOMPOSITION MAY YIELD CARBON DIOXIDE, CARBON MONOXIDE.
         4] AVOID EXPOSURE TO HIGH DUST CONCENTRATIONS.
         5] MOVE CONTAINERS FROM AREA IF WITHOUT RISK, COOL EXPOSED CONTAINERS.
         6] DIKE AREA FOR CONTROL AND CONTAINMENT TO PREVENT ENTRY INTO SEWERS, DRAIN, AND WATER WAYS.

LARGE SPILL/NO FIRE/RESCUE: WEAR RUBBER OR NEOPRENE BOOTS, GLOVES.

SPILL CONTROL AND CONTAINMENT:
   HOUSEHOLD SPILL: SWEEP UP POWDER. DAMP MOP UP RESIDUE.
   LARGE SPILL: AVOID CREATING A DUST. GENTLY SWEEP UP POWDER AND PUT IN A DISPOSAL CONTAINER. RINSE AREA WITH
   WATER. PICK UP WITH ABSORBENT. PUT IN A DISPOSAL CONTAINER. SEAL AND REMOVE FROM THE WORK AREA.

HEALTH HAZARD INFORMATION:
   WHEN USING THIS PRODUCT WEAR: NO SPECIAL REQUIREMENT. Inhalation may cause coughing, sore throat, wheezing, and
   shortness of breath. Eye contact will cause stinging, tearing, swelling, and redness. Ingestion may result in
   abdominal cramps, nausea, and vomiting.

   A physician should be contacted if anyone develops any signs or symptoms and suspects that they are caused by
   exposure to this product.

FIRST AID:
   Eye Exposure:
   Flush with water for 15 minutes while lifting the upper and lower eye lids. Contact lenses should not be worn
   when working with this product. Get medical attention.

   Skin Exposure:
   Wash with soap and water.If irritation develops, get medical attention.

   Breathing:
   If irritation develops, move victim to fresh air.

   Swallowing:
   If the victim is conscious, give 2-3 glasses of water to drink. Immediately contact the local poison control
   center for advice. Keep the victim warm and at rest. Get medical attention.

PRODUCT: TIDE WITH BLEACH, ULTRA (LAUNDRY GRANULES)

INCOMPATIBILITY: NONE KNOWN

INGREDIENTS: SODIUM PERBORATE   TLV: CAS#: 10486-00-7
             SODIUM CARBONATE   TLV: CAS#:   497-19-8
             SODIUM SULFATE   TLV: CAS#: 7757-82-6
             SODIUM SILICATE   TLV: CAS#: 1344-09-8
             ANIONIC SURFACTANTS, SODIUM PHOSPHATE, ALUMINOSILICATE, ENZYME, PERFUME

IGNI TEMP: NA FP: NA LEL: NA UEL: NA VP: NA VD: NA SG: NA PS: POWDER APPEAR: WHITE - BLUE SPECKS
     ODOR: PERFUMED   PH FACTOR: NA

HAZARD RATINGS: H: F: R:

FIRE FIGHTING:
  HIGH EXPANSION FOAM, LOW EXPANSION FOAM, ALCOHOL FOAM; DRY CHEMICAL, CARBON DIOXIDE, WATER FOG.
  PROTECTIVE CLOTHING, RUBBER GLOVES, AND BREATHING APPARATUS.
WARNING: 1] STRUCTURAL PROTECTIVE CLOTHING IS PERMEABLE, REMAIN CLEAR OF SMOKE, WATER FALL OUT AND WATER RUN OFF.
         2] KEEP OUT OF THE REACH OF CHILDREN.
         3] THERMAL DECOMPOSITION MAY YIELD CARBON DIOXIDE, CARBON MONOXIDE.
         4] MOVE CONTAINERS FROM AREA IF WITHOUT RISK, COOL EXPOSED CONTAINERS.
         5] DIKE AREA FOR CONTROL AND CONTAINMENT TO PREVENT ENTRY INTO SEWERS, DRAIN, AND WATER WAYS.

LARGE SPILL/NO FIRE/RESCUE: WEAR RUBBER OR NEOPRENE BOOTS, GLOVES.

SPILL CONTROL AND CONTAINMENT:
  HOUSEHOLD SPILL: SWEEP UP POWDER. PUT IN A DISPOSAL CONTAINER. DAMP MOP UP RESIDUE. RINSE OUT IN THE SINK.
  LARGE SPILL: AVOID CREATING A DUST. SCOOP UP POWDER. PUT IN A DISPOSAL CONTAINER. RINSE AREA WITH WATER. PICK UP
  WITH ABSORBENT. PUT IN A DISPOSAL CONTAINER. SEAL AND REMOVE FROM THE WORK AREA.

HEALTH HAZARD INFORMATION:
  WHEN USING THIS PRODUCT WEAR: NO SPECIAL REQUIREMENT. Inhalation may cause coughing, sore throat, wheezing, and
  shortness of breath. Eye contact will cause stinging, tearing, itching, swelling, and redness. Skin contact may
  cause irritation. Ingestion may result in nausea, vomiting, and diarrhea.

  A physician should be contacted if anyone develops any signs or symptoms and suspects that they are caused by
  exposure to this product.

FIRST AID:
  Eye Exposure:
  Flush with water for 15 minutes while lifting the upper and lower eye lids. Contact lenses should not be worn
  when working with this product. Get medical attention.

  Skin Exposure:
  Wash with water. If irritation develops, get medical attention.

  Breathing:
  If irritation develops, move the victim to fresh air.

  Swallowing:
  If the victim is conscious, give 2-3 glasses of milk or water to drink. Immediately contact the local poison
  control center for advice. Keep the victim warm and at rest. Get medical attention.

PRODUCT: TILEX

INCOMPATIBILITY: TOILET BOWL CLEANERS; AMMONIA; RUST REMOVERS; VINEGAR; ACIDS; PRODUCTS CONTAINING AMMONIA.

INGREDIENTS: SODIUM HYDROXIDE   .5-2% TLV: 2 mg/m3   CAS#: 1310-73-2
            SODIUM HYPOCHLORITE   2-5% TLV: CAS#: 7681-52-9

IGNI TEMP: NA FP: NA LEL: NA UEL: NA VP: NA VD: NA SG: 1.034 PS: LIQUID APPEAR: CLEAR YELLOW
    ODOR: CHLORINE   PH FACTOR: 12.4-12.8

HAZARD RATINGS: H: F: R: 2 0 1   HAZARD CLASS   8

FIRE FIGHTING:
  HIGH EXPANSION FOAM, LOW EXPANSION FOAM, ALCOHOL FOAM; DRY CHEMICAL, CARBON DIOXIDE, WATER FOG.
  PROTECTIVE CLOTHING, RUBBER GLOVES, AND BREATHING APPARATUS.
WARNING: 1] STRUCTURAL PROTECTIVE CLOTHING IS PERMEABLE, REMAIN CLEAR OF SMOKE, WATER FALL OUT AND WATER RUN OFF.
         2] KEEP OUT OF THE REACH OF CHILDREN.
         3] THERMAL DECOMPOSITION RELEASES SODIUM CHLORATE A CORROSIVE/OXIDIZER
         4] DO NOT MIX WITH PRODUCTS LISTED ABOVE AS INCOMPATIBLE. HAZARDOUS GASES WILL BE GENERATED.
         5] MOVE CONTAINERS FROM AREA IF WITHOUT RISK, COOL EXPOSED CONTAINERS.
         6] DIKE AREA FOR CONTROL AND CONTAINMENT TO PREVENT ENTRY INTO SEWERS, DRAIN, AND WATER WAYS.

LARGE SPILL/NO FIRE/RESCUE: WEAR RUBBER OR NEOPRENE BOOTS, GLOVES.

SPILL CONTROL AND CONTAINMENT:
  HOUSEHOLD SPILL: WIPE UP WITH AN ABSORBENT CLOTH. RINSE OUT IN THE SINK. WIPE UP RESIDUE WITH A DAMP CLOTH.
  LARGE SPILL: PICK UP WITH ABSORBENT. RINSE AREA WITH WATER. PICK UP WITH ABSORBENT. PUT ALL ABSORBENT IN A
  DISPOSAL CONTAINER. SEAL AND REMOVE FROM THE WORK AREA.

HEALTH HAZARD INFORMATION:
  WHEN USING THIS PRODUCT WEAR: RUBBER GLOVES. Skin contact causes irritation. Eye contact causes moderate
  stinging, tearing, swelling, and redness. Ingestion may cause abdominal cramps, nausea, and vomiting. Inhalation
  of vapors or mist may cause irritation of the nose, throat, and lungs.

  A physician should be contacted if anyone develops any signs or symptoms and suspects that they are caused by
  exposure to this product.

FIRST AID:
  Eye Exposure:
  Flush with water for 15 minutes while lifting the upper and lower eye lids. Contact lenses should not be worn
  when working with this product. Get medical attention.

  Skin Exposure:
  Wash with soap and water. If irritation develops, get medical attention.

  Breathing:
  If irritation develops, move the victim to fresh air.

  Swallowing:
  If the victim is conscious, give 2-3 glasses of water to drink. Immediately contact the local poison control
  center for advice. Keep the victim warm and at rest. Get medical attention.

PRODUCT: TOP JOB - LIQUID HOUSEHOLD CLEANER (NON-PHOSPHATE)

INCOMPATIBILITY: CHLORINE BLEACHES; HALOGENS; SODIUM HYPOCHLORITE.

```
INGREDIENTS: ANIONIC SURFACTANTS      TLV:          CAS#:
             BUTYL DIGLYCOL           TLV:          CAS#:
             SODIUM CITRATE           TLV:          CAS#:  6132-04-3
             SODIUM CARBONATE         TLV:          CAS#:   497-19-8
             AMMONIA                  TLV:          CAS#:  7664-41-7
```

IGNI TEMP: NA FP: 200F LEL: NA UEL: NA VP: NA VD: NA SG: 1.2 PS: LIQUID APPEAR: AQUA COLORED
    ODOR: LAVENDER   PH FACTOR: NA

HAZARD RATINGS: H: F: R:

FIRE FIGHTING:
    HIGH EXPANSION FOAM, LOW EXPANSION FOAM, ALCOHOL FOAM; DRY CHEMICAL, CARBON DIOXIDE, WATER FOG.
    PROTECTIVE CLOTHING, RUBBER GLOVES, AND BREATHING APPARATUS.
WARNING: 1] STRUCTURAL PROTECTIVE CLOTHING IS PERMEABLE, REMAIN CLEAR OF SMOKE, WATER FALL OUT AND WATER RUN OFF.
         2] KEEP OUT OF THE REACH OF CHILDREN.
         3] CONTAINERS WILL MELT AND LEAK IN A FIRE.
         4] PRODUCT MUST BE PREHEATED FOR IGNITION TO OCCUR.
         5] THERMAL DECOMPOSITION MAY YIELD CARBON DIOXIDE, CARBON MONOXIDE.
         6] MOVE CONTAINERS FROM AREA IF WITHOUT RISK, COOL EXPOSED CONTAINERS.
         7] DIKE AREA FOR CONTROL AND CONTAINMENT TO PREVENT ENTRY INTO SEWERS, DRAIN, AND WATER WAYS.

LARGE SPILL/NO FIRE/RESCUE: WEAR RUBBER OR NEOPRENE BOOTS, GLOVES.

SPILL CONTROL AND CONTAINMENT:
    HOUSEHOLD SPILL: WIPE UP WITH AN ABSORBENT CLOTH. RINSE OUT IN THE SINK. WIPE UP RESIDUE WITH A DAMP CLOTH.
    RINSE OUT IN THE SINK.
    LARGE SPILL: PICK UP WITH ABSORBENT. RINSE AREA WITH WATER. PICK UP WITH ABSORBENT. PUT ALL ABSORBENT IN A
    DISPOSAL CONTAINER. SEAL AND REMOVE FROM THE WORK AREA.

HEALTH HAZARD INFORMATION:
    WHEN USING THIS PRODUCT WEAR: NO SPECIAL REQUIREMENT. Eye contact will cause mild stinging, tearing, and
    redness. Skin contact will cause mild irritation. Ingestion will cause mild abdominal cramps, nausea, and
    vomiting. Inhalation will cause irritation of the nose, throat, and lungs.

    A physician should be contacted if anyone develops any signs or symptoms and suspects that they are caused by
    exposure to this product.

FIRST AID:
    Eye Exposure:
    Flush with water for 15 minutes while lifting the upper and lower eye lids. Contact lenses should not be worn
    when working with this product. Get medical attention.

    Skin Exposure:
    Wash with soap and water. If irritation develops, get medical attention.

    Breathing:
    If irritation develops, move victim to fresh air. If it persists, get medical attention.

    Swallowing:
    If the victim is conscious, give 2-3 glasses of milk or water to drink. Immediately contact the local poison
    control center for advice. Keep the victim warm and at rest. Get medical attention.

PRODUCT: TOP JOB SPRAY - ALL PURPOSE CLEANER

INCOMPATIBILITY: CHLORINE BLEACHES; HALOGENS; SODIUM HYPOCHLORITE.

INGREDIENTS:		TLV:	CAS#:
ANIONIC SURFACTANTS		TLV:	CAS#:
BUTYL DIGLYCOL		TLV:	CAS#:
SODIUM CITRATE		TLV:	CAS#: 6132-04-3
SODIUM CARBONATE		TLV:	CAS#: 497-19-8
AMMONIA		TLV:	CAS#: 7664-41-7

IGNI TEMP: NA FP: 200F LEL: NA UEL: NA VP: NA VD: NA SG: 1.03 PS: LIQUID APPEAR: AQUA COLORED
    ODOR: LAVENDER   PH FACTOR: NA

HAZARD RATINGS: H: F: R:

FIRE FIGHTING:
  HIGH EXPANSION FOAM, LOW EXPANSION FOAM, ALCOHOL FOAM; DRY CHEMICAL, CARBON DIOXIDE, WATER FOG.
  PROTECTIVE CLOTHING, RUBBER GLOVES, AND BREATHING APPARATUS.
WARNING: 1] STRUCTURAL PROTECTIVE CLOTHING IS PERMEABLE, REMAIN CLEAR OF SMOKE, WATER FALL OUT AND WATER RUN OFF.
        2] KEEP OUT OF THE REACH OF CHILDREN.
        3] CONTAINERS WILL MELT AND LEAK IN A FIRE.
        4] PRODUCT MUST BE PREHEATED FOR IGNITION TO OCCUR.
        5] THERMAL DECOMPOSITION MAY YIELD CARBON DIOXIDE, CARBON MONOXIDE.
        6] MOVE CONTAINERS FROM AREA IF WITHOUT RISK, COOL EXPOSED CONTAINERS.
        7] DIKE AREA FOR CONTROL AND CONTAINMENT TO PREVENT ENTRY INTO SEWERS, DRAIN, AND WATER WAYS.

LARGE SPILL/NO FIRE/RESCUE: WEAR RUBBER OR NEOPRENE BOOTS, GLOVES.

SPILL CONTROL AND CONTAINMENT:
  HOUSEHOLD SPILL: WIPE UP WITH AN ABSORBENT CLOTH. RINSE OUT IN THE SINK. WIPE UP RESIDUE WITH A DAMP CLOTH.
  RINSE OUT IN THE SINK.
  LARGE SPILL: PICK UP WITH ABSORBENT. RINSE AREA WITH WATER. PICK UP WITH ABSORBENT. PUT ALL ABSORBENT IN A
  DISPOSAL CONTAINER. SEAL AND REMOVE FROM THE WORK AREA.

HEALTH HAZARD INFORMATION:
  WHEN USING THIS PRODUCT WEAR: NO SPECIAL REQUIREMENT. Eye contact will cause mild stinging, tearing, and
  redness. Skin contact will cause mild irritation. Ingestion will cause mild abdominal cramps, nausea, and
  vomiting. Inhalation will cause irritation of the nose, throat, and lungs.

  A physician should be contacted if anyone develops any signs or symptoms and suspects that they are caused by
  exposure to this product.

FIRST AID:
Eye Exposure:
Flush with water for 15 minutes while lifting the upper and lower eye lids. Contact lenses should not be worn
when working with this product. Get medical attention.

Skin Exposure:
Wash with soap and water. If irritation develops, get medical attention.

Breathing:
If irritation develops, move victim to fresh air. If it persists, get medical attention.

Swallowing:
If the victim is conscious, give 2-3 glasses of milk or water to drink. Immediately contact the local poison
control center for advice. Keep the victim warm and at rest. Get medical attention.

PRODUCT: TSP POWDER CLEANER

INCOMPATIBILITY: ACIDS.

INGREDIENTS: TRISODIUM PHOSPHATE   TLV: CAS#: 7601-54-9

IGNI TEMP: NA FP: NA LEL: NA UEL: NA VP: NA VD: NA SG: NA PS: POWDER APPEAR: WHITE GRANULAR
     ODOR: NONE   PH FACTOR: NA

HAZARD RATINGS: H: F: R:

FIRE FIGHTING:
   HIGH EXPANSION FOAM, LOW EXPANSION FOAM, ALCOHOL FOAM; DRY CHEMICAL, CARBON DIOXIDE, WATER FOG.
   PROTECTIVE CLOTHING, RUBBER GLOVES, AND BREATHING APPARATUS.
WARNING: 1] STRUCTURAL PROTECTIVE CLOTHING IS PERMEABLE, REMAIN CLEAR OF SMOKE, WATER FALL OUT AND WATER RUN OFF.
         2] KEEP OUT OF THE REACH OF CHILDREN.
         3] THERMAL DECOMPOSITION MAY YIELD CARBON DIOXIDE, CARBON MONOXIDE.
         4] MOVE CONTAINERS FROM AREA IF WITHOUT RISK, COOL EXPOSED CONTAINERS.
         5] DIKE AREA FOR CONTROL AND CONTAINMENT TO PREVENT ENTRY INTO SEWERS, DRAIN, AND WATER WAYS.

LARGE SPILL/NO FIRE/RESCUE: WEAR RUBBER OR NEOPRENE BOOTS, GLOVES.

SPILL CONTROL AND CONTAINMENT:
   HOUSEHOLD SPILL: SWEEP UP POWDER. PUT IN A DISPOSAL CONTAINER. DAMP MOP AREA TO PICK UP RESIDUE.
   LARGE SPILL: AVOID CREATING A DUST. GENTLY SWEEP UP AND PUT IN A DISPOSAL CONTAINER. RINSE AREA WITH WATER. PICK
   UP WITH ABSORBENT. PUT IN A DISPOSAL CONTAINER. SEAL AND REMOVE FROM THE WORK AREA.

HEALTH HAZARD INFORMATION:
   WHEN USING THIS PRODUCT WEAR: NO SPECIAL REQUIREMENT. Eye contact will cause stinging, tearing, itching,
   swelling, and redness. Skin contact may cause irritation. Ingestion may cause abdominal cramps, nausea, and
   vomiting.

   A physician should be contacted if anyone develops any signs or symptoms and suspects that they are caused by
   exposure to this product.

FIRST AID:
   Eye Exposure:
   Flush with water for 15 minutes while lifting the upper and lower eye lids. Contact lenses should not be worn
   when working with this product. Get medical attention.

   Skin Exposure:
   Wash with water. If irritation develops, get medical attention.

   Breathing:
   Not expected to be a problem.

   Swallowing:
   Immediately contact the local poison control center for advice. Keep the victim warm and at rest. Get medical
   attention.

PRODUCT: TWINKLE SILVER POLISH

INCOMPATIBILITY: NONE KNOWN

INGREDIENTS: NONE LISTED BY MANUFACTURER NON-HAZARDOUS UNDER OSHA 1910.1200

IGNI TEMP: NA FP: NA LEL: NA UEL: NA VP: NA VD: NA SG: 1.21 PS: PASTE APPEAR: LIGHT BLUE ODOR: SOAPY
  PH FACTOR: 9.7

HAZARD RATINGS: H: F: R:

FIRE FIGHTING:
  HIGH EXPANSION FOAM, LOW EXPANSION FOAM, ALCOHOL FOAM; DRY CHEMICAL, CARBON DIOXIDE, WATER FOG.
  PROTECTIVE CLOTHING, RUBBER GLOVES, AND BREATHING APPARATUS.
WARNING: 1] STRUCTURAL PROTECTIVE CLOTHING IS PERMEABLE, REMAIN CLEAR OF SMOKE, WATER FALL OUT AND WATER RUN OFF.
         2] KEEP OUT OF THE REACH OF CHILDREN.
         3] HAZARDOUS DECOMPOSITION PRODUCTS ARE NOT KNOWN.
         4] MOVE CONTAINERS FROM AREA IF WITHOUT RISK, COOL EXPOSED CONTAINERS.
         5] DIKE AREA FOR CONTROL AND CONTAINMENT TO PREVENT ENTRY INTO SEWERS, DRAIN, AND WATER WAYS.

LARGE SPILL/NO FIRE/RESCUE: WEAR RUBBER OR NEOPRENE BOOTS, GLOVES.

SPILL CONTROL AND CONTAINMENT:
  HOUSEHOLD SPILL: SCOOP OR SCRAPE UP AND PUT IN A   DISPOSAL CONTAINER. WIPE UP RESIDUE WITH A DAMP CLOTH. RINSE
  OUT IN THE SINK.
  LARGE SPILL: SCOOP OR SCRAPE UP PASTE. PUT IN A DISPOSAL CONTAINER. RINSE AREA WITH WATER. PICK UP WITH
  ABSORBENT. PUT IN A DISPOSAL CONTAINER. SEAL AND REMOVE FROM THE WORK AREA.

HEALTH HAZARD INFORMATION:
  WHEN USING THIS PRODUCT WEAR: WEAR RUBBER LATEX GLOVES. Eye contact will cause irritation. May cause skin
  irritation.

  A physician should be contacted if anyone develops any signs or symptoms and suspects that they are caused by
  exposure to this product.

FIRST AID:
  Eye Exposure:
  Flush with water for 15 minutes while lifting the upper and lower eye lids. Contact lenses should not be worn
  when working with this product. Get medical attention.

  Skin Exposure:
  Wash skin with soap and water. If irritation develops, get medical attention.

  Breathing:
  Not expected to be a problem.

  Swallowing:
  If the victim is conscious, give 2-3 glasses of milk or water to drink. Immediately   contact the local poison
  control center for advice. Keep the victim warm and at rest. Get medical attention.

PRODUCT: ULTRAGESIC LOTION

INCOMPATIBILITY: NONE KNOWN

INGREDIENTS: METHYL SALICYLATE   10-16%   TLV: CAS#: 119-36-8
            STEARIC ACID   1-3%   TLV: CAS#: 57-11-4
            MENTHOL   2-4%   TLV: CAS#: 89-78-1
            WATER   TLV: CAS#: 7732-18-5
            EMULSIFIERS   1-2%   TLV: CAS#:

IGNI TEMP: NA FP: NA LEL: NA UEL: NA VP: NA VD: NA SG: 1.0 PS: LIQUID APPEAR: WHITE ODOR: MENTHOL
   PH FACTOR: 6.0-7.5

HAZARD RATINGS: H: F: R:

FIRE FIGHTING:
   HIGH EXPANSION FOAM, LOW EXPANSION FOAM, ALCOHOL FOAM; DRY CHEMICAL, CARBON DIOXIDE, WATER FOG.
   PROTECTIVE CLOTHING, RUBBER GLOVES, AND BREATHING APPARATUS.
WARNING: 1] STRUCTURAL PROTECTIVE CLOTHING IS PERMEABLE, REMAIN CLEAR OF SMOKE, WATER FALL OUT AND WATER RUN OFF.
       2] KEEP OUT OF THE REACH OF CHILDREN.
       3] CONTAINERS WILL MELT AND LEAK IN A FIRE.
       4] THERMAL DECOMPOSITION MAY YIELD CARBON DIOXIDE, CARBON MONOXIDE.
       5] MOVE CONTAINERS FROM AREA IF WITHOUT RISK, COOL EXPOSED CONTAINERS.
       6] DIKE AREA FOR CONTROL AND CONTAINMENT TO PREVENT ENTRY INTO SEWERS, DRAIN, AND WATER WAYS.

LARGE SPILL/NO FIRE/RESCUE: WEAR RUBBER OR NEOPRENE BOOTS, GLOVES.

SPILL CONTROL AND CONTAINMENT:
   HOUSEHOLD SPILL: WIPE UP WITH AN ABSORBENT CLOTH. RINSE OUT IN THE SINK. WIPE UP RESIDUE WITH A DAMP CLOTH.
   RINSE OUT IN THE SINK.
   LARGE SPILL: PICK UP WITH ABSORBENT. RINSE AREA WITH WATER. PICK UP WITH ABSORBENT. PUT ALL ABSORBENT IN A
   DISPOSAL CONTAINER. SEAL AND REMOVE FROM THE WORK AREA.

HEALTH HAZARD INFORMATION:
   WHEN USING THIS PRODUCT WEAR: NO SPECIAL REQUIREMENT. Eye contact may cause stinging, tearing, itching,
   swelling, and redness. Skin contact may cause irritation.

   A physician should be contacted if anyone develops any signs or symptoms and suspects that they are caused by
   exposure to this product.

FIRST AID:
   Eye Exposure:
   Flush with water for 15 minutes while lifting the upper and lower eye lids. Contact lenses should not be worn
   when working with this product. Get medical attention.

   Skin Exposure:
   Wash with soap and water. If irritation develops, get medical attention.

   Breathing:
   Not expected to be a problem.

   Swallowing:
   Immediately contact the local poison control center for advice. Keep the victim warm and at rest. Get medical
   attention.

PRODUCT: VANISH BLUE AUTOMATIC TOILET BOWL CLEANER

INCOMPATIBILITY: NONE KNOWN

INGREDIENTS: NONIONIC SURFACTANT under 5% TLV: CAS#:

IGNI TEMP: NA FP: NA LEL: NA UEL: NA VP: 17.5 mmHg VD: NA SG: 1.0 PS: LIQUID   APPEAR: DARK BLUE
    ODOR: NON-DESCRIPTIVE   PH FACTOR: 9.0

HAZARD RATINGS: H: F: R: 1 0 0

FIRE FIGHTING:
   HIGH EXPANSION FOAM, LOW EXPANSION FOAM, ALCOHOL FOAM; DRY CHEMICAL, CARBON DIOXIDE, WATER FOG.
   PROTECTIVE CLOTHING, RUBBER GLOVES, AND BREATHING APPARATUS.
WARNING: 1] STRUCTURAL PROTECTIVE CLOTHING IS PERMEABLE, REMAIN CLEAR OF SMOKE, WATER FALL OUT AND WATER RUN OFF.
             2] KEEP OUT OF THE REACH OF CHILDREN.
             3] THERMAL DECOMPOSITION PRODUCTS ARE UNKNOWN.
             4] MOVE CONTAINERS FROM AREA IF WITHOUT RISK, COOL EXPOSED CONTAINERS.
             5] DIKE AREA FOR CONTROL AND CONTAINMENT TO PREVENT ENTRY INTO SEWERS, DRAIN, AND WATER WAYS.

LARGE SPILL/NO FIRE/RESCUE: WEAR RUBBER OR NEOPRENE BOOTS, GLOVES.

SPILL CONTROL AND CONTAINMENT:
   HOUSEHOLD SPILL: WIPE UP DRY WITH AN ABSORBENT CLOTH. RINSE OUT IN THE SINK. WIPE UP RESIDUE WITH A DAMP CLOTH.
   RINSE OUT IN THE SINK.
   LARGE SPILL: PICK UP WITH ABSORBENT. RINSE AREA WITH WATER. PICK UP WITH ABSORBENT. SEAL AND REMOVE FROM THE
   WORK AREA.

HEALTH HAZARD INFORMATION:
   WHEN USING THIS PRODUCT WEAR: NO SPECIAL REQUIREMENT. Eye contact will cause stinging, tearing, itching,
   swelling, and redness.

   A physician should be contacted if anyone develops any signs or symptoms and suspects that they are caused by
   exposure to this product.

FIRST AID:
   Eye Exposure:
   Flush with water for 15 minutes while lifting the upper and lower eye lids. Contact lenses should not be worn
   when working with this product. Get medical attention.

   Skin Exposure:
   Wash with soap and water. If irritation develops, get medical attention.

   Breathing:
   Not expected to be a problem.

   Swallowing:
   If the victim is conscious, give 2-3 glasses of milk or water to drink. Immediately contact the local poison
   control center for advice. Keep the victim warm and at rest. Get medical attention.

PRODUCT: VANISH BOWL FRESHENER

INCOMPATIBILITY: NONE KNOWN

INGREDIENTS: ALL ARE NON-HAZARDOUS UNDER OSHA 1910.1200

IGNI TEMP: NA FP: NA LEL: NA UEL: NA VP: NA VD: NA SG: 1.3 PS: SOLID APPEAR: VARIOUS COLORS
    ODOR: VARIOUS   PH FACTOR: 7.0

HAZARD RATINGS: H: F: R: 0 0 0

FIRE FIGHTING:
    HIGH EXPANSION FOAM, LOW EXPANSION FOAM, ALCOHOL FOAM; DRY CHEMICAL, CARBON DIOXIDE, WATER FOG.
    PROTECTIVE CLOTHING, RUBBER GLOVES, AND BREATHING APPARATUS.
WARNING: 1] STRUCTURAL PROTECTIVE CLOTHING IS PERMEABLE, REMAIN CLEAR OF SMOKE, WATER FALL OUT AND WATER RUN OFF.
         2] KEEP OUT OF THE REACH OF CHILDREN.
         3] THERMAL DECOMPOSITION PRODUCTS ARE UNKNOWN.
         4] MOVE CONTAINERS FROM AREA IF WITHOUT RISK, COOL EXPOSED CONTAINERS.
         5] DIKE AREA FOR CONTROL AND CONTAINMENT TO PREVENT ENTRY INTO SEWERS, DRAIN, AND WATER WAYS.

LARGE SPILL/NO FIRE/RESCUE: WEAR RUBBER OR NEOPRENE BOOTS, GLOVES.

SPILL CONTROL AND CONTAINMENT:
    HOUSEHOLD SPILL: SWEEP UP MATERIAL. PUT IN A DISPOSAL CONTAINER. WIPE UP RESIDUE WITH A DAMP CLOTH. RINSE OUT IN
    THE SINK.
    LARGE SPILL: SCOOP UP MATERIAL. PUT IN A DISPOSAL CONTAINER. RINSE AREA WITH WATER. PICK UP WITH ABSORBENT. PUT
    IN A DISPOSAL CONTAINER. SEAL AND REMOVE FROM THE WORK AREA.

HEALTH HAZARD INFORMATION:
    WHEN USING THIS PRODUCT WEAR: NO SPECIAL REQUIREMENT. Product is not expected to cause any health effects.

    A physician should be contacted if anyone develops any signs or symptoms and suspects that they are caused by
    exposure to this product.

FIRST AID:
Eye Exposure:
Flush with water for 15 minutes while lifting the upper and lower eye lids. Contact lenses should not be worn
when working with this product. Get medical attention.

Skin Exposure:
Wash with soap and water. If irritation develops, get medical attention.

Breathing:
Not expected to be a problem.

Swallowing:
If the victim is conscious, give 2-3 glasses of milk or water to drink. Immediately contact the local poison
control center for advice. Keep the victim warm and at rest. Get medical attention.

PRODUCT: VANISH CLEAR DROP-INS - TANK AUTOMATIC BOWL CLEANER

INCOMPATIBILITY: STRONG OXIDIZERS.

INGREDIENTS: N-ALKYL DIMETHYL
             AMMONIUM CHLORIDE under 14% TLV: CAS#: 68391-01-5

IGNI TEMP: NA FP: NA LEL: NA UEL: NA VP: NA VD: NA SG: 1.2 PS: SOLID APPEAR: SQUARE WHITE DISK   ODOR:
   PH FACTOR: 6.2

HAZARD RATINGS: H: F: R: 3 0 0   HAZARD CLASS 8

FIRE FIGHTING:
   HIGH EXPANSION FOAM, LOW EXPANSION FOAM, ALCOHOL FOAM; DRY CHEMICAL, CARBON DIOXIDE, WATER FOG.
   PROTECTIVE CLOTHING, RUBBER GLOVES, AND BREATHING APPARATUS.
WARNING: 1] STRUCTURAL PROTECTIVE CLOTHING IS PERMEABLE, REMAIN CLEAR OF SMOKE, WATER FALL OUT AND WATER RUN OFF.
         2] KEEP OUT OF THE REACH OF CHILDREN.
         3] HAZARDOUS DECOMPOSITION YIELDS TOXIC NITROUS OXIDES, AMMONIA VAPORS.
         4] DO NOT HANDLE OR REMOVE THE DISK ONCE PUT IN THE TANK.
         5] AVOID CONTACT WITH THE PRODUCT AND WET MATERIAL IN THE TANK. IT CAN CAUSE IRRITATION OF THE EYES AND
            SKIN.
         6] MOVE CONTAINERS FROM AREA IF WITHOUT RISK, COOL EXPOSED CONTAINERS.
         7] DIKE AREA FOR CONTROL AND CONTAINMENT TO PREVENT ENTRY INTO SEWERS, DRAIN, AND WATER WAYS.

LARGE SPILL/NO FIRE/RESCUE: WEAR RUBBER OR NEOPRENE BOOTS, GLOVES.

SPILL CONTROL AND CONTAINMENT:
   HOUSEHOLD SPILL: WEAR RUBBER GLOVES. SCOOP UP BROKEN DISK AND PARTS. PUT IN A DISPOSAL CONTAINER. WIPE UP ANY
   RESIDUE WITH A DAMP CLOTH. RINSE OUT IN THE SINK.
   LARGE SPILL: SCOOP UP MATERIAL. PUT IN A DISPOSAL CONTAINER. RINSE AREA WITH WATER. PICK UP WITH ABSORBENT. PUT
   IN A DISPOSAL CONTAINER. SEAL AND REMOVE FROM THE WORK AREA.

HEALTH HAZARD INFORMATION:
   WHEN USING THIS PRODUCT WEAR: RUBBER GLOVES. PRODUCT IS A CORROSIVE. Skin contact may cause irritation and
   burns. Eye contact will cause pain, redness, swelling, and tissue damage. Ingestion will cause burns to the
   mouth, tongue, throat, and stomach. Spontaneous vomiting may occur.

   A physician should be contacted if anyone develops any signs or symptoms and suspects that they are caused by
   exposure to this product.

FIRST AID:
   Eye Exposure:
   Flush with water for 15 minutes while lifting the upper and lower eye lids. Contact lenses should not be worn
   when working with this product. Get medical attention.

   Skin Exposure:
   Flush skin with water. Wash with soap and water. If irritation develops, get medical attention.

   Breathing:
   Not expected to be a problem.

   Swallowing:
   If the victim is conscious, rinse out the mouth, give 2-3 glasses of milk or water to drink. Immediately contact
   the local poison control center for advice. Keep the victim warm and at rest. Get medical attention.

PRODUCT: VANISH CRYSTAL TOILET BOWL CLEANER

INCOMPATIBILITY: METALS; CHLORINE COMPOUNDS e.g. BLEACH.

INGREDIENTS: SODIUM BISULFATE   under 70% TLV: CAS#: 7681-38-1

IGNI TEMP: NA FP: NA LEL: NA UEL: NA VP: NA VD: NA SG: 2.0 PS: CRYSTALS APPEAR: WHITE AND BLUE
     ODOR: WINTERGREEN   PH FACTOR: under 1

HAZARD RATINGS: H: F: R: 3 0 0 NO WATER   HAZARD CLASS 8

FIRE FIGHTING:
   HIGH EXPANSION FOAM, LOW EXPANSION FOAM, ALCOHOL FOAM; DRY CHEMICAL, CARBON DIOXIDE, WATER FOG.
   PROTECTIVE CLOTHING, RUBBER GLOVES, AND BREATHING APPARATUS.
WARNING: 1] STRUCTURAL PROTECTIVE CLOTHING IS PERMEABLE, REMAIN CLEAR OF SMOKE, WATER FALL OUT AND WATER RUN OFF.
         2] KEEP OUT OF THE REACH OF CHILDREN.
         3] IF INVOLVED IN A FIRE USE DRY SAND, EARTH, SALT, SODA ASH, LIME.
         4] REACTS WITH WATER PRODUCING HEAT AND TOXIC FUMES OF SULFURIC ACID.
         5] MOVE CONTAINERS FROM AREA IF WITHOUT RISK, COOL EXPOSED CONTAINERS.
         6] DIKE AREA FOR CONTROL AND CONTAINMENT TO PREVENT ENTRY INTO SEWERS, DRAIN, AND WATER WAYS.

LARGE SPILL/NO FIRE/RESCUE: WEAR NON-SEALED CHEMICAL PROTECTIVE CLOTHING, RUBBER BOOTS, GLOVES, BREATHING
                           APPARATUS.

SPILL CONTROL AND CONTAINMENT:
   HOUSEHOLD SPILL: WEAR RUBBER GLOVES. SWEEP UP CRYSTALS. PUT IN THE TOILET BOWL. WIPE UP ANY RESIDUE USING A
   SOLUTION OF (2 TABLESPOONS OF BAKING SODA TO A GLASS OF WATER) ON A CLOTH. RINSE OUT IN THE SINK. REWIPE AREA
   USING A DAMP CLOTH. RINSE OUT IN THE SINK.
   LARGE SPILL: AVOID CREATING A DUST. SCOOP UP CRYSTALS. PUT IN A DISPOSAL CONTAINER. RINSE AREA WITH WATER. PICK
   UP WITH ABSORBENT. PUT IN A DISPOSAL CONTAINER. SEAL AND REMOVE FROM THE WORK AREA.

HEALTH HAZARD INFORMATION:
   WHEN USING THIS PRODUCT WEAR: NO SPECIAL REQUIREMENT. Product is corrosive to the eyes and skin, throat, and
   stomach. Eye contact will cause pain, stinging and tissue damage. Skin contact may cause burns. Ingestion will
   cause burns to the mouth, tongue, throat, and stomach, with difficulty swallowing.

   A physician should be contacted if anyone develops any signs or symptoms and suspects that they are caused by
   exposure to this product.

FIRST AID:
   Eye Exposure:
   Flush with water for 15 minutes while lifting the upper and lower eye lids. Contact lenses should not be worn
   when working with this product. Get medical attention.

   Skin Exposure:
   Remove contaminated clothing. Flush with water for 15 minutes. If irritation develops, get medical attention.

   Breathing:
   Not expected to be a problem.

   Swallowing:
   Do not induce vomiting. If the victim is conscious, rinse out the mouth, give victim 2-3 glasses of milk or
   water to drink. Immediately contact the local poison control center for advice. Keep the victim warm and at
   rest. Get medical attention.

PRODUCT: VANISH DISINFECTANT THICK HEAVY DUTY TOILET BOWL CLEANER

INCOMPATIBILITY: METALS; CHLORINE COMPOUNDS e.g. BLEACH; HOUSEHOLD CLEANERS AMMONIA; ALUMINUM; PRODUCTS WITH
               AMMONIA.

INGREDIENTS: HYDROCHLORIC ACID   under 10% TLV: 5 ppm (ceil) CAS#: 7647-01-0

IGNI TEMP: NA FP: NA LEL: NA UEL: NA VP: 20 mmHg VD: 1 SG: 1.04 PS: LIQUID   APPEAR: THICK BLUE
   ODOR: SPICY MINT PH FACTOR: under 1

HAZARD RATINGS: H: F: R: 3 0 0   HAZARD CLASS   8

FIRE FIGHTING:
   HIGH EXPANSION FOAM, LOW EXPANSION FOAM, DRY CHEMICAL, WATER FOG.
   PROTECTIVE CLOTHING, RUBBER GLOVES, AND BREATHING APPARATUS.
WARNING: 1] STRUCTURAL PROTECTIVE CLOTHING IS PERMEABLE, REMAIN CLEAR OF SMOKE, WATER FALL OUT AND WATER RUN OFF.
         2] KEEP OUT OF THE REACH OF CHILDREN.
         3] THERMAL DECOMPOSITION PRODUCTS ARE NOT KNOWN.
         4] MOVE CONTAINERS FROM AREA IF WITHOUT RISK, COOL EXPOSED CONTAINERS.
         5] DIKE AREA FOR CONTROL AND CONTAINMENT TO PREVENT ENTRY INTO SEWERS, DRAIN, AND WATER WAYS.

LARGE SPILL/NO FIRE/RESCUE: WEAR NON SEALED CHEMICAL PROTECTIVE CLOTHING, RUBBER BOOTS, GLOVES, BREATHING
                           APPARATUS.

SPILL CONTROL AND CONTAINMENT:
   HOUSEHOLD SPILL: WEAR RUBBER GLOVES. SCOOP UP AS MUCH AS POSSIBLE. PUT IN TOILET. MIX A SOLUTION OF
   (2 TABLESPOONS OF BAKING SODA IN A GLASS OF WATER). SOAK A CLOTH IN IT. WIPE UP REMAINING PRODUCT. RINSE OUT IN
   THE SINK. WIPE UP WITH A DAMP CLOTH. RINSE OUT IN THE SINK.
   LARGE SPILL: CONTAIN AND NEUTRALIZE WITH SODA ASH OR LIME. SCOOP UP AND PUT IN A DISPOSAL CONTAINER. RINSE AREA
   WITH WATER. PICK UP WITH ABSORBENT. PUT IN A DISPOSAL CONTAINER. SEAL AND REMOVE FROM THE WORK AREA.

HEALTH HAZARD INFORMATION:
   WHEN USING THIS PRODUCT WEAR: CORROSIVE: DAMAGING TO EYES, SKIN, MOUTH, THROAT, STOMACH TISSUE. Skin contact
   will cause severe pain and damage to the tissue. Eye contact will cause severe pain and damage to the tissue.
   Inhalation may cause coughing and burning sensation in the nasal passage. Ingestion may cause burns to the
   mouth, tongue, throat, and stomach. Spontaneous vomiting may occur.

   A physician should be contacted if anyone develops any signs or symptoms and suspects that they are caused by
   exposure to this product.

FIRST AID:
   Eye Exposure:
   Flush with water for 15 minutes while lifting the upper and lower eye lids. Contact lenses should not be worn
   when working with this product. Get medical attention.

   Skin Exposure:
   Remove contaminated clothing. Flush with water. If irritation develops, get medical attention.

   Breathing:
   If irritation develops, move the victim to fresh air.

   Swallowing:
   Do not induce vomiting. If victim is conscious, rinse out the mouth, give 2-3 glasses of milk or water to drink.
   Immediately contact the local poison control center for advice. Keep the victim warm and at rest. Get medical
   attention.

PRODUCT: VANISH DROP-INS - SOLID AUTOMATIC TOILET BOWL CLEANER (BLUE)

INCOMPATIBILITY: NONE KNOWN

INGREDIENTS: ALL ARE NON-HAZARDOUS UNDER OSHA 1910.1200 STANDARD

IGNI TEMP: NA FP: NA LEL: NA UEL: NA VP: NA VD: NA SG: 1.3 PS: SOLID APPEAR: DARK BLUE   ODOR: WINTERGREEN
    PH FACTOR: 6.1

HAZARD RATINGS: H: F: R: 0 0 0

FIRE FIGHTING:
    HIGH EXPANSION FOAM, LOW EXPANSION FOAM, ALCOHOL FOAM; DRY CHEMICAL, CARBON DIOXIDE, WATER FOG.
    PROTECTIVE CLOTHING, RUBBER GLOVES, AND BREATHING APPARATUS.
WARNING: 1] STRUCTURAL PROTECTIVE CLOTHING IS PERMEABLE, REMAIN CLEAR OF SMOKE, WATER FALL OUT AND WATER RUN OFF.
         2] KEEP OUT OF THE REACH OF CHILDREN.
         3] HAZARDOUS DECOMPOSITION PRODUCTS ARE UNKNOWN.
         4] MOVE CONTAINERS FROM AREA IF WITHOUT RISK, COOL EXPOSED CONTAINERS.
         5] DIKE AREA FOR CONTROL AND CONTAINMENT TO PREVENT ENTRY INTO SEWERS, DRAIN, AND WATER WAYS.

LARGE SPILL/NO FIRE/RESCUE: WEAR RUBBER OR NEOPRENE BOOTS, GLOVES.

SPILL CONTROL AND CONTAINMENT:
    HOUSEHOLD SPILL: SWEEP UP MATERIAL. PUT IN A DISPOSAL CONTAINER. WIPE UP RESIDUE WITH A DAMP CLOTH. RINSE OUT IN
    THE SINK. WASH HANDS WITH SOAP AND WATER.
    LARGE SPILL: SCOOP UP MATERIAL. PUT IN A DISPOSAL CONTAINER. RINSE AREA WITH WATER. PICK UP WITH ABSORBENT. PUT
    IN A DISPOSAL CONTAINER. SEAL AND REMOVE FROM THE WORK AREA.

HEALTH HAZARD INFORMATION:
    WHEN USING THIS PRODUCT WEAR: NO SPECIAL REQUIREMENT. Eye contact may cause stinging, tearing, itching,
    swelling, and redness. Skin contact may cause irritation. Ingestion may cause nausea and vomiting.

    A physician should be contacted if anyone develops any signs or symptoms and suspects that they are caused by
    exposure to this product.

FIRST AID:
    Eye Exposure:
    Flush with water for 15 minutes while lifting the upper and lower eye lids. Contact lenses should not be worn
    when working with this product. Get medical attention.

    Skin Exposure:
    Wash with soap and water. If irritation develops, get medical attention.

    Breathing:
    Not expected to be a problem.

    Swallowing:
    If the victim is conscious, give 2-3 glasses of milk or water to drink. Immediately contact the local poison
    control center for advice. Keep the victim warm and at rest. Get medical attention.

PRODUCT: VANISH DROP-INS - SOLID AUTOMATIC TOILET BOWL CLEANER (GREEN)

INCOMPATIBILITY: NONE KNOWN

INGREDIENTS: ALL ARE NON-HAZARDOUS UNDER OSHA 1910.1200 STANDARD

IGNI TEMP: NA FP: NA LEL: NA UEL: NA VP: NA VD: NA SG: 1.3 PS: SOLID APPEAR: DARK GREEN ODOR: WINTERGREEN
    PH FACTOR: 8.0

HAZARD RATINGS: H: F: R: 0 0 0

FIRE FIGHTING:
  HIGH EXPANSION FOAM, LOW EXPANSION FOAM, ALCOHOL FOAM; DRY CHEMICAL, CARBON DIOXIDE, WATER FOG.
  PROTECTIVE CLOTHING, RUBBER GLOVES, AND BREATHING APPARATUS.
WARNING: 1] STRUCTURAL PROTECTIVE CLOTHING IS PERMEABLE, REMAIN CLEAR OF SMOKE, WATER FALL OUT AND WATER RUN OFF.
         2] KEEP OUT OF THE REACH OF CHILDREN.
         3] HAZARDOUS DECOMPOSITION PRODUCTS ARE UNKNOWN.
         4] MOVE CONTAINERS FROM AREA IF WITHOUT RISK, COOL EXPOSED CONTAINERS.
         5] DIKE AREA FOR CONTROL AND CONTAINMENT TO PREVENT ENTRY INTO SEWERS, DRAIN, AND WATER WAYS.

LARGE SPILL/NO FIRE/RESCUE: WEAR RUBBER OR NEOPRENE BOOTS, GLOVES.

SPILL CONTROL AND CONTAINMENT:
  HOUSEHOLD SPILL: SWEEP UP MATERIAL. PUT IN A DISPOSAL CONTAINER. WIPE UP RESIDUE WITH A DAMP CLOTH. RINSE OUT IN
  THE SINK. WASH HANDS WITH SOAP AND WATER.
  LARGE SPILL: SCOOP UP MATERIAL. PUT IN A DISPOSAL CONTAINER. RINSE AREA WITH WATER. PICK UP WITH ABSORBENT. PUT
  IN A DISPOSAL CONTAINER. SEAL AND REMOVE FROM THE WORK AREA.

HEALTH HAZARD INFORMATION:
  WHEN USING THIS PRODUCT WEAR: NO SPECIAL REQUIREMENT. Eye contact may cause stinging, tearing, itching,
  swelling, and redness. Skin contact may cause irritation. Ingestion may cause nausea and vomiting.

  A physician should be contacted if anyone develops any signs or symptoms and suspects that they are caused by
  exposure to this product.

FIRST AID:
  Eye Exposure:
  Flush with water for 15 minutes while lifting the upper and lower eye lids. Contact lenses should not be worn
  when working with this product. Get medical attention.

  Skin Exposure:
  Wash with soap and water. If irritation develops, get medical attention.

  Breathing:
  Not expected to be a problem.

  Swallowing:
  If the victim is conscious, give 2-3 glasses of milk or water to drink. Immediately contact the local poison
  control center for advice. Keep the victim warm and at rest. Get medical attention.

PRODUCT: VANISH DROP-INS - SOLID AUTOMATIC TOILET BOWL CLEANER (GREEN) (CANADIAN)

INCOMPATIBILITY: OXIDIZERS.

INGREDIENTS: GLYCERIN   UNDER   10%   TLV: 10 mg/m3   CAS#: 56-81-5

IGNI TEMP: NA FP: NA LEL: NA UEL: NA VP: NA VD: NA SG: 1.3 PS: SOLID APPEAR: DARK GREEN ODOR: NONE
    PH FACTOR: 8.0

HAZARD RATINGS: H: F: R: 1 1 0

FIRE FIGHTING:
   HIGH EXPANSION FOAM, LOW EXPANSION FOAM, ALCOHOL FOAM; DRY CHEMICAL, CARBON DIOXIDE, WATER FOG.
   PROTECTIVE CLOTHING, RUBBER GLOVES, AND BREATHING APPARATUS.
WARNING: 1] STRUCTURAL PROTECTIVE CLOTHING IS PERMEABLE, REMAIN CLEAR OF SMOKE, WATER FALL OUT AND WATER RUN OFF.
         2] KEEP OUT OF THE REACH OF CHILDREN.
         3] HAZARDOUS DECOMPOSITION PRODUCTS ARE UNKNOWN.
         4] MOVE CONTAINERS FROM AREA IF WITHOUT RISK, COOL EXPOSED CONTAINERS.
         5] DIKE AREA FOR CONTROL AND CONTAINMENT TO PREVENT ENTRY INTO SEWERS, DRAIN, AND WATER WAYS.

LARGE SPILL/NO FIRE/RESCUE: WEAR RUBBER OR NEOPRENE BOOTS, GLOVES.

SPILL CONTROL AND CONTAINMENT:
   HOUSEHOLD SPILL: SWEEP UP MATERIAL. PUT IN A DISPOSAL CONTAINER. WIPE UP RESIDUE WITH A DAMP CLOTH. RINSE OUT IN
   THE SINK. WASH HANDS WITH SOAP AND WATER.
   LARGE SPILL: SCOOP UP MATERIAL. PUT IN A DISPOSAL CONTAINER. RINSE AREA WITH WATER. PICK UP WITH ABSORBENT. PUT
   IN A DISPOSAL CONTAINER. SEAL AND REMOVE FROM THE WORK AREA.

HEALTH HAZARD INFORMATION:
   WHEN USING THIS PRODUCT WEAR: NO SPECIAL REQUIREMENT. Eye contact may cause stinging, tearing, itching,
   swelling, and redness. Skin contact may cause irritation. Ingestion may cause nausea and vomiting.

   A physician should be contacted if anyone develops any signs or symptoms and suspects that they are caused by
   exposure to this product.

FIRST AID:
   Eye Exposure:
   Flush with water for 15 minutes while lifting the upper and lower eye lids. Contact lenses should not be worn
   when working with this product. Get medical attention.

   Skin Exposure:
   Wash with soap and water. If irritation develops, get medical attention.

   Breathing:
   Not expected to be a problem.

   Swallowing:
   If the victim is conscious, give 2-3 glasses of milk or water to drink. Immediately contact the local poison
   control center for advice. Keep the victim warm and at rest. Get medical attention.

PRODUCT: VANISH EXTRA STRENGTH FRAGRANCE BOWL FRESHENER & DEODORIZER

INCOMPATIBILITY: NONE KNOWN

INGREDIENTS: ALL ARE NON-HAZARDOUS UNDER OSHA 1910.1200 STANDARD

IGNI TEMP: NA FP: NA LEL: NA UEL: NA VP: NA VD: NA SG: 1.3 PS: SOLID APPEAR: COLORED ROD
    ODOR: STRONG PERFUME   PH FACTOR: 7.0

HAZARD RATINGS: H: F: R: 0 0 0

FIRE FIGHTING:
    HIGH EXPANSION FOAM, LOW EXPANSION FOAM, ALCOHOL FOAM; DRY CHEMICAL, CARBON DIOXIDE, WATER FOG.
    PROTECTIVE CLOTHING, RUBBER GLOVES, AND BREATHING APPARATUS.
WARNING: 1] STRUCTURAL PROTECTIVE CLOTHING IS PERMEABLE, REMAIN CLEAR OF SMOKE, WATER FALL OUT AND WATER RUN OFF.
         2] KEEP OUT OF THE REACH OF CHILDREN.
         3] HAZARDOUS DECOMPOSITION PRODUCTS ARE UNKNOWN.
         4] MOVE CONTAINERS FROM AREA IF WITHOUT RISK, COOL EXPOSED CONTAINERS.
         5] DIKE AREA FOR CONTROL AND CONTAINMENT TO PREVENT ENTRY INTO SEWERS, DRAIN, AND WATER WAYS.

LARGE SPILL/NO FIRE/RESCUE: WEAR RUBBER OR NEOPRENE BOOTS, GLOVES.

SPILL CONTROL AND CONTAINMENT:
    HOUSEHOLD SPILL: SWEEP UP MATERIAL. PUT IN A DISPOSAL CONTAINER. WIPE UP RESIDUE WITH A DAMP CLOTH. RINSE OUT IN
    THE SINK. WASH HANDS WITH SOAP AND WATER.
    LARGE SPILL: SCOOP UP MATERIAL. PUT IN A DISPOSAL CONTAINER. RINSE AREA WITH WATER. PICK UP WITH ABSORBENT. PUT
    IN A DISPOSAL CONTAINER. SEAL AND REMOVE FROM THE WORK AREA.

HEALTH HAZARD INFORMATION:
    WHEN USING THIS PRODUCT WEAR: NO SPECIAL REQUIREMENT. Eye contact may cause stinging, tearing, itching,
    swelling, and redness. Skin contact may cause irritation. Ingestion may cause nausea and vomiting.

    A physician should be contacted if anyone develops any signs or symptoms and suspects that they are caused by
    exposure to this product.

FIRST AID:
    Eye Exposure:
    Flush with water for 15 minutes while lifting the upper and lower eye lids. Contact lenses should not be worn
    when working with this product. Get medical attention.

    Skin Exposure:
    Wash with soap and water. If irritation develops, get medical attention.

    Breathing:
    Not expected to be a problem.

    Swallowing:
    If the victim is conscious, give 2-3 glasses of milk or water to drink. Immediately contact the local poison
    control center for advice. Keep the victim warm and at rest. Get medical attention.

PRODUCT: VANISH FOAMIN' TOILET BOWL CLEANER

INCOMPATIBILITY: NONE KNOWN

INGREDIENTS: PROPANE   TLV: 1000 ppm   CAS#:   74-98-6
            BUTANE   TLV: 800 ppm   CAS#:   106-97-8

IGNI TEMP: NA FP: 20F LEL: NA UEL: NA VP: 17 mmHg VD: 1 SG: 1.0 PS: LIQUID APPEAR: FOAM   ODOR: FLORAL
   PH FACTOR: 11.8

HAZARD RATINGS: H: F: R:

FIRE FIGHTING:
   HIGH EXPANSION FOAM, LOW EXPANSION FOAM, ALCOHOL FOAM; DRY CHEMICAL, CARBON DIOXIDE, WATER FOG.
   PROTECTIVE CLOTHING, RUBBER GLOVES, AND BREATHING APPARATUS.
WARNING: 1] STRUCTURAL PROTECTIVE CLOTHING IS PERMEABLE, REMAIN CLEAR OF SMOKE, WATER FALL OUT AND WATER RUN OFF.
        2] KEEP OUT OF THE REACH OF CHILDREN.
        3] REMOVE ALL IGNITION SOURCES IF WITHOUT RISK.
        4] CONTAINERS WILL EXPLODE IN A FIRE OR IF HEATED ABOVE 120F.
        5] THERMAL DECOMPOSITION PRODUCTS ARE UNKNOWN.
        6] MOVE CONTAINERS FROM AREA IF WITHOUT RISK, COOL EXPOSED CONTAINERS.
        7] DIKE AREA FOR CONTROL AND CONTAINMENT TO PREVENT ENTRY INTO SEWERS, DRAIN, AND WATER WAYS.

LARGE SPILL/NO FIRE/RESCUE: WEAR RUBBER OR NEOPRENE BOOTS, GLOVES.

SPILL CONTROL AND CONTAINMENT:
   HOUSEHOLD SPILL: WIPE UP DRY WITH AN ABSORBENT CLOTH. RINSE OUT IN THE SINK. WIPE UP RESIDUE WITH A DAMP CLOTH.
   RINSE OUT IN THE SINK.
   LARGE SPILL: PICK UP WITH ABSORBENT. RINSE AREA WITH WATER. PICK UP WITH ABSORBENT. PUT ALL ABSORBENT IN A
   DISPOSAL CONTAINER. SEAL AND REMOVE FROM THE WORK AREA.

HEALTH HAZARD INFORMATION:
   WHEN USING THIS PRODUCT WEAR: NO SPECIAL REQUIREMENT. Eye contact will cause stinging, tearing, itching, and
   redness. Ingestion may result in nausea, vomiting, and diarrhea.

   A physician should be contacted if anyone develops any signs or symptoms and suspects that they are caused by
   exposure to this product.

FIRST AID:
   Eye Exposure:
   Flush with water for 15 minutes while lifting the upper and lower eye lids. Contact lenses should not be worn
   when working with this product. Get medical attention.

   Skin Exposure:
   Wash with soap and water. If irritation develops, get medical attention.

   Breathing:
   Not expected to be a problem.

   Swallowing:
   If the victim is conscious, give 2-3 glasses of milk or water to drink. Immediately contact the local poison
   control center for advice. Keep the victim warm and at rest. Get medical attention.

PRODUCT: VANISH   GREEN AUTOMATIC TOILET CLEANER

INCOMPATIBILITY: NONE KNOWN

INGREDIENTS: NONIONIC SURFACTANT under 5% TLV: CAS#:

IGNI TEMP: NA FP: NA LEL: NA UEL: NA VP: 17,2 mmHg VD: 1 SG: 1.0 PS: LIQUID   APPEAR: DARK GREEN
    ODOR: PINE   PH FACTOR: 8.5

HAZARD RATINGS: H: F: R: 1 0 0

FIRE FIGHTING:
  HIGH EXPANSION FOAM, LOW EXPANSION FOAM, ALCOHOL FOAM; DRY CHEMICAL, CARBON DIOXIDE, WATER FOG.
  PROTECTIVE CLOTHING, RUBBER GLOVES, AND BREATHING APPARATUS.
WARNING: 1] STRUCTURAL PROTECTIVE CLOTHING IS PERMEABLE, REMAIN CLEAR OF SMOKE, WATER FALL OUT AND WATER RUN OFF.
         2] KEEP OUT OF THE REACH OF CHILDREN.
         3] HAZARDOUS DECOMPOSITION PRODUCTS ARE UNKNOWN.
         4] MOVE CONTAINERS FROM AREA IF WITHOUT RISK, COOL EXPOSED CONTAINERS.
         5] DIKE AREA FOR CONTROL AND CONTAINMENT TO PREVENT ENTRY INTO SEWERS, DRAIN, AND WATER WAYS.

LARGE SPILL/NO FIRE/RESCUE: WEAR RUBBER OR NEOPRENE BOOTS, GLOVES.

SPILL CONTROL AND CONTAINMENT:
  HOUSEHOLD SPILL: WIPE UP WITH AN ABSORBENT CLOTH. RINSE OUT IN THE SINK. WIPE UP RESIDUE WITH A DAMP CLOTH.
  RINSE OUT IN THE SINK. WASH HANDS WITH SOAP AND WATER.
  LARGE SPILL: PICK UP WITH ABSORBENT. RINSE AREA WITH WATER. PICK UP WITH ABSORBENT. PUT ALL ABSORBENT IN A
  DISPOSAL CONTAINER. SEAL AND REMOVE FROM THE WORK AREA.

HEALTH HAZARD INFORMATION:
  WHEN USING THIS PRODUCT WEAR: NO SPECIAL REQUIREMENT. Eye contact may cause stinging, tearing, itching,
  swelling, and redness. Skin contact may cause irritation. Ingestion may cause nausea and vomiting.

  A physician should be contacted if anyone develops any signs or symptoms and suspects that they are caused by
  exposure to this product.

FIRST AID:
  Eye Exposure:
  Flush with water for 15 minutes while lifting the upper and lower eye lids. Contact lenses should not be worn
  when working with this product. Get medical attention.

  Skin Exposure:
  Wash with soap and water. If irritation develops, get medical attention.

  Breathing:
  Not expected to be a problem.

  Swallowing:
  If the victim is conscious, give 2-3 glasses of milk or water to drink. Immediately contact the local poison
  control center for advice. Keep the victim warm and at rest. Get medical attention.

PRODUCT: VANISH   SOLID AUTOMATIC TOILET BOWL CLEANER

INCOMPATIBILITY: NONE KNOWN

INGREDIENTS: ALL ARE NON HAZARDOUS UNDER OSHA 1910.1200 STANDARD

IGNI TEMP: NA FP: NA LEL: NA UEL: NA VP: NA VD: NA SG: 1.4 PS: SOLID APPEAR: BLUE   ODOR: NON DESCRIPTIVE
    PH FACTOR: 6.0

HAZARD RATINGS: H: F: R: 0 0 0

FIRE FIGHTING:
   HIGH EXPANSION FOAM, LOW EXPANSION FOAM, ALCOHOL FOAM; DRY CHEMICAL, CARBON DIOXIDE, WATER FOG.
   PROTECTIVE CLOTHING, RUBBER GLOVES, AND BREATHING APPARATUS.
WARNING: 1] STRUCTURAL PROTECTIVE CLOTHING IS PERMEABLE, REMAIN CLEAR OF SMOKE, WATER FALL OUT AND WATER RUN OFF.
            2] KEEP OUT OF THE REACH OF CHILDREN.
            3] HAZARDOUS DECOMPOSITION PRODUCTS ARE UNKNOWN.
            4] MOVE CONTAINERS FROM AREA IF WITHOUT RISK, COOL EXPOSED CONTAINERS.
            5] DIKE AREA FOR CONTROL AND CONTAINMENT TO PREVENT ENTRY INTO SEWERS, DRAIN, AND WATER WAYS.

LARGE SPILL/NO FIRE/RESCUE: WEAR RUBBER OR NEOPRENE BOOTS, GLOVES.

SPILL CONTROL AND CONTAINMENT:
   HOUSEHOLD SPILL: SWEEP UP MATERIAL. PUT IN A DISPOSAL CONTAINER. WIPE UP RESIDUE WITH A DAMP CLOTH. RINSE OUT IN
   THE SINK. WASH HANDS WITH SOAP AND WATER.
   LARGE SPILL: SCOOP UP MATERIAL. PUT IN A DISPOSAL CONTAINER. RINSE AREA WITH WATER. PICK UP WITH ABSORBENT. PUT
   IN A DISPOSAL CONTAINER. SEAL AND REMOVE FROM THE WORK AREA.

HEALTH HAZARD INFORMATION:
   WHEN USING THIS PRODUCT WEAR: NO SPECIAL REQUIREMENT. Eye contact may cause stinging, tearing, itching,
   swelling, and redness. Skin contact may cause irritation. Ingestion may cause nausea and vomiting.

   A physician should be contacted if anyone develops any signs or symptoms and suspects that they are caused by
   exposure to this product.

FIRST AID:
   Eye Exposure:
   Flush with water for 15 minutes while lifting the upper and lower eye lids. Contact lenses should not be worn
   when working with this product. Get medical attention.

   Skin Exposure:
   Wash with soap and water. If irritation develops, get medical attention.

   Breathing:
   Not expected to be a problem.

   Swallowing:
   If the victim is conscious, give 2-3 glasses of milk or water to drink. Immediately contact the local poison
   control center for advice. Keep the victim warm and at rest. Get medical attention.

PRODUCT: VIDAL SASSOON AEROSOL PROTEIN ENRICHED HAIR SPRAY

INCOMPATIBILITY: OXIDIZERS.

INGREDIENTS: ISOBUTANE   TLV: CAS#: 75-28-5
            SD ALCOHOL 40   OCTYL ACRYLAMIDE
            AMINOMETHYL PROPANOL DIMETHICONE COPOLYOL   ACRYLATES
            BUTYLAMINO ETHYL METHACRYLATE COPOLYMER   CYCLOMETHICONE
            MYRISTOYL HYDROLIZED ANIMAL PROTEIN   OCTYLSALICYLATE
            PANTHENOL   KERATIN AMINO ACIDS FRAGRANCE

IGNI TEMP: NA FP: -117F LEL: 1.8 UEL: 8.4 VP: NA VD: 2 SG: .8 PS: LIQUID   APPEAR: CLEAR MIST
    ODOR: PERFUME   PH FACTOR: NA

HAZARD RATINGS: H: F: R: 1 4 0

FIRE FIGHTING:
  HIGH EXPANSION FOAM, LOW EXPANSION FOAM, ALCOHOL FOAM; DRY CHEMICAL, CARBON DIOXIDE, WATER FOG.
  PROTECTIVE CLOTHING, RUBBER GLOVES, AND BREATHING APPARATUS.
WARNING: 1] STRUCTURAL PROTECTIVE CLOTHING IS PERMEABLE, REMAIN CLEAR OF SMOKE, WATER FALL OUT AND WATER RUN OFF.
         2] KEEP OUT OF THE REACH OF CHILDREN.
         3] REMOVE ALL IGNITION SOURCES IF WITHOUT RISK.
         4] THERMAL DECOMPOSITION MAY YIELD CARBON DIOXIDE, CARBON MONOXIDE.
         5] MOVE CONTAINERS FROM AREA IF WITHOUT RISK, COOL EXPOSED CONTAINERS.
         6] DIKE AREA FOR CONTROL AND CONTAINMENT TO PREVENT ENTRY INTO SEWERS, DRAIN, AND WATER WAYS.

LARGE SPILL/NO FIRE/RESCUE: WEAR RUBBER OR NEOPRENE BOOTS, GLOVES.

SPILL CONTROL AND CONTAINMENT:
  HOUSEHOLD SPILL: WIPE UP WITH AN ABSORBENT CLOTH. RINSE OUT IN THE SINK. WIPE UP RESIDUE WITH A DAMP CLOTH.
  RINSE OUT IN THE SINK.
  LARGE SPILL: PICK UP WITH ABSORBENT. RINSE AREA WITH WATER. PICK UP WITH ABSORBENT. PUT ALL ABSORBENT IN A
  DISPOSAL CONTAINER. SEAL AND REMOVE FROM THE WORK AREA.

HEALTH HAZARD INFORMATION:
  WHEN USING THIS PRODUCT WEAR: NO SPECIAL REQUIREMENT. Eye contact will cause stinging, tearing, itching, and
  redness. Ingestion will result in nausea and vomiting. Inhalation may cause irritation of the nose.

  A physician should be contacted if anyone develops any signs or symptoms and suspects that they are caused by
  exposure to this product.

FIRST AID:
  Eye Exposure:
  Flush with water for 15 minutes while lifting the upper and lower eye lids. Contact lenses should not be worn
  when working with this product. Get medical attention.

  Skin Exposure:
  Wash with soap and water. If irritation develops, get medical attention.

  Breathing:
  If irritation develops, move victim to fresh air.

  Swallowing:
  If the victim is conscious, give 1-2 glasses of milk or water to drink. Immediately contact the local poison
  control center for advice. Keep the victim warm and at rest. Get medical attention.

PRODUCT: VIDAL SASSOON COLORIFIC GELS

INCOMPATIBILITY: NONE KNOWN

INGREDIENTS: WATER   TLV: CAS#: 7732-18-5
             ETHYL ALCOHOL   TLV: CAS#:   64-17-5
             TITANIUM DIOXIDE   TLV: CAS#: 13463-67-7
             TRIETHANOLAMINE   TLV: CAS#:   102-71-6
             POLYACRYLAMIDO METHYL PROPANE SULFONIC ACID
             FERRIC FERROCYANIDE, IRON OXIDE, CARMINE, FRAGRANCE, KATHON, GETEARETH-9, POYLQUATERIUM-4, MICA

IGNI TEMP: NA FP: 143F LEL: NA UEL: NA VP: NA VD: 1.6 SG: 1.02 PS: GEL APPEAR: VARIOUS COLORS
   ODOR: PERFUME   PH FACTOR: NA

HAZARD RATINGS: H: F: R:

FIRE FIGHTING:
  HIGH EXPANSION FOAM, LOW EXPANSION FOAM, ALCOHOL FOAM; DRY CHEMICAL, CARBON DIOXIDE, WATER FOG.
  PROTECTIVE CLOTHING, RUBBER GLOVES, AND BREATHING APPARATUS.
WARNING: 1] STRUCTURAL PROTECTIVE CLOTHING IS PERMEABLE, REMAIN CLEAR OF SMOKE, WATER FALL OUT AND WATER RUN OFF.
         2] KEEP OUT OF THE REACH OF CHILDREN.
         3] REMOVE ALL IGNITION SOURCES IF WITHOUT RISK.
         4] HAZARDOUS DECOMPOSITION PRODUCTS ARE UNKNOWN.
         5] MOVE CONTAINERS FROM AREA IF WITHOUT RISK, COOL EXPOSED CONTAINERS.
         6] DIKE AREA FOR CONTROL AND CONTAINMENT TO PREVENT ENTRY INTO SEWERS, DRAIN, AND WATER WAYS.

LARGE SPILL/NO FIRE/RESCUE: WEAR RUBBER OR NEOPRENE BOOTS, GLOVES.

SPILL CONTROL AND CONTAINMENT:
  HOUSEHOLD SPILL: WIPE UP WITH AN ABSORBENT CLOTH. RINSE OUT IN THE SINK. WIPE UP RESIDUE WITH A DAMP CLOTH.
  RINSE OUT IN THE SINK.
  LARGE SPILL: PICK UP WITH ABSORBENT. RINSE AREA WITH WATER. PICK UP WITH ABSORBENT. PUT ALL ABSORBENT IN A
  DISPOSAL CONTAINER. SEAL AND REMOVE FROM THE WORK AREA.

HEALTH HAZARD INFORMATION:
  WHEN USING THIS PRODUCT WEAR: NO SPECIAL REQUIREMENT. Eye contact will cause stinging, tearing, itching, and
  redness. Ingestion may result in nausea, vomiting, and diarrhea.

  A physician should be contacted if anyone develops any signs or symptoms and suspects that they are caused by
  exposure to this product.

FIRST AID:
  Eye Exposure:
  Flush with water for 15 minutes while lifting the upper and lower eye lids. Contact lenses should not be worn
  when working with this product. Get medical attention.

  Skin Exposure:
  If irritation develops discontinue use. If persistent, get medical attention.

  Breathing:
  Not expected to be a problem.

  Swallowing:
  If the victim is conscious, give 1-2 glasses of milk or water to drink. Immediately contact the local poison
  control center for advice. Keep the victim warm and at rest. Get medical attention.

PRODUCT: VIDAL SASSOON COLORIFIC MOUSSES

INCOMPATIBILITY: NONE KNOWN

INGREDIENTS: WATER   TLV: CAS#: 7732-18-5
             ISOBUTANE   TLV: CAS#:   75-28-5
             PROPYLENE GLYCOL   TLV: CAS#: 4254-15-3
             TETRASODIUM EDTA   TLV: CAS#: 10378-23-1
             METHYL PARABEN   TLV: CAS#:   99-76-3
             PROPYL PARABEN   TLV: CAS#:   94-13-3
             AMINO-METHYL PROPANOL, PEG-150 DISTEARETH, STEARETH-21, OCTYL SALICYLATE, COCO-BETAINE

IGNI TEMP: NA FP: -117F LEL: 1.8 UEL: 8.4 VP: NA VD: NA SG: 1 PS: LIQUID APPEAR: FOAM   ODOR: PERFUME
    PH FACTOR: NA

HAZARD RATINGS: H: F: R: 1 4 0

FIRE FIGHTING:
  HIGH EXPANSION FOAM, LOW EXPANSION FOAM, ALCOHOL FOAM; DRY CHEMICAL, CARBON DIOXIDE, WATER FOG.
  PROTECTIVE CLOTHING, RUBBER GLOVES, AND BREATHING APPARATUS.
WARNING: 1] STRUCTURAL PROTECTIVE CLOTHING IS PERMEABLE, REMAIN CLEAR OF SMOKE, WATER FALL OUT AND WATER RUN OFF.
         2] KEEP OUT OF THE REACH OF CHILDREN.
         3] REMOVE ALL IGNITION SOURCES IF WITHOUT RISK.
         4] HAZARDOUS DECOMPOSITION PRODUCTS ARE UNKNOWN.
         5] MOVE CONTAINERS FROM AREA IF WITHOUT RISK, COOL EXPOSED CONTAINERS.
         6] DIKE AREA FOR CONTROL AND CONTAINMENT TO PREVENT ENTRY INTO SEWERS, DRAIN, AND WATER WAYS.

LARGE SPILL/NO FIRE/RESCUE: WEAR RUBBER OR NEOPRENE BOOTS, GLOVES.

SPILL CONTROL AND CONTAINMENT:
  HOUSEHOLD SPILL: WIPE UP WITH AN ABSORBENT CLOTH. RINSE OUT IN THE SINK. WIPE UP RESIDUE WITH A DAMP CLOTH.
  RINSE OUT IN THE SINK.
  LARGE SPILL: PICK UP WITH ABSORBENT. RINSE AREA WITH WATER. PICK UP WITH ABSORBENT. PUT ALL ABSORBENT IN A
  DISPOSAL CONTAINER. SEAL AND REMOVE FROM THE WORK AREA.

HEALTH HAZARD INFORMATION:
  WHEN USING THIS PRODUCT WEAR: NO SPECIAL REQUIREMENT. Eye contact will cause stinging, tearing, itching, and
  redness. Ingestion may result in nausea, vomiting, and diarrhea.

  A physician should be contacted if anyone develops any signs or symptoms and suspects that they are caused by
  exposure to this product.

FIRST AID:
  Eye Exposure:
  Flush with water for 15 minutes while lifting the upper and lower eye lids. Contact lenses should not be worn
  when working with this product. Get medical attention.

  Skin Exposure:
  If irritation develops, discontinue use. If it persists, get medical attention.

  Breathing:
  Not expected to be a problem.

  Swallowing:
  If the victim is conscious, give 1-2 glasses of milk or water to drink. Immediately contact the local poison
  control center for advice. Keep the victim warm and at rest. Get medical attention.

PRODUCT: VIDAL SASSOON EXTRA BODY MOUSSE

INCOMPATIBILITY: NONE KNOWN

INGREDIENTS: WATER   TLV: CAS#: 7732-18-5
             PROPANE   TLV: CAS#:   74-98-6
             ISOBUTANE   TLV: CAS#:   75-28-5
             BUTANE   TLV: CAS#:   106-97-8
             PROPYLENE GLYCOL   TLV: CAS#: 4254-15-3
             CITRIC ACID   TLV: CAS#:   77-92-9
             DISODIUM EDTA   TLV: CAS#: 6381-92-6
             POYLQUATERIUM-4 LAURAMINE OXIDE   OCTYL SALICYLATE
             PANTHENOL   SILK AMINO ACID   ANIMAL KERATIN AMINO ACID

IGNI TEMP: NA FP: 20F LEL: NA UEL: NA VP: NA VD: 2 SG: 1 PS: LIQUID APPEAR: FOAM   ODOR: PERFUME
   PH FACTOR: NA

HAZARD RATINGS: H: F: R: 1 4 0

FIRE FIGHTING:
   HIGH EXPANSION FOAM, LOW EXPANSION FOAM, ALCOHOL FOAM; DRY CHEMICAL, CARBON DIOXIDE, WATER FOG.
   PROTECTIVE CLOTHING, RUBBER GLOVES, AND BREATHING APPARATUS.
WARNING: 1] STRUCTURAL PROTECTIVE CLOTHING IS PERMEABLE, REMAIN CLEAR OF SMOKE, WATER FALL OUT AND WATER RUN OFF.
         2] KEEP OUT OF THE REACH OF CHILDREN.
         3] REMOVE ALL IGNITION SOURCES IF WITHOUT RISK.
         4] HAZARDOUS DECOMPOSITION PRODUCTS ARE UNKNOWN.
         5] MOVE CONTAINERS FROM AREA IF WITHOUT RISK, COOL EXPOSED CONTAINERS.
         6] DIKE AREA FOR CONTROL AND CONTAINMENT TO PREVENT ENTRY INTO SEWERS, DRAIN, AND WATER WAYS.

LARGE SPILL/NO FIRE/RESCUE: WEAR RUBBER OR NEOPRENE BOOTS, GLOVES.

SPILL CONTROL AND CONTAINMENT:
   HOUSEHOLD SPILL: WIPE UP WITH AN ABSORBENT CLOTH. RINSE OUT IN THE SINK. WIPE UP RESIDUE WITH A DAMP CLOTH.
   RINSE OUT IN THE SINK.
   LARGE SPILL: PICK UP WITH ABSORBENT. RINSE AREA WITH WATER. PICK UP WITH ABSORBENT. PUT ALL ABSORBENT IN A
   DISPOSAL CONTAINER. SEAL AND REMOVE FROM THE WORK AREA.

HEALTH HAZARD INFORMATION:
   WHEN USING THIS PRODUCT WEAR: NO SPECIAL REQUIREMENT. Eye contact will cause stinging, tearing, itching, and
   redness. Ingestion may result in nausea, vomiting, and diarrhea.

   A physician should be contacted if anyone develops any signs or symptoms and suspects that they are caused by
   exposure to this product.

FIRST AID:
   Eye Exposure:
   Flush with water for 15 minutes while lifting the upper and lower eye lids. Contact lenses should not be worn
   when working with this product. Get medical attention.

   Skin Exposure:
   If irritation develops discontinue use. If persistent, get medical attention.

   Breathing:
   Not expected to be a problem.

   Swallowing:
   If the victim is conscious, give 2-3 glasses of milk or water to drink. Immediately contact the local poison
   control center for advice. Keep the victim warm and at rest. Get medical attention.

PRODUCT: WINDEX ENVIRO - REFILL GLASS CLEANER BLUE

INCOMPATIBILITY: NONE KNOWN

INGREDIENTS: ISOPROPYL ALCOHOL   3% TLV: 400 ppm   CAS#: 67-63-0
             2-BUTOXYETHANOL   2% TLV: 25 ppm(skin) CAS#: 111-76-2

IGNI TEMP: NA FP: 129F LEL: NA UEL: NA VP: 17.6 mmHg VD: 1.2 SG: .99 PS: LIQUID APPEAR: CLEAR BLUE
   ODOR: AMMONIA   PH FACTOR: 11

HAZARD RATINGS: H: F: R: 0 2 0

FIRE FIGHTING:
   HIGH EXPANSION FOAM, LOW EXPANSION FOAM, ALCOHOL FOAM; DRY CHEMICAL, CARBON DIOXIDE, WATER FOG.
   PROTECTIVE CLOTHING, RUBBER GLOVES, AND BREATHING APPARATUS.
WARNING: 1] STRUCTURAL PROTECTIVE CLOTHING IS PERMEABLE, REMAIN CLEAR OF SMOKE, WATER FALL OUT AND WATER RUN OFF.
         2] KEEP OUT OF THE REACH OF CHILDREN.
         3] REMOVE ALL IGNITION SOURCES IF WITHOUT RISK.
         4] UNPLUG ELECTRICAL APPLIANCES BEFORE USING PRODUCT.
         5] HAZARDOUS DECOMPOSITION PRODUCTS ARE UNKNOWN.
         6] MOVE CONTAINERS FROM AREA IF WITHOUT RISK, COOL EXPOSED CONTAINERS.
         7] DIKE AREA FOR CONTROL AND CONTAINMENT TO PREVENT ENTRY INTO SEWERS, DRAIN, AND WATER WAYS.

LARGE SPILL/NO FIRE/RESCUE: WEAR RUBBER OR NEOPRENE BOOTS, GLOVES.

SPILL CONTROL AND CONTAINMENT:
   HOUSEHOLD SPILL: WIPE UP WITH AN ABSORBENT CLOTH. RINSE OUT IN THE SINK. WIPE UP RESIDUE WITH A DAMP CLOTH.
   RINSE OUT IN THE SINK.
   LARGE SPILL: PICK UP WITH ABSORBENT. RINSE AREA WITH WATER. PICK UP WITH ABSORBENT. PUT ALL ABSORBENT IN A
   DISPOSAL CONTAINER. SEAL AND REMOVE FROM THE WORK AREA.

HEALTH HAZARD INFORMATION:
   WHEN USING THIS PRODUCT WEAR: NO SPECIAL REQUIREMENT. Eye contact will cause stinging, tearing, itching, and
   redness. Ingestion may result in nausea and vomiting.

   A physician should be contacted if anyone develops any signs or symptoms and suspects that they are caused by
   exposure to this product.

FIRST AID:
   Eye Exposure:
   Flush with water for 15 minutes while lifting the upper and lower eye lids. Contact lenses should not be worn
   when working with this product. Get medical attention.

   Skin Exposure:
   Rinse off with water.

   Breathing:
   Not expected to be a problem.

   Swallowing:
   If the victim is conscious, give 2-3 glasses of milk or water to drink. Immediately contact the local poison
   control center for advice. Keep the victim warm and at rest. Get medical attention.

PRODUCT: WINDEX GLASS CLEANER (AEROSOL)

INCOMPATIBILITY: NONE KNOWN

INGREDIENTS: ISOPROPYL ALCOHOL   3% TLV: 400 ppm   CAS#: 67-63-0
             2-BUTOXYETHANOL   2% TLV: 25 ppm(skin) CAS#: 111-76-2
             ISOBUTANE   3% TLV: CAS#: 75-28-5
             PROPANE   1% TLV: 1000 ppm   CAS#: 74-98-6

IGNI TEMP: NA FP: 131F LEL: NA UEL: NA VP: 17.6 mmHg VD: 1.2 SG: .99 PS: LIQUID APPEAR: CLEAR
    ODOR: AMMONIA   PH FACTOR: 11

HAZARD RATINGS: H: F: R: 0 2 0

FIRE FIGHTING:
  HIGH EXPANSION FOAM, LOW EXPANSION FOAM, ALCOHOL FOAM; DRY CHEMICAL, CARBON DIOXIDE, WATER FOG.
  PROTECTIVE CLOTHING, RUBBER GLOVES, AND BREATHING APPARATUS.
WARNING: 1] STRUCTURAL PROTECTIVE CLOTHING IS PERMEABLE, REMAIN CLEAR OF SMOKE, WATER FALL OUT AND WATER RUN OFF.
         2] KEEP OUT OF THE REACH OF CHILDREN.
         3] REMOVE ALL IGNITION SOURCES IF WITHOUT RISK.
         4] CONTAINERS WILL EXPLODE IN A FIRE OR IF HEATED ABOVE 120F
         5] UNPLUG ELECTRICAL APPLIANCES BEFORE USING PRODUCT.
         6] HAZARDOUS DECOMPOSITION PRODUCTS ARE UNKNOWN.
         7] MOVE CONTAINERS FROM AREA IF WITHOUT RISK, COOL EXPOSED CONTAINERS.
         8] DIKE AREA FOR CONTROL AND CONTAINMENT TO PREVENT ENTRY INTO SEWERS, DRAIN, AND WATER WAYS.

LARGE SPILL/NO FIRE/RESCUE: WEAR RUBBER OR NEOPRENE BOOTS, GLOVES.

SPILL CONTROL AND CONTAINMENT:
  HOUSEHOLD SPILL: WIPE UP WITH AN ABSORBENT CLOTH. RINSE OUT IN THE SINK. WIPE UP RESIDUE WITH A DAMP CLOTH.
  RINSE OUT IN THE SINK.
  LARGE SPILL: PICK UP WITH ABSORBENT. RINSE AREA WITH WATER. PICK UP WITH ABSORBENT. PUT ALL ABSORBENT IN A
  DISPOSAL CONTAINER. SEAL AND REMOVE FROM THE WORK AREA.

HEALTH HAZARD INFORMATION:
  WHEN USING THIS PRODUCT WEAR: NO SPECIAL REQUIREMENT. Eye contact will cause stinging, tearing, itching, and
  redness. Ingestion may result in nausea and vomiting.

  A physician should be contacted if anyone develops any signs or symptoms and suspects that they are caused by
  exposure to this product.

FIRST AID:
  Eye Exposure:
  Flush with water for 15 minutes while lifting the upper and lower eye lids. Contact lenses should not be worn
  when working with this product. Get medical attention.

  Skin Exposure:
  Rinse off with water.

  Breathing:
  Not expected to be a problem.

  Swallowing:
  If the victim is conscious, give 2-3 glasses of milk or water to drink. Immediately contact the local poison
  control center for advice. Keep the victim warm and at rest. Get medical attention.

PRODUCT: WINDEX GLASS CLEANER (BLUE)

INCOMPATIBILITY: NONE KNOWN

INGREDIENTS: ISOPROPYL ALCOHOL   3% TLV: 400 ppm   CAS#: 67-63-0
            2-BUTOXYETHANOL   2% TLV: 25 ppm(skin) CAS#: 111-76-2

IGNI TEMP: NA FP: 129F LEL: NA UEL: NA VP: 17.6 mmHg VD: 1.2 SG: .99 PS: LIQUID APPEAR: CLEAR BLUE
    ODOR: AMMONIA   PH FACTOR: 11

HAZARD RATINGS: H: F: R: 0 2 0

FIRE FIGHTING:
  HIGH EXPANSION FOAM, LOW EXPANSION FOAM, ALCOHOL FOAM; DRY CHEMICAL, CARBON DIOXIDE, WATER FOG.
  PROTECTIVE CLOTHING, RUBBER GLOVES, AND BREATHING APPARATUS.
WARNING: 1] STRUCTURAL PROTECTIVE CLOTHING IS PERMEABLE, REMAIN CLEAR OF SMOKE, WATER FALL OUT AND WATER RUN OFF.
        2] KEEP OUT OF THE REACH OF CHILDREN.
        3] REMOVE ALL IGNITION SOURCES IF WITHOUT RISK.
        4] UNPLUG ELECTRICAL APPLIANCES BEFORE USING PRODUCT.
        5] HAZARDOUS DECOMPOSITION PRODUCTS ARE UNKNOWN.
        6] MOVE CONTAINERS FROM AREA IF WITHOUT RISK, COOL EXPOSED CONTAINERS.
        7] DIKE AREA FOR CONTROL AND CONTAINMENT TO PREVENT ENTRY INTO SEWERS, DRAIN, AND WATER WAYS.

LARGE SPILL/NO FIRE/RESCUE: WEAR RUBBER OR NEOPRENE BOOTS, GLOVES.

SPILL CONTROL AND CONTAINMENT:
  HOUSEHOLD SPILL: WIPE UP WITH AN ABSORBENT CLOTH. RINSE OUT IN THE SINK. WIPE UP RESIDUE WITH A DAMP CLOTH.
  RINSE OUT IN THE SINK.
  LARGE SPILL: PICK UP WITH ABSORBENT. RINSE AREA WITH WATER. PICK UP WITH ABSORBENT. PUT ALL ABSORBENT IN A
  DISPOSAL CONTAINER. SEAL AND REMOVE FROM THE WORK AREA.

HEALTH HAZARD INFORMATION:
  WHEN USING THIS PRODUCT WEAR: NO SPECIAL REQUIREMENT. Eye contact will cause stinging, tearing, itching, and
  redness. Ingestion may result in nausea and vomiting.

  A physician should be contacted if anyone develops any signs or symptoms and suspects that they are caused by
  exposure to this product.

FIRST AID:
  Eye Exposure:
  Flush with water for 15 minutes while lifting the upper and lower eye lids. Contact lenses should not be worn
  when working with this product. Get medical attention.

  Skin Exposure:
  Rinse off with water.

  Breathing:
  Not expected to be a problem.

  Swallowing:
  If the victim is conscious, give 2-3 glasses of milk or water to drink. Immediately contact the local poison
  control center for advice. Keep the victim warm and at rest. Get medical attention.

PRODUCT: WINDEX LEMON GLASS CLEANER (YELLOW)

INCOMPATIBILITY: NONE KNOWN

INGREDIENTS: ISOPROPYL ALCOHOL   3% TLV: 400 ppm   CAS#: 67-63-0
             2-BUTOXYETHANOL   2% TLV: 25 ppm(skin) CAS#: 111-76-2

IGNI TEMP: NA FP: 129F LEL: NA UEL: NA VP: 17.6 mmHg VD: 1.2 SG: .99 PS: LIQUID APPEAR: CLEAR YELLOW
     ODOR: LEMON/AMMONIA   PH FACTOR: 11

HAZARD RATINGS: H: F: R: 0 2 0

FIRE FIGHTING:
  HIGH EXPANSION FOAM, LOW EXPANSION FOAM, ALCOHOL FOAM; DRY CHEMICAL, CARBON DIOXIDE, WATER FOG.
  PROTECTIVE CLOTHING, RUBBER GLOVES, AND BREATHING APPARATUS.
WARNING: 1] STRUCTURAL PROTECTIVE CLOTHING IS PERMEABLE, REMAIN CLEAR OF SMOKE, WATER FALL OUT AND WATER RUN OFF.
         2] KEEP OUT OF THE REACH OF CHILDREN.
         3] REMOVE ALL IGNITION SOURCES IF WITHOUT RISK.
         4] UNPLUG ELECTRICAL APPLIANCES BEFORE USING PRODUCT.
         5] HAZARDOUS DECOMPOSITION PRODUCTS ARE UNKNOWN.
         6] MOVE CONTAINERS FROM AREA IF WITHOUT RISK, COOL EXPOSED CONTAINERS.
         7] DIKE AREA FOR CONTROL AND CONTAINMENT TO PREVENT ENTRY INTO SEWERS, DRAIN, AND WATER WAYS.

LARGE SPILL/NO FIRE/RESCUE: WEAR RUBBER OR NEOPRENE BOOTS, GLOVES.

SPILL CONTROL AND CONTAINMENT:
  HOUSEHOLD SPILL: WIPE UP WITH AN ABSORBENT CLOTH. RINSE OUT IN THE SINK. WIPE UP RESIDUE WITH A DAMP CLOTH.
  RINSE OUT IN THE SINK.
  LARGE SPILL: PICK UP WITH ABSORBENT. RINSE AREA WITH WATER. PICK UP WITH ABSORBENT. PUT ALL ABSORBENT IN A
  DISPOSAL CONTAINER. SEAL AND REMOVE FROM THE WORK AREA.

HEALTH HAZARD INFORMATION:
  WHEN USING THIS PRODUCT WEAR: NO SPECIAL REQUIREMENT. Eye contact will cause stinging, tearing, itching, and
  redness. Ingestion may result in nausea and vomiting.

  A physician should be contacted if anyone develops any signs or symptoms and suspects that they are caused by
  exposure to this product.

FIRST AID:
  Eye Exposure:
  Flush with water for 15 minutes while lifting the upper and lower eye lids. Contact lenses should not be worn
  when working with this product. Get medical attention.

  Skin Exposure:
  Rinse off with water.

  Breathing:
  Not expected to be a problem.

  Swallowing:
  If the victim is conscious, give 2-3 glasses of milk or water to drink. Immediately contact the local poison
  control center for advice. Keep the victim warm and at rest. Get medical attention.

PRODUCT: WINDEX PROFESSIONAL STRENGTH MULTI-SURFACE AND GLASS CLEANER

INCOMPATIBILITY: NONE KNOWN

INGREDIENTS: 2-BUTOXYETHANOL   4% TLV: 25 ppm(skin) CAS#: 111-76-2

IGNI TEMP: NA FP: 212F LEL: NA UEL: NA VP: 17.4 mmHg VD: 1.2 SG: .99 PS: LIQUID APPEAR: CLEAR BLUE
    ODOR: AMMONIA   PH FACTOR: 11

HAZARD RATINGS: H: F: R: 0 1 0

FIRE FIGHTING:
  HIGH EXPANSION FOAM, LOW EXPANSION FOAM, ALCOHOL FOAM; DRY CHEMICAL, CARBON DIOXIDE, WATER FOG.
  PROTECTIVE CLOTHING, RUBBER GLOVES, AND BREATHING APPARATUS.
WARNING: 1] STRUCTURAL PROTECTIVE CLOTHING IS PERMEABLE, REMAIN CLEAR OF SMOKE, WATER FALL OUT AND WATER RUN OFF.
         2] KEEP OUT OF THE REACH OF CHILDREN.
         3] REMOVE ALL IGNITION SOURCES IF WITHOUT RISK.
         4] UNPLUG ELECTRICAL APPLIANCES BEFORE USING PRODUCT.
         5] HAZARDOUS DECOMPOSITION PRODUCTS ARE UNKNOWN.
         6] MOVE CONTAINERS FROM AREA IF WITHOUT RISK, COOL EXPOSED CONTAINERS.
         7] DIKE AREA FOR CONTROL AND CONTAINMENT TO PREVENT ENTRY INTO SEWERS, DRAIN, AND WATER WAYS.

LARGE SPILL/NO FIRE/RESCUE: WEAR RUBBER OR NEOPRENE BOOTS, GLOVES.

SPILL CONTROL AND CONTAINMENT:
  HOUSEHOLD SPILL: WIPE UP WITH AN ABSORBENT CLOTH. RINSE OUT IN THE SINK. WIPE UP RESIDUE WITH A DAMP CLOTH.
  RINSE OUT IN THE SINK.
  LARGE SPILL: PICK UP WITH ABSORBENT. RINSE AREA WITH WATER. PICK UP WITH ABSORBENT. PUT ALL ABSORBENT IN A
  DISPOSAL CONTAINER. SEAL AND REMOVE FROM THE WORK AREA.

HEALTH HAZARD INFORMATION:
  WHEN USING THIS PRODUCT WEAR: NO SPECIAL REQUIREMENT. Eye contact will cause stinging, tearing, itching, and
  redness. Ingestion may result in nausea and vomiting.

  A physician should be contacted if anyone develops any signs or symptoms and suspects that they are caused by
  exposure to this product.

FIRST AID:
Eye Exposure:
Flush with water for 15 minutes while lifting the upper and lower eye lids. Contact lenses should not be worn
when working with this product. Get medical attention.

Skin Exposure:
Rinse off with water.

Breathing:
Not expected to be a problem.

Swallowing:
If the victim is conscious, give 2-3 glasses of milk or water to drink. Immediately contact the local poison
control center for advice. Keep the victim warm and at rest. Get medical attention.

PRODUCT: WINDEX PROFESSIONAL STRENGTH MULTI-SURFACE & GLASS CLEANER

INCOMPATIBILITY: NONE KNOWN

INGREDIENTS: ISOPROPYL ALCOHOL   7% TLV: 400 ppm   CAS#: 67-63-0
             2-BUTOXYETHANOL   3% TLV: 25 ppm(skin) CAS#: 111-76-2

IGNI TEMP: NA FP: 122F LEL: NA UEL: NA VP: 1 mmHg VD: 1.0 SG: .983 PS: LIQUID APPEAR: CLEAR BLUE
     ODOR: AMMONIA   PH FACTOR: 11

HAZARD RATINGS: H: F: R: 1 2 0

FIRE FIGHTING:
    HIGH EXPANSION FOAM, LOW EXPANSION FOAM, ALCOHOL FOAM; DRY CHEMICAL, CARBON DIOXIDE, WATER FOG.
    PROTECTIVE CLOTHING, RUBBER GLOVES, AND BREATHING APPARATUS.
WARNING: 1] STRUCTURAL PROTECTIVE CLOTHING IS PERMEABLE, REMAIN CLEAR OF SMOKE, WATER FALL OUT AND WATER RUN OFF.
         2] KEEP OUT OF THE REACH OF CHILDREN.
         3] REMOVE ALL IGNITION SOURCES IF WITHOUT RISK.
         4] UNPLUG ELECTRICAL APPLIANCES BEFORE USING PRODUCT.
         5] HAZARDOUS DECOMPOSITION PRODUCTS ARE UNKNOWN.
         6] MOVE CONTAINERS FROM AREA IF WITHOUT RISK, COOL EXPOSED CONTAINERS.
         7] DIKE AREA FOR CONTROL AND CONTAINMENT TO PREVENT ENTRY INTO SEWERS, DRAIN, AND WATER WAYS.

LARGE SPILL/NO FIRE/RESCUE: WEAR RUBBER OR NEOPRENE BOOTS, GLOVES.

SPILL CONTROL AND CONTAINMENT:
    HOUSEHOLD SPILL: WIPE UP WITH AN ABSORBENT CLOTH. RINSE OUT IN THE SINK. WIPE UP RESIDUE WITH A DAMP CLOTH.
    RINSE OUT IN THE SINK.
    LARGE SPILL: PICK UP WITH ABSORBENT. RINSE AREA WITH WATER. PICK UP WITH ABSORBENT. PUT ALL ABSORBENT IN A
    DISPOSAL CONTAINER. SEAL AND REMOVE FROM THE WORK AREA.

HEALTH HAZARD INFORMATION:
    WHEN USING THIS PRODUCT WEAR: NO SPECIAL REQUIREMENT. Eye contact will cause stinging, tearing, itching, and
    redness. Ingestion may result in nausea and vomiting.

    A physician should be contacted if anyone develops any signs or symptoms and suspects that they are caused by
    exposure to this product.

FIRST AID:
Eye Exposure:
Flush with water for 15 minutes while lifting the upper and lower eye lids. Contact lenses should not be worn
when working with this product. Get medical attention.

    Skin Exposure:
    Rinse off with water.

    Breathing:
    Not expected to be a problem.

    Swallowing:
    If the victim is conscious, give 2-3 glasses of milk or water to drink. Immediately contact the local poison
    control center for advice. Keep the victim warm and at rest. Get medical attention.

PRODUCT: WINDEX VINEGAR GLASS CLEANER (GREEN)

INCOMPATIBILITY: NONE KNOWN

INGREDIENTS: 2-BUTOXYETHANOL   10% TLV: 25 ppm(skin) CAS#: 111-76-2

IGNI TEMP: NA FP: 212F LEL: NA UEL: NA VP: 17.3 mmHg VD: 1.2 SG: .99 PS: LIQUID APPEAR: CLEAR GREEN
    ODOR: GREEN APPLE   PH FACTOR: NA

HAZARD RATINGS: H: F: R: 0 1 0

FIRE FIGHTING:
  HIGH EXPANSION FOAM, LOW EXPANSION FOAM, ALCOHOL FOAM; DRY CHEMICAL, CARBON DIOXIDE, WATER FOG.
  PROTECTIVE CLOTHING, RUBBER GLOVES, AND BREATHING APPARATUS.
WARNING: 1] STRUCTURAL PROTECTIVE CLOTHING IS PERMEABLE, REMAIN CLEAR OF SMOKE, WATER FALL OUT AND WATER RUN OFF.
         2] KEEP OUT OF THE REACH OF CHILDREN.
         3] REMOVE ALL IGNITION SOURCES IF WITHOUT RISK.
         4] UNPLUG ELECTRICAL APPLIANCES BEFORE USING PRODUCT.
         5] HAZARDOUS DECOMPOSITION PRODUCTS ARE UNKNOWN.
         6] MOVE CONTAINERS FROM AREA IF WITHOUT RISK, COOL EXPOSED CONTAINERS.
         7] DIKE AREA FOR CONTROL AND CONTAINMENT TO PREVENT ENTRY INTO SEWERS, DRAIN, AND WATER WAYS.

LARGE SPILL/NO FIRE/RESCUE: WEAR RUBBER OR NEOPRENE BOOTS, GLOVES.

SPILL CONTROL AND CONTAINMENT:
  HOUSEHOLD SPILL: WIPE UP WITH AN ABSORBENT CLOTH. RINSE OUT IN THE SINK. WIPE UP RESIDUE WITH A DAMP CLOTH.
  RINSE OUT IN THE SINK.
  LARGE SPILL: PICK UP WITH ABSORBENT. RINSE AREA WITH WATER. PICK UP WITH ABSORBENT. PUT ALL ABSORBENT IN A
  DISPOSAL CONTAINER. SEAL AND REMOVE FROM THE WORK AREA.

HEALTH HAZARD INFORMATION:
  WHEN USING THIS PRODUCT WEAR: NO SPECIAL REQUIREMENT. Eye contact will cause stinging, tearing, itching, and
  redness. Ingestion may result in nausea and vomiting.

  A physician should be contacted if anyone develops any signs or symptoms and suspects that they are caused by
  exposure to this product.

FIRST AID:
  Eye Exposure:
  Flush with water for 15 minutes while lifting the upper and lower eye lids. Contact lenses should not be worn
  when working with this product. Get medical attention.

  Skin Exposure:
  Rinse off with water.

  Breathing:
  Not expected to be a problem.

  Swallowing:
  If the victim is conscious, give 2-3 glasses of milk or water to drink. Immediately contact the local poison
  control center for advice. Keep the victim warm and at rest. Get medical attention.

PRODUCT: WISK NON-PHOSPHATE POWER SCOOP POWDER LAUNDRY DETERGENT

INCOMPATIBILITY: NONE KNOWN

INGREDIENTS: SODIUM CARBONATE   1%   TLV: CAS#: 497-19-8

IGNI TEMP: NA FP: NA LEL: NA UEL: NA VP: NA VD: NA SG: NA PS: POWDER APPEAR: BLUE   ODOR:
   PH FACTOR: 13

HAZARD RATINGS: H: F: R: 2 0 0

FIRE FIGHTING:
   HIGH EXPANSION FOAM, LOW EXPANSION FOAM, ALCOHOL FOAM; DRY CHEMICAL, CARBON DIOXIDE, WATER FOG.
   PROTECTIVE CLOTHING, RUBBER GLOVES, AND BREATHING APPARATUS.
WARNING: 1] STRUCTURAL PROTECTIVE CLOTHING IS PERMEABLE, REMAIN CLEAR OF SMOKE, WATER FALL OUT AND WATER RUN OFF.
         2] KEEP OUT OF THE REACH OF CHILDREN.
         3] THERMAL DECOMPOSITION PRODUCTS ARE NOT KNOWN.
         4] MOVE CONTAINERS FROM AREA IF WITHOUT RISK, COOL EXPOSED CONTAINERS.
         5] DIKE AREA FOR CONTROL AND CONTAINMENT TO PREVENT ENTRY INTO SEWERS, DRAIN, AND WATER WAYS.

LARGE SPILL/NO FIRE/RESCUE: WEAR RUBBER OR NEOPRENE BOOTS, GLOVES.

SPILL CONTROL AND CONTAINMENT:
   HOUSEHOLD SPILL: SWEEP UP POWDER. PUT IN DISPOSAL CONTAINER. DAMP MOP UP RESIDUE. RINSE OUT IN THE SINK.
   LARGE SPILL: AVOID CREATING A DUST. SWEEP UP POWDER. PUT IN A DISPOSAL CONTAINER. RINSE AREA WITH WATER. PICK UP
   WITH ABSORBENT. PUT IN A DISPOSAL CONTAINER. SEAL AND REMOVE FROM THE WORK AREA.

HEALTH HAZARD INFORMATION:
   WHEN USING THIS PRODUCT WEAR: NO SPECIAL REQUIREMENT. Eye contact will cause stinging, tearing, itching, and
   redness. Accumulation of fluid in the eye lid tissue and possibly the eye itself will be involved. Prolonged
   skin contact may cause a rash. Ingestion may result in nausea, vomiting, and diarrhea. Prolonged inhalation may
   cause irritation of the nasal passage.

   A physician should be contacted if anyone develops any signs or symptoms and suspects that they are caused by
   exposure to this product.

FIRST AID:
   Eye Exposure:
   Flush with water for 15 minutes while lifting the upper and lower eye lids. Contact lenses should not be worn
   when working with this product. Get medical attention.

   Skin Exposure:
   Remove contaminated clothing and flush skin with water.

   Breathing:
   If irritation develops, move the victim to fresh air.

   Swallowing:
   If the victim is conscious, give 2-3 glasses of milk or water to drink. Immediately contact the local poison
   control center for advice. Keep the victim warm and at rest. Get medical attention.

PRODUCT: ZACT TOOTH PASTE

INCOMPATIBILITY: NONE KNOWN

INGREDIENTS: DICALCIUM PHOSPHATE DIHYDRATE TLV: CAS#: 7757-93-9
             SORBITOL   TLV: CAS#:   50-74-4
             PROPYLENE GLYCOL   TLV: CAS#: 4354-15-3
             ETHYL PARABEN   TLV: CAS#:   120-47-8
             WATER   TLV: CAS#: 7732-18-5
             SODIUM LAURYL SULFATE   TLV: CAS#:   151-21-3
             BUTYL PARABEN   TLV: CAS#:   94-26-8
             TRIMAGNESIUM PHOSPHATE, DIHYDROCHLORIDE, SODIUM SACCHARIN

IGNI TEMP: NA FP: NA LEL: NA UEL: NA VP: NA VD: NA SG: NA PS: GEL APPEAR: THICK ODOR: NA   PH FACTOR: NA

HAZARD RATINGS: H: F: R:

FIRE FIGHTING:
  HIGH EXPANSION FOAM, LOW EXPANSION FOAM, ALCOHOL FOAM; DRY CHEMICAL, CARBON DIOXIDE, WATER FOG.
  PROTECTIVE CLOTHING, RUBBER GLOVES, AND BREATHING APPARATUS.
WARNING: 1] STRUCTURAL PROTECTIVE CLOTHING IS PERMEABLE, REMAIN CLEAR OF SMOKE, WATER FALL OUT AND WATER RUN OFF.
         2] KEEP OUT OF THE REACH OF CHILDREN.
         3] THERMAL DECOMPOSITION MAY YIELD CARBON DIOXIDE, CARBON MONOXIDE.
         4] MOVE CONTAINERS FROM AREA IF WITHOUT RISK, COOL EXPOSED CONTAINERS.
         5] DIKE AREA FOR CONTROL AND CONTAINMENT TO PREVENT ENTRY INTO SEWERS, DRAIN, AND WATER WAYS.

LARGE SPILL/NO FIRE/RESCUE: WEAR RUBBER OR NEOPRENE BOOTS, GLOVES.

SPILL CONTROL AND CONTAINMENT:
  HOUSEHOLD SPILL: WIPE UP WITH AN ABSORBENT CLOTH. RINSE OUT IN THE SINK. WIPE UP RESIDUE WITH A DAMP CLOTH.
  RINSE OUT IN THE SINK.
  LARGE SPILL: PICK UP WITH ABSORBENT. RINSE AREA WITH WATER. PICK UP WITH ABSORBENT. PUT ALL ABSORBENT IN A
  DISPOSAL CONTAINER. SEAL AND REMOVE FROM THE WORK AREA.

HEALTH HAZARD INFORMATION:
  WHEN USING THIS PRODUCT WEAR: NO SPECIAL REQUIREMENT. Not expected to have any ill effects. Eye contact may
  cause irritation.

  A physician should be contacted if anyone develops any signs or symptoms and suspects that they are caused by
  exposure to this product.

FIRST AID:
  Eye Exposure:
  Flush with water for 15 minutes while lifting the upper and lower eye lids. Contact lenses should not be worn
  when working with this product. Get medical attention.

  Skin Exposure:
  Not expected to be a problem.

  Breathing:
  Not expected to be a problem.

  Swallowing:
  Immediately contact the local poison control center for advice. Keep the victim warm and at rest. Get medical
  attention.

PRODUCT: ZEST DEODORANT BEAUTY BAR - TOILET SOAP

INCOMPATIBILITY: NONE KNOWN

INGREDIENTS:

SODIUM SULFATE	CAS#: 7757-82-6	POTASSIUM CHLORIDE	CAS#: 7447-40-7
SODIUM CHLORIDE	CAS#: 7647-14-5	SODIUM TALLOWATE	CAS#:
LAURIC ACID	CAS#: 143-07-7	SODIUM COCOATE	CAS#:
TITANIUM DIOXIDE	CAS#: 13463-67-7	MAGNESIUM TALLOWATE	CAS#:
TETRASODIUM EDTA	CAS#: 10378-23-1	MAGNESIUM COCOATE	CAS#:
POTASSIUM SULFATE	CAS#: 7778-80-5	TRICLOCARBEN	CAS#:
SODIUM SILICATE	CAS#: 1344-09-8	FRAGRANCE	CAS#:

IGNI TEMP: NA FP: NA LEL: NA UEL: NA VP: NA VD: NA SG: 1.14 PS: SOLID APPEAR: BAR
    ODOR: FLORAL - SPICE - SCENT FREE   PH FACTOR: NA

HAZARD RATINGS: H: F: R:

FIRE FIGHTING:
    HIGH EXPANSION FOAM, LOW EXPANSION FOAM, ALCOHOL FOAM; DRY CHEMICAL, CARBON DIOXIDE, WATER FOG.
    PROTECTIVE CLOTHING, RUBBER GLOVES, AND BREATHING APPARATUS.
WARNING: 1] STRUCTURAL PROTECTIVE CLOTHING IS PERMEABLE, REMAIN CLEAR OF SMOKE, WATER FALL OUT AND WATER RUN OFF.
         2] KEEP OUT OF THE REACH OF CHILDREN.
         3] HAZARDOUS DECOMPOSITION PRODUCTS ARE NOT KNOWN.
         4] MOVE CONTAINERS FROM AREA IF WITHOUT RISK, COOL EXPOSED CONTAINERS.
         5] DIKE AREA FOR CONTROL AND CONTAINMENT TO PREVENT ENTRY INTO SEWERS, DRAIN, AND WATER WAYS.

LARGE SPILL/NO FIRE/RESCUE: WEAR RUBBER OR NEOPRENE BOOTS, GLOVES.

SPILL CONTROL AND CONTAINMENT:
    HOUSEHOLD SPILL: SCRAPE UP RESIDUE. PUT IN A DISPOSAL CONTAINER. WIPE UP RESIDUE WITH A DAMP CLOTH. RINSE OUT IN
    THE SINK.
    LARGE SPILL: SCOOP UP PRODUCT. PUT IN A DISPOSAL CONTAINER. RINSE AREA WITH WATER. PICK UP WITH ABSORBENT. PUT
    IN A DISPOSAL CONTAINER. SEAL AND REMOVE FROM THE WORK AREA.

HEALTH HAZARD INFORMATION:
    WHEN USING THIS PRODUCT WEAR: NO SPECIAL REQUIREMENT. Eye contact will cause stinging, tearing, itching, and
    redness. Ingestion may result in nausea, vomiting, and diarrhea.

    A physician should be contacted if anyone develops any signs or symptoms and suspects that they are caused by
    exposure to this product.

FIRST AID:
Eye Exposure:
Flush with water for 15 minutes while lifting the upper and lower eye lids. Contact lenses should not be worn
when working with this product. Get medical attention.

Skin Exposure:
If rash develops, discontinue use.

Breathing:
Not expected to be a problem.

Swallowing:
If the victim is conscious, give 2-3 glasses of water to drink. Immediately contact the local poison control
center for advice. Keep the victim warm and at rest. Get medical attention.

Milton Keynes UK
Ingram Content Group UK Ltd.
UKHW051942071024
449327UK00026B/2137